How does the variability of the Sun affect the climate on Earth and other conditions in our atmosphere? The 143rd IAU Colloquium, held in Boulder, Colorado, brought together international experts to investigate this connection. This timely volume presents their articles and together these clearly illustrate just how variations in the energy output of the Sun affect the terrestrial climate, radiative environment, and chemistry of the upper atmosphere.

This survey provides a complete and up-to-date review of observations, theoretical interpretations, and empirical and physical models of variations in the energy output of the Sun and solar-type stars. In particular, it shows how this variability is related to magnetic activity on the surface of the star and how long-term modulations are driven by the stellar interior.

Together, these articles clearly demonstrate the link between changes in the radiative output of the Sun and registered climatic anomalies – essential reading for researchers and graduate students.

The Sun as a Variable Star

IAU Astronomical Union
Union Astronomique International

The following Colloquia of the International Astronomical Union are published for the Union by Cambridge University Press.

82. Cepheids. *Edited by Barry F. Madore.* 0 521 30091 6. 1985

91. History of Oriental Astronomy. *Edited by G. Swarup, A. K. Bag and K. S. Shukla.* 0 521 34659 2. 1987

92. Physics of Be Stars. *Edited by A. Slettebak and T. P. Snow.* 0 521 33078 5. 1987

101. Supernova Remnants and the Interstellar Medium. *Edited by R. S. Roger and T. L. Landecker.* 0 521 35062 X. 1988

105. The Teaching of Astronomy. *Edited by Jay M. Pasachoff and John R. Percy.* 0 521 35331 9. 1990

106. Evolution of Peculiar Red Giant Stars. *Edited by Hollis Johnson and Ben Zuckerman.* 0 521 36617 8. 1989

111. The Use of Pulsating Stars in Fundamental Problems of Astronomy. *Edited by Edward G. Schmidt.* 0 521 37023 X. 1989

136. Stellar Photometry – Current Techniques and Future Developments. *Edited by C. J. Butler and I. Elliott.* 0 521 41866 6. 1993

139. Stellar Pulsation and Pulsating Variable Stars. *Edited by James M. Nemec and Jaymie M. Matthews.* 0 521 44382 2. 1993

143. The Sun as a Variable Star. *Edited by J. M. Pap, C. Fröhlich, H. S. Hudson and S. K. Solanki.* 0 521 42006 7. 1994

145. Supernovae and Supernova Remnants. *Edited by Richard McCray.* 0 521 46080 8. 1994

147. The Equation of State in Astrophysics. *Edited by Gilles Chabrier and Evry Schatzman.* 0 521 47260 1. 1994

The Sun as a Variable Star
Solar and Stellar Irradiance Variations

Proceedings of IAU Colloquium No. 143
held in Boulder, Colorado, USA
June 20–25, 1993

Edited by
J. M. PAP
Jet Propulsion Laboratory,
California Institute of Technology
Pasadena, California, USA

C. FRÖHLICH
World Radiation Centre, PMOD
Davos, Switzerland

H. S. HUDSON
University of Hawaii,
Honolulu, Hawaii, USA

S. K. SOLANKI
ETH–Institute of Astronomy,
Zürich, Switzerland

CAMBRIDGE
UNIVERSITY PRESS

Published by the Press Syndicate of the University of Cambridge
The Pitt Building, Trumpington Street, Cambridge CB2 1RP
40 West 20th Street, New York, NY 10011-4211, USA
10 Stamford Road, Oakleigh, Melbourne 3166, Australia

First published 1994

Printed in Great Britain at the University Press, Cambridge

A catalogue record for this book is available from the British Library

Library of Congress cataloguing in publication data available

ISBN 0 521 42006 7 hardback

Dedicated to the memory of

Ernest Gurtovenko

Contents

Participants xiii
Group Photograph xvii
Preface xix
Acknowledgements xxi
Statement from the Scientific Organizing Committee xxiii

Introduction

A Historical Perspective on Measurements of Solar Irradiance
V. Gaizauskas 1

Solar Monitoring has a Past and Present: Does it have a Future?
R. C. Willson 4

Solar Variability and its Terrestrial Effects
G. L. Withbroe and W. Kalkofen 11

General Reviews on Observations of Solar and Stellar Irradiance Variability from Space and from the Ground

Irradiance Variations of Stars
G. W. Lockwood 20

Irradiance Observations of the Sun
C. Fröhlich 28

Solar Absolute Reference Spectrum
H. Neckel 37

The Solar Spectral Irradiances from X-ray to Radio Wavelengths
O. R. White 45

Observational Programs for Solar and Stellar Irradiance Variability

Irradiance Observations of SMM, Spacelab 1, UARS, and ATLAS Experiments
R. C. Willson 54

Total Solar Irradiance Observations from the EURECA and ATLAS Experiments
D. Crommelynck, V. Domingo, A. Fichot, and R. Lee 63

Irradiance Observations in Near-UV, Visible and Infrared Spectral Bands from Measurements Carried out during ATLAS-1 and EURECA-1 Missions
G. Thuillier, M. Herse, D. Labs, P. C. Simon, D. Gillotay, and H. Mandel. 70

Irradiance Observations from the UARS/SUSIM and ATLAS/SUSIM Experiments
G. Brueckner, L. E. Floyd, P. A. Lund, D. K. Prinz, and M. E. VanHoosier 72

Irradiance Observations from the UARS/SOLSTICE Experiment
G. J. Rottman, T. N. Woods, O. R. White, and J. London 73

Middle Ultraviolet Solar Spectral Irradiance Measurements, 1985–1992, from the SBUV/2 and SSBUV Instruments
R. P. Cebula, E. Hilsenrath, and M. T. Deland 81

The Variable Structure of the X-ray Corona as Viewed by *Yohkoh*
K. Shibata 89

Summary of Future Space Observations at Different Wavelengths
V. Domingo 101

Photometric Variations of Solar Type Stars
R. R. Radick 109

Photometric Observations of the Sun
G. A. Chapman 117

Brightness Observations of the Sun
J. R. Kuhn 130

Observations of Global Solar Magnetic and Velocity Fields
J. T. Hoeksema 138

Variability of Solar and Stellar Irradiance Related to the Network, Active Regions (Sunspots and Plages), and Large-Scale Magnetic Structures

The Variability of Stellar Magnetic Flux, Related Surface Structures and Activity
S. Saar 147

Variations of the Coronal Radiation in X-ray Related to Coronal Holes, Active Region Loop Systems, Bright Points
E. Antonucci 159

Variations of Coronal Radiations at Optical Wavelengths
R. C. Altrock 172

Slowly-Varying Microwave Emissions from the Solar Corona
K. F. Tapping and K. L. Harvey 182

Empirical Models of Solar Total and Spectral Irradiance Variability

Spatial Resolution of Solar Total Irradiance Variability: The *YOHKOH* White-Light Observations
H. S. Hudson 196

The Sun in Time: Detecting and Modelling Magnetic Inhomogenities on Solar-Type Stars
J. D. Dorren and E. F. Guinan 206

Irradiance Models Based on Solar Magnetic Fields
K. L. Harvey 217

Irradiance Effects of Small-Scale Magnetic Fields on the Sun
S. K. Solanki 226

Estimating Solar Forcing of Climate Change during the Maunder Minimum
J. Lean, A. Skumanich, O. R. White, and D. Rind 236

Solar Rotation, Irradiance Changes and Climate
E. Nesme-Ribes, D. Sokoloff, and R. Sadourny 244

Solar and Stellar Oscillations, Irradiance Variations, and their Interpretation

What can we Learn from Oscillation Studies about Irradiance and Radius Changes?
D. Gough 252

Theoretical Interpretation of Magnetic Activity
E. N. Parker 264

Theoretical Interpretation of Solar and Stellar Irradiance Variations
H. C. Spruit 270

Convection and Irradiance Variations
P. A. Fox and S. Sofia 280

The Response of the Earth's Atmosphere to Solar Irradiance Variations and Sun-Climate Connections

Solar Variability Traced by Cosmogenic Isotopes
J. Beer, S. T. Baumgartner, B. Dittrich-Hannen, J. Hauenstein, P. Kubik, C. Lukasczyk, W. Mende, R. Stellmacher, and M. Suter 291

Solar Forcing of Global Climate Change
P. E. Damon and J. L. Jirikowic 301

Response of the Middle Atmosphere to Solar Variability – Model Simulations
T. Y. W. Huang and G. Y. Brasseur 315

A Probable Connection between Solar and Atmospheric Decadal Variability
K. Labitzke and H. Loon 330

Solar Activity and Global Temperature
E. Friis-Christensen and K. Lassen 339

List of Contributed (Poster) Papers 348

Participants

Joe Allen	NOAA/National Geophysical Data Center (U.S.)
Richard F. Altrock	USAF Phillips Lab (U.S.)
Pavel Ambrož	Astronomical Inst. (Czech Republic)
Bo Andersen	Norwegian Space Center (Norway)
Ester Antonucci	Univ. of Torino (Italy)
Steve Arendt	NCAR/High Altitude Obs. (U.S.)
Markus Aschwanden	Univ. of Maryland (U.S.)
Nathalie Audard	Obs. de la Côte d'Azur (France)
Eugene H. Avrett	Smithsonian Astrophysical Obs. (U.S.)
Thomas Ayres	Univ. of Colorado (CASA) (U.S.)
Kurt Bachmann	National Center for Atmospheric Research (U.S.)
Taeil Bai	Stanford Univ. (U.S.)
Sallie Baliunas	Harvard-Smithsonian (U.S.)
Tünde Baranyi	Heliophys. Obs., Hungarian Acad. of Sci. (Hungary)
Jürg Beer	EAWAG-ETH-Zürich (Switzerland)
Gena Bisnovatyi-Kogan	Space Research Center (Russia)
Valentina Bobova	Krimeon (Ukraine)
Pat Bornmann	NOAA Space Environment Lab. (U.S.)
Dave Bouwer	NOAA Space Environment Lab. (U.S.)
Axel Brandenburg	National Center for Atmospheric Research (U.S.)
Peter Brandt	Kiepenheuer Inst. (Germany)
Guy Brasseur	National Center for Atmospheric Research (U.S.)
Pal Brekke	Univ. of Oslo (Norway)
Guenter Brueckner	Naval Research Lab. (U.S.)
Konstantine Burlov-Vasiljev	Main Astronomical Obs. (Ukraine)
John Butler	Armagh Obs. (N. Ireland)
Bruno Caccin	Univ. of Roma (Italy)
Christina Cadavid	SFO/California State Univ. (U.S.)
Cini G. Castagnoli	Univ. of Torino (Italy)
Richard Cebula	Hughes Corp. (U.S.)
Guido Ceppatelli	Arcetri Astrophysical Obs. (Italy)
Kwing L. Chan	Applied Research Corp. (U.S.)
Gary Chapman	SFO/California State Univ. (U.S.)
Xiezhen Chen	Purple Mountain Obs. (China)
Helen Coffey	NOAA/National Geophysical Data Center (U.S.)
Dominique Crommelynck	Royal Meteor. Inst. (Belgium)
Pierre Cugnon	Observatoire Royal de Belgique (Belgium)
Paul Damon	Univ. of Arizona (U.S.)
Matthew Deland	Hughes Corp. (U.S.)
Pierre Demarque	Yale Univ. (U.S.)
Vladimir Dermendjiev	Inst. of Astronomy (Bulgaria)
Giuliana De Toma	Univ. of Trieste (Italy)
Vicente Domingo	ESTEC/ESA (The Netherlands
Richard F. Donnelly	NOAA Space Environment Lab. (U.S.)
Frank Eparvier	Univ. of Colorado, LASP (U.S.)
Laurent Eyer	Observatoire de Genève (Switzerland)
Joan Feynman	Jet Propulsion Lab. (U.S.)
Alan Fiala	U.S. Naval Obs. (U.S.)
Massimo Fofi	Osservatorio Astronomico di Roma (Italy)

Peter Foukal	CRI, Inc. (U.S.)
Peter Fox	NCAR/High Altitude Obs. (U.S.)
Eigil Friis-Christensen	Danish Meterological Inst. (Denmark)
Claus Fröhlich	World Radiation Center, PMOD (Switzerland)
Tim Fuller-Rowler	NOAA Space Environment Lab. (U.S.)
Victor Gaizauskas	Herzberg Inst. (Canada)
Dennis Gallager	Univ. of Colorado (U.S.)
Jorge Gameiro	Centro de Astrofisica, Univ. do Porto (Portugal)
Howard Garcia	NOAA Space Environment Lab. (U.S.)
Elena Gavryuseva	Inst. de Astrofisica de Canarias (Spain)
Didier Gillotay	Aeronomy Inst. (Belgium)
Peter Gilman	National Center for Atmospheric Research (U.S.)
Irina Glushneva	Sternberg Astronomical Inst. (Russia)
Douglas Gough	Univ. of Cambridge (U.K.)
David Gray	Univ. of W. Ontario (Canada)
Madhulika Guhathakurta	USRA/GSFC (U.S.)
Edward Guinan	Villanova Univ. (U.S.)
Ernest Gurtovenko	Main Astronomical Obs. (Ukraine)
Joyce Guzik	Los Alamos National Lab. (U.S.)
Peggy Hartsell	Univ. Colorado, APAS/LASP (U.S.)
John Harvey	National Solar Obs. (U.S.)
Karen Harvey	Solar Physics Research Corp. (U.S.)
Donald F. Heath	Hughes Corp. (U.S.)
Roger Helizon	Jet Propulsion Lab. (U.S.)
John Hickey	Eppley Lab. (U.S.)
Ernest Hildner	NOAA Space Environment Lab. (U.S.)
Ernest Hilsenrath	NASA/Goddard Space Flight Center (U.S.)
Walter Hoegy	NASA/Goddard Space Flight Center (U.S.)
Todd Hoeksema	Stanford Univ. (U.S.)
Douglas Hoyt	Research & Data System Corp. (U.S.)
Hugh Hudson	Univ. of Hawaii (U.S.)
Gor Ivanov-Kholodny	IZMIRAN (Russia)
Rekha Jain	Univ. of St. Andrews (Scotland)
Antonio Jimenez	Inst. de Astrofisica de Canarias (Spain)
John Jirikowic	Univ. of Arizona (U.S.)
Philip Judge	NCAR/High Altitude Observatory (U.S.)
Wolfgang Kalkofen	NASA/Headquarters (U.S.)
R. Kariyappa	Indian Inst. of Astrophysics (India)
Iraida Kim	Sternberg Astronomical Inst. (Russia)
Diana Kjurkchieva	Shoumen Univ. (Bulgaria)
Nathan Kleeorin	Ben-Gurion Univ. of Negev (Israel)
Ron Kroll	Univ. of Arizona (U.S.)
Jeffrey Kuhn	NSO/Sacramento Peak (U.S.)
Donald Kurtz	Univ. of Cape Town (South Africa)
H. Lee Kyle	NASA/Goddard Space Flight Center (U.S.)
Erkki Kyröiä	Finnish Meteorological Isnt. (Finland)
Karin Labitzke	Inst. für Meteorologie (Germany)
Dietrich Labs	Landessternwarte Heidelberg Labs. (Germany)
Massimo Landini	Univ. of Florence (Italy)
Eugene Lavely	Univ. of Colorado (U.S.)

John Lawrence	SFO/California State Univ. (U.S.)
Judith Lean	Naval Research Lab. (U.S.)
Robert Lee	NASA/Langley Research Center (U.S.)
Egil Leer	Univ. of Oslo (Norway)
Torben Leifsen	Univ. of Oslo (Norway)
William Livingston	National Solar Obs. (U.S.)
Jeffrey Linsky	Univ. of Colorado, JILA (U.S.)
Wesley Lockwood	Lowell Obs. (U.S.)
Julius London	Univ. of Colorado, APAS (U.S.)
András Ludmány	Heliophys. Obs., The Hung. Acad. of Sci. (Hungary)
Günter Lustig	Inst. für Astronomie, Univ. Graz (Austria)
Elena Makarova	Sternberg Astronomical Inst. (Russia)
John Martin	National Physical Lab. (U.K.)
Yuri Matvejev	Main Astronomical Obs. (Ukraine)
Patrick McIntosh	NOAA Space Environment Lab. (U.S.)
John McKinnon	NOAA/National Geophysical Data Center (U.S.)
Ann Mecherikunnel	NASA/Goddard Space Flight Center (U.S.)
Werner Mende	Inst. for Hydroecology (Germany)
David Mkrtichian	Astronomical Obs. (Ukraine)
Richard Muller	Obs. du Pic du Midi (France)
Heinz Neckel	Hamburg Obs. (Germany)
John Neff	Univ. of Iowa (U.S.)
Elizabeth Nesme-Ribes	CNRS, Observatoire de Paris (France)
Jun Nishikawa	Hiraiso Solar Terr. Res. Cntr. (Japan)
Jeyasingh Nithianandam	Naval Research Lab. (U.S.)
A. Nusinov	Inst. of Applied Geophys. (Russia)
Atila Özgüz	Kandilli Obs. (Turkey)
Judit Pap	Jet Propulsion Lab. (U.S.)
Eugene Parker	Univ. of Chicago (U.S.)
Lucio Paterno	Catania Univ. (Italy)
Dean Pesnell	Applied Research Corp. (U.S.)
Klaus Pflug	Astrophysical Inst. Potsdam (Germany)
Vladimir Piotrovitch	St. Petersburg Univ. (Russia)
Janine Provost	Obs. de la Côte d'Azur (France)
Lawrence Puga	NOAA Space Environment Lab. (U.S.)
Richard Radick	USAF Phillips Lab. (U.S.)
George Reid	NOAA/ERL (U.S.)
Harold Reitsema	Ball Aerospace Sys. Group (U.S.)
Edward Rhodes	Univ. of South California (U.S.)
Alberto Righini	Univ. of Florence (Italy)
Richard Robinson	NASA/Goddard Space Flight Center (U.S.)
Juan Roederer	Univ. of Alaska (U.S.)
Yuri Romanov	Astronomical Obs., Odessa St. Univ. (Ukraine)
José Romero	World Radiation Center, PMOD (Switzerland)
Gary Rottman	NCAR/High Altitude Obs. (U.S.)
Jean Rozelot	OCA/CERGA (France)
V. Rusin	Astronomical Inst. (Slovakia)
Alexander Ruzmaikin	SFO/California State Univ. (U.S.)
Steve H. Saar	Harvard Univ. (U.S.)
Kenneth Schatten	NASA/GSFC-NSF (U.S.)

Barry Schlesinger	Hughes Corp. (U.S.)
Ed Schmahl	Univ. of Maryland (U.S.)
Hans Schober	Inst. für Astronomie (Austria)
Giuseppe Severino	Osservatorio Astronomico-Napoli (Italy)
Kazunari Shibata	National Astronomical Obs. (Japan)
Jennifer Shiplay	Los Alamos National Lab. (U.S.)
Paul Simon	Aeronomy Inst. (Belgium)
K. R. Sivaraman	Indian Inst. of Astrophysics (India)
Sabatino Sofia	Yale Univ. (U.S.)
Sami Solanki	ETH-Inst. of Astronomy (Switzerland)
Stan Solomon	Univ. of Colorado, LASP (U.S.)
Willie Soon	Harvard-Smithsonian (U.S.)
Henk Spruit	Max Planck Inst. of Astrophysics (Germany)
Jürgen Staude	Astrophysical Inst. Potsdam (Germany)
Robin Stebbins	Univ. of Colorado, JILA (U.S.)
Klaus Strassmeier	Univ. of Vienna (Austria)
Tuula Summanen	Finnish Meteorological Inst. (Finland)
Kenneth Tapping	Dominion Radio Astrophysical Obs. (Canada)
Edward Terez	Simferopol St. Univ. (Ukraine)
Gerard Thuillier	Service d'Aeronomie du CNRS (France)
Gethyn Timothy	CSSA, Stanford Univ. (U.S.)
Antonin Tlamicha	Astronomical Inst. (Czech Republic)
Kent Tobiska	Telos/Jet Propulsion Lab. (U.S.)
Fumihiko Tomita	Hiraiso Solar Terr. Res. Cntr. (Japan)
Maria Torelli	Osservatorio Astronomico-Napoli (Italy)
Michael VanHoosier	Naval Research Lab. (U.S.)
Harry Van Loon	National Center for Atmospheric Research (U.S.)
Anne Vigouroux	Obs. de la Côte d'Azur (France)
Istvan Vince	Astronomical Obs. (Yugoslavia)
William Wagner	NASA/Headquarters (U.S.)
Christoph Wehrli	World Radiation Center, PMOD (Switzerland)
Oran R. White	NCAR/High Altitude Obs. (U.S.)
Dan Wilkinson	NOAA (U.S.)
Richard C. Willson	Jet Propulsion Lab. (U.S.)
George L. Withbroe	NASA/Headquarters (U.S.)
Tom Woods	NCAR/High Altitude Obs. (U.S.)
John Worden	National Center for Atmospheric Research (U.S.)
Fei Wu	Univ. of Colorado (U.S.)
Jiaxiang Zhang	Purple Mountain Obs. (China)
Qizhou Zhang	Harvard Univ. (U.S.)
Harold Zirin	Caltech/Big Bear Solar Observatory (U.S.)

Preface

The IAU Colloquium No. 143 *The Sun as a Variable Star: Solar and Stellar Irradiance Variations* was held on June 20 – 25, 1993 at the Clarion Harvest House, Boulder, Colorado, USA. The main objective of this Colloquium was to review the most recent results on the observations, theoretical interpretations, empirical and physical models of the variations observed in solar and stellar irradiances. A special emphasis of the Colloquium was to discuss the results gained on the climatic impact of solar irradiance variability.

Study of changes in solar and stellar irradiances has been of high interest for a long time. Determining the absolute value of the luminosity of stars with different ages is a crucial question for the theory of stellar evolution and energy production in stellar interiors. Observations of the temporal changes of solar and stellar irradiances – in the entire spectral band and at different wavelengths – provide an additional tool for studying the physical processes below the photosphere and in the solar-stellar atmospheres. Since the Sun's radiative output is the main driver of the physical processes within the Earth's atmosphere, study of irradiance changes is an extremely important issue for climatic studies as well. Climatic models show that small, but persistent changes in solar irradiance may influence the Earth's climate. Furthermore, to understand the human effect on global climatic change, the role of irradiance variations (as a significant source of the natural climate changes) in terrestrial and climatic processes must be revealed.

The Colloquium was a historical meeting since this was the first time when a conference sponsored by the International Astronomical Union was entirely devoted to irradiance variations and their climatic impact. 200 scientists from 30 countries participated in this Colloquium. It was divided into six sessions as defined by their key topics: (1) General Reviews on Observations of Solar and Stellar Irradiance Variability; (2) Observational Programs for Solar and Stellar Irradiance Variability; (3) Variability of Solar and Stellar Irradiance Related to the Network, Active Regions (Sunspots and Plages), and Large-Scale Magnetic Structures; (4) Empirical Models of Solar Total and Spectral Irradiance Variability; (5) Solar and Stellar Oscillations, Irradiance Variations and their Interpretation; and (6) The Response of the Earth's Atmosphere to Solar Irradiance Variations and Sun-Climate Connections. A special 1-day session of the *"Solar Electromagnetic Radiation Study for Solar Cycle 22"* (SOLERS22) was held on June 25, 1993, where the five working groups discussed their progress and future plans on measuring the absolute value of total solar and spectral irradiances and studying their temporal variations.

There were 38 invited talks and 110 contributed poster papers presented at the Colloquium. These papers have demonstrated that the solar energy output changes on different time scales: the short-term (from minutes to months) variations are related to surface modulations mainly caused by the evolution of active regions; the solar-cycle-related long-term variations are directly linked with the evolution of magnetic fields over the activity cycle; while the secular variations over centuries are associated with long-term internal modulations. Although considerable information exists on solar-stellar irradiance variations, their physical origin is not well-understood. The lack of adequate physical models of irradiance variations for predicting the solar-induced climatic changes led to extensive discussions of the consequences of the planned delay or even a possible termination of irradiance observations performed in space. Based on these discussions, the Scientific Organizing Committee released a resolution that addressed this issue to the leaders of the Space Agencies concerned. The resolution was also forwarded to the General Secretary of the International Astronomical Union by the President of IAU Commission 10.

The proceedings of the Colloquium have been printed in two volumes. This volume contains the invited papers in the order of their presentation at the Colloquium within the listed key topics. The contributed papers have been published separately in a special issue of Solar Physics (Vol. 152, 1994) that will also appear in book format by Kluwer Academic Publishers.

This volume has been edited with a LaTeX formatting program released on September 30, 1993 by Alison Woollatt. Ferenc Váradi provided his technical expertise in the LaTeX typesetting. The editors invested considerable effort into making the proceedings uniform throughout. We are responsible for any errors that may have been introduced in this process.

Judit M. Pap
Pasadena, CA, USA

Claus Fröhlich
Davos, Switzerland

Hugh S. Hudson
Honolulu, HI, USA

Sami K. Solanki
Zürich, Switzerland

March, 1994

Acknowledgements

Many people and many organizations helped in organizing the Colloquium and making the proceedings ready for publication. We are sincerely thankful to Dr. Jacqueline Bergeron, IAU General Secretary, and Dr. Immo Appenzeller, IAU Assistant General Secretary, for their help and assistance throughout the planning process of the meeting. We would like to take the opportunity to express our gratitude to Dr. Richard C. Willson (JPL, Solar Irradiance Monitoring Group) and Dr. Ernest Hildner (NOAA Space Environment Laboratory) for their continuous support.

We would also like to express our appreciation and thanks to the members of the Scientific Organizing Committee for their support. The Scientific Organizing Committee consisted of Sallie Baliunas, Richard F. Donnelly, Peter Foukal, Claus Fröhlich (Co-Chair), V. Gaizauskas, Ernest Gurtovenko, Tadashi Hirayama, Hugh S. Hudson (Co-Chair), Julius London, Judit M. Pap (Chair), Paul Simon, Henk Spruit, Béla Szeidl, Jean-Claude Vial, and Oran R. White. Here we are very sorry to report that Dr. Ernest Gurtovenko died on January 20, 1994. Dr. Gurtovenko headed the Solar Physics Department of the Main Astronomical Observatory of the Ukrainian Academy of Sciences for years and served on many national and international committees. His work gave a significant contribution to the models of solar radiation and he built the space telescope *DIFOS* to be launched on *Coronas* in 1994 for studying the global oscillations of the solar brightness. His work as a member of the Scientific Organizing Committee provided a significant help to scientists from the former Soviet Union to attend the Colloquium.

We also wish to express our gratitude to the members of the Local Organizing Committee: Pat Bornmann, Peter Fox, Howard Garcia (Chair), Patrick McIntosh, Larry Puga, Gary Rottman, Andy Skumanich, Kent Tobiska (Co-Chair), and Tom Woods for their help and assistance. We are especially thankful to Kent Tobiska, Pat McLane, Helga Mycroft, Liana The, Mary Eberle, Pam Bergstedt, Judy McGarvey, Gwen Dickenson, and Gregory Dickenson for all their time and hard work in handling the numerous details necessary for the planning and smooth running of the meeting. The staff of the Clarion Harvest House provided a pleasant environment for the meeting. We also express our gratitude to Liana The, Ferenc Váradi and Virginia White, for their help and time during the process of putting together the conference proceedings.

Financial support provided by the National Aeronautics and Space Administration, NASA Jet Propulsion Laboratory, California Institute of Technology, NOAA Space Environment Laboratory, and Ball Aerospace made it possible to carry out this meeting. Travel supports provided by the International Astronomical Union, National Science Foundation, International Science Foundation, and the Department of Energy helped more than 30 scientists from different countries to attend the Colloquium. We are very grateful for the financial assistance of the above organizations.

The Sun as a Variable Star: Solar and Stellar Irradiance Variations

Statement from the Scientific Organizing Committee
Colloquium No. 143 of the International Astronomical Union

It is now established that the terrestrial climate, radiative environment, and upper atmospheric chemistry are strongly influenced by the varying luminosity of the Sun. The human consequences are such that quantitative study over the solar magnetic cycle is imperative for societal planning. The solar variability together with the accumulation of anthropogenic pollutants determine the human milieu of the future. We note that the World Climate Research Program has expounded a full scientific study of these problems.

The scientific presentations at the IAU Colloquium No. 143 make it clear that the forthcoming termination of essential scientific measurements of solar irradiance places these studies in serious jeopardy. One of the most important aspects of the study is the long-term variation of the Sun, which we have enough reason to believe to be significant. However, the present understanding of the basic physical mechanisms of irradiance variations does not allow us to develop adequate irradiance models. Moreover, the limitations on absolute calibration of instruments, the large short-term variability of the Sun, and the inadequacy of theory conspire to make it impossible to establish long-term trends if there are gaps in the observations. Therefore we request the IAU to draw the attention of the major national space agencies named below to these concerns. The problems we have noted can be addressed by implementing the following specific recommendations:

- *NASA*

  (1) High precision observations of solar total and spectral irradiance from UARS and ATLAS are required for one full solar 11-year cycle to achieve progress in understanding key issues of climate change and ozone variation.

  (2) Future flight opportunities on EOS for continuous monitoring of the solar radiative outputs urgently need to be maintained, to study the longer-term trends of potential importance to global change. Failure to achieve overlapping continuous measurement sets from successive satellites will reduce the usefulness of the excellent data already obtained (since 1980) addressing these issues.

- *ESA*

Considering the results obtained from measurements of solar total and spectral irradiance of the Sun from the ultraviolet to the infrared, a second flight of EURECA would provide a unique opportunity to extend the measurements carried out by UARS and EURECA-1, to fill the gaps before and after the launch of SOHO. This proposal is in line with the general objectives of the European Community with respect to the Global Change Program.

- *ISAS and NASDA*

The contribution of *Yohkoh* to studies of solar global properties are recognized, and the participants of IAU Colloquium No. 143 recommend and encourage continued analysis of *Yohkoh* data, and of the development of future Japanese solar and applications satellites that can extend our research knowledge in these important directions.

INTERNATIONAL ASTRONOMICAL UNION

A Historical Perspective on Measurements of Solar Irradiance

By V. GAIZAUSKAS

Herzberg Institute of Astrophysics, National Research Council of Canada, Ottawa, Canada K1A 0R6

Recent measurements made from platforms in space prove beyond question that the radiant energy received from the Sun at the Earth, once called the 'solar constant', fluctuates over a wide range of amplitudes and time scales. The source of that variability and its impact on our terrestrial environment pose major challenges for modern science. We are confronted with a tangled web of facts which requires the combined ingenuity of solar, stellar, planetary and atmospheric scientists to unravel. This brief overview draws attention to key developments during the past century which shaped our concepts about sources of solar variability and their connection with solar activity.

Commission 10 of the International Astronomical Union (IAU) has a long and distinguished association with the subject of this conference on "The Sun as a Variable Star: Solar and Stellar Irradiance Variations". At the time of the first General Assembly of the IAU, held in Rome in 1922, Commission 10 was responsible for 'Solar Radiation'. Its President was Charles Greeley Abbot of the Smithsonian Astrophysical Observatory (Abbot 1922). He is still renowned for directing an epic series of daily measurements of the 'solar constant' from several excellent mountain-top sites for about 30 years. In those early days, the goal of the Commission was to validate the measured variability in the 'solar constant' in terms of solar activity. Subsequent Transactions of the IAU reflect Abbot's vexation with the magnitude of any real variation in the Sun's output. His report to the 1925 General Assembly strikes a plaintive note: "Solar variations exceeding 3% occur rarely, but evidence is accumulating in yet unpublished meteorological researches that solar changes of as little as 1/2 of one percent are not negligible. Hence the problem of securing satisfying accuracy in solar constant measurements is exceedingly difficult, considering that we must observe at the bottom of a turbid atmospheric sea" (Abbot 1925).

From the IAU's beginnings, the motivation behind the quest for variations in solar luminosity was as much to establish a link with variations in terrestrial climate as with solar activity. The Sun's internal source of energy did not enter into the reports of the solar commissions in the 1920's; the great work on the internal constitution of the stars was just then underway. Not until the next decade would nuclear fusion be proposed as the means of energy generation. And it would take four decades before models for stellar evolution were placed on a secure foundation. Those great conceptual advances would have been impossible without accurate measurements of the luminous outputs of the Sun and of a myriad of stars.

Nuclear fusion in stars leads to their variability in luminosity on time scales determined by the exhaustion of nuclear fuels at each stage in the process for converting mass into radiant energy. These time scales are many orders of magnitude greater than the period of sunspot activity. The awesome panoply of coronal and chromospheric activity that we can witness almost every day represents just a minuscule loss from the Sun's vast reserves of energy.

And so, by the time they where in a position to measure solar irradiance directly with instruments placed above Abbot's 'turbid sea', solar scientists realized that any variations must be extremely small. Some even argued that there might be no variation at all in the total luminous output. Whatever radiation was blocked by sunspots might reappear elsewhere in order to preserve a perfect balance – as a brightening of the photosphere around spots, as enhanced emission in faculae, or as waves dissipating energy in the corona. These arguments were supported by lingering suspicions about the validity of Abbot's correlations made into the 1950's between measured changes in the 'solar constant' and weather patterns. Recognition also came in the 1950's and 1960's that solar emissions from UV to X-ray wavelengths were far in excess of black-body values extrapolated from photospheric wavelengths and were, besides, highly variable.

These considerations no doubt weighed heavily in the design of the self-calibrating cavity radiometers flown on the Nimbus-7 satellite (ERB experiment, Hickey *et al.* 1980) and on the Solar Maximum Mission (SMM) satellite (ACRIM experiment, Willson 1981). For the first time, the sensitivity for measuring the total solar irradiance was sufficiently high that dips caused by spots and peaks caused by facular excesses were easily detected and compared successfully with models. Stability of the ACRIM instrument was such that intensity fluctuations due to the 5-minute p-mode oscillations were also recognized. All doubt about the ability of sunspots to block energy flow through the photosphere has been removed; no evidence has since been found to suggest that the contribution of faculae to the solar luminosity cancels that of sunspots. The most important discovery of the Nimbus-7/ERB and SMM/ACRIM irradiance observations was the long-term variation of total solar irradiance over the solar cycle with an amplitude of 0.1% (Willson & Hudson 1988). The trend in the variation of the luminosity suggests that it is directly correlated with the population of active regions and the bright magnetic network. Thus one of the long-term goals set out by the original Commission 10 over 70 years ago has been achieved.

But there is no shortage of remaining problems. All one has to do is to glance at the topics covered in this conference. For example, even after the long-term trend in the irradiance values from ACRIM have been corrected for the sunspot deficits and the facular excesses in active regions, an additional irradiance component remains with a solar-cycle time scale. We tend to forget that, even when there are no sunspots at the minimum phase of the solar cycle, the total magnetic flux of the Sun is still about one-quarter of its mean value at the peak of the cycle and is still variable on a small scale.

We are still far from resolving problems related to atmospheric physics in the presence of a variable solar output at UV wavelengths. These problems have ceased to be merely academic. Public concern about the varying distribution of ozone in Earth's middle atmosphere is high and growing. To assess the effect of variable UV output from the Sun, we require precise solar UV spectrophotometry at the 1% level. In order to assess long-term trends in global warming, more research is needed into the fossil records preserved in ancient sediments and glacial deposits. When, for how long, and by how much did solar activity change significantly from the levels we experience today?

And we must turn to the study of other stars like the Sun to give us insight to the Sun's past and future behavior. It has been pointed out that the ACRIM record shows the total luminosity of the Sun increasing with sunspot activity; yet younger Sun-like stars seem to be anti-correlated with their activity as measured by chromospheric outputs. On the other hand the same stars over a stellar rotation period show an anti-correlation between chromospheric and photospheric outputs. It would seem that the Sun has evolved from an early stage, where its variability was determined by dark spots, to the present one, where it is dominated by faculae. Conversely, problems of general interest to stellar evolution can be answered from solar studies of those phenomena which we now know to be responsible for the variability in the Sun's radiative output.

There is much to celebrate in recent achievements in research on solar-stellar variability. But it is no time to rest on laurels. All the topics I have skimmed over rely on access to data banks of uniformly high quality and high precision. Many of the exciting problems uncovered by recent research demand long-term continuity for key data bases. Yet economic realities are threatening the viability of the kind of long-term research discussed in these proceedings. Commitment is now very difficult to extract from funding agencies for programs of synoptic measurements made either in space or at ground level. Given the public awareness about global warming and ozone depletion, we have to be vigorous in promoting those areas of solar research which address those concerns. We have to take great care that key data bases continue to accumulate. As President of Commission 10 I welcome any resolutions on these matters from this meeting to forward to the Executive Committee of the IAU for formal adoption at next year's General Assembly. And I invite you to consider if a new working group, as proposed by the Scientific Organizing Committee of this Colloquium, is needed in the forthcoming reorganization of the solar commissions of the IAU in order to focus attention on these concerns and to propose courses of action.

REFERENCES

ABBOT, C. G. 1922 *Trans. Int. Astron. Union* **1**, 28–30.

ABBOT, C. G. 1925 *Trans. Int. Astron. Union* **2**, 27–39.

HICKEY, J. R., STOWE, L.L., JACOBOWITZ, H., PELLEGRINO, P., MASCHHOFF, R. H., HOUSE, F. & VONDER HAAR, T. H. 1980 Initial solar irradiance determination from Nimbus-7 cavity radiometer measurements. *Science* **208**, 281–282.

RISE: 1990 Radiative Inputs of the Sun to Earth. *NSF Research Plan*, Washington.

WILLSON, R. C. 1981 Solar total irradiance observations by active cavity radiometers. *Science* **74**, 217–229.

WILLSON, R. C. & HUDSON, H. S. 1988 Solar luminosity variations in solar cycle 21. *Nature* **332**, 810–813.

Solar Monitoring has a Past and Present: Does it have a Future?

By RICHARD C. WILLSON

Solar Irradiance Monitoring Group, Earth and Space Sciences Division, Jet Propulsion Laboratory, California Institute of Technology, Pasadena, CA 91109, USA

Intrinsic variations of total solar irradiance have, after nearly a century of sustained effort, been demonstrated by flight experiments during solar cycles 21 and 22. This accomplishment has been the result of a collaborative effort on behalf of several groups of researchers in both the United States and Europe. The overriding concern at this point of time is whether the integrity of the precision long-term solar total irradiance database, which began at the maximum of solar cycle 21 and continues in the present, will be lost in the late 1990's, prior to the inception of experiments planned for the NASA Earth Observing System beginning in 2002. Experiments are not currently in place to prevent an unbridgeable discontinuity in the precision total solar irradiance database.

---

1. Introduction

The desire to monitor total solar irradiance (TSI), the energy supplied to the Earth by the Sun, has been a goal of mankind for more than a century. Interest in possible variability in this quantity arose from a realization that it is a primary determining factor for the Earth's climate. At any time in the Earth's history, the total solar irradiance, atmospheric chemical composition, and the distribution of oceans and land masses act in combination to determine the radiative balance, and hence the climate, of the biosphere.

In recent years the science of paleoclimatology has made great advances in understanding past climate changes and their causal mechanisms. On the timescale of billions of years, solar luminosity is thought to have increased about 30% during the evolutionary cycle of the Sun. About a billion years before present, the increase in total solar irradiance combined with changes in the composition of the Earth's atmosphere to provide a climate that could support life (Newkirk 1982). The largest climate changes during the existence of life on Earth were most likely of non-solar origin. Some were caused by volcanic-driven sea floor spreading that moved the continental land masses and changed the topology of the ocean basins. The resulting alterations of ocean circulation, the primary heat exchange mechanism for the Earth's climate, resulted in major climate changes. Others may have been caused by the impact of extra-terrestrial objects such as asteroids or large meteorites. The most recent large-scale climate change began about 65 million years ago and continues into the present.

On timescales between 19 and 100 hundred thousand years, periodic variations in the Earth's orbit, the so-called Milankovich effects, have induced an effective variation of the total solar irradiance of the Earth. The record of resulting climate variations has been inferred from the stratigraphy of benthic formanifera and the ratio of oxygen isotopes (16/18 amu) found in cores of sea floor sediment. While this is not an example of intrinsic solar variability and its climatic effect, it is a dramatic example of the climate impact of subtle variations of solar irradiance (Hays *et al.* 1976).

The magnetic 'activity' cycle of the Sun is known to have varied in periodicity and magnitude. Variations over the past ten thousand years are chronicled by the varying abundances of 'cosmogenic' isotopes, notably carbon 14 in living and fossil organic materials and beryllium 10 in Greenland ice cores. Periodicities near 11, 88, 208 and 2300 years have been found (Damon & Sonnet 1991; Beer *et al.* 1990). During the past millennia, variations of solar activity are correlated with periods of significant climate change. While TSI variations were suspected as the cause of these changes, direct observations linking TSI and solar activity variation was not available until recently (Willson & Hudson 1988).

2. Attempts to measure total solar irradiance variability

Attempts were made to detect TSI variability from ground-based observatories, most notably by the Smithsonian Astrophysical Observatory, over the first half of the 20th century (Abbot & Fowle 1908; Hoyt 1979). During the following 30 years, experiments with the same objective were conducted using aircraft, high-altitude balloons, sounding rocket and space flight platforms (Fröhlich 1976). All were unable to detect total irradiance variations that were unambiguously solar in origin. Failure of the intra-atmospheric experiments was caused by uncertainties of atmospheric attenuation for the TSI that masked solar variations from even high flying aircraft and balloons. Failure of extra atmospheric experiments on rockets and early satellites was due to limitations in accuracy and precision of early solar monitoring flight instrumentation (Willson 1984).

Two developments following the end of the Smithsonian ground-based program facilitated the modern exploration of solar total irradiance variations: the development of new, more accurate and precise TSI detectors, and the availability of space flight platforms capable of supporting sustained observations.

The first of a new generation of total solar irradiance measurement instrumentation, so-called electrically self-calibrating cavity (ESCC) detectors, was developed at the Jet Propulsion Laboratory (JPL) in the mid 1960's as part of its effort to understand the thermal behavior of interplanetary spacecraft. Several versions of these new "pyrheliometers" were subsequently developed at JPL for different applications: the Thermal Control Flux Monitor (TCFM) for the Mariner 1969 mission (Plamondon 1969), the Practical Cavity Radiometer (PACRAD) for laboratory measurements (Kendall & Berdahl 1970) and the Active Cavity Radiometer (ACR) for flight observations of TSI on balloon, rocket, shuttle and satellite platforms (Willson 1971, 1984). Similar detectors were developed at other laboratories during the early 1970's: at the Physical Meteorological Observatory at Davos (PMOD – Switzerland) (Brusa & Fröhlich 1972), the National Bureau of Standards (NBS – USA) (Geist 1972), the Eppley Laboratory (EL – USA) (Hickey & Karoli 1974) and the Royal Meteorological Institute of Belgium (RMIB) (Crommelynck 1973). An experiment was conducted in 1975 to establish a TSI radiation scale (the World Radiation Reference) embodying the performance of five of these new sensors: three from JPL and one each from the PMOD and RMIB (Brusa & Fröhlich 1975).

3. Observations of total solar irradiance variability

The first implementation of ESCC pyrheliometry in a space flight environment was by the Jet Propulsion Laboratory's TCFM experiments on the Mariner 6 and 7 spacecraft in 1969. A non-ESCC solar detector was included in the Nimbus-6/ERB experiment in 1975. Both were unable to detect intrinsic solar variability due to limitations in instrument accuracy, calibration, precision and sensor degradation (Plamondon 1969).

The first long-term solar monitoring utilizing an ESCC sensor was the Earth Radiation Budget (ERB) experiment on the NOAA Nimbus-7 spacecraft. The ERB solar database, beginning in late 1978 and continuing to early 1993, is the longest currently available (Kyle *et al.* 1993). Limitations imposed on ERB solar observations by the absence of solar pointing on the Nimbus platform sustained a noise level in the ERB results that inhibited recognition of intrinsic solar variability until subsequent detection by JPL's Active Cavity Radiometer Irradiance Monitor I (ACRIM I) experiment on the NASA Solar Maximum Mission (SMM) in 1980 (Willson *et al.* 1981). The mutually corroborative function of the ACRIM I and ERB results has played an important role in verifying solar variability on solar activity cycle timescales.

A series of shorter-term TSI experiments has been flown on or deployed from the Space Shuttle to provide comparison experiments for satellite monitors. The Spacelab 1 and ATLAS flights between 1983 and 1993 employed two different TSI experiments, as has the shuttle-deployed EURECA platform that operated for 10 months starting in mid 1992 (Willson 1994; Fröhlich 1994; Crommelynck *et al.* 1994). The shuttle ACRIM experiment has demonstrated a capability of sustaining satellite TSI experiments with less than 300 parts-per-million (ppm) uncertainty, a precision comparable or superior to other methods (including radiometers operating at cryogenic temperatures) except for overlapping, long duration satellite experiments.

4. Results of total solar irradiance variability observations

The most significant finding from the precision TSI database thus far is on solar cycle timescales: a direct correlation of luminosity and solar activity (Willson & Hudson 1988, 1991). With an 0.1% peak-to-peak amplitude during solar cycle 21, it agrees in sense with that predicted from the coincidence of the "Little Ice Age" climate anomaly and the "Maunder Minimum" of solar activity during the 16th and 17th centuries.

Solar cycle TSI variation is predicted with varying degrees of success by regression models using the precision TSI database and 'proxies' of solar activity, such as the Zürich sunspot number, the 10.7 cm microwave flux, He I 1083 nm flux and the 'core-to-wing ratio' of Mg II. The use of the He I model led to the initial realization of the primary role of faculae and the bright network in the solar cycle TSI variation (Foukal & Lean 1988, 1990; Livingston *et al.* 1988; Fröhlich & Pap 1989; Willson & Hudson 1991).

As useful as these statistical 'proxy' models have been, however, it is not surprising in view of their essentially non-physical nature that significant errors are found in some model predictions. Underestimation of TSI has been a characteristic flaw near the maxima of solar cycles 21 and 22. Use of models based on proxies with long time series, such as the sunspot area, has tempted some modelers to make tenuous extrapolations into the past, regardless of the highly uncertain character of early proxy observations.

An inverse relationship between sunspot area and total irradiance has been found on the solar rotational timescale (27 days) with deficits in total irradiance of as much as -0.3% (Willson *et al.* 1981; Willson 1982). There is growing evidence that most of the missing flux is balanced by excess facular radiation on the active region timescale (months) with the rest re-distributed through the bright network on the solar cycle timescale.

On the shortest timescales, solar global oscillations of low degree have been detected in the ACRIM I total irradiance data, including pressure modes (timescales of minutes – the so-called 5-minute oscillations or 'p-modes') (Hudson & Woodard 1983; Woodard & Hudson 1983) and possible gravity modes (timescales of hours to days) (Fröhlich 1987). Interpretation of the 5-minute oscillation results from the ACRIM I experiment has placed an upper limit on the interior rotation rate of the Sun, and therefore on solar oblateness, providing support for the relativistic interpretation of the precession of the perihelion of Mercury. The p-mode oscillations are concentrated in the convection zone. Therefore the depth within the Sun for which their analysis can provide new physical insight is limited. Should gravity mode oscillations be verified in TSI data, however, information on physical processes extending to the solar core would be accessible.

Timescales shorter than a year do not appear to be of direct climatological interest, but contain information on solar variability that provided much new insight into the physics of the Sun. Continuous TSI monitoring, particularly by satellites with a high duty cycle of solar pointing during each orbit, can provide the observations that will facilitate future solar models that can predict TSI variability with sufficient precision to anticipate corresponding climate variations.

5. Present and planned total solar irradiance monitoring

The precision TSI database is being sustained at present by two experiments: the NASA ERBS experiment launched in 1984, and the JPL/NASA Upper Atmosphere Research Satellite (UARS)/ACRIM II experiment launched in 1991. The Nimbus-7/ERB experiment, after more than 14 years of operation, ceased acquiring TSI data in early 1993. The next planned satellite TSI experiment is the ESA SOHO/VIRGO (1995 launch), to be followed by three ACRIM experiments in the NASA Earth Observation System CHEM platforms (beginning in 2002). Annual underflights aboard the Space Shuttle by ACRIM and CROM instrumentation will continue at least through 1994.

6. Sustaining the total solar irradiance database

The detection of solar luminosity variability proportional to magnetic activity during solar cycles 21 and 22 provides the impetus to extend the irradiance database with maximum precision. While the 0.1% peak-to-peak variation during these cycles may be of minor climatological interest, the direct correlation between activity and luminosity implicates TSI variations as the causal mechanism for known frequencies of climate changes. If that is the case there must be other luminosity variabilities with longer periodicities and proportionately larger amplitudes that have significant climatic implications. Subtle trends in the total irradiance of as little as ± 0.5 percent per century could eventually produce the extreme range of climates known to have existed in the past, from warm periods without permanent ice, to the great ice ages (Eddy 1976).

Accumulation of a century-long database capable of detecting TSI trends of climatological significance, with the continuity required for development of a physically meaningful solar TSI model, will necessarily be composed of the results of many individual solar monitoring experiments. If these experiments last between three and ten years each, the results of from ten to twenty will have to be related with adjacent precisions of 50 to less than 20 ppm to resolve a century of change at the 0.5% level with a 10:1 signal-to-noise ratio. A careful measurement strategy will be required to sustain the required precision for the TSI database since the absolute uncertainty of current satellite instrumentation is inadequate for this purpose.

The absolute uncertainty (relative to S.I. units) of the current generation of flight TSI instrumentation, which operates at ambient temperatures, is about 1000 ppm in the laboratory and about 2000 ppm in flight experiments. The absolute uncertainty of a new generation of TSI sensors operating near the temperature of liquid Helium approaches 100 ppm in the laboratory (Martin & Fox 1994). These cryogenic sensors face some significant hurdles in the effort to transform them into space flight experiments, however. They must use much smaller apertures (0.3 cm diameter) than their laboratory versions to minimize solar input that would otherwise prevent the space flight qualified stirling cycle 'refrigerators' from maintaining temperatures below 20 °K. Such small apertures are difficult to make and measure accurately and they are much more susceptible to contamination than the large-apertured (0.8 cm diameter) ambient temperature sensors. A realistic expectation for their eventual in-flight performance would be in the several hundred ppm uncertainty range. Clearly, the absolute uncertainty of neither the ambient or cryogenic temperature TSI sensor technology is adequate to sustain a contiguous long-term database.

A relative precision of less than 50 ppm is achievable for the data of overlapped solar monitors with the current state-of-the-art ambient temperature instrumentation, given a sufficient number of overlapping comparisons (Willson 1994). The principal source of uncertainty for the 'overlap strategy' is degradation of solar monitoring sensors over multi-year missions. The ACRIM experiments have demonstrated that calibration of such degradation is possible at the level of a 50 ppm per decade or less (Willson & Hudson 1991).

It is clear that the only available monitoring approach capable of supporting the contiguous TSI database is one that relates successive solar monitoring experiments at a precision level defined by the operation of the instrumentation in the space flight environment. The method of choice is the overlapping and comparison of successive solar monitoring experiments: a so-called 'overlap strategy'. Failing that, relating successive TSI monitors using results from a 'third party' experiment that overlaps both would be next best, but with inherent additional uncertainty.

A logical role for cryogenic instrumentation, at least during its developmental stage, is as a shuttle-based underflight experiment. A relatively inexpensive experiment based on the use of expendable cryogens could be used to test the technology and provide periodic calibrations of ambient temperature satellite TSI monitors with enhanced accuracy.

The 'overlap strategy' was to have begun with the SMM/ACRIM I and UARS/ACRIM II experiments. Unfortunately the SMM mission ended in late 1989, two years before UARS could be launched (the result of the Space Shuttle hiatus following the Challenger loss). The relationship between the ACRIM I and ACRIM II experiments has instead been established using a 'third party' overlap strategy using the results of comparisons of both ACRIM I and ACRIM II with the Nimbus-7/ERB experiment. The ratio of ACRIM II to ACRIM I is 1.001759 with a statistical uncertainty smaller than 10 ppm, a demonstration of the capability of the 'overlap strategy'.

The Nimbus-7/ERB experiment ceased acquiring data in early 1993. The outlook for current TSI monitoring experiments by ACRIM II and ERBS must be viewed with a degree of pessimism. ERBS has been in flight nearly a decade, it could fail at any time without surprising anyone. Several problems have already surfaced with the UARS satellite: the batteries are deteriorating faster than anticipated, and the primary solar panel drive system has failed. Neither may last into the SOHO/VIRGO timeframe of 1995 and beyond. SOHO is assumed to be at least a 3-year minimum mission, but its follow-on, EOS/ACRIM experiment, is not scheduled until 2002 in current plans. The probability of a TSI data gap in the mid or late 1990's or early 2000's is high.

Additional solar monitoring experiments are needed between 1994 and 2003 to guarantee continuity of the high-precision TSI database. Failure to implement them may produce an unbridgeable gap in the long-term TSI database at the level of precision required to provide a contiguous, long-term TSI database for climate and solar physics. If a gap in monitoring occurs it will become necessary to start the high precision TSI database over again in 2003, with the relationship between it and the current database unknown with useful precision.

7. Recommendations for saving the total solar irradiance database

A combination of the continuation of one and the addition of a TSI Mission comprised of a series of new, low cost flight experiment activities during the mid-to-late 1990's could provide a high probability of preserving the contiguous TSI database. The first would be the continuation of the annual flights of the shuttle TSI comparison instrumentation beyond its currently planned third and final flight in the ATLAS series of missions in 1994. The ATLAS/ACRIM experiment can sustain a year-to-year precision of better than 300 ppm. This is far short of the 20 – 50 ppm desired, but much superior to the 2000 ppm of ambient TSI experiment accuracies, and comparable to the probable performance of flight versions of the cryogenic TSI experiments.

The first in the series of new flight experiments would have to be launched soon to guarantee overlap with UARS/ACRIM II and SOHO/VIRGO. New small satellite and TSI instrumentation developments make this a potentially feasible, science-effective and cost-effective possibility. Ideally, two TSI satellites should be launched initially. Following comparison, one of the two small satellites would continue monitoring on a full time basis and the other would be saved (except for periodic re-comparisons) as backup for the first. Ground spares would be launched as replacements upon failure of one of the first two satellites, replacing the reserve satellite capability and adding redundancy. Such a TSI mission approach could provide an optimum long-term monitoring modality.

Acknowledgments. This paper was completed by the Jet Propulsion Laboratory, California Institute of Technology, under a contract with the National Aeronautics and Space Administration.

REFERENCES

ABBOT, C.G. & FOWLE, F. I. 1908 *Annals of the Astrophysical Observatory of the Smithsonian Institution*, **II**.

BRUSA, R.W. & FRÖHLICH, C. 1972 Entwicklung eines neuen absolutradiometers. *Technical Note 1.*, World Radiation Center, Davos, Switzerland.

BRUSA, R.W. & FRÖHLICH, C. 1975 *Scientific discussions IPC IV*, World Radiation Center, Davos, Switzerland.

BEER, J., BLINOV, A., BONANI, G., FINKEL, R.C., HOFMANN, H.J., LEHMANN, B., OESCHGER, H., SIGG, A., SCHWANDER, I., STAFFELBACH, T., STAUFFER, T., SUTER, M. & WOLFLI, W. 1990 Use of ^{10}Be in polar ice to trace the 11-year cycle of solar activity: Information on cosmic ray history. *Nature* **347**, 164–166.

CROMMELYNCK, D. 1973 Instrumentation theory of absolute radiometry. *Pub. A* **81**, Royal Meteorological Institute of Belgium, Brussels, Belgium.

CROMMELYNCK, D., DOMINGO, V., FICHOT, A. & LEE, R. 1994 Irradiance observations from the EURECA and ATLAS experiments. In *The Sun as a Variable Star: Solar and Stellar Irradiance Variations* (ed. J.M. Pap, C. Fröhlich, H.S. Hudson & S.K. Solanki). Cambridge University Press, in press.

DAMON, P. & SONNET, C. P. 1991 Solar and terrestrial components of the atmospheric ^{14}C variation spectrum. In *The Sun in Time* (ed. C. P. Sonnett, M. S. Giampapa & M. S. Matthews). pp. 360–388. Univ. of Arizona Press, Tucson, AZ, USA.

EDDY, J.A. 1977 Historical evidence for the existence of the solar cycle. In *Solar Output and Its Variation* (ed. O.R. White). pp. 51–71, Univ. of Colorado Press, Boulder, CO, USA.

FRÖHLICH, C. 1977 Contemporary measures of the solar constant. In *Solar Output and Its Variation* (ed. O.R. White). pp. 93–110. Univ. of Colorado Press, Boulder, CO, USA.

FRÖHLICH, C. 1987 Solar oscillations and helioseismology from ACRIM/SMM irradiance data. In *New and Exotic Phenomena* (ed. O. Fackler & J. T. T. Van). pp. 395–402. Gif-sur-Yvette: Editions Frontieres.

FRÖHLICH, C. & PAP, J. M. 1989 Multi-spectral analysis of total solar irradiance variations. *Astron. Astrophysics* **220**, 272–280.

FRÖHLICH, C. 1994 Reviews of space observations of total solar irradiance. In *The Sun as a Variable Star: Solar and Stellar Irradiance Variations*, (ed. J.M. Pap, C. Fröhlich, H.S. Hudson & S.K. Solanki). Cambridge Univ. Press, in press.

GEIST, J. 1972 Fundamental principles of absolute radiometry and the philosophy of this NBS program. *Tech. Note* 591-1, USNBS, Gaithersburg, MD.

HAYS, J.D., IMBRIE, J. & SHACKLETON, N. J. 1976 Variations in the Earth orbit: pacemaker of the ice ages. *Science* **194**, 1121–1132.

HICKEY, J. R. & KAROLI, A. R. 1974 Radiometric calibrations for the Earth Radiation Budget experiment. *Appl. Opt.* **13**, 523–533.

HOYT, D.V. 1979 The Smithsonian Astrophysical Observatory solar constant program. *Rev. of Geophysics and Space Physics* **17**, 427–458.

HUDSON, H.S. & WOODARD, M. F. 1983 Solar surface granulation and variations of total solar irradiance. *Bull. Amer. Astron. Soc.* **15**, 715.

KENDALL, J. M. & BERDAHL, M. 1970 Two blackbody radiometers of high accuracy. *J. Appl. Opt.* **9**, 1082–1091.

KYLE, H. L., HOYT, D. V., HICKEY, J. R., MASCHOFF, R. H. & VALLETTE, B. J. 1993 Nimbus-7 Earth Radiation Budget calibration history–Part I: the solar channels. *NASA Reference Publication* **1316**, Goddard Space Flight Center, Greenbelt, MD.

LIVINGSTON, W., WALLACE, L. & WHITE, O. R. 1988 Spectrum line intensity as a surrogate for solar irradiance variation. *Science* **240**, 1765–1767.

MARTIN, J. E. & FOX, N. P. 1994 Cryogenic absolute radiometer, CSAR. *Solar Physics*, in press.

NEWKIRK, G. A. 1982 The nature of solar variability. In *Panel on Solar Variability, Weather and Climate, Studies in Geophysics* pp. 33–47. National Academy Press, Washington, D.C., USA.

PLAMONDON, J.A. 1969 TCFM solar observations on Mariner 69. *JPL Space Programs Summary* **3**, pp. 162. Jet Propulsion Laboratory, Pasadena, CA, USA.

WILLSON, R.C. 1971 The active cavity radiometric scale, the international pyrheliometric scale and the solar constant. *J. Geophys. Res.* **74**, 4325.

WILLSON, R.C. 1982 Solar irradiance variations and solar activity. *J. Geophys. Res.* **86** 4319–4326.

WILLSON, R.C. 1984 Measurements of solar total irradiance and its variability. *Space Science Reviews* **38**, 203–242.

WILLSON, R.C. 1994 Irradiance observations from the SMM, UARS and ATLAS ACRIM experiments. In *The Sun as a Variable Star: Solar and Stellar Irradiance Variations* (ed. J. M. Pap, C. Fröhlich, H. S. Hudson & S. K. Solanki). Cambridge University Press, in press.

WILLSON, R.C., GULKIS, S., JANSSEN, M., HUDSON, H. S. & CHAPMAN, G. A. 1981 Observations of solar irradiance variability. *Science* **211**, 700–702.

WILLSON, R.C., HUDSON, H. S., FRÖHLICH, C. & BRUSA, R. 1986 Long-term downward trend in total solar irradiance. *Science* **234**, 1114–1117.

WILLSON, R.C. & HUDSON, H. S. 1988 Solar luminosity variations in solar cycle 21. *Nature* **332**, 810–812.

WILLSON, R.C. & HUDSON, H. S. 1991 The Sun's luminosity over a complete solar cycle. *Nature* **351**, 42–44.

WOODARD, M.F. & HUDSON, H. S. 1983 Frequencies, amplitudes and linewidths of solar oscillations from solar total irradiance observations. *Nature* **318**, 449–450.

Solar Variability and its Terrestrial Effects

By GEORGE L. WITHBROE[1] AND WOLFGANG KALKOFEN[1,2]

[1]Space Physics Division, NASA Headquarters, Washington, DC 20546, USA

[2]Harvard-Smithsonian Center for Astrophysics, Cambridge, MA 02138, USA

The solar output in radiation and particles varies on an eleven-year time scale, affecting terrestrial temperatures and perhaps even climate. We examine some of the data that support this contention.

1. Introduction

The Sun varies on many time scales. They may be as short as seconds or less for flares; weeks, for phenomena linked to the solar rotation; decades, for phenomena associated with the activity cycle and the magnetic cycle; or centuries and longer, for variations on time scales of ice ages and warm interglacial periods.

While we may experience the effects of events of the shortest to intermediate time scales, we must rely on the historical and geological records to tell us about the longest. For the shortest time scales we observe the explosive phenomena directly in the outermost layers of the Sun, both from the ground in visible light and, much more dramatically, from space in very high energy radiation and relativistic particles; for the longest time scales, we can observe the behavior of stars that are solar twins in most respects yet differ from the Sun in the level of their luminosity variation and, thus, exhibit behavior that the Sun may have shown in the past or may show in the future.

In this introductory paper we will merely touch upon some of the topics associated with these time scales. In Section 2 we discuss the variation in the solar output in relation to the eleven-year solar cycle, in Section 3 we consider the effects of solar variability on terrestrial surface temperatures and climate, and in Section 4 we compare the Sun to other stars; in the final section we draw some conclusions.

2. Solar variability

The most conspicuous phenomenon of solar variability is the appearance and disappearance of dark spots on the face of the Sun, which occurs at the rhythm of ten to twelve years (Figure 1) and an average (during the last three centuries) of eleven years. Sunspots mark the waxing and waning of solar activity. At the same time, the polarity of the solar magnetic field reverses, so the magnetic cycle is twice as long as the activity cycle. As is evident from Figure 1, which shows the sunspot number (a measure of activity that depends on the number of individual spots and the number of groups of spots), the length of the cycle is not sharply defined, nor is the level of activity. Another indicator of the solar cycle is the radio flux at a wavelength of 10.7 cm. Its smoothed variation matches very well the behavior of the sunspot number, and is another convenient measure of solar activity. There is a hint that a rapid rise of activity leads to a short cycle and a high maximum and, thus, to a relation between cycle length and activity level at solar maximum (cf. also Figure 5).

Near solar maximum, the level of emission from the hot outermost layers of the Sun rises. This can be seen in Figure 2, which shows the variation in the flux at several wavelengths late during the rising phase towards the maximum of cycle 21. The dominant

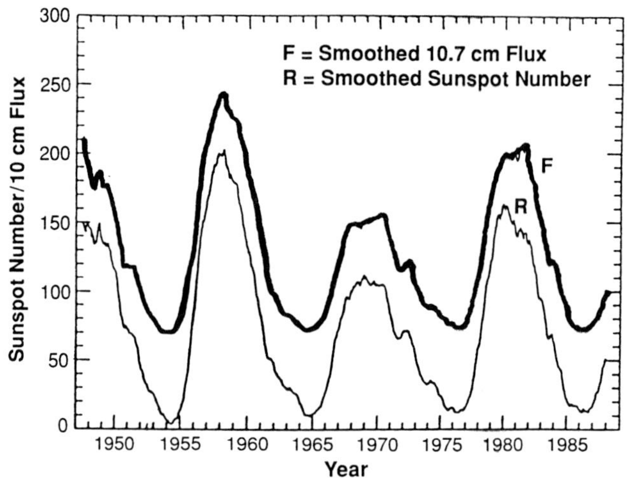

FIGURE 1. Smoothed (13-month running mean) sunspot number, R, and solar 10.7 cm radio flux for solar cycles 19, 20, and 21 (Withbroe 1990).

variation is that of the 27-day solar rotation, but a distinct increase in the emission maxima is also evident. Most of the emission occurs in the ultraviolet part of the spectrum, as shown in the right-hand panels of the figure. At Lyman-α, for example, the increase in emission at solar maximum over that at minimum may be as much as a factor of two, while at X-ray wavelengths it can be more than two orders of magnitude. It is interesting to note that, for this rotational modulation of the emission, the maxima of the short wavelength radiation from the hot corona coincide approximately with the minima in the total irradiance (cf. Lean 1991, Figure 14). Thus, sunspots cause a short-term deficit in total irradiance when they are directly facing the Earth. Even though the extra emission at short wavelengths also peaks at that time, it is insufficient to compensate this sunspot deficit. Nevertheless, despite this short-term shortfall in solar output, total irradiance averaged over a solar rotation is maximal at sunspot maximum (cf. Figure 7). The source of the excess radiation is unknown.

The frequency of optical flares follows the sunspot number (cf. Figure 3), increasing by more than an order of magnitude at sunspot maximum over that at minimum. Flares are explosive events occurring in the very hot, tenuous outer layers of the solar atmosphere. They generate copious amounts of high-energy photons and particles, including relativistic elementary particles with energies beyond 10 MeV. Large flares, and especially large X-class flares which emit prodigious quantities of radiation, tend to occur in the declining phase of the solar cycle beyond maximum. The frequency of proton events (cf. Figure 3) follows this behavior, as do geomagnetic storms, which are responses of the Earth's magnetosphere to coronal mass ejections and high-speed solar wind streams from low-latitude coronal holes. Geomagnetic storms can cause catastrophic failures of electric power grids. In the upper atmosphere of the Earth, where high-energy radiation and particles from flares are absorbed, they modify the photochemistry as well as the temperature and density structure of the atmosphere. Large flares affect satellites in two

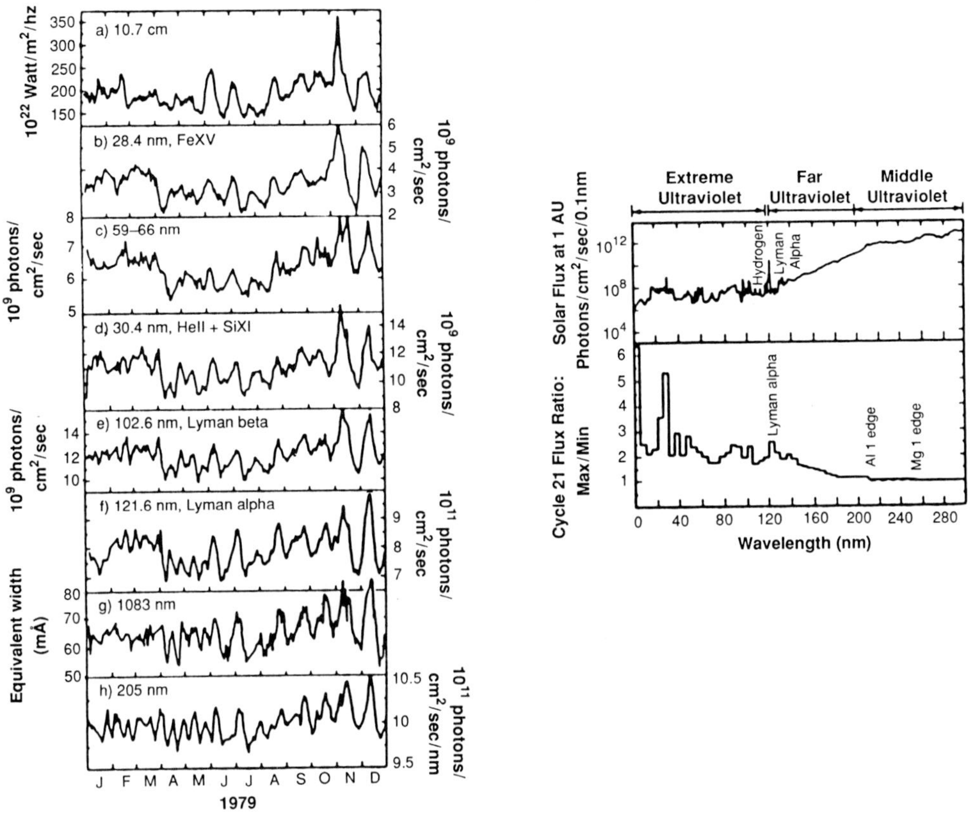

FIGURE 2. Left: Solar variability in 1979 at 10.7 cm in the radio band, 10,830 Å in the infrared, and at several UV and EUV wavelengths. Right: The solar UV spectrum, and an estimate of the spectral variability shortward of 2900 Å during cycle 21 (Lean 1987).

ways: by exposing them to intense, penetrating radiation, which could be dangerous to astronauts and biological systems, and by limiting the life times of satellites in low earth orbit through atmospheric heating and increased drag.

3. Terrestrial surface temperatures and climate

As is evident from Figure 1, the solar cycle is irregular, both in length and in intensity. We may ask what the extent of the variability of the Sun has been in the past, and what it might be for a star like the Sun in the future. Considering the first question, we see in Figure 4 the sunspot number from 1600 to the present time (Lean 1991). The solid line represents monthly means and the dashed line and crosses, yearly means. The data clearly show the alternation between maxima and minima in the sunspot number. While the length of the activity cycle is quite well defined by the data from the last three centuries as close to 11 years, the amplitude of the sunspot number varied by as much as a factor of five. Yet sunspots never failed to appear at the expected time of solar maximum. It is different for the preceding half century, the so-called Maunder Minimum, when sunspots were either few in number or completely absent. The cycle length is therefore undefined in the data plotted. However, Nesme-Ribes reports at this

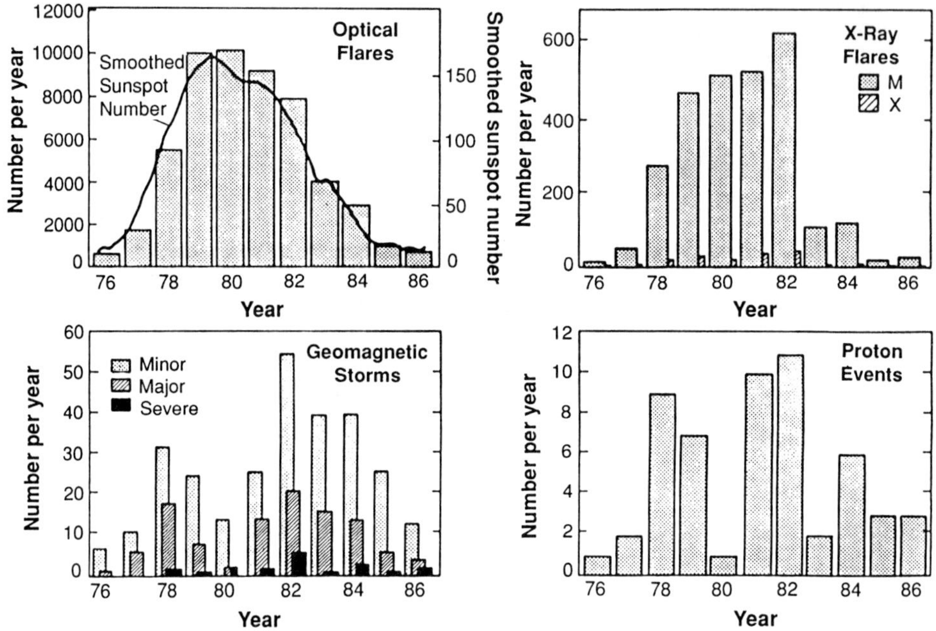

FIGURE 3. Frequency of flares, geomagnetic storms and proton events during maximum of cycle 21.

conference that the length of the cycle during the Maunder Minimum was 9 or 10 years. This combination of a short cycle length and very low activity is surprising in view of the usual result that a short cycle length implies an active maximum.

According to Eddy (1976), "The 11 year sunspot cycle may not have been operative before about 1700. Instead, solar activity appears to have been characterized by extended periods of either enhanced activity or inactivity. Thus during the 290 years when the Sun has been observed most intensely its behavior may have been unusually regular and benign." The quality of this conclusion parallels the finding (Anklin *et al.* 1993) from the Greenland Ice Core Project (GRIP) that abrupt and large climatic oscillations can occur during both glacial and interglacial periods, and that the past 8000 years have been strangely stable. Even though the periods discussed in the two papers are different and a definite connection between solar activity and climate has not been definitely established, circumstantial evidence suggests that the similarity of conclusions is not accidental. It is interesting to note here that approximately one third of the solar-like stars appear to be in a state analogous to the Maunder Minimum (cf Baliunas & Jastrow 1990), which matches the fraction of time the Sun has been in a state of minimal activity with very low sunspot numbers during the last millenium (cf Figure 6). Thus, both the past behavior of the Sun and the present behavior of solar-like stars suggest that the stability in solar behavior we have witnessed during the last several centuries is unusual.

A relation between solar activity and climate is suggested by the data covering a century on sunspot number, cycle length and terrestrial surface temperatures displayed in Figure 5. There is an obvious correspondence between cycle length and land temperature in the northern hemisphere and, similarly, between sunspot number and global sea temperature. Apart from the impression that the (inverse) cycle length peaks earlier than does the sunspot number, one would conclude that a short cycle length corresponds to a strong maximum in sunspot number. It is interesting to note again the apparent contra-

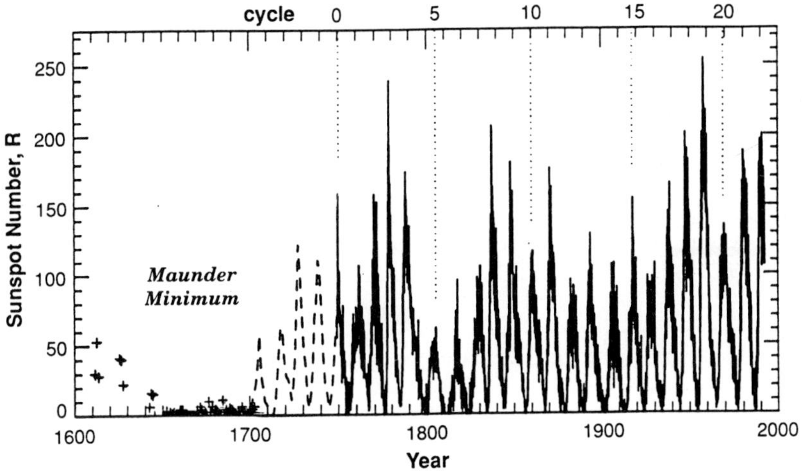

FIGURE 4. Sunspot number from 1600 to the present time.

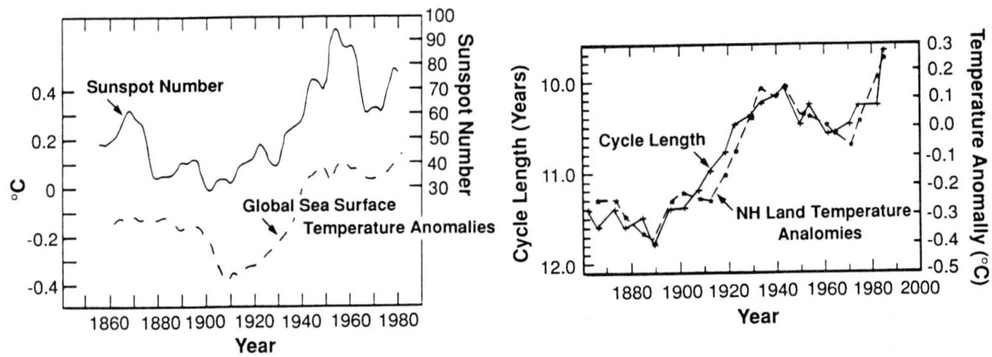

FIGURE 5. Comparison (left) of 11 year running mean of the sunspot number with global average sea surface temperature anomalies (Reid 1991), and (right) length of solar cycle, from smoothed sunspot extrema, with northern hemisphere temperature anomalies (Friis-Christensen & Lassen 1991).

diction that the cycle length appears to have been unusually short during the Maunder Minimum. The data displayed in Figure 5 would have suggested land and sea temperatures higher than usual at that time. But it is known from the historical record as well as from Dutch landscape paintings that temperatures were uncommonly low, giving the period the name of Little Ice Age. Other observations showing a close relation between solar activity and climate are, for example, the amount of rainfall in several nothern latitude bands over the last century (cf. Noyes 1982 Ch. 10): The rainfall was correlated with the sunspot number in the latitude band from 70° to 80°, anticorrelated from 60° to 70°, and changed correlation in the band from 50° to 60°.

Figure 6 makes again the case for a connection between solar activity and climate in northern Europe. Several measures of solar activity over the last millenium are indicated: ^{14}C variation (lower line and open circles), a qualitative measure of the severity of winter (upper line, shifted right to allow for the carbon isotope to be assimilated into tree rings), the sunspot number since 1700, and naked-eye sunspot observations (dark circles); the

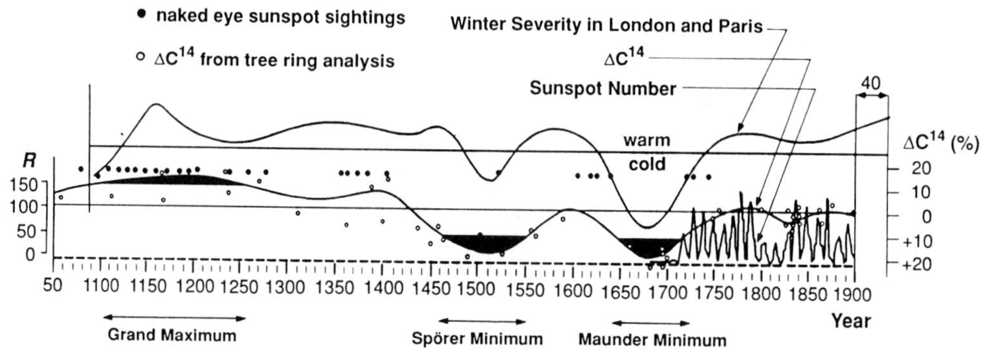

FIGURE 6. Relation between the severity of winter and solar activity (from Eddy 1976, 1977).

Maunder and Spörer Minima and the Medieval Maximum are indicated by shading. Qualitatively, the general correspondence between climate, as perceived in London and Paris, and solar activity, as indicated by organic carbon deposits and sunspot number, is evident.

4. Solar and stellar variability

A means by which the Sun can influence climate is the variation of the total irradiance. The question is whether the measured irradiance variation is sufficiently large. As Figure 5 showed, solar activity correlates with terrestrial surface temperature. However, it is generally believed that the current amplitude of the variation of total irradiance of 0.1% can have only a negligible effect on climate. Figure 7 displays total solar irradiance over approximately one cycle, as well as sunspot number. It is noteworthy that although both ERB and ACRIM agree on the variation of 0.1%, they differ on the level of the irradiance by 0.3%. This worrisome fact would not have been discovered without overlapping coverage, which points to the necessity of having a second instrument commence measurement before the first one ends operation.

The expectation that an irradiance change of 0.1% cannot affect climate appears to be inconsistent with the implications of the obvious correspondence between the low level of solar activity during the Maunder Minimum and the low temperatures experienced in many parts of the world. However, it is thought that solar irradiance during the Maunder Minimum was depressed not by the usual 0.1%, but by 0.3%, and during an extended period of time (see also papers in this volume by Lean *et al.* and Nesme-Ribes *et al.*). This raises the question of the range of luminosity variation that can be expected from a star like the Sun.

Figure 8 gives the histogram of magnetic activity for solar-analog stars as measured by calcium emission. An HK index greater than 0.17 corresponds to the active Sun, and smaller than 0.17 to the quiet Sun. The Sun thus appears to show average behavior as far as chromospheric emission is concerned. The same conclusion is reached by Lockwood (1994) who notes that in the Mount Wilson sample of stars, the Sun fits into the middle of the distribution of calcium indices. However, he also reports on the luminosity varation of solar-type stars observed at Lowell Observatory. In this sample, the Sun joins a minority

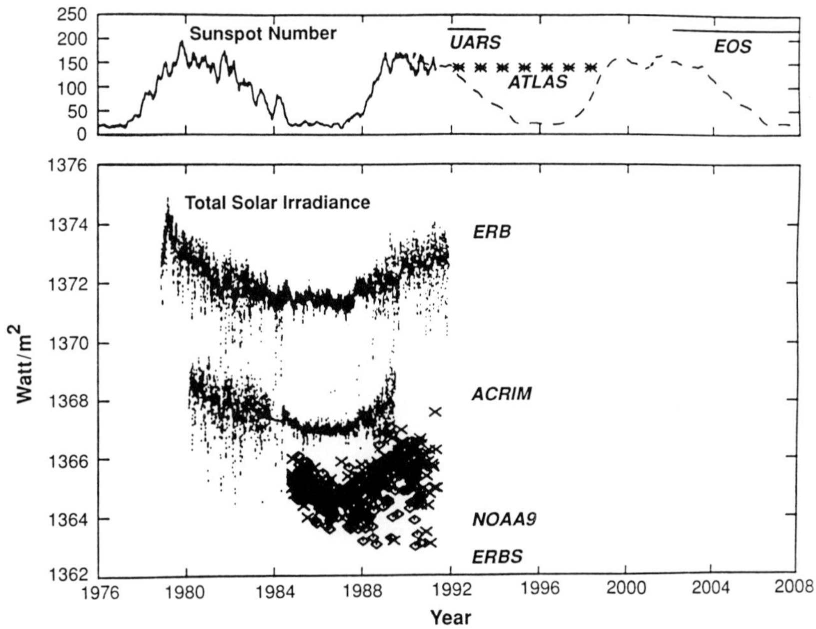

FIGURE 7. Solar activity variation measured by sunspot number (top), and by the ERB radiometer on the Nimbus-7 satellite and ACRIM on SMM, and by the ERBE program on NOAA9 and ERBS (Lean, private communication).

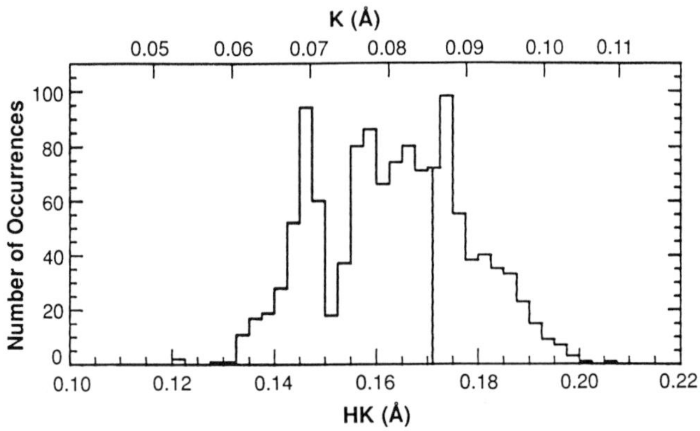

FIGURE 8. Frequency distribution of solar-like stars as determined from emission in the H and K lines of Ca II, and in the K line alone (Baliunas & Jastrow 1990).

of stars with a low luminosity variation of 0.1%. Most of the stars vary by 0.3%, and some by as much as 1%. These results suggest that the larger luminosity variation of 0.3% inferred for the Sun during the Maunder Minimum would be typical behavior for stars that are like the Sun in mass, composition and rotation period and, therefore, in the activity of their magnetic dynamo. Thus, a substantially larger solar variation than the current 0.1% is consistent with the stellar observations.

It may be worth noting that the study of solar-analog stars may be essential in pointing up differences in various features, such as chromospheric and coronal emission, luminosity, and phase relations between maximal brightness and maximal color index (cf Gray 1994). Such differences may help elucidate the physical mechanism underlying solar and stellar variability.

5. Conclusions

Solar activity and global temperatures appear to be highly correlated. However, according to studies with general circulation models the current level of irradiance variation is too low to change climate. A higher level of solar variability appears possible and, as the study of solar-type stars indicates, even likely. The societal consequences of a change in climate are expected to be enormous. In order to assess the relative importance of solar variability and of anthropogenic causes it is imperative that we investigate and understand the influence of the Sun alone. This requires a long-term program of monitoring total and spectral irradiance and an effort to study the physical mechanism of solar and stellar luminosity variation. The stars are important in this context as well since different phenomena observed in stars may help us identify the unknown source of excess radiation in the Sun at activity maximum and point to the underlying cause of solar as well as stellar variability.

Acknowledgments. We are grateful to Drs. Judith Lean and William Wagner for helpful discussions and critical reading of the manuscript.

REFERENCES

ANKLIN, M. ET AL. 1993 Climate instability during the last interglacial period recorded in the GRIP ice core. *Nature* **364**, 203–207.

BALIUNAS, S. & JASTROW, R. 1990 Evidence for long-term brightness changes of solar-type stars. *Nature* **348**, 520–523.

EDDY, J. A. 1976 The Maunder Minimum. *Science* **192**, 1189–1202.

EDDY, J. A. 1977 Historical evidence for the existence of the solar cycle. In *The Solar Output and its Variation* (ed. O.R. White). pp. 51–71. Assoc. Univ. Press, Boulder, CO.

FRIIS-CHRISTENSEN, E. & LASSEN, K. 1991 Length of the solar cycle: an indicator of solar activity closely associated with climate. *Science* **254**, 698–700.

GRAY, D. 1993 Variations during stellar cycles. Contributed paper presented at *IAU Colloquium No. 143 "The Sun as a Variable Star: Solar and Stellar Irradiance Variations"*, June 20-25, 1993, Boulder, Colorado.

LEAN, J. 1987 Solar ultraviolet irradiance variations: A review. *J. Geophys. Res.* **92**, 839–868.

LEAN, J. 1991 Variations in the Sun's radiative output. *Rev. Geophys.* **29**, 505–535.

LEAN, J., SKUMANICH, A., WHITE, O.R. & RIND, D. 1994 Estimating solar forcing of climate change during the Maunder Minimum. In *The Sun as a Variable Star: Solar and Stellar Irradiance Variations* (ed. J.M. Pap, C. Fröhlich, H.S. Hudson & S.K. Solanki). Cambridge University Press, in press.

LOCKWOOD, G.W. 1994 Irradiance variations of stars. In *The Sun as a Variable Star: Solar and Stellar Irradiance Variations* (ed. J.M. Pap, C. Fröhlich, H.S. Hudson & S.K. Solanki. Cambridge University Press, in press.

NESME-RIBES, E., SOKOLOFF, D. & SADOURNY, R. 1994 Solar rotation, irradiance changes and climate. In *The Sun as a Variable Star: Solar and Stellar Irradiance Variations* (ed. J.M. Pap, C. Fröhlich, H.S. Hudson & S.K. Solanki). Cambridge University Press, in press.

NOYES, R. W. 1982 *The Sun, Our Star*, Harvard Univ. Press, Cambridge, MA.

REID, G. C. 1991 Solar total irradiance variation and the global sea surface temperature record. *J. Geophys. Res.* **96**, 2835–2844.

WITHBROE, G. L. 1990 *Adv. in Astronaut. Sc.*, Vol. **71**.

Irradiance Variations of Stars

By G. WESLEY LOCKWOOD

Lowell Observatory, 1400 West Mars Hill Road, Flagstaff, AZ 86001, USA

Using a nine-year time series of photometric observations of 33 stars similar in temperature and mass to the Sun, but covering a wide range of age and mean chromospheric activity, we found that a majority varies from year to year, some by as much as several percent. We describe the methodology, circumstances of the observations, and photometric results. Stars most similar in mean chromospheric activity to the Sun varied by amounts several times greater than the Sun over a comparable time interval. Thus, the Sun's present low level of variability, as measured from 1980 to 1989 by the Solar Maximum Mission, appears unusual.

1. Introduction

Throughout the early decades of this century, a number of scientists were convinced that the solar constant varied by up to several percent and that these variations were associated with fluctuations in the Earth's climate (e.g., Clayton 1923; Huntington 1923; Abbott 1963). This enduring if elusive idea has been explored anew in recent conferences (e.g., McCormac 1983; Schatten & Arking 1990).

Intrigued by measurements of the solar constant which seemed to leave open the question of solar variability, astronomers at the Lowell Observatory began monitoring the steadiness of the sunlight reflected from the planets Uranus and Neptune in 1949. The Air Force Cambridge Research Laboratory sponsored this study until 1966 with the goal of improving weather forecasts by uncovering the supposed connection to solar variability.

By comparing the brightness of the planets with stars nearby in the sky, making differential rather than absolute measurements, the Lowell observers avoided the worst complications of direct solar observation—the variability and spectral composition of the transmission of the Earth's atmosphere, and the need for an absolutely calibrated (and stable!) radiometer. They found no plausible evidence for intrinsic variations in planetary brightness attributable to solar changes. However, their related studies of solar-type stars provided benchmark stellar data for the ongoing quest for solar variations. To standardize the planetary photometry, they carefully measured a small group of 16 solar-type stars many nights each year from 1955 to 1966. In a report summarizing their results, Jerzykiewicz & Serkowski (1966) concluded that ". . . this long series of photoelectric observations has taught us more about the variations of solar-type stars than about the Sun itself. The observations . . . indicate that for none of these stars does the standard deviation of the yearly mean magnitude exceed 0.008, and for [three stars] this deviation is less than 0.004 mag."

Their observations thus capped the possible variability of sunlike stars at the 1% level but left open the question of smaller fluctuations. Our recent work reveals a rich domain of variability below 1% populated by stars whose photometric behavior mimics the solar paradigm in exaggerated form.

2. Differential photometric observations of solar-type stars

With vastly improved equipment, we undertook a photometric study of sunlike stars in 1981, using an 0.5-m reflecting telescope reserved for long-term photometric programs. At first, we concentrated on young, active stars in the Hyades open cluster. To our

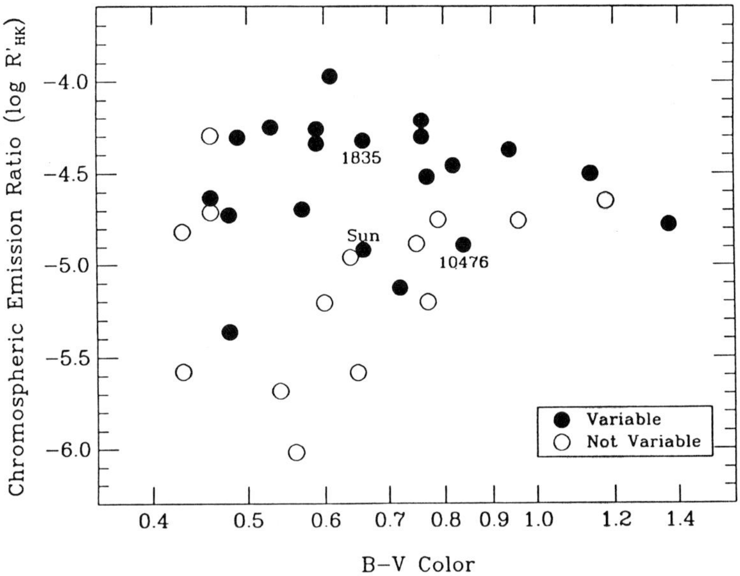

FIGURE 1. Properties of the program stars and the Sun. The symbols indicate whether we have found the stars to be variable or not. The ordinate, log R'_{HK}, the mean chromospheric emission ratio, is an estimate of the fraction of the stars' total output appearing as chromospheric emission (Noyes *et al.* 1984). The $B--V$ color, expressed as the magnitude difference between broadband filters at 0.44 and 0.55 microns, is a measure of photospheric temperature.

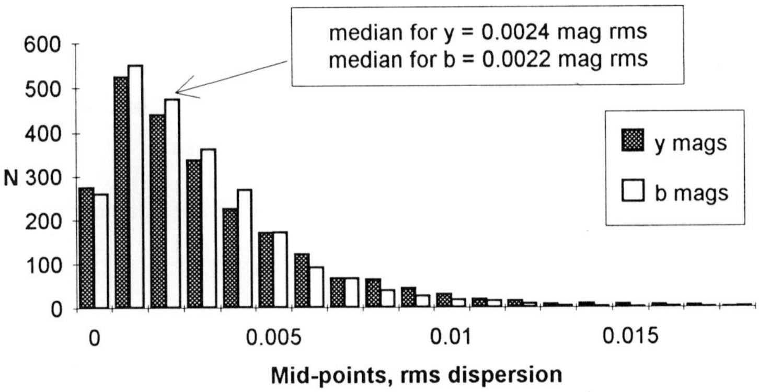

FIGURE 2. Distribution of the repeatability of differential b and y measurements made in a y,b,b,y sequence.

surprise, we found that many were variable by up to several percent from night to night and from year to year (Radick *et al.* 1987). Further, brightness and Ca II emission are inversely correlated, strongly suggesting an association with active regions.

In 1984, we began measuring ordinary main sequence field stars, concentrating on 33 F, G, and K type dwarf and subgiant stars originally observed in Ca II H and K at Mount Wilson Observatory (Wilson 1978). These stars bracket the Sun in age, mass, temperature, and mean chromospheric activity (Figure 1). Some show cyclic activity mimicking the Sun (Baliunas & Vaughan 1985); others appear quiescent, perhaps in Maunder minima (Baliunas & Jastrow 1990).

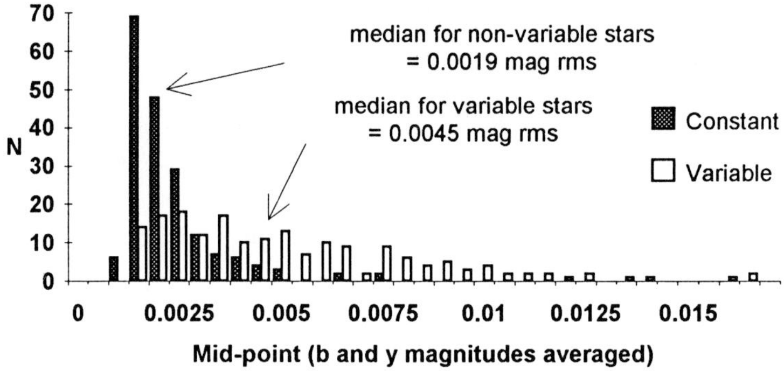

FIGURE 3. Distribution of the observed magnitude dispersion over an observing season for non-variable and variable stars.

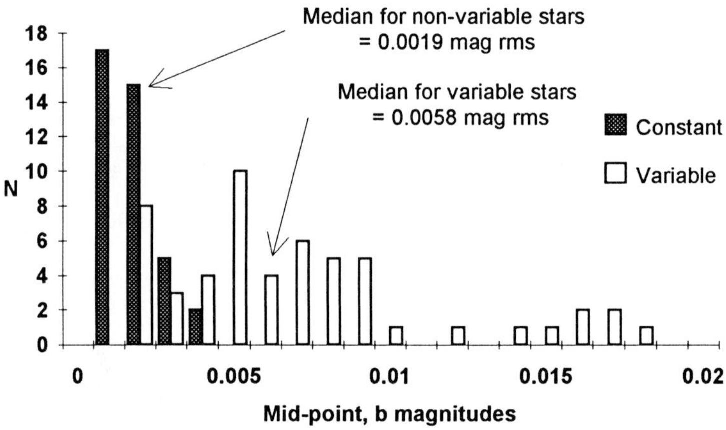

FIGURE 4. Distribution of the dispersion of annual mean b magnitudes over 9 seasons for non-variable and variable stars.

The stars are grouped into trios or quartets comprised of one or two program stars and two adjacent (within a few degrees) comparison stars of similar color and visual magnitude. A nightly observation takes about 1/2 hour and involves cycling through the group two times through each of two filters, b (472 nm) and y (551 nm). This produces four independent sets of differential magnitudes. We typically observe each group about 10 times per season.

From observations on ~1000 nights over the past decade, we have mapped out the slow year-to-year variations, which range from 0.1 – 3% rms (Skiff & Lockwood 1986; Lockwood & Skiff 1988, 1990). The variations tend to be smooth rather than random, and some appear to be cyclic. The amplitude of variation decreases with increasing stellar age, just as chromospheric activity does (Lockwood *et al.* 1992). Despite an impressive improvement in precision over the Lowell study three decades ago, we are still unable, by a factor of three or so, to detect variations in stars as small as the 0.1% peak-to-peak change in total solar irradiance seen in cycle 21 (Willson & Hudson 1991). Nevertheless, useful comparisons are possible.

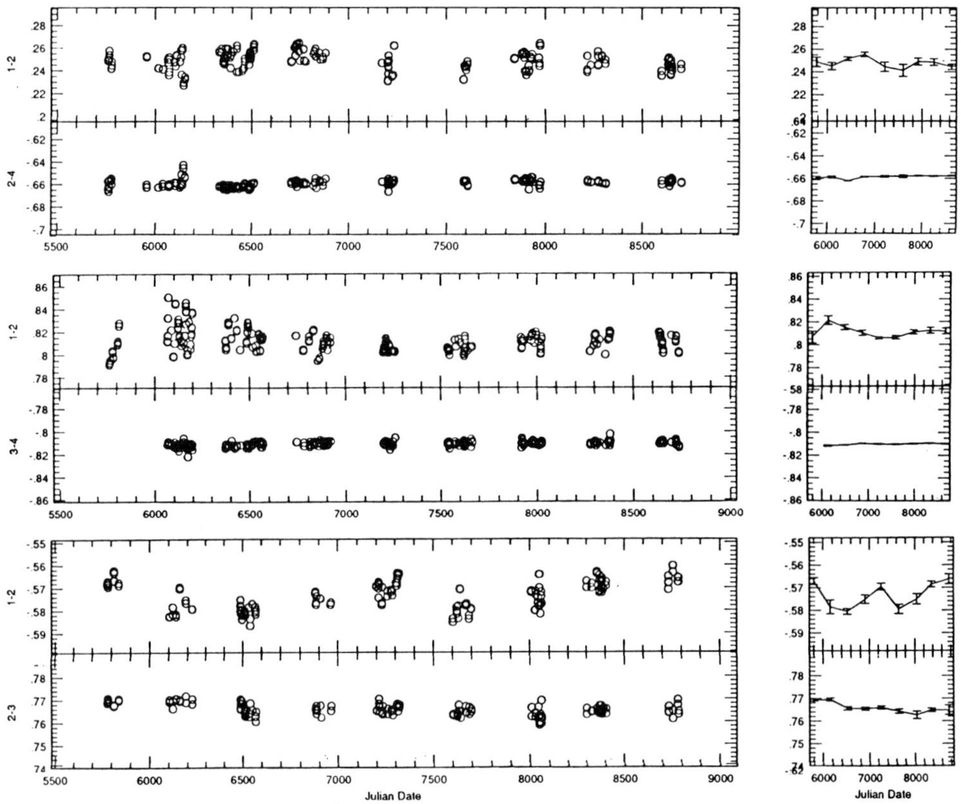

FIGURE 5. Light curves (b and y averaged) for three stars whose variations appear cyclic. Below each differential light curve we show the corresponding light curves for the comparison star pair. The significance of the variation can therefore be judged by comparing the two light curves. (*left*) the individual nightly data points. (*right*) annual mean magnitudes and 95% confidence intervals. The program stars are: (*top*) HD35296; (*center*) HD82885; (*bottom*) HD115383. The comparison star pair for HD115383 show a secular trend, so one of these stars must also be variable. HD82885 showed strong rotational modulation during the second observing season, accounting for the large scatter in that year (Skiff & Lockwood 1986).

3. Observational errors

Owing to skill and attentiveness to sky conditions, the observer (Brian A. Skiff) seldom makes a measurement that must be rejected later. Nonetheless, acceptable nights vary in quality and some observations are better than others. The histograms of measurement dispersion (Figure 2) show the distribution of error over half-hour observing intervals.

Two points are noteworthy: (1) the distribution of noise is virtually identical in b and y. It arises in similar proportions from scintillation, Poisson noise, sky transparency variations, and guiding errors (cf. Table 2 in Lockwood & Skiff 1988). (2) The median dispersion is 0.002 mag rms and few observations are worse than 0.010 mag (1%). The tail of the distribution comes mostly from observations on poor nights. Assuming four observations per night (two each in b and y) and 10+ nights per season, the predicted observational error in annual mean magnitudes is < 0.001 mag.

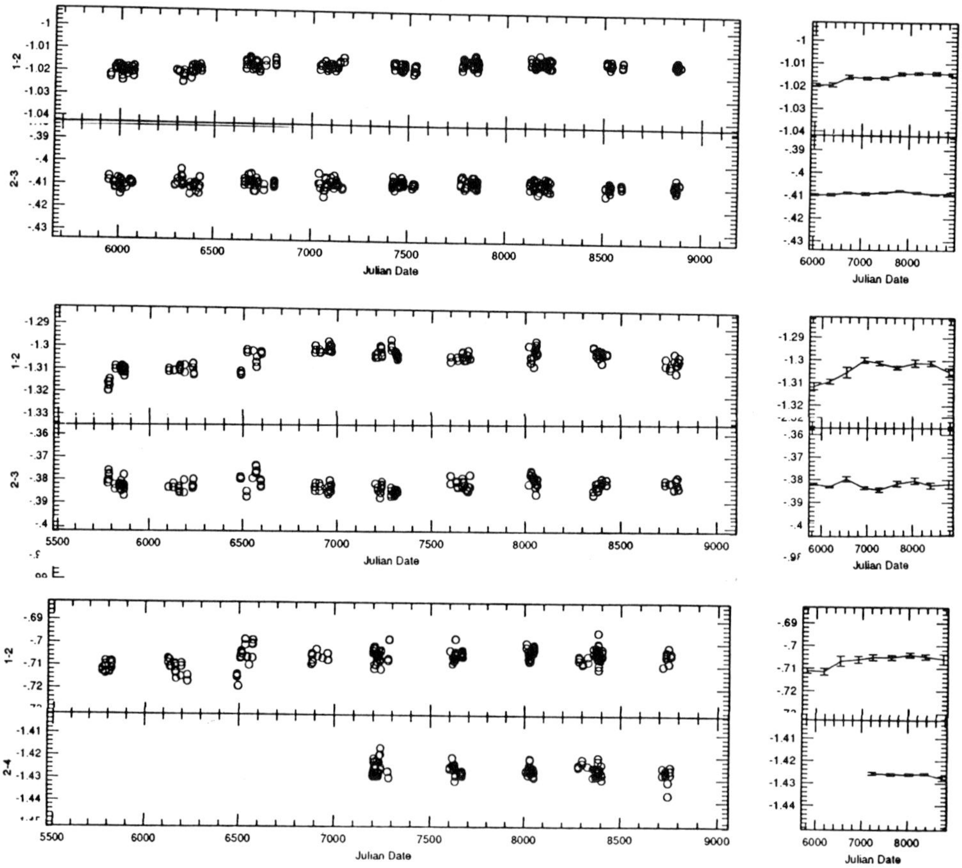

FIGURE 6. Light curves for three stars whose brightness varied monotonically (*top*) HD10476; (*middle*) HD120136; (*bottom*) HD114710. The mean chromospheric activity levels of these three stars are only slightly higher than the Sun's. HD10476 is a close solar analog with a distinct ~10 year chromospheric activity cycle (see paper by Radick, this volume). One of the comparison stars for HD120136 is unstable; it may be a δ Scuti variable.

4. Detecting variable stars

Observing stars in trios (or quartets) yields three (or six) parallel time series of differential magnitudes for pairwise star combinations. By testing the significance of the correlation between pairs of light curves, we determine which stars in a group are variable. Sometimes a comparison star is variable—an unfortunate but not unexpected complication—hence the need for multiple comparison stars. The threshold of detectable variability for a program star depends critically on the inherent stability of its comparison stars, a property which differs from group to group. The best comparison stars are stable to < 0.001 mag.

Corroborating evidence for variability comes from the tendency of variability to recur season after season. Further, because of the blackbody distribution of stellar radiation, intrinsic variations in b are greater than in y, while observational error shows no such color dependence.

Figure 3 shows histograms of magnitude dispersion over the duration of single observing seasons (months) for variable and non-variable stars. The histograms overlap because of

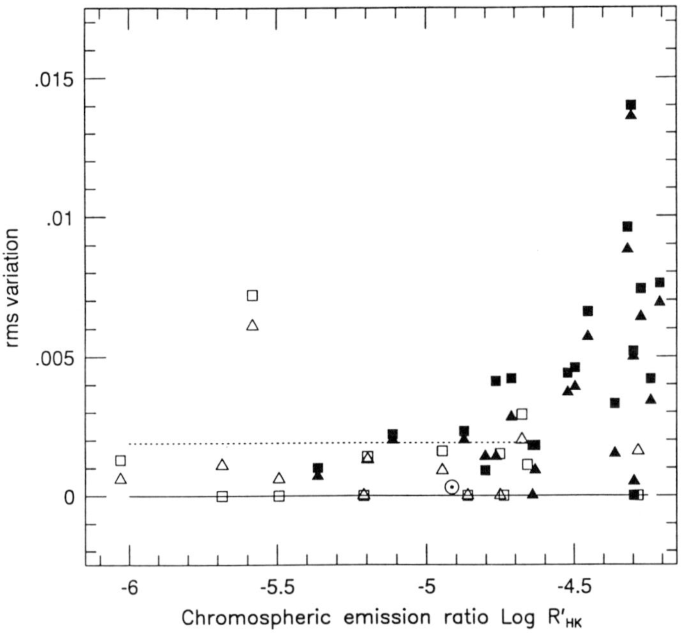

FIGURE 7. The rms dispersion of annual mean magnitudes, b (*squares*) and y (*triangles*), as a function of the chromospheric emission ratio, log R'_{HK}. Filled symbols indicate stars whose intrinsic variability is decisively established by statistical tests. Open symbols indicate stars whose observed fluctuations do not exceed the fluctuations of their particular comparison stars. The star near (−5.6, 0.007) may be variable but its comparison star pair is so unreliable that no decision is possible. The active star HD129333 is omitted because its large variations lie off scale. The dotted line indicates the median value of measurement error and intrinsic variability for comparison stars. Plotted values falling below this line have been significantly adjusted and represent, for those stars which are not demonstrably variable, upper limits to intrinsic variability.

differing thresholds of detectable variability from group to group. For non-variable stars, the median dispersion is 0.0019 mag. Since the median dispersion for detected variable stars is more than twice this large, the distinction between variables and non-variables is often clear cut. Under the most favorable circumstances when the comparison stars are exceptionally stable, we can detect short-term variability as small as ~0.0015 mag rms.

Similarly, in Figure 4 we show the distribution of the dispersion of annual mean magnitudes over the 9-year duration of the program. The median value for non-variable stars was only 0.0019 mag, and the smallest detected variation was 0.0015 mag.

5. Some light curves

The two principal data products from our program are light curves, which display data in an easily comprehended visual format, and formal statistics quantifying the amount and significance of variability. Figures 5 and 6 show some sample light curves for stars showing evidence of photometric cycles and monotonic trends, respectively. The light curves display individual nightly points on the left-hand panels and annual mean brightness levels with 95% confidence error bars on the right-hand panels. To simplify the presentation we have averaged data for b and y. The light curves appear in pairs with the top panel showing the differential magnitude for program star minus comparison star ("signal") and the lower panel showing analogous data for the comparison stars ("noise").

The average levels of chromospheric activity for the three stars in Figure 5 are all higher than the Sun's. Their light variations are correlated inversely with Ca II H and K variation as is typical for young, active stars (Lockwood *et al.* 1992).

Figure 6 shows three stars whose annual mean brightness changed monotonically over nine years.

6. Photometric variability and mean chromospheric activity

We found previously that photometric variability is a strong function of mean chromospheric variability, and hence, stellar age (e.g., Lockwood & Skiff 1990; Lockwood *et al.* 1992). Figure 7 shows an updated version of the relationship between the rms dispersion in annual mean b and y magnitudes and the chromospheric emission ratio, log R'_{HK}. This ratio, expressed in the notation of Noyes *et al.* (1984) but calculated using a different prescription for correcting for the non-magnetic component of the emission, estimates the fraction of a star's output arising from the chromosphere. Different symbols indicate the presence or absence of detected photometric variability. The intrinsic stellar variability is almost always greater in b than in y. A regression of σ_b on σ_y yields $\sigma_b/\sigma_y = 1.07 \pm 0.03$.

The plotted values in Figure 7, have been adjusted slightly downward to correct for measurement errors and variability of the comparison stars. We subtracted the observed variance of the comparison star magnitudes, or in a few cases where the comparison stars are clearly suspect, we used estimated typical values. In most cases the change is trivial, but it does make a difference for some stars whose variability is either small or undetectable, that is, those on the left side of the diagram. For them, the values plotted represent a statistically unbiased estimate of the upper limit of possible intrinsic variability.

We included the Sun in Figure 7 based on its observed variation from 1980–1989 (Willson & Hudson 1991), adjusting the plotted value slightly upward to account for the distinction between variation in total irradiance and the variation in the visual continuum. Several stars located near the Sun on this diagram are demonstrably variable by much larger amounts than the Sun, a factor of four on the average. Thus, the Sun's recent photometric variability appears deficient relative to some similar stars.

In reaching this conclusion we recognize several deficiencies that can be at least partially remedied by further solar and stellar observations. Foremost and most easily rectified is the limited duration of present data—just over one 11-year cycle for the Sun and probably less than a cycle length for many of the stars. We know that our photometric observations may not sample the full range of chromospheric variation, as shown for example in the illustrations in the paper by Radick in this volume. It seems equally likely that the full range of solar irradiance variation over activity cycles of differing amplitude is not exhibited in the available data. Secondly, as solar variability lies well below the level at which we can reliably detect stellar variability, we can only point to those stars whose variability grossly exceeds that of the Sun as examples of what the Sun's behavior might be like at some other time. We have no real population statistics to lead us to a picture of normal stellar behavior, only the tantalizing indication that the Sun has potential for greater variations that we have observed thus far.

REFERENCES

ABBOTT, C. G. 1963 Solar Variation and Weather. *Smithsonian Misc. Coll.* **146**, No. 3.

BALIUNAS, S. L., & JASTROW, R. 1990 Evidence for long-term brightness changes of solar-type stars. *Nature* **345**, 520–522.

BALIUNAS, S. L. & VAUGHAN, A. H. 1985 Stellar activity cycles. *Ann. Rev. Astron. & Astrophys.* **23**, 379.

CLAYTON, H. H. 1923 *Solar Relations to Weather and Life.* The Clayton Weather Service: Canton, MA.

HUNTINGTON, E. 1923 *Earth and Sun.* Yale University Press: New Haven.

JERZYKIEWICZ, M., & SERKOWSKI, K. 1966 The Sun as a Variable Star. *Lowell Observatory Bulletin* (No. 137) **6**, 295–323.

LOCKWOOD, G. W., & SKIFF, B. A. 1988 Luminosity variations of stars similar to the Sun. Air Force Geophysics Lab Report AFGL-TR-88-0221, 101 pp. AFGL: Hanscom Air Force Base.

LOCKWOOD, G. W. & SKIFF, B. A. 1990 Some insights on solar variability from precision stellar astronomical photometry. In *Climate Impact of Solar Variability* (ed. K. H. Schatten & A. Arking). NASA Conference Publ. 3086. pp. 8–15.

LOCKWOOD, G. W., SKIFF, B. A., BALIUNAS, S. L., & RADICK, R. 1992 Long-term solar brightness changes estimted from a survey of Sun-like stars. *Nature* **360**, 653–655.

MCCORMAC, B. M. 1983 *Weather and Climate Responses to Solar Variation.* Colorado Associated University Press: Boulder.

NOYES, R. W., HARTMANN, L. W., BALIUNAS, S. L., DUNCAN, D. K., & VAUGHAN, A. H. 1984 Rotation, convection, and magnetic activity in lower main-sequence star. *Astrophys. J.* **279**, 763–777.

RADICK, R. R., LOCKWOOD, G. W., THOMPSON, D. T., DUNCAN, D. K., & BAGGETT, W. E. 1987 The activity, variability, and rotation of lower main-sequence Hyades stars. *Astrophys. J.* **321**, 459–472.

SCHATTEN, K. H., & ARKING, A. ED. 1990 *Climate Impact of Solar Variability. NASA Conference Publ.* 3086.

SKIFF, B. A., & LOCKWOOD, G. W. 1986 The photometric variability of solar-type stars. V. The standard stars 10 and 11 Leonis Minoris. *Publ. Astron. Soc. Pacific* **98**, 338–441.

WILLSON, R. C. & HUDSON, H. S. 1991 The Sun's luminosity over a complete solar cycle. *Nature* **351**, 42–44.

WILSON, O. C. 1978 Chromospheric variations in main-sequence stars. *Astrophys. J.* **226**, 379–396.

Irradiance Observations of the Sun

By CLAUS FRÖHLICH

Physikalisch-Meteorologisches Observatorium Davos, World Radiation Center
CH-7260 Davos Dorf, Switzerland

Measurements of the total solar irradiance during the last 14 years from satellites show variations over time scales from minutes to years and decades. The most important variance is in the range from days to several months and is related to the photospheric features of solar activity: decreasing the irradiance during the appearance of sunspots, and increasing it by faculae and the bright magnetic network. Long-term modulation by the 11-year activity cycle is observed conclusively with the irradiance being higher during solar maximum. The accuracy of the determined variability and its interpretation in terms of manifestations of activity related features on the photosphere is discussed. Besides the direct influence of the spots, faculae and magnetic network more profound changes in the thermal transport seem to influence the behaviour of the solar photospheric radiation on the solar cycle and longer time scales.

1. Introduction

The irradiance from the Sun at the mean Sun-Earth distance, integrated over all wavelengths, hence total irradiance, is traditionally called "solar constant" S although it has been shown to vary on time scales from minutes to decades (see e.g. Fröhlich 1993). 'All wavelengths' means essentially the energetically important range from 200 nm to 5μm containing 99.9 percent of S. The measurements from satellites discussed here (until the end of 1992) have been performed by the radiometer HF of the Earth Radiation Budget Experiment (ERB) on the NIMBUS-7 satellite since November 16, 1978 (Hoyt *et al.* 1992), by ACRIM I on the Solar Maximum Mission Satellite (SMM) from February 14, 1980 until June 1, 1989 (e.g. Willson & Hudson 1991), by ACRIM II on the Upper Atmospheric Research Satellite (UARS) since October 1991 (Willson 1994) and by PMO6/SOVA2 on the European Retrievable Carrier (EURECA) since August 11, 1992 (Romero *et al.* 1994). For comparison, the rocket and balloon experiments by JPL and PMOD/WRC are included in the discussion. These experiments were used in 1986 to prove that the downward trend of the solar irradiance during the declining phase of the cycle 21 was real and not instrumental (Willson *et al.* 1986). The selection of the experimental data is somewhat arbitrary and represents rather the author's familiarity with them than other criteria.

2. How accurate is the variability?

Time series of the measured irradiances are plotted in Figure 1 and illustrate the variability on all time scales. The differences among the experiments are due to their absolute calibrations which are accurate to 'only' about $\pm 0.2\%$ and do not reflect the precision and stability of the instruments which are obviously much better. The time series of the two ACRIM instruments demonstrate also the difficulty to bridge gaps; again due to the limited absolute accuracy: Willson (1994) has to adjust the ACRIM II measurements in order to get a homogeneous data set covering the whole period of ACRIM I and II. The HF time series is essentially uninterrupted and can be used as reference, although it cannot be proven that the sensitivity of HF has no long-term trend. The comparison between HF and ACRIM I and II is shown in Figure 2. There are obvious

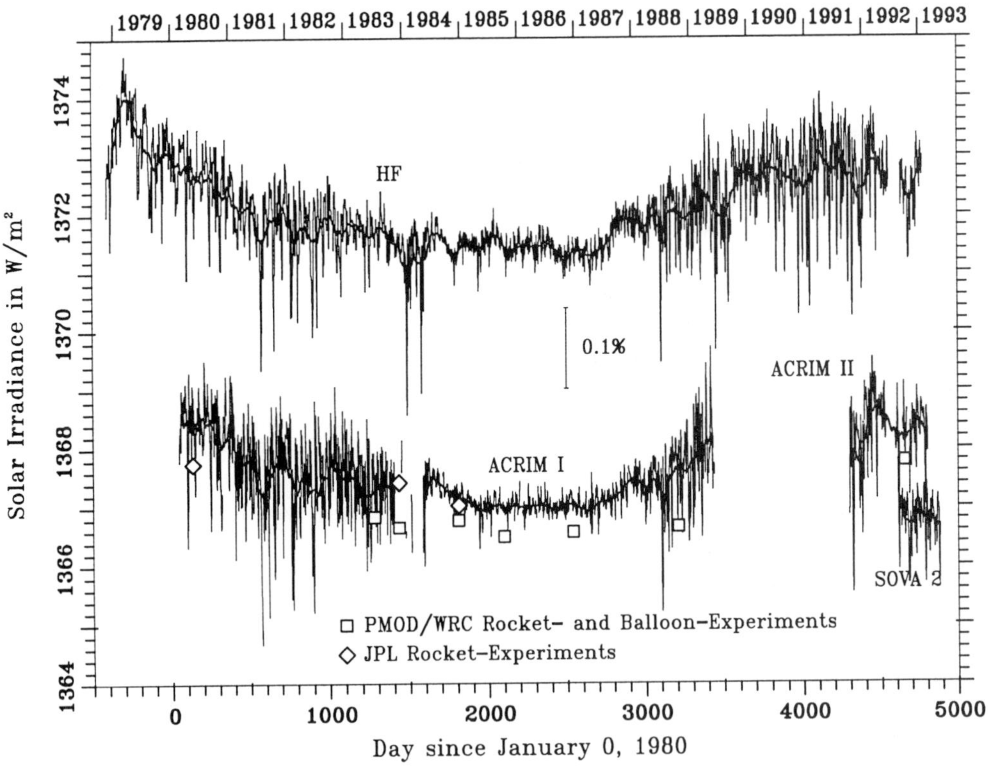

FIGURE 1. Time series of ACRIM I and II, HF and PMO6/SOVA 2 total solar irradiance measurements from November 1978 to beginning of 1993. The individual points represent daily mean values, the heavy line the 81-day running mean. Furthermore, the results of the rocket and balloon experiments of JPL and PMOD/WRC are plotted.

departures of up to 500 ppm (about half the solar cycle's amplitude) in these results which ask for explanation. During period from the end of 1980 until spring 1984 the SMM satellite was not pointing properly, thus ACRIM I had to be operated in a different mode and much less data have been acquired; this obviously increases the noise. During the solar minimum e.g. NIMBUS-7 shows much larger variations than ACRIM I (see also Figure 1). It is obvious from comparison of both time series that ACRIM (disregarding the 'non-pointing' period of SMM) shows much lower noise than HF; the latter is more noisy due to higher contributions from instrumental and operational effects, but also due to the fact that the observation time per orbit is much shorter. The modulation during the solar minimum, which is sensed by HF with a greater amplitude, is part of a more or less periodic signal with a period of about 300 days persisting throughout the measurement period. Besides the possibility that this is an interesting solar feature (see also Section 3), the ratio of its amplitude as measured by HF and ACRIM varies with time. This could be due to a modulation of the sensitivity by e.g. the distance change on the Earth' orbit around the Sun and a related temperature effect in one or both radiometers which is not fully accounted for by the applied corrections. The short observation and exposure time of HF may hide such an effect, whereas for ACRIM the 60 minutes of observation allow to analyze the orbital behaviour of the instrument in detail. There are other differences, especially during the period of ACRIM II where a gradual increase of the latter relative to HF is observed. These may still be due to some residuals of the 'inverse degradation' observed in ACRIM II (Willson 1994) but the applied correction would need to be increased by about 50% which seems unlikely. The decreasing ratio may also be due to a degradation of the HF cavity's absorptance during the high solar fluxes encountered before their rapid drop in spring 1992. These still not fully understood uncertainties underline the importance for having more than one and better three experiments simultaneously in space in order to cross-check the results and to tie new experiments into existing data basis.

The results of the rocket and balloon experiments were able to confirm the decrease of the solar radiation during the declining phase of cycle 21. A major drawback of these measurements is that they are rather short – a few minutes for a rocket and generally a few hours for a balloon flight – and a truly simultaneous comparison with an instrument on board an orbiting satellite may not be possible. The latter is a prerequisite for accurate comparison due to the short term variations of solar irradiance. Together with the limited accuracy of a radiometer working in a rapidly varying thermal environment during a rocket experiment or the limitations of the atmospheric corrections still needed for a balloon flight, an overall precision and repeatability of somewhat less than 0.1% can be achieved. This was sufficient for the downward trend, but is not enough to settle the above mentioned differences between space experiments or to assess solar irradiance variability over periods of more than a few years accurately. These uncertainties have to be kept in mind when the results of modelling solar variations are discussed.

3. Influence of solar activity on irradiance variations

The dominant feature in the time series is the 'dip', a negative excursion of a few days' length and a depth ranging up to a few tenths of a percent of the irradiance. These dips result from sunspot groups rotating past the central meridian and have been noticed in data sets as early as in the ones from Mariner VI and VII. But only the prominent dips observed by ACRIM in April and May 1980 convinced the community that sunspots can produce such large depressions. Willson *et al.* (1981) described them in terms of the Photometric Sunspot Index, the PSI function, similar to the models of sunspot darkness noted earlier by Foukal & Vernazza (1979) and Hoyt & Eddy (1982). Hudson & Willson (1982) define this photometric index as the sum of the projected areas of the sunspots multiplied by a factor α taking into account the umbra/penumbra area ratio and the effective temperature of the sunspot relative to the photosphere, in the simplest way possible. An improved PSI calculation has been presented by Fröhlich *et al.* (1994). The major improvement is due to screening of the observations from outliers which improves the homogeneity of the data set substantially, at least for the period after December 1981 when NOAA started to report data from several stations instead of one to two stations only. Moreover, these calculations take into account the area dependence of the contrast α and calculate 'true' daily means for each observation using the latitude dependent surface rotation of the spots. The correlation between the newly calculated PSI and total solar irradiance using bi-variate spectral analysis yields a major improvement due to the screening of the data (an increase in coherence of about 5-10%) and a minor one due to the more sophisticated methods (1-2%) as shown by Fröhlich *et al.* (1994). Furthermore, this study has been performed for different phases of solar cycles 21 and 22. This analysis shows that the gain, the factor by which PSI has to be multiplied to yield the observed irradiance change, is time dependent. It changes from about 0.6 in 1980 to 1.1 in 1990 and cannot be interpreted by a change of the contrast relative to the quiet Sun (as it is normally defined and determined by direct photometry), but rather as a change of the contrast between the spots and their surrounding as seen in

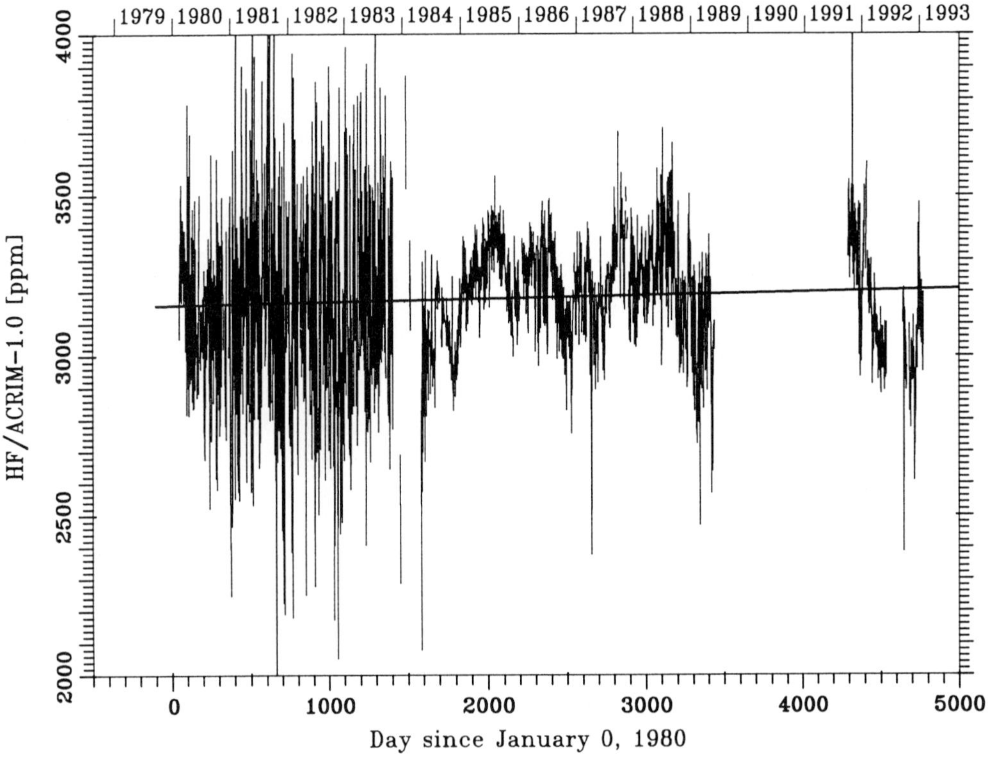

FIGURE 2. Comparison of ACRIM I & II and HF total solar irradiance measurements from February 1980 to end of 1992.

total irradiance (integrated over the solar disk). Part of this change is possibly due to a change in the spot to facular area ratio. But, independent of its interpretation, this time dependent factor can be used to calibrate the PSI, which then is used to subtract the effect of sunspots from the irradiance observations. The result of the subtraction of the calibrated PSI is shown in Figure 3 for both the ACRIM and HF data. The differences between the two data sets with a quasi-period of 300 days were already discussed in section 2 and Figure 2 and are clearly seen during the minimum and the rising part of cycle 22.

Several ideas and models have been put forward to explain these quasi-periodic variations and the solar cycle modulation. One approach is to account for the variations in terms of the effects that magnetic flux tubes seem to have on the radiation and convection in the relatively shallow photospheric layers that emit most of the Sun's luminosity. A relatively straightforward approach to both the 6 month and 300 days quasi-periodic and the 11-year variations has been proposed by Foukal & Lean (1988) and Willson & Hudson (1988). In these studies, it was shown that the residual irradiance variations of $S + PSI$ correlate quite well with the time-series of properly scaled He I index or the 10.7 cm radio flux. This is not surprising since facular area variations were previously shown to account also for shorter term variations in these residuals. Thus, one may conclude that the 6 and 10 month variations are caused by the tendency of major complexes of activity to persist for about this number of months or solar rotations. This time scale in persistence of solar activity episodes has been documented before in studies of the He I index time series by Harvey (1984). These variations are also found by the multivariate

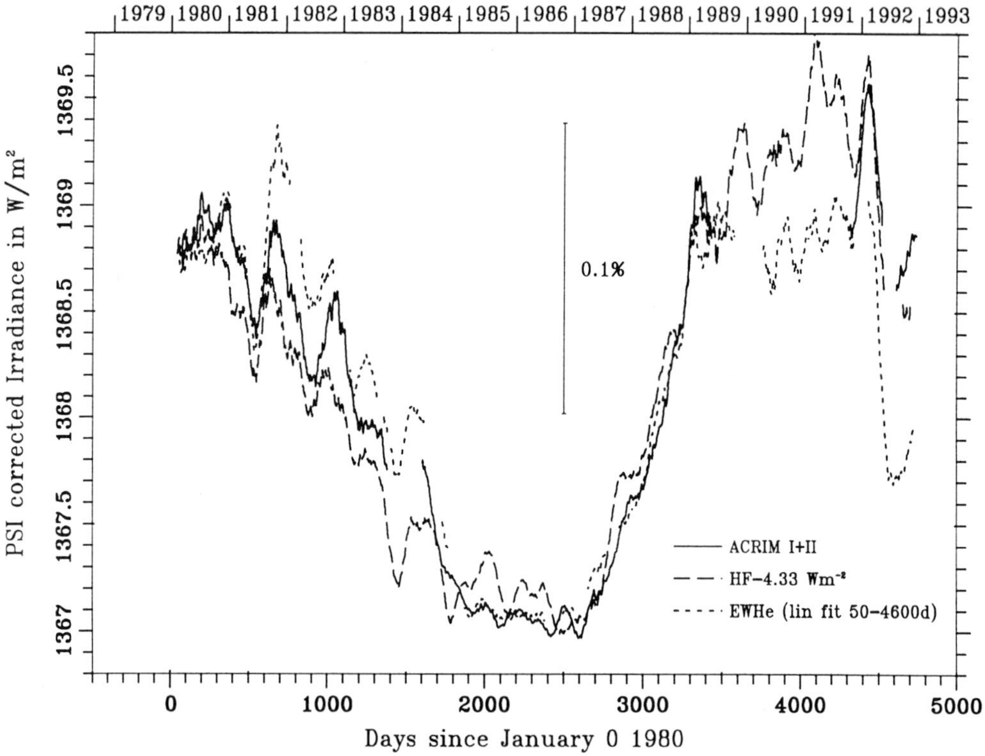

FIGURE 3. Solar irradiance measured by ACRIM (full line) and HF (long dashed line) with the effect of the sunspots removed ($S + PSI$, 81-day running mean). The short dashed line represents the result of the linear regression of the He I index with ($S + PSI$) for the time period Feb. 1980–Aug. 1992.

spectral analysis of the 9 year time series of ACRIM with PSI and He I (Pap & Fröhlich 1992; Fröhlich 1993). The part of the power density explained by He I has a large peak of about 80% at a period of 190 days and a broader peak with about 65% between periods of 250–450 days. This indicates that the 6 month and the 300 day quasi-periodic modulation is indeed due to the bright network and faculae. The solar cycle modulation has to be associated with some kind of a slow change in the solar atmosphere. The simplest explanation is a slow variation in the amount of the bright magnetic network outside active regions as shown by e.g. Foukal *et al.* (1991). This was mainly supported by the success of the modelling of $S + PSI$ by the He I index until about 1989, the end of the ACRIM I record (see e.g. Figure 6 of Fröhlich 1993). But, when the fit is either performed over the full period (1980–1992) or limited to the period of the declining phase of cycle 21 and the result used for the whole period, the model values fall short of the irradiance by as much as 25% around 1991. In the case of the fit over the whole period the model further overestimates the irradiance during the declining phase of cycle 21, as shown in Figure 3.

4. More detailed analysis of the solar cycle variation

In order to get more insight in the behaviour of the relationship between irradiance and He I the full period has been divided into the following intervals: 6 Feb 1980 – 13

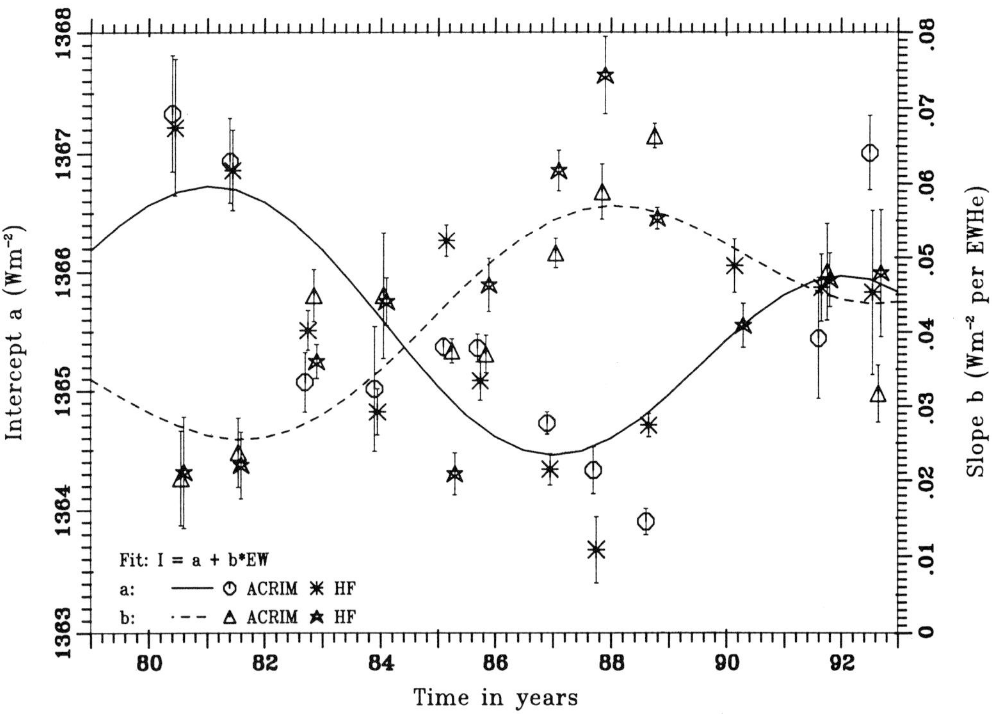

FIGURE 4. Linear regression analysis of irradiance against He I for different time intervals with $\pm 1\sigma$ error bars: the full line represent the results for the intercept a (left hand scale); the broken line the ones for the slope b (right hand scale).

Nov 1980; 1 Jan 1981 – 31 Dec 1981; 1 Jan 1982 – 30 Jun 1983; 1 Jul 1983 – 30 Jun 1984; 1 Jul 1984 – 31 Dec 1985; 1 Feb 1985 – 31 Mar 1987; 1 Jan 1986 – 31 Dec 1987; 1 Apr 1987 – 31 Mar 1988; 1 Jan 1988 – 31 May 1989; 1 Jun 1989 – 31 Dec 1990; 1 Jan 1991 – 1 May 1992; 1 Jan 1992 – 25 Feb 1993. The choice of these intervals is mainly dictated by the availability of data in the sense that gaps are excluded and they coincide with the periods analyzed for the *PSI* analysis (Fröhlich *et al.* 1994). A linear regression analysis between irradiance I and He I EW is performed according to $I = a + b \cdot EW$. The results for the slope b and the intercept a are shown in Figure 4 for each interval. The full and broken lines represent a fit of the slope and intercept respectively with an 11 year period sine-wave superimposed on a linear trend. Interesting enough, the slope b is increasing with time in a very similar manner as the contrast in the *PSI* representation of the spot influence: it is low in 1980/81, increasing until 1988 and slightly decreasing towards 1992 (*PSI* contrast has a much lesser decrease). The intercept does the opposite and weakens the slope effect to some extent. As we are dealing with a factor of 3.5 between minimum and maximum slope, however, the weakening due to the out-of-phase change of the intercept does not influence the net effect very much. It is interesting to note, that Harvey (1994) has observed a similar difference between magnetic flux from the disk outside sunspots and He I. This may indicate that the He I index should be replaced by the more direct surrogate: the magnetic flux outside spots. Further investigations in this direction are planned. The irradiance reconstructed using the fitted slope and intercept is shown in Figure 5. The fit is quite good although some discrepancy seem

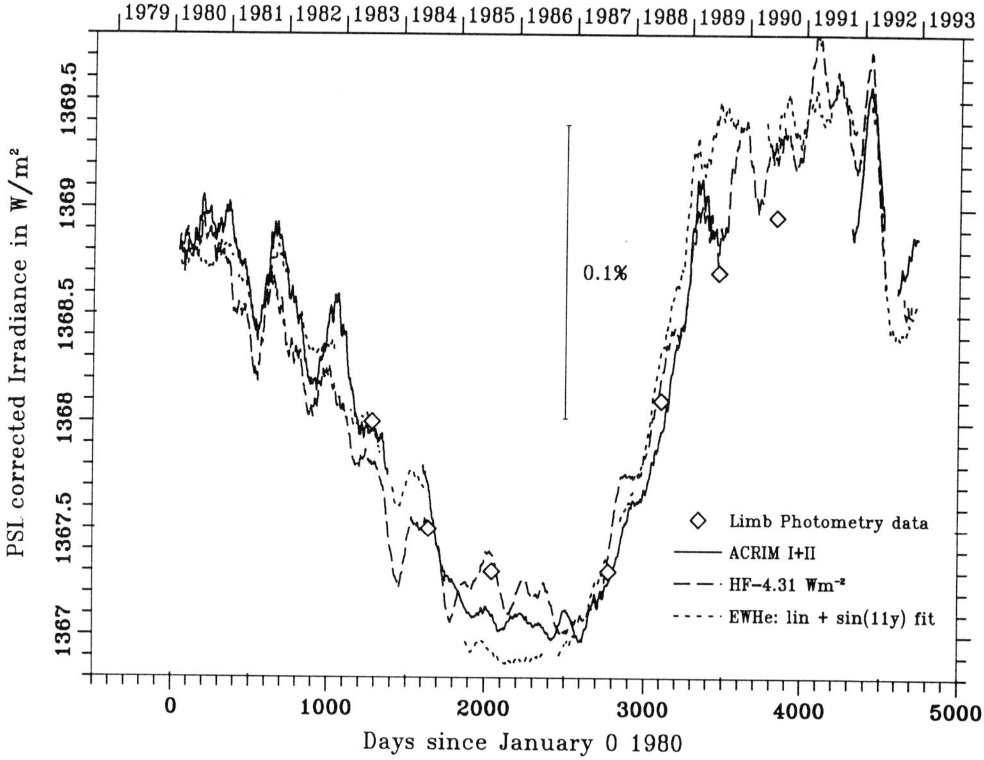

FIGURE 5. Solar irradiance measured by ACRIM (full line) and HF (long dashed line) with the effect of the sunspots removed ($S + PSI$, 81-day running mean). The short dashed line represents the result of the linear regression of He I line with ($S + PSI$) using the time varying regression parameters from Figure 4.

to be too large to be accounted for by uncertainties in the index or irradiance. Major discrepancies occurred at solar minimum and around and after the maximum of cycle 22. The discrepancy in 1980, noted earlier by Foukal & Lean (1988) has disappeared.

Kuhn *et al.* (1988) and Kuhn & Libbrecht (1991) have reported observations of the limb brightness which can be used to explain the total irradiance variation of the solar cycle. The observations are broad-band, two-color photometric measurements of the brightness distribution in a narrow annulus 20 arcsec wide, just inside the solar limb. The solar limb flux, observed as a function of latitude, can be divided into a 'facular' and 'temperature' part based on the assumptions that the 'temperature' part is constant over the 4-month observing summer period and that the 'facular' part shows up as intermittent bright regions. The component of excess brightness moves toward the equator between 1983 and 1985, and then reappears at relatively high latitudes in 1987 and again moves to the equator while it becomes brighter. The excess brightness responsible for the limb flux is due to features which are not resolved by this observation, and it could be due to the bright network in and outside active regions. The contribution of the limb flux (including the facular component) can account for the total irradiance decrease after 1983, and its increase since 1987 (Kuhn & Libbrecht 1991) as shown in Figure 5. However, there are systematic differences as e.g. the solar cycle amplitude which is smaller from the limb data than the observed one in irradiance. This could be corrected by using a slightly different limb darkening function for the translation of the limb data to full disk observations. This point should be investigated further as it may give a hint of what features are producing the solar cycle irradiance variation.

5. Conclusions

The past 15 years have seen the introduction of radiometers with sufficient precision to measure a variety of small variations of the total solar irradiance. These variations are interesting to solar physicists in several ways. In addition, the presence of a distinct 11-year modulation of total irradiance suggests that longer-term variations may be significant for the Earth's climate.

It is also important to note that the available time series is based on at least one instrument covering the full period of 14 years. Without this instrument the other time series might have been in trouble to fit new to old data with gaps of several years. The lesson learned from this is that at least one instrument should be in space to monitor the solar irradiance accurately. The understanding of the underlying physics can only be improved by comparing models with reliable continuous data.

For most of the variations physical explanations are available by e.g. Hudson & Willson (1982), Spruit (1988), Schatten & Mayr (1992). In order to better understand the reasons for the observed temporal changes of the model parameters for sunspots and other magnetic influences, however, more simultaneous studies of the detailed features on the solar disk from ground and space together with continued solar constant observations from space are needed. The Solar Heliospheric Observatory, SOHO, the next ESA/NASA solar mission, will certainly yield a major contribution to this issue. However, for the time between now and SOHO's launch in 1995 it is hoped that the existing measurements of ACRIM II will last long enough to have at least some time overlap for direct comparisons.

Acknowledgments. Part of this work has been supported by the Swiss National Science Foundation which is greatly acknowledged. Thanks are extended to J.Romero and Ch.Wehrli, Physikalisch-Meteorologisches Observatorium Davos, and J.Pap, Jet Propulsion Laboratory, Pasadena, for helpful discussions and R.C.Willson, Jet Propulsion Laboratory, Pasadena, for suggestions to improve the text and for providing data before publication.

REFERENCES

FOUKAL, P.V. & VERNAZZA, J. 1979 The effect of magnetic fields on solar luminosity. *Astrophys.J.* **234**, 707–715.

FOUKAL, P.V. & LEAN, J. 1988 Magnetic modulation of solar luminosity by photospheric activity. *Astrophys.J.* **328**, 347–357.

FOUKAL, P.V., HARVEY, K. & HILL, F. 1991 Do changes in the photospheric network cause the 11 year variation of total solar irradiance?. *Astrophys.J.* **383**, L89–L92.

FRÖHLICH, C. 1993 Changes of total solar irradiance. In *Interactions Between Global Climate Subsystems, The Legacy of Hann* pp. 123–129. Geophysical Monograph 75: IUGG 15.

FRÖHLICH, C., PAP, J. & HUDSON, H.S. 1994 Improvement of the Photospheric Sunspot Index and changes of the effective sunspot influence with time. *Solar Phys.* , in press.

HARVEY, J. 1984 Helium 10830 Å irradiance: 1975-1983. In *Solar Irradiance Variations on Active Region Time Scales* (ed. B.J. LaBonte, G.A. Chapman, H.S. Hudson & R.C. Willson). pp. 197–212. NASA Conf. Publ. CP-2310: Washington, DC, USA.

HARVEY, K. 1994 Irradiance models based on solar magnetic field observations. In *The Sun as a Variable Star: Solar and Stellar Irradiance Variations* (ed. J.M. Pap, C. Fröhlich, H.S. Hudson & S.K. Solanki). Cambridge University Press, in press.

HOYT, D.V. & EDDY, J.A. 1982 An Atlas of variations in the solar constant caused by sunspot blocking and facular emissions from 1874 to 1981. *NCAR/TN194 + STR*, NCAR, Boulder, CO., USA

HOYT, D.V., KYLE, H.L., HICKEY, J.R. & MASCHHOFF, R.H. 1992 The Nimbus-7 solar total irradiance: A new algorithm for its derivation. *J.Geoph.Res.* **97**, 51–63.

HUDSON, H.S. & WILLSON, R.C. 1982 Sunspots & solar variability. In *Physics of Sunspots* (ed. L. Cram & J. Thomas). pp. 434–445. Sacramento Peak Obs: Sunspot, NM, USA.

KUHN, J.R., LIBBRECHT, K.G. & DICKE, R. 1988 The surface temperature of the Sun and changes in the solar constant. *Science* **242**, 908–911.

KUHN, J.R. & LIBBRECHT, K.G. 1991 Nonfacular solar luminosity variations. *Astrophys.J.* **381**, L35–L37.

PAP, J.M. & FRÖHLICH, C. 1992 Multi-variate spectral analysis of short term irradiance variations. In *Proceedings of the Workshop on the Solar Electromagnetic Radiation Study for Solar Cycle 22* (ed. R.F. Donnelly). pp. 62–75. NOAA ERL SEL, Boulder, CO, USA.

ROMERO, J., WEHRLI, C. & FRÖHLICH, C. 1994 Solar total irradiance variability from SOVA 2 on board EURECA. *Solar Phys.* , in press.

SCHATTEN, K.H. & MAYR, H.G. 1992 On the structure and contrast of sunspots and faculae. In *Proceedings of the Workshop on the Solar Electromagnetic Radiation Study for Solar Cycle 22* (ed. R.F. Donnelly). pp. 142–153. NOAA ERL SEL, Boulder, CO, USA.

SPRUIT, H.C. 1988 Influence of magnetic activity on the solar luminosity and radius. In *Solar Radiative Output Variations* (ed. P.V. Foukal). pp. 143–155. Cambridge Research and Instrumentation Inc., Cambridge, MA, USA.

WILLSON, R.C. 1984 Measurements of solar total irradiance and its variability. *Space Sci. Rev.* **38**, 203–242.

WILLSON, R.C. 1994 Irradiance observations from the Spacelab 1, SMM, UARS and ATLAS experiments. In *The Sun as a Variable Star: Solar and Stellar Irradiance Variations* (ed. J.M. Pap, C. Fröhlich, H.S. Hudson & S.K. Solanki). Cambridge University Press, in press.

WILLSON, R.C., GULKIS, S., JANSSEN, M., HUDSON, H.S. & CHAPMAN, G.A. 1981 Observations of solar irradiance variability. *Science* **211**, 700–702.

WILLSON, R.C., HUDSON, H.S., FRÖHLICH, C. & BRUSA, R.W. 1986 Long-term downward trend in total solar irradiance. *Science* **234**, 1114–1116.

WILLSON, R.C. & HUDSON, H.S. 1988 Solar luminosity variations in solar cycle 21. *Nature* **332**, 332–334.

WILLSON, R.C. & HUDSON, H.S. 1991 The Sun's luminosity over a complete solar cycle. *Nature* **351**, 42–44.

Solar Absolute Reference Spectrum

By HEINZ NECKEL

Hamburg Observatory, Gojenbergsweg 112, D–21029 Hamburg, Germany

Requirements and problems in establishing an absolute solar spectrum, which could serve equally well all concerned sciences (e.g. solar physics, atmospheric physics, astrophysics) as a qualified reference, are illustrated by means of a high resolution spectral atlas with absolute intensity and irradiance scales. Also the organizational aspects are pointed out: the desire for one worldwide reference spectrum involves its recommendation by a competent team of scientists and the agreement of the relevant, official organizations.

1. Introduction

In the first half of this century inquiries for an absolute solar reference spectrum could be easily answered: the work of Abbot and his co-workers had provided spectral irradiance data of unique and undoubted quality. Even without official recommendation, these data were adopted for most hand and text books and were used by scientists and engineers everywhere on the world.

Since then, many new absolute measurements of solar radiation were made. They often differed, however, in important aspects (e.g. in bandwidth) and yielded – at least partly – very inconsistent results. To be able to relate all relevant data to a common base, and also to satisfy all those who are interested in absolute radiation data, the recommendation of a generally accepted reference spectrum appears to be highly desirable. I think it is this desire which let the organizers of this colloquium propose the topic of this contribution.

About 20 years ago there was already one attempt to establish something like a 'solar absolute reference spectrum'. The procedure to establish this reference was exemplary, and it is worthwhile to quote the relevant passages from Thekaekara (1975): *'An ad hoc committee ... was ... appointed ... to study this question. The Committee made a detailed survey of all available information and recommended standard values for the solar constant and solar spectrum. ... These values have been formally adopted by the American Society for Testing and Materials (ASTM) as the "Standard Specification for Solar Constant and Air Mass Zero Solar Spectral Irradiance". ...'*. Actually, this 'NASA/ASTM Standard of Total and Spectral Solar Irradiance' was primarily based on the results from the GALILEO experiment aboard the NASA Convair 990 aircraft. Curiously, it ranked among the least reliable irradiance data which have been published 'after Abbot', and therefore never played the role of a generally accepted reference. Nevertheless, the quoted 'recipe' how to establish a reference spectrum is still valid, but with associations like IAU (International Astronomical Union) and IAMAP (International Association of Meteorology and Atmospheric Physics) as the responsible organizations.

The more important problem is, of course, the realization of an absolute spectrum which is reliable enough for recommendation and adoption as a worldwide reference. I will demonstrate the requirements – and the involved problems – by means of a special, unpublished spectral atlas, which gives the solar intensities in absolute values. The main characteristic of this atlas is that it actually deals with two spectra, one of the disk-center and one of the disk-averaged radiation. Clearly, also the desire for a reference spectrum concerns not only the flux, but the intensity as well. Therefore, any official agreement should refer to a comparable set of such two consistent spectra.

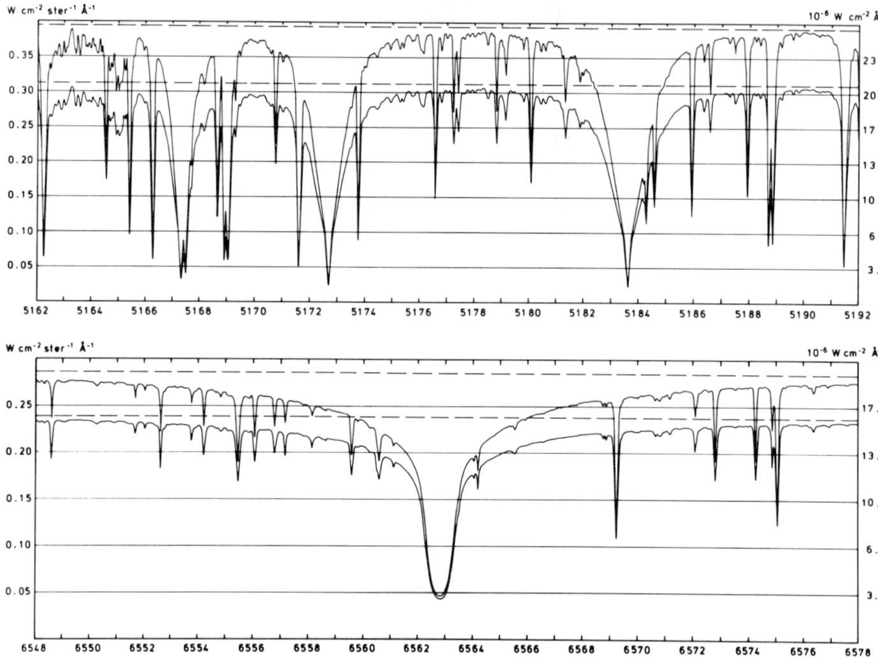

FIGURE 1. Two extracts from an absolutely scaled solar spectrum atlas for disk-center (upper curve) and disk-averaged (lower curve) intensity (left scale) Right scale: irradiance at 1 AU (lower curve). Dashed lines indicate the levels of the corresponding (quasi) continua.

First, I will describe the final product and point to its specialities, and then make a few comments on the problems with respect to the mixing of its 'ingredients'.

2. Absolute spectra of disk center and disk-averaged radiation

Figure 1 displays two extracts from the atlas, which actually extends from 329 to 1251 nm and is stored on magnetic tape. The left scales are in W cm^{-2} sr^{-1} $Å^{-1}$ and provide for both, disk center and disk-averaged spectra, the absolute intensities. The right scales are in μW cm^{-2} $Å^{-1}$ and refer only to the lower spectra on each panel, which represent the spectral irradiance at 1 AU. The atlas includes also the quasi-continuum levels (dashed lines), which are defined by the highest intensities in the two overall spectra (Figure 2).

The spectra are based on Fourier Transform Spectra (FTS), which were obtained by Brault and his co-workers at the Kitt Peak National Solar Observatory (Brault 1979; the disk spectra are copies from the same original data which were used by Kurucz *et al.* 1984 to prepare their Solar Flux Atlas). The main characteristics of these spectra are:

(1) The spectral resolution, which depends solely on the maximum distance of travel of the retro-reflectors, was for all spectra about 350000 (10 mÅ at 350 nm).

(2) The dispersion (= distance between two successive recorded spectral data) changes from 4 mÅ at 330 nm to 20 mÅ at 1250 nm.

(3) Scattered light is unimportant, since it is unmodulated and not detected as signal.

(4) If the bandwidth is limited to the upper third of the available range, the intensity scale is perfectly linear even with a nonlinear detector.

(5) There may be, however, minor errors in the zero points of the intensity scale, which

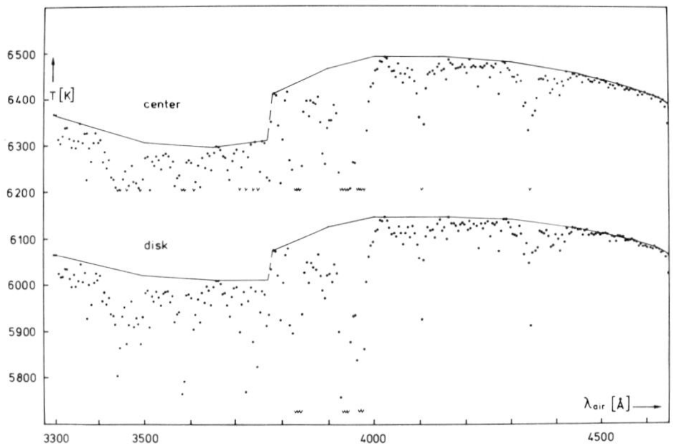

FIGURE 2. Radiation temperatures from intensity maxima in 5 Å wide spectral intervals of disk center and disk-veraged spectra. Polygonal tracks indicate the levels of the adopted 'quasi-continuum'.

can be caused by errors of the height of the central peaks in the interferograms due to noise.

(6) After each solar spectrum the (continuous) spectrum of a reference lamp was recorded and any further treatment, e.g. the absolute calibration of the intensity scales, referred to the ratio spectra Sun/lamp.

The intensity scale was calibrated by means of the absolute data of Neckel & Labs (1984) for spectral bands being precisely 20.0 Å wide. It should be noted that the data for both disk-center and disk-averaged intensities are based on the same observations of the disk center. These were later supplemented by special observations of the center-to-limb variation of the 20 Å bands by Neckel & Labs (1984), which provided then the correct ratios of disk-averaged to disk-center intensities F/I (Neckel & Labs 1984 for $\lambda < 660$ nm and Pierce & Slaughter 1977 or Pierce *et al.* 1977 for $\lambda > 660$ nm).

Figure 3 displays the resulting calibration curves and the residuals resulting from the least square fits. The overall spectral distribution of the curves reflects basically the spectrum of the reference lamp. The scatter of the residuals is usually about 0.5 % (1σ).

The wavelength-scale refers to wavelengths in air, is corrected for radial velocity and instrumental alignment errors and is in the system of the Kitt Peak Tables of Pierce & Breckinridge (1973). The remaining wavelength errors range from 0.2 mÅ at 330 nm to 2 mÅ at 1250 nm. So, the atlas spectra can be used as reference for intensities as well as for wavelength differences and line profiles.

3. Consistency with other solar and stellar data

Within the limits set by intrinsic variations of the intensity distribution across the solar disk (see below), the atlas ratios of disk-averaged to disk-center intensity agree at continuum wavelengths with those which follow from the limb darkening data of Pierce & Slaughter (1977) for $\lambda < 730$ nm, of Pierce *et al.* (1977) for $\lambda > 740$ nm, and of Neckel & Labs (1994) for $330 < \lambda < 1099$ nm. The complete integral of the irradiance spectrum (330 to 1250 nm) is 1.06 kW m^{-2} and corresponds to a solar constant of about 1.37 kW m^{-2}. The irradiance spectrum also agrees almost perfectly with the most recent model prediction of Kurucz (1992) where the opacities resulted from atomic and

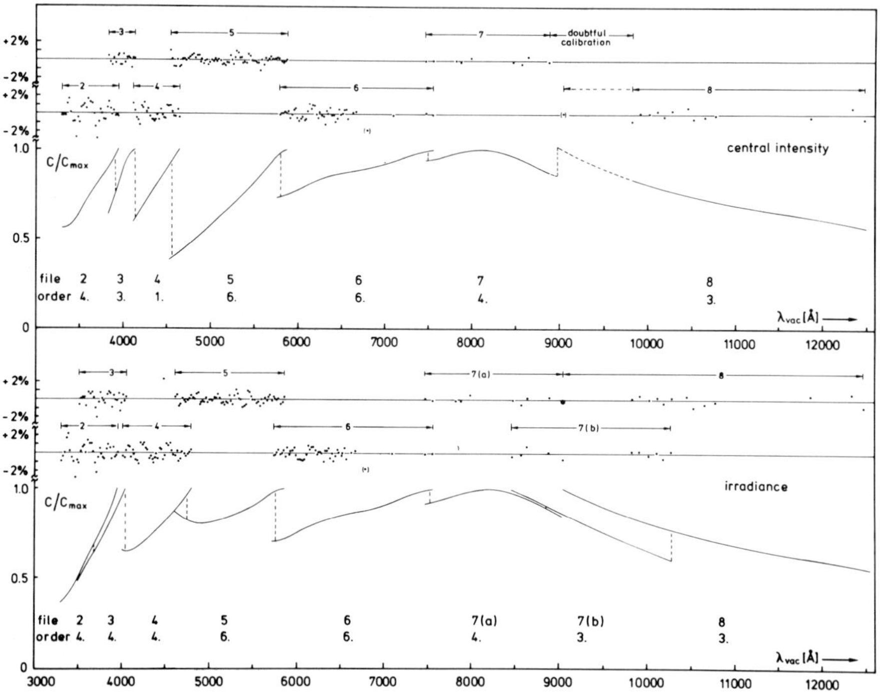

FIGURE 3. Calibration curves for Kitt Peak FTS spectra (ratio Sun to reference lamp), in units of their maximum values. Polynomials of order 1 to 6 were fitted to the ratios between observed absolute intensity integrals being exactly 20 Å wide and the corresponding FTS integrals. The residuals of the least square fits are plotted above the curves. Their standard deviations range between 0.3 and 1.1 %, the average is 0.5%.

molecular line data of some 58 million lines with a mixing-length to scale-height ratio of 1.25, and micro-turbulent velocity of 1.5 km/s. This agreement is obvious from Figure 3 which was adopted from Kurucz (1992).

A comparison of the irradiance spectrum with the absolute flux of a solar type star (16 Cyg B) is given in Figure 5. The upper plot displays the residuals with respect to the overall, $1/\lambda$-proportional gradient shown in plot 2, which indicates that 16 Cyg B is slightly cooler than the Sun ($\approx$ 40 K). Besides this gradient are no other significant systematic differences. Only at a few wavelengths one or the other (but never both) stellar value deviates by more than 0.02 magnitudes – or 2 % – from the solar data. It should be noted that this Sun-star comparison avoids any passband mismatches, since the solar data refer precisely to the same passband that were used by the stellar observers. Even the errors in the wavelength settings of the stellar observations could be evaluated precisely. Assuming errors of 1 Å for the Hardorp (1980) data and between 0 and 4 Å for the Taylor (1984) data reduced not only the scatter in the Sun-star differences significantly, but improved also very clearly the agreement between both stellar data sets. This is a good example for the advantage of using a high resolution spectrum as reference.

The Sun-star comparison provided also rather precise data for the Sun's colors (U–B, B–V), which agree well with those of other solar type stars and correspond precisely to the Sun's spectral type as shown in Table 1. In the spectral band which defines the visual magnitudes 'V', the ratio solar irradiance to flux of 16 Cyg B is 1.52×10^{13} (magnitude

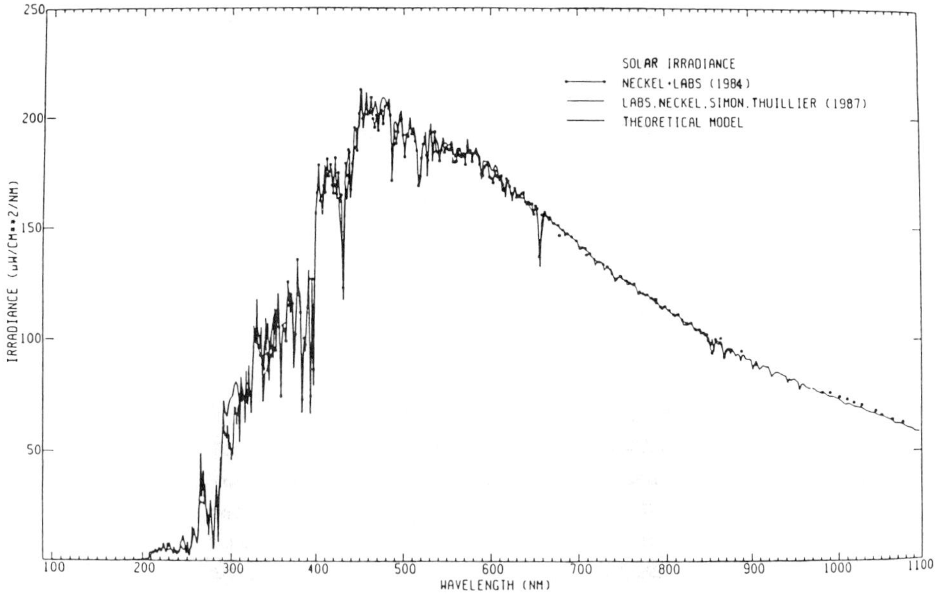

FIGURE 4. Predicted solar irradiance from Kurucz (1992) compared to the observed data of Labs *et al.* (1987) for $\lambda < 360$ nm and Neckel & Labs (1984) for $\lambda > 330$ nm. The latter data provided the absolute scale for the atlas irradiance spectrum.

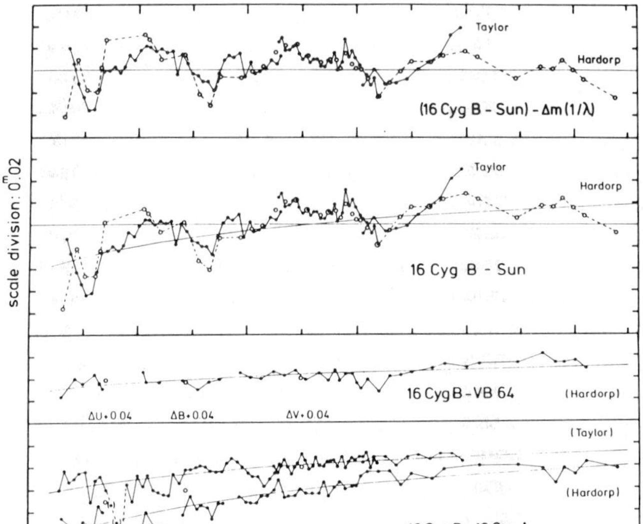

FIGURE 5. Comparison of the flux of the solar type star 16 Cyg B with the solar irradiance spectrum and with the fluxes of two other 'solar analogs'. The stellar data are from Hardorp (1980) and Taylor (1984) and are in the system defined by the absolute flux data of the standard star Vega provided by Hayes (1985). For details see Neckel (1986).

Star	T_{eff}	U − B	B − V	Spectral type
16 Cyg B	5736	0.204	0.658	G3 V
VB 64	5758	0.200	0.654	G2+ V
Sun	5777	0.195	0.650	G2 V
16 Cyg A	5804	0.190	0.644	G1.5 Vb

TABLE 1. Consistent spectral characteristics of Sun and solar type stars. Stellar colors are from Blanco *et al.* (1966) or Nicolet (1978); MK spectral types from Keenan & Yorka (1985) (N.B.: all three stars and the Sun are MK standards).

difference $\Delta V = 32.95$). With $V_{16CygB} = 6.20$ (and $V_{Vega} = 0.03$) one gets for the Sun's apparent visual magnitude $V_{Sun} = -26.75$, one of the familiar text book values.

4. Problems and restrictions

The main restriction for preparing a high resolution absolute reference spectrum comes, of course, from the Earth's atmosphere, which not only fixes the lower wavelength limit (here: 329 nm; actually, the atlas range could be extended down to about 300 nm), but also blends the red and infrared spectral region with terrestrial lines and absorption bands or even blocks completely wide spectral regions. As far as ground-based observations are concerned, the precise determination of atmospheric extinction is another, special problem. While absolute broad band measurements of the solar irradiance from space seem to be already a matter of standard routine, high resolution spectra of FTS quality and absolute intensity measurements will for some time remain a matter of ground based observations. Another problem arises from the Sun's variability, which prevents the provision of homogeneous data and limits so the accuracy of any final spectral reference. With respect to the wavelength scale it should be noted that in high resolution spectra like the Kitt Peak FTS not only line profiles and central line depths, but also even the wavelengths vary on time scales of the order of hours and days (disk center: 500 m/s, full disk: 50 m/s; see e.g. Neckel & Labs 1990). Time scales between minutes and weeks are found also for variations of the intensity distribution across the disk. In particular, for wide spectral bands including strong absorption lines the ratios disk-averaged to disk-center intensity (F/I) may undergo variations in the order of 1 − 2 % (occasionally even more!) of the disk center intensity (Neckel & Labs 1987, 1994). Clearly, such variations of F/I indicate mainly local intensity variations, e. g. at disk center, and do not necessarily point to variations of the overall flux. Therefore, absolute fluxes may be a better base for the construction of absolute reference spectra than absolute intensities. In any case, ideal reference spectra would have to be derived from data which either all refer to the same instant of time, or are all representative averages for the same, sufficiently large time interval, possibly near activity minimum. A look at Table 2 reveals that the absolutely calibrated Kitt Peak FTS spectra are far from such ideal combination: the observing days of the contributing data are distributed randomly over a period of about 20 years.

After all, one has to realize that a homogeneous, absolute solar spectrum of high spectral resolution, which includes the major spectral region of interest (200 nm − 10 μm corresponding to 99.9 % of all radiation) is not yet available and can not be constructed from available observational data. Therefore, any agreement on a reference spectrum must include one or another compromise, which has to be based on different data coming from different observations (ground based and space borne) made at different times. But such an agreement should not only refer to the spectral reference, but to all sorts of solar

FTS(center)		**FTS(disk)**	
λ (nm)	dates	λ (nm)	dates
329– 391	1979/12/28	329– 368	1981/ 6/21
391– 414	1980/ 2/15	368– 404	/ 6/22
414– 456	/ 2/15	404– 475	1980/11/23
456– 580	/ 3/30	475– 576	1981/ 3/24
580– 749	1981/ 5/12	576– 753	/ 3/25
749– 897	/ 5/13	753–1028	/ 3/25
897–1251	/ 5/13	1028–1251	/ 5/11

I(center)			**F/I**		
λ (nm)	dates	# of days	λ (nm)	dates	ref.
329– 413	1964/4/26–5/13	4	319– 660	1981/4/14–30	NL
401– 657	1961/9/1–10/14	8	660– 730	1974/3–1975/1	PS
639–1248	1962/9/9–10/ 2	10	730–1251	1974/9–1976/3	PSW

TABLE 2. Observation days for atlas data collection. The references NL, PS and PSW refer to Neckel & Labs (1984), Pierce & Slaughter (1977) and Pierce *et al.* (1977), respectively.

and stellar radiation data, including solar total irradiance, stellar colors, limb darkening, model predictions, etc. One basic requirement for the spectral reference is its internal consistency, which can be easily tested by means of the Kitt Peak FTS or any other comparable high resolution spectra (see Figure 3).

REFERENCES

BLANCO, V. M., DEMERS, S., DOUGLASS, G, G. & FITZGERALD, M.P. 1966 Photoelectric catalog. *Publ. U. S. Naval Obs.*, pp. 21.

BRAULT, J. W. 1979 Solar Fourier Transform spectroscopy. In *Proceedings of JOSO Workshop* pp. 33. Osservazioni e memorie dell'Osservatorio Astrofisico di Arcetri 106.

Hardorp, J. 1980 The Sun amoung the stars III: Energy distribution of 16 G-type stars and the solar flux calibration *Astron. Astrophys.* **91**, 221.

HAYES, D. S. 1985 Stellar absolute fluxes and energy distributions from 0.32 to 4.0 μm. In *Calibration of Fundamental Stellar Quantities, IAU Symp. No 111* (ed. D.S. Hayes, L.E. Pasinetti & A.D. Philip). pp. 225. D. Reidel Publ. Co..

KEENAN, P. C. & YORKA, A. R. 1985 1986 revised MK spectral standards: Stars G0 and later. *Standard Star Newsletter* **6**, 2.

KURUCZ, R. L. 1992 Model atmospheres for population synthesis. In *The Stellar Populations of Galaxies, IAU Symp. No 149* (ed. B. Barbuy & A. Renzini). pp. 225. Kluwer Academic Publishers.

KURUCZ, R. L., FURENLID, I., BRAULT, J. & TESTERMANN, L. 1984 *Solar Flux Atlas from 296 to 1300 nm* Harvard University Press.

LABS, D., NECKEL, H., SIMON, P. C. & THUILLIER, G. 1987 Ultraviolet solar irradiance measurements from 200 to 358 nm during Spacelab 1 mission. *Solar Phys.* **107**, 203.

NECKEL, H. 1986 The absolute energy distributions of the Sun, of the solar analogs 16 Cyg B, Hyades VB 64, 16 Cyg A, and of the standard stars α Lyr and 29 Psc. *Astron. Astrophys.* **159**, 175.

NECKEL, H. & LABS, D. 1984 The solar radiation between 3300 and 12500 Å. *Solar Phys.* **90**, 205.

NECKEL, H. & LABS, D. 1987 Asymmetry and variations of solar limb darkening along the diameter defined by diurnal motion in April 1981 (and later). *Solar Phys.* **110**, 139.

NECKEL, H. & LABS, D. 1990 Variations of 'wavelengths' and 'bisector indices' of 70 solar spectral lines between 3300 and 3960 Å in Kitt Peak FTS spectra. *Solar Phys.* **126**, 207.

NECKEL, H. & LABS, D. 1994 Solar limb darkening 1986–1990 ($\lambda\lambda$ 303 to 1099 nm). *Solar Phys.*, submitted.

NICOLET, B. 1978 Catalog of homogeneous data in the UBV photoelectric photometric system. *Astron. Astrophys. Suppl.* **34**, 1.

PIERCE, A. K. & BRECKINRIDGE, J. B. 1973 *The Kitt Peak Table of Photographic Solar Spectrum Wavelengths* Kitt Peak Nat. Obs. Contribution No. 559.

PIERCE, A. K. & SLAUGHTER, C. D. 1977 Solar limb darkening I: $\lambda\lambda$ 3033-7297 nm. *Solar Phys.* **51**, 25.

PIERCE, A. K., SLAUGHTER, C. D. & WEINBERGER, D. 1977 Solar limb darkening in the interval 7404–24018 Å. *Solar Phys.* **52**, 179.

TAYLOR, B. J. 1984 Energy distribution for 1 solar-type stars and synthetic UBV photometry of the Sun. *Astrophys. J. Suppl.* **54**, 167.

THEKAEKARA, M. P. 1975 The total and spectral solar irradiance and its possible variations. In *The Solar Constant and the Earth's Atmosphere* (ed. H. Zirin & J. Walter). pp. 232. Boulder, CO., USA.

The Solar Spectral Irradiances from X–Ray to Radio Wavelengths

By ORAN R. WHITE

High Altitude Observatory, National Center for Atmospheric Research,
Boulder, CO 80307, USA

Sources of new measurements of the solar EUV, UV, and visible spectrum are presented together with discussion of formation of the solar spectrum as a problem in stellar atmospheres. Agreement between the data and a modern synthetic spectrum shows that observed radiative variability is a minor perturbation on a photosphere in radiative equilibrium and LTE. Newly observed solar variability in 1992 defines a magnetic episode on the Sun closely associated with changes in both spectral irradiances and the total irradiance. This episode offers the opportunity to track the relationship between radiation and magnetic flux evolution.

1. Introduction

Our understanding of the Sun's spectrum has improved substantially in the last two years because of progress in both theoretical synthesis of the spectrum and new observations. It is now possible to compute a synthetic solar spectrum that closely matches the existing measurements in the visible. New high precision UV irradiance observations beginning in 1991 will give a detailed time history of the UV solar spectrum between 115 nm and 410 nm. A new experiment to measure the photospheric spectrum from 180 nm to 3200 nm referenced to a black body has just been flown by Thuillier and his colleagues. This is a major step toward knowing the absolute spectrum from the vacuum UV to the infrared. In 1992 the radiative output of the Sun apparently changed in response to a large scale decrease in the total magnetic flux in the photosphere, and this episode will be discussed as an example of the thermal and radiative consequences due to evolution of solar magnetism.

Here I address these topics to show not just progress but promise in our grasp of the solar spectrum and its variation. First I give a brief summary of new observations of spectral irradiance observations as well as discuss the formation of the solar spectrum as a problem in stellar atmospheres. As background for the empirical picture, the review by Lean (1991) and the book, *Solar Radiation Flux*, by E.A. Makarova, A.V. Kharitonov & T.V. Kazachevskaya (1991) offer current compendia of measurements and their analyses. Individual measurements mentioned here are described more fully by the experimenters themselves in presentations at this meeting.

1.1. *New observations of spectral irradiance*

Figure 1 shows examples of irradiance measurements from the two extremes of the solar spectrum, X-ray and radio wavelengths. The first, 10.7 cm radio flux data, is our standard radiometric index of solar activity measured for the full solar disk (see Tapping & Harvey 1994). The second, the X-ray irradiance between .5 and 3 nm is a new irradiance measurement constructed from 10 arcsec resolution solar images recorded continuously from space by the SXT instrument on the *Yohkoh* satellite (see Shibata 1994). The radio flux record shows solar variation over solar cycle 22, while the new X-ray record shows the change in solar radiative output from the chromosphere, transition region, and corona in 1991–93 as the Sun evolved from solar maximum in the declining phase of cycle 22.

Our picture of the visible solar spectrum comes from several observers as discussed by

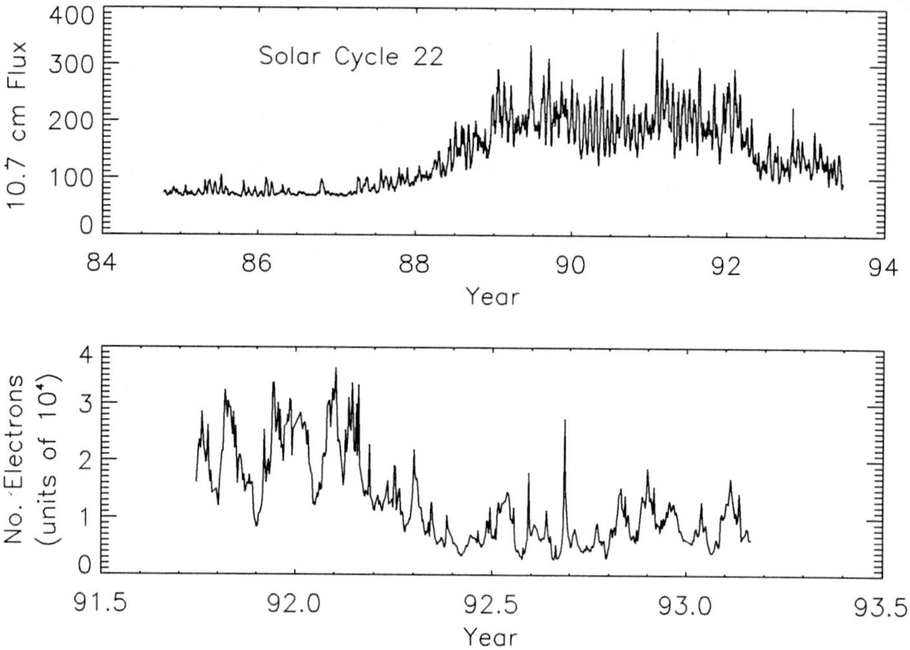

FIGURE 1. Upper panel: the 10.7 cm solar radio flux (10^{-22} W/m^2) in solar cycle 22 (Tapping & Harvey 1994). Lower panel: daily mean X-ray irradiance from the *Yohkoh*/SXT X-ray imaging experiment, .5 to 2 nm (AlMg filter), data courtesy of L. Acton, G. Slater, and the *Yohkoh* project.

Makarova *et al.* (1994). Makarova and her colleagues show the precision of the mean of the best observations to be ~3%, which is still insufficient for detection of spectral variability corresponding to the .1% change in total irradiance over the solar cycle. At wavelengths below 300 nm, measurements come from a variety of satellites such as the Nimbus 7, Atmospheric Explorer E(AE–E), Solar Mesosphere Explorer (SME), Upper Atmosphere Research Satellite (UARS), the NOAA series (NOAA–9 and NOAA–11), and the Atlas series of shuttle flights. Occasional rocket-borne experiments that are recovered and recalibrated provide checks on the space instruments. The results from the many experiments before 1991 can be found in the review by Lean (1991).

The measurement picture changed in 1991 with the flight of the Solar Ultraviolet Spectral Irradiance Monitor (SUSIM) (Brueckner *et al.* 1994), the Solar Stellar Intercomparison Experiment (SOLSTICE) (Rottman *et al.* 1994), and the European Recoverable Carrier (EURECA) (Thuillier *et al.* 1994; Romero *et al.* 1994) instrument packages for comparison with the SBUV experiments on NOAA–9 and NOAA–11. These experiments are expected to give the highest accuracy and precision measurements of the solar UV spectrum between 115 and 420 nm ever made as well as new measurements of the solar spectrum from 180 nm to infrared. We now have for the first time five instruments recording the time history of the solar spectrum continuously from space. Data from these instruments are still in a preliminary state of analysis at this writing, but the early results suggest precisions of better than 1 percent at 300 nm in the SOLSTICE measurements. Table 1 summarizes these new experiments to measure spectral irradiance at radio, visible, and ultraviolet wavelengths. The Pioneer–Venus measurements discussed

Obs. S/C	Experimenter Inst.	wavelength (nm)	op. date	type of measurement
Pioneer–Venus	Hoegy[1]	10 – 130	12/1/78–10/15/92	integrated EUV flux
San Marco	ASSI[2]	20 – 200	3/25/88–12/6/88	ref. spectrum
UARS	SUSIM[3]	115 – 420	10/16/91 to date	spectral time series
UARS	SOLSTICE[4]	115 – 420	10/3/91 to date	spectral time series
ATLAS1	SUSIM[3]	115 – 420	4/18/93–4/17/93	ref. spectrum
ATLAS1	Thuillier[5]	180 – 3200		ref. spectrum
ATLAS2	SUSIM[3]	115 – 420	3/24/92–4/2/92	ref. spectrum
EURECA	Thuillier[5]	180 – 3200	8/10/92–6/21/93	spectral time series
DRAO	Tapping[6]	10.7 cm	1947 to date	radio flux series
TOYOKAWA		3.2, 8, 15, 30 cm	Jan. 75 to date	radio flux series
NOAA	SBUV2[7]	175 – 190	1986 to date	spectral time series

Refs: [1]Pesnell & Hoegy (1994); [2]Schmidtke et al. (1992); [3]Brueckner et al. (1994); [4]Rottman et al. (1994); [5]Thuillier et al. (1994); [6]Tapping & Harvey (1994); [7]Cebula & Deland (1992).

TABLE 1. Summary of new spectral irradiance observations.

by Pesnell & Hoegy (1994) are unique since they are the only source of EUV flux time series since Hinteregger's AE–E experiments.

Photometrically calibrated imaging systems to record full disk solar images are the next step in understanding variation in the radiative output of the Sun, and Table 2 lists the systems producing data. Nishikawa (1990) and Chapman (1994) show that well calibrated images can yield relative irradiances that follow ACRIM I measurements. The San Fernando Observatory (SFO) system is designed specifically to determine the sources of variability due to sunspots and faculae. The images are made in the CaII K line and over a broad band at 672.3 nm (see Chapman 1994). Although designed for helioseismology, the National Solar Observatory (NSO) High L instrument yields very high quality images of the full disk in a 10Å band centered on the K line (see Harvey 1993, private communication). The *Yohkoh* images contain information about the irradiance for soft X-rays and, therefore, describe the variability of the radiative output of the solar corona and flares. The Big Bear Solar Observatory (BBSO) archive contains both H alpha and CaII K full disk images that serve as reference data (see Marquette 1992).

2. What is the solar spectrum?

We tend to think of the solar spectrum as a Planck function at a particular temperature, but the variation of opacity with wavelength plus the existence of temperature gradients in the outer solar atmosphere naturally lead to departures from a Planck function. A stellar spectrum can be thought of as a wavelength mapping of the source function at $\tau_\lambda \approx 2/3$. Figure 2 shows the solar spectrum together with a Planck function at 5770°K. The large departures from the Planck function below 100 nm arise from the

Obs. S/C	Experimenter Inst.	wavelength (nm)	op. date
NAO	Nishikawa[1]	545, 770	Dec. 87 – Apr. 88
YOHKOH	SXT[2]	.5 – 3	Aug. 91 to date
SFO	CFDT1, 2[3]	393.1 – 393.5	10/1/91 to date
SFO	CFDT1, 2[3]	672.3	10/1/91 to date
NSO	High L[4]	392.8 – 393.8	3/2/93 to date
BBSO	CaII K[5]	393.5	1987 to date

Refs: [1]Nishikawa (1990); [2]Shibata (1994); [3]Chapman (1994); [4]Harvey (1993); [5]Marquette (1992)

TABLE 2. Summary of photometric imager observations

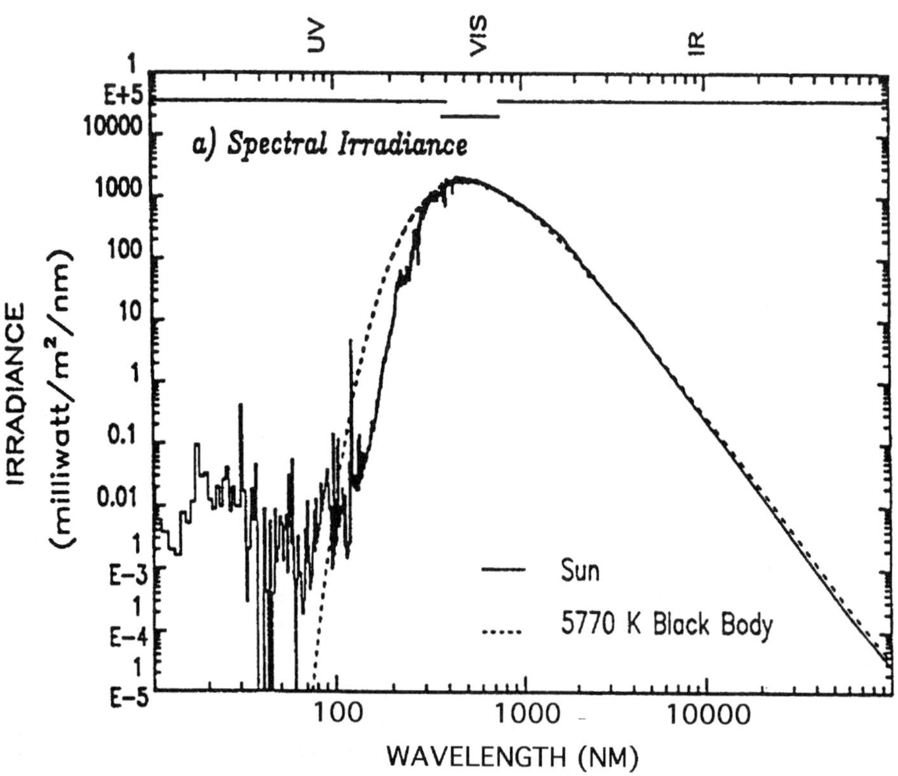

FIGURE 2. The solar radiative energy distribution compared to black body Planck function at 5770°K, from Lean (1991).

existence of the chromosphere and corona while the departure between 150 nm and 450 nm arises from an increased photospheric opacity at these wavelengths.

The problem of how a star adjusts its emergent spectrum to carry the necessary flux is addressed in the theory of stellar atmospheres. In the case of a spherically symmetric photosphere where the energy flux is carried wholly by radiation; specification of the radiative flux, the chemical composition, the mass of the star and the opacity as function of temperature, density, and wavelength is sufficient to compute an emergent spectrum

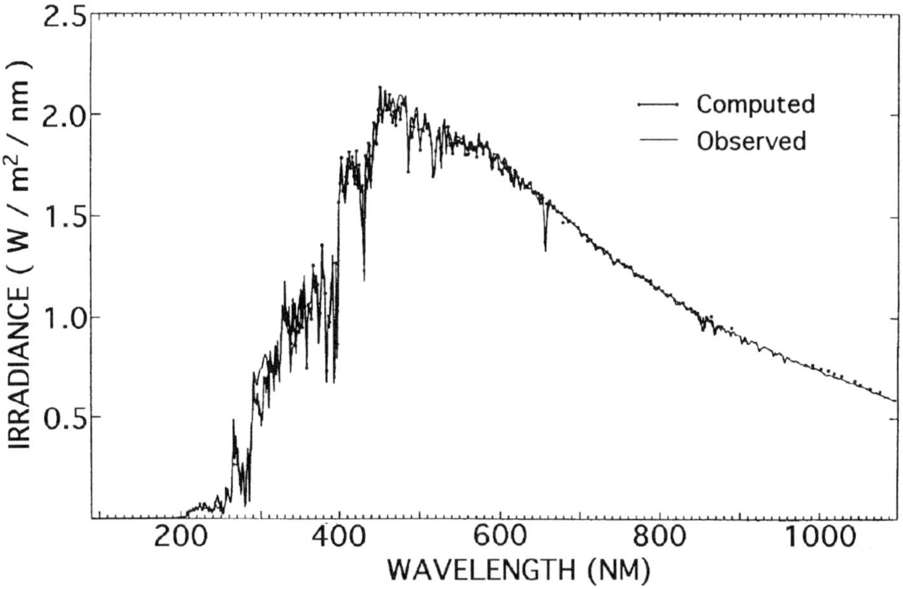

FIGURE 3. Comparison of synthetic solar spectrum computed by Kurucz (1991) with the measured spectrum from Neckel & Labs (1984).

using a modern LTE stellar atmospheres code such as ATLAS developed by Kurucz (1991). Kurucz's contribution is the specification of some 58 million atomic and molecular transitions contributing to the opacity. The solar spectrum computed by Kurucz is shown in Figure 3 together with the empirical spectrum given by Neckel & Labs (1984). Obviously, the shape of the computed spectrum follows the details of observed spectra well, which shows Kurucz' success in specifying free-bound absorption edges and overlapping spectral lines contributing to the opacity below 450 nm. The effective temperature for this synthetic spectrum is 5771°K.

The answer to my question 'What is the solar spectrum?' is 'In terms of the energy carrying capacity of the visible photosphere, the Sun's spectrum is formed in a quiet photosphere in LTE and radiative equilibrium.' Observationally, the Sun's total radiative output varies by only 1/1000th of the area under the spectral energy distribution curve shown in Figure 3.

The cause of radiative variability on the Sun is, to the first order, a small perturbation on the energy-carrying capacity of the outer layers where the bulk of the radiative energy originates. As stated by H. Spruit in his presentation at this colloquium: the Sun's radiative variability appears to arise from changes in the surface emissivity due to the presence of magnetic fields, not from a uniform global fluctuation in the luminosity of the star.

3. How to improve physical understanding of variability of the spectrum?

A promising approach is the work of Avrett (1992) and Kurucz (1993) in the calculation of synthetic solar spectra for different stellar models corresponding to different types and mixes of the structures associated with solar activity. Such calculations will give us estimates of the change in 'color' of the solar spectrum, particularly below 450 nm where

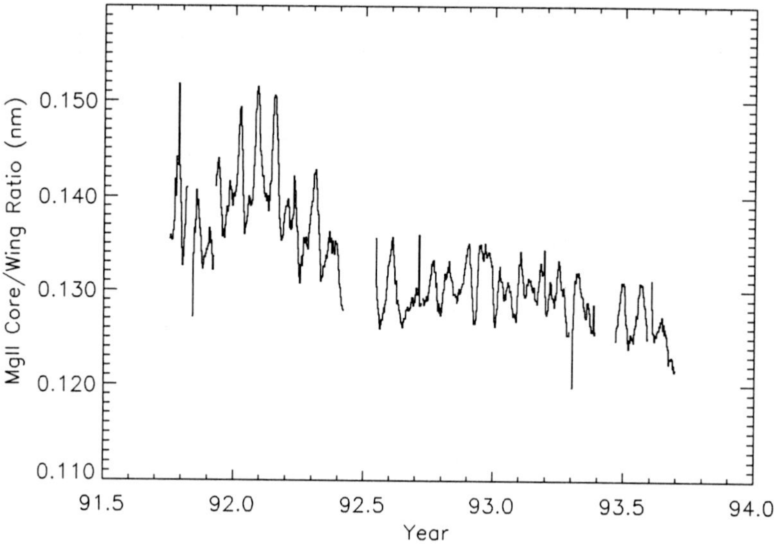

FIGURE 4. Plot of the MgII core/wing ratio for 1991 to 1993 from UARS/SOLSTICE spectral irradiance measurements. The core/wing ratio is a relative intensity measurement of the integrated chromospheric emission in the cores of the 280 nm MgII doublet. From Rottman *et al.* (1994).

the effect of bound-bound line and free-bound continuum opacities change the shape of the spectral energy distribution.

Current measurements of spectral irradiances between 115 nm and 410 nm address the variability of this spectral region. The UARS measurements run from 3 October 1991 to date and now cover a range of solar activity from solar maximum levels to the descending phase of cycle 22, i.e., a period when the solar surface becomes less complicated. The interplay between these new observations and the spectrum synthesis will give insight into the cause of variation in the Sun's UV spectral energy distribution.

However, these approaches do not address the basic problem of the change in surface emissivity with the emergence and disappearance of magnetic flux in the photosphere. The question of the thermodynamics of evolving flux tubes and their effect on the radiative output is the core of the spectral variability problem as we see it today. The UARS and ground-based measurements in 1992 show a decline in chromospheric emission associated with a decrease in apparent magnetic field strength. This observed change offers the opportunity to see directly the relationship between radiative variability and solar magnetism on an almost global scale.

As an illustration of the radiative change observed in 1992, Figure 4 shows a plot of the MgII core/wing ratio as a function of time from the UARS (SOLSTICE) database. Beginning in early 1992 the MgII index and other chromospheric line emissions decline to a lower level by June 1992 and remain in this state since that time. Data from the National Solar Observatory (NSO) and Mt. Wilson Observatory (MWO) show that this radiative change occurred when the total apparent magnetic flux decreased by about a factor of two. Furthermore, the magnetic flux change occurs principally in the southern solar hemisphere where the number of active regions decreased markedly during the spring of 1992.

I illustrate this change in magnetism in Figure 5 from K. Harvey (1994) where she plots the total apparent magnetic flux in active regions for each Carrington rotation between

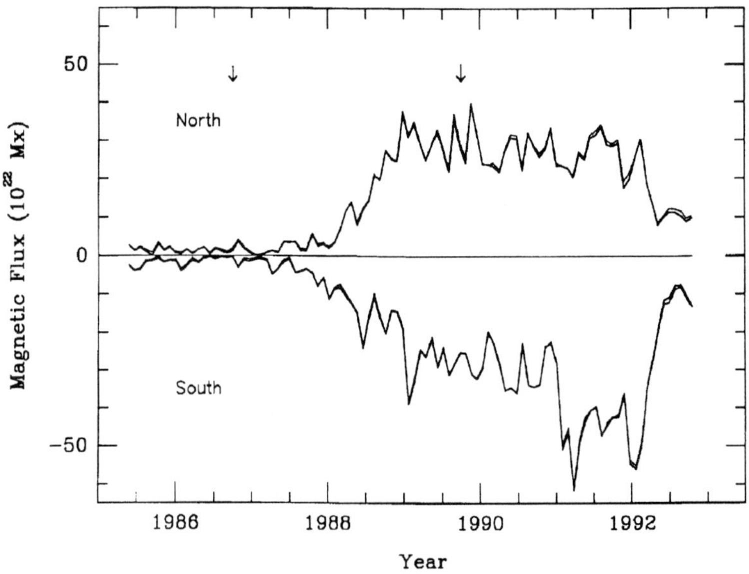

FIGURE 5. The total mean magnetic flux in active regions during solar cycle 22. Note the rapid decline in the total solar magnetism in 1992. From K. Harvey (1994). Arrows mark solar cycle minimum and maximum.

1985.5 to 1992.7. This plot shows that the magnetism in active regions decreased in both hemispheres to low levels almost globally around the Sun. This episode began in a period of high solar activity preceded by a rapid emergence of magnetic flux in the southern hemisphere during the last three months of 1991. Thus, we see the radiative consequences of the emergence of flux in several active regions in the southern hemisphere followed by the temporary disappearance of the active region flux in both hemispheres in a period of nine months between October 1991 and June 1992.

What of the change in the integrated spectrum during this time? Four sets of total irradiance measurements are available during parts of this period: Earth Radiation Budget (Nimbus 7/ERB) from D. Hoyt and Lee Kyle (Hoyt *et al.* 1992), Active Cavity Radiometer (UARS/ACRIM II) from R.C. Willson (Willson 1994), and Solar Variability Experiment (EURECA/SOVA1 and SOVA2) from Crommelynk (Crommelynk *et al.* 1994) and from C. Fröhlich (Romero *et al.* 1994). Only the ERB data are available for detailed study at this time, and they are shown as the middle curve in Figure 6, together with the sunspot deficit computed from Photometric Sunspot Index (PSI) data supplied by J. Pap and C. Fröhlich. Correction of the ERB data for the effects of sunspots using the PSI time series gives the upper curve in Figure 6. This curve is then the estimate of the run of 'facular excess' on the Sun from the beginning of 1991 to January 1993 when the PSI database stops. This bright facular contribution began increasing by $\sim$1 Watt/m^2 at the very end of 1991, reached maximum near the first of February, and then steadily declined at the rate of 2 Watt/m^2 per 6 months. We observe the decline for six months until the facular excess can no longer be computed with existing data.

I cautiously ask: Does this short history of the facular excess show the effect of thermal shadowing by large-scale flux tube aggregates lying below the visible surface late in 1991 followed by the rise and fall in facular excess as thermal energy carried by these magnetic aggregates appears and then steadily declines during the first six months of 1992?

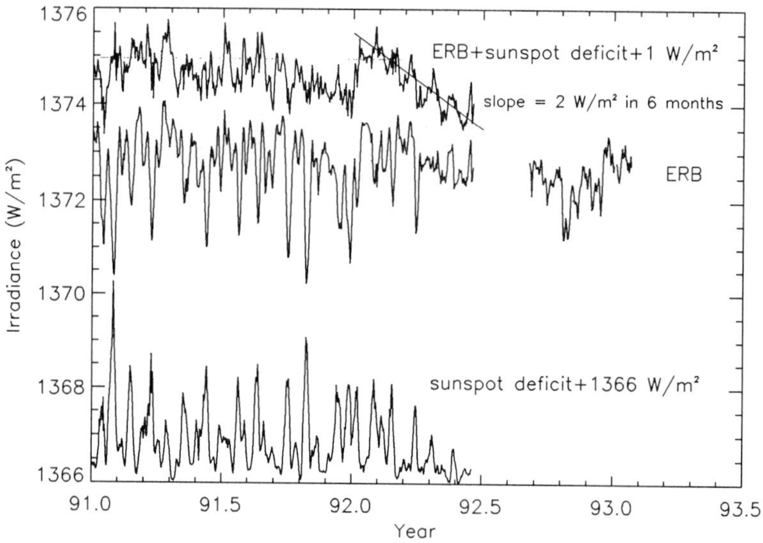

FIGURE 6. Comparison of the total irradiance variation from ERB in 1991 to 1993 with the sunspot deficit, and the total irradiance corrected with PSI. Upper curve: ERB+sunspot deficit+ 1 W/m^2, middle curve: ERB data uncorrected, lower curve: sunspot deficit +1366 W/m^2. ERB data courtesy of Hoyt and Kyle. PSI data courtesy of Pap and Fröhlich.

Acknowledgments. I am privileged to have many generous colleagues who contributed both data and ideas to this paper. I particularly appreciate contributions from Dick Donnelly, Doug Hoyt, Peter Fox, Judit Pap, Claus Fröhlich, Dick Willson, Karen Harvey, Jack Harvey, Roger Ulrich, Loren Acton, Greg Slater, Tom Woods, Gary Rottman, Bernard Durney, B.C. Low and Gary Chapman. This research is supported in part by contract NASW–4721. The National Center for Atmospheric Research is sponsored by the National Science Foundation.

REFERENCES

AVRETT, E. H. 1992 Temporal variations of near-UV, visible, and infrared spectral irradiance from a theoretical viewpoint. In *Proceedings of the Workshop on the Solar Electromagnetic Radiation Study for Solar Cycle 22* (ed. R. Donnelly). pp. 20–42. NOAA ERL SEL, Boulder, CO, USA.

BRUECKNER, G. E., FLOYD, L.E., LEAN, J.L., LUND, P.A., PRINZ, D.K. & VAN HOOSIER, M.E. 1994 Irradiance observations from UARS/SUSIM and ATLAS/SUSIM In *The Sun as a Variable Star: Solar and Stellar Irradiance Variations* (ed. J.M. Pap, C. Fröhlich, H. S. Hudson & S.K. Solanki). Cambridge University Press, in press.

CEBULA, R. P. & DELAND, M. T. 1992 The SBUV–2 monitors and the NOAA–9 and NOAA–11 satellites. In *Proceedings of the Workshop on the Solar Electromagnetic Radiation Study for Solar Cycle 22* (ed. R.F. Donnelly). pp. 239–249. NOAA ERL SEL, Boulder, CO, USA.

CHAPMAN, G. 1994 Full-disk ground-based photometric observations of the Sun. In *The Sun as a Variable Star: Solar and Stellar Irradiance Variations* (ed. J.M. Pap, C. Fröhlich, H. S. Hudson & S. K. Solanki). Cambridge University Press, in press.

CROMMELYNCK, D., DOMINGO, V., FICHOT, A. & LEE, R. 1994 Irradiance observations from the EURECA and ATLAS experiments. In *The Sun as a Variable Star: Solar and Stellar Irradiance Variations* (ed. J.M. Pap, C. Fröhlich, H. S. Hudson & S. K. Solanki). Cambridge University Press, in press.

HARVEY, J. W. 1993 Private communication.

HARVEY, K. & TAPPING, K.F. 1994 Irradiance models based of solar magnetic field observations. *Solar Phys.*, in press.

PESNELL, W.D. & HOEGY, W.R. 1994 Solar cycle variations in the extreme ultraviolet . *Solar Phys.*, in press.

HOYT, D. V., KYLE, H.I., J.R. HICKEY & MASCHHOFF, R.H. 1993 The Nimbus 7 total irradiance: A new algorithm for its derivations. *J. Geophys. Res.* **97, A1**, 5–63.

KURUCZ, R. L 1991 *New Lines, New Models, New Colors, Precision Photometery: Astrophysics of the Galaxy* (ed. A.G.D. Philip, A.R. Upgren & K.A. Janes). L. Davis Press, Schnectady, NY.

LEAN, J. 1991 Variations in the Sun's radiative output. *Rev. of Geophys.* **29, 4**, 505–535.

MAKAROVA, E. A., KHARITONOV, A.V. & KAZACHEVSKOYA, T.V. 1991 *Solar Radiation Flux*, Naulka Publishers, USSR.

MAKAROVA, E. A., KAZACHEVSKOYA, T.V. & KHARITONOV, A.V. 1991 On the variability of some characteristics of solar flux. *Solar Phys.*, in press.

MARQUETTE, W. 1992 Calcium plage observations at Big Bear Solar Observatory. In *Proceedings of the Workshop on the Solar Electromagnetic Radiation Study for Solar Cycle 22* (ed. R. Donnelly). pp. 154–159. NOAA ERL SEL, Boulder, CO., USA.

NECKEL, H. & LABS, D. 1984 The solar radiation between 3300 Å and 12,500 Å. *Solar Physics* **909**, 205–258.

NISHIKAWA, J. 1990 Estimation of total irradiance variations with the CCD solar surface photometer. *Astrophys. J.* **359**, 235–345.

ROMERO, J., WEHRLI, C. & FRÖHLICH, C. 1994 Solar total irradiance variability from SOVA2 on board EURECA. *Solar Phys.*, in press.

ROTTMAN, G. J., WOODS, T.N., WHITE, O.R. & LONDON, J. 1994 Irradiance observations from the UARS/SOLSTICE experiment. In *The Sun as a Variable Star: Solar and Stellar Irradiance Variations* (ed. J. M. Pap, C. Fröhlich, H. S. Hudson & S. M. Solanki). Cambridge University Press, in press.

SHIBATA, K. 1994 The variable structure of the X-ray corona as viewed by *Yohkoh*. In *The Sun as a Variable Star: Solar and Stellar Irradiance Variations* (ed. J.M. Pap, C. Fröhlich, H. S. Hudson & S. K. Solanki). Cambridge University Press, in press.

SCHMIDTKE, G., WOODS, T.N., WORDEN, J., ROTTMAN, G.J., DOLLS, H., WITA, C. & SOLOMON, S.C. 1992 Solar EUV irradiance from the San Marco ASSI: A reference spectrum. *Geophys. Res. Letters* **19, 21**, 2175–2178.

TAPPING, K. F. & HARVEY, K. 1994 Variations of coronal radiations at radio wavelengths. In *The Sun as a Variable Star: Solar and Stellar Irradiance Variations* (ed. J.M. Pap, C. Fröhlich, H. S. Hudson & S.K. Solanki). Cambridge University Press, in press.

THUILLIER, G., HERSÉ, M., LABS, D., SIMON, P.C., GILLOTAY, D. & MANDEL, G. 1994 Irradiance observations in near-UV, visible, and near infrared spectral bands from measurements carried out during ATLAS–1 and EURECA–1 missions. In *The Sun as a Variable Star: Solar and Stellar Irradiance Variations* (ed. J. M. Pap, C. Fröhlich, H. S. Hudson & S.K. Solanki). Cambridge University Press, in press.

WILLSON, R. C. 1994 Irradiance observations from the Spacelab 1, SMM, UARS, and ATLAS experiments. In *The Sun as a Variable Star: Solar and Stellar Irradiance Variations* (ed. J. Pap, C. Fröhlich, H. S. Hudson & S. Solanki). Cambridge University Press, in press.

Irradiance Observations of SMM, Spacelab 1, UARS, and ATLAS Experiments

By RICHARD C. WILLSON

Solar Irradiance Monitoring Group, Earth and Space Sciences Division, Jet Propulsion Laboratory, California Institute of Technology, Pasadena, CA 91109, USA

Detection of intrinsic solar variability on the total flux level was made using results from the first Active Radiometer Irradiance Monitor (ACRIM) experiment, launched on the Solar Maximum Mission (SMM) in early 1980. ACRIM I, specifically designed to start the high precision total solar irradiance database as part of the U.S. Climate Research Program, produced high precision results throughout the 9.75 years of the Solar Maximum Mission. The second ACRIM experiment was flown aboard the Space Shuttle as part of the NASA/ESA Spacelab 1 Mission in late 1983. Its primary function has been to provide a comparison with ACRIM I that could be used to relate its observations with future satellite solar monitors, should they and ACRIM I fail to overlap in time. The second ACRIM satellite solar monitoring experiment (ACRIM II) has provided high precision total solar irradiance observations since its launch as part of the Upper Atmosphere Research Satellite (UARS) mission in late 1991 and continues at present. The shuttle ACRIM instrumentation has been flown on the ATLAS 1 and 2 missions in 1992 and 1993, providing comparisons with the UARS/ACRIM II.

1. The SMM/ACRIM I experiment

The ACRIM I experiment on the Solar Maximum Mission provided the first unambiguous evidence of intrinsic total solar irradiance (TSI) variability (Willson 1980) (see Figure 1). The first clear evidence was a TSI 'dip' near 1980 day 100 (Willson *et. al.* 1981; Willson & Hudson 1981). This was caused by two large sunspot groups in an early stage of their evolution. The TSI increases on either side of this dip were caused by faculae and plage surrounding the sunspots, which are the first and last components of the groups to be seen when they rotated onto and then off our side of the Sun. These two groups were identifiable for six solar rotations. On each successive rotation the ratio of facular area to sunspot area increased within these two groups. The impact of this can be clearly seen during the second dip near day 125 at which time the effect of the sunspot area was significantly offset by the presence of relatively enhanced facular area (Willson 1982, 1984).

The solar cycle TSI variation of about 0.1% peak-to-peak can be clearly seen in Figure 2 which shows the entirety of the ACRIM I results for the 9.75 years mission. The occurrence of the TSI minimum near the time of solar activity minimum (September 1986), as reckoned by the reversal of polarity of active regions, confirms the solar cycle character of the TSI variation (Willson & Hudson 1988, 1991). One of the most important aspects of the discovery of a direct proportionality of TSI and solar activity is that its direction is that required for solar forcing of climate change. This makes solar forcing a possible explanation of the 'Spörer, Maunder and Dalton' climate minima that occurred from the 14th through the 19th centuries and the 'Medieval Climate Maximum' of the 12th century (Eddy 1977; Anderson 1991).

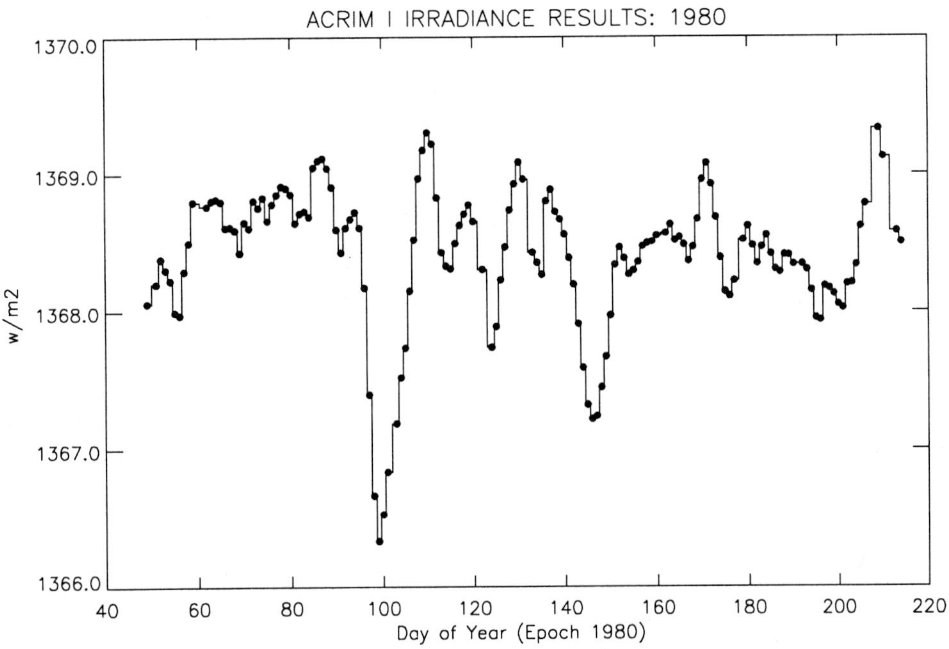

FIGURE 1. Results from the SMM/ACRIM I experiment. Daily mean irradiance values are shown for the first six months of observation.

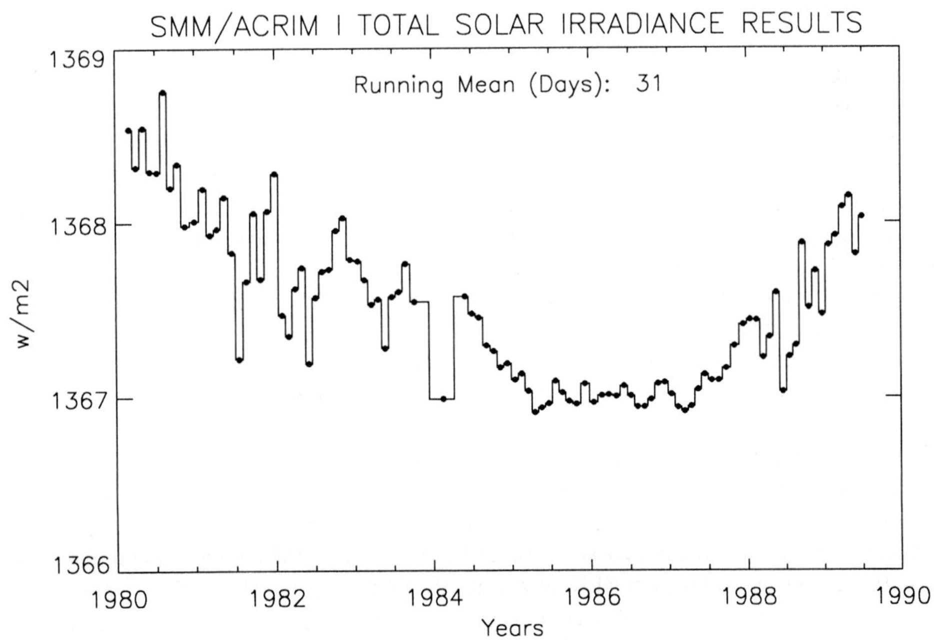

FIGURE 2. Results from the SMM/ACRIM I experiment. 31-day running means smoothing is used to clarify the long term solar cycle trend.

Comparisons	Ratio to ACRIM	Standard Error (σ)	Number of Samples
Nimbus7/ERB - SMM/ACRIM I	1.003138	0.000004	2714
Nimbus7/ERB - SMM/ACRIM II	1.004946	0.000003	225
ACRIM I/ACRIM II (ERB Comparisons)	1.001802	0.000002	

TABLE 1. Comparisons of SMM/ACRIM I and UARS/ACRIM II with Nimbus-7/ERB experiment. Demonstration of the 'overlap strategy' capability for preserving the precision of the total solar irradiance database.

2. The UARS/ACRIM II experiment

The ACRIM II experiment was launched in September 1991, nearly two years after the demise of the Solar Maximum Mission. The results to date are shown in Figure 3. The launch delay of several years, caused by the Challenger accident, prevented the planned direct on-orbit comparisons between ACRIM I and II. The importance of relating their results with precision through direct in-flight comparison cannot be understated since the inherent absolute accuracy of flight TSI sensors is inadequate to sustain a useful long term database. Fortunately, we have been able to relate their results using the next best approach – by comparing both ACRIM I and ACRIM II with the contemporary observations by the Nimbus-7/ERB experiment (Willson 1992). These comparisons are summarized in Table 1 and shown in Figure 4. The benefit of having redundant TSI monitoring experiments in place is clear from this, it has saved the continuity of the TSI database. Without the benefit of these overlapping comparisons, the relationship of future TSI observations to the past could not be established with precision smaller than the solar cycle variation.

From Table 1 it can be seen that the ratio of ACRIM I/ACRIM II can be established with an uncertainty of less than 5 ppm, assuming that only random errors affect the results. That this is not quite the case can be clearly seen in Figure 4. Systematic differences are seen in the plot of the ratio of Nimbus-7/ERB to ACRIM I in the form of an approximately sinusoidal variation nearly centered on their 9.45 years set of comparisons. Similarly, a systematic downward trend can be seen in the ratio of Nimbus-7/ERB to ACRIM II.

The causes of systematic differences between ACRIM and ERB are not known with certainty. They may be due to uncorrected thermal effects of the ERB instrumentation during the 1980–1990 period and to uncorrected degradation of ERB's sensitivity during 1991–1992. Other evidences of shorter term ERB sensitivities to thermal effects can be observed in the results as divergences from the means in Figure 4 that significantly exceed the standard errors. The revision of the ERB results by Hoyt & Kyle (1990) in which earlier database problems due to pointing errors and thermal drifting were significantly reduced, concluded that residual thermal effects remained. A component of the systematic differences is probably contributed by ACRIM I misperformance as well, although by virtue of its more stable spacecraft thermal environment and much higher frequency of electrical self-calibration, the contributions are likely to be much smaller than ERB's.

Similar arguments implicate degradation of ERB's sensitivity as the source of the systematic downward trend relative to the ACRIM II results. By 1992 ERB had been in orbit more than 14 years, and although its total amount of direct solar exposure is only

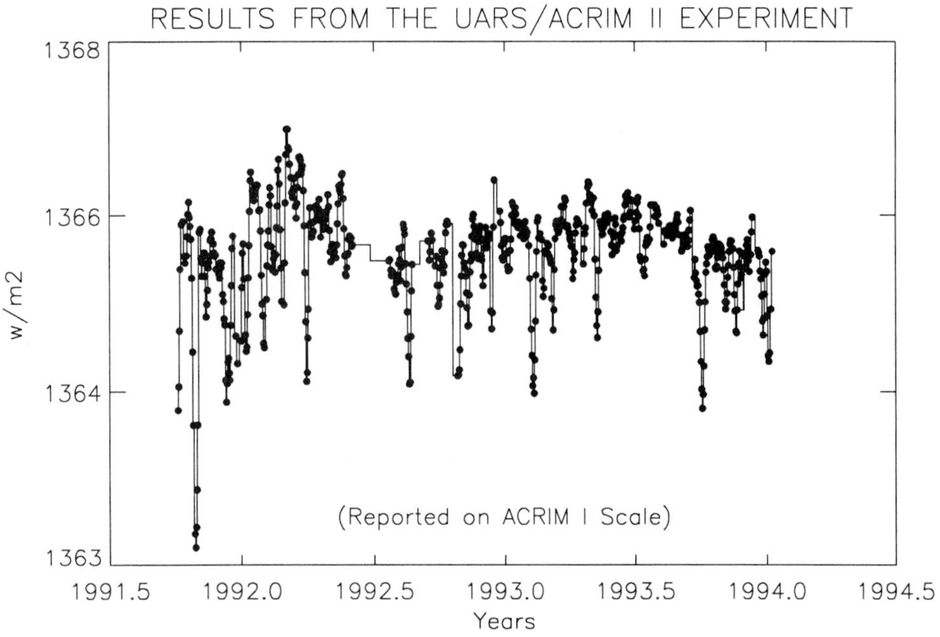

FIGURE 3. Results from the UARS/ACRIM II experiment. Daily mean irradiances are shown reported on the scale of the SMM/ACRIM I experiment (see Table 1).

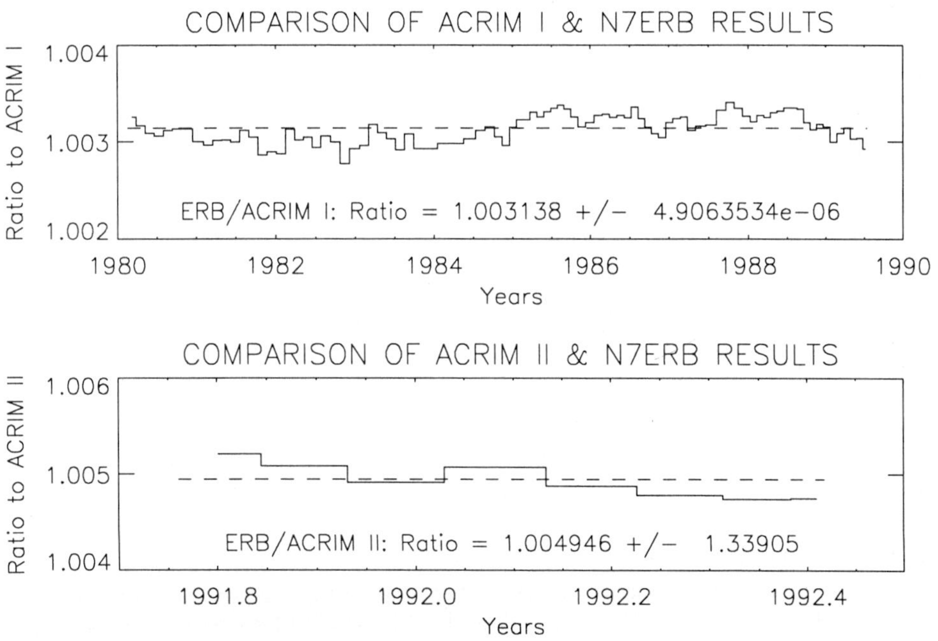

FIGURE 4. Comparisons of SMM/ACRIM I, UARS/ACRIM II and the Nimbus-7/ERB experiments. Results are shown as percentage variation.

about 10% of that of the ACRIM's per orbit, degradation should be expected to be a factor in ERB results after that length of time. ERB has no degradation self-calibration capability. The ACRIM experiments have a multi-sensor degradation calibration ability that has demonstrated an inverse relationship of detector sensitivity and exposure to solar direct flux. Therefore this type of relative trend suggests sensitivity degradation of the ERB, particularly during this period of exceptionally high solar activity when enhanced levels of high energy solar flux and particles are present, as the cause of the relative trend.

A more realistic estimate of the uncertainties of the ERB/ACRIM comparisons is being pursued through minimization of the systematic effects by modeling. Preliminary results indicate the residual uncertainty of the ACRIM I/ACRIM II ratio will be greater than the 2 ppm of Table 1 and smaller than 50.

3. Spacelab 1/ACRIM, ATLAS/ACRIM and contemporary total solar irradiance experiments

Several TSI experiments have been conducted since launch of the UARS/ACRIM II. The first was the ATLAS1/ACRIM in March 1992, the second the ATLAS2/ACRIM experiment in April. Two ESA TSI experiments, the PMOD (Brusa & Fröhlich 1972) and CROM (Crommelynck 1981) were included on the EURECA platform which was launched by the Shuttle in 1992 and retrieved in 1993 (see also Crommelynck *et. al.* 1994; Fröhlich 1994; Romero *et. al.* 1994).

The shuttle-based ACRIM experiment flown on the ATLAS 1 and 2 missions was also flown on the Spacelab 1 mission in December 1983. The principal purpose of the shuttle ACRIM experiment was to make underflight comparisons with satellite TSI monitoring experiments so that if the latter failed to overlap the precision of the database could be sustained through mutual inter comparisons with the shuttle ACRIM. The results of comparisons with satellite experiments are shown in Table 2. The Spacelab 1/ACRIM data were acquired during the SMM/ACRIM I mission. The ATLAS 1 and 2/ACRIM results were acquired during the UARS/ACRIM II experiment.

In the interest of understanding the reliability of the shuttle ACRIM results, the ACRIM II observations are scaled to ACRIM I as discussed previously, to facilitate comparison of Spacelab 1 and ATLAS measurements. Sensor ACR502 was included in the ACRIM instrument on all three shuttle flights, the ACR504 on two. The weighted means are dominated by the ATLAS results because of their smaller uncertainties. This is caused by far smaller number of observations acquired during the Spacelab 1 mission when near total failure of the shuttle command-and-data system (RAU) occurred. The spread of the observations from the flights (±242 ppm for ACR502 and ±152 ppm for ACR504) exceeds their statistical uncertainties by a large margin, indicating the clear intervention of sources of systematic error. It has been observed that the clean cavities of the ACRIM sensors provided for integration in each mission are contaminated when returned for recalibration after each flight. Contamination is believed to be the principal cause of the variation of results from flight-to-flight. The largest spread is between the ACR502 result of ATLAS 1 and the other two, indicating that its second flight result could be anomalously high. This may be resolved by the ATLAS-3/ACRIM experiment in late 1994.

It was hoped that the shuttle ACRIM series would be able to provide mission-to-mission precision better than 100 ppm instead of the approximately 200 ppm observed from the results to date. While larger than expected, this bound of uncertainty is an order of magnitude superior to the absolute uncertainty of the current generation of sensors, that

Mission	Sensor ACR502 Average (% var)	Uncertainty Std. Error (σ) ± (% var)	Sensor ACR504 Average (% var)	Uncertainty Std. Error (σ) ± (% var)
Spacelab 1 (12/83)	0.3679	0.0098	0.4479	0.0088
ATLAS 1 (3/92)	0.4164	0.0028	Not Flown	–
ATLAS 2 (4/93)	0.3680	0.0006	0.4174	0.0007
Weighted Mean (All flights)	0.3701	0.0006	0.4176	0.0007
Variation (±)	0.0242		0.0152	
Weighted Mean (ATLAS 1 & 2)	0.3701	0.0006	–	–
Variation (±)	0.0242			
Weighted Mean (SL 1 & ATLAS 2)	0.3680	0.0010	0.4176	0.0011
Variation (±)	0.0022		0.0152	

TABLE 2. Results from the Space Shuttle ACRIM experiment on the Spacelab 1, ATLAS 1 and ATLAS 2 missions. Results are shown for ACR sensors 502 and 504 as their percentage variation relative to the radiation scale defined by the SMM/ACRIM I instrument.

operate at 'ambient temperature' and comparable to the absolute uncertainty of new cryogenic TSI sensors currently under development. From this standpoint, the shuttle ACRIM experiment is a valuable insurer of the long term TSI database. Should overlap of satellite TSI monitoring experiments fail to occur in the future, the shuttle ACRIM comparisons could be used to sustain the TSI database precision at a useful level, even if less precisely than had been hoped.

The ESA EURECA platform was launched from the Shuttle in mid–1992 and retrieved about nine months later. Two TSI monitors resided on the platform: the PMOD developed by Fröhlich (see Romero *et al.* 1994) and CROM developed by Crommelynck (see Crommelynck *et al.* 1994). A comparison of preliminary daily mean results obtained simultaneously by the PMOD, CROM and the UARS/ACRIM II experiments is shown in Figure 5 (Fröhlich 1993; Crommelynck 1993). The EURECA data shown begins shortly after its deployment. The results from all three experiments depict essentially the same principal TSI variations, including the 'sunspot deficit' effect of large sunspot groups in the early and late part of the period. The ACRIM II and PMOD results are most closely matched in both SI units and the shape of their time series. The CROM utilizes a different operational mode and differs significantly in design from the PMOD and ACRIM which probably explains the differences of shape of its time series. Both sets of EURECA results demonstrate higher initial decreases in flux, relative to ACRIM II. This is probably due to high rates of degradation of their cavity absorbing surfaces in the early observations, a common feature of the initial performance of TSI sensors in space. The dashed lines of Figure 5 occur during periods when the UARS spacecraft was inoperational due to failure of the solar array's pointing mechanism and no simultaneous results were available for comparison.

These same data are presented in Figure 6 as the percentage difference from the UARS/ACRIM II results. A general downward trend in both the relative sets of results probably indicates uncorrected degradation in the PMOD and CROM sensors accumu-

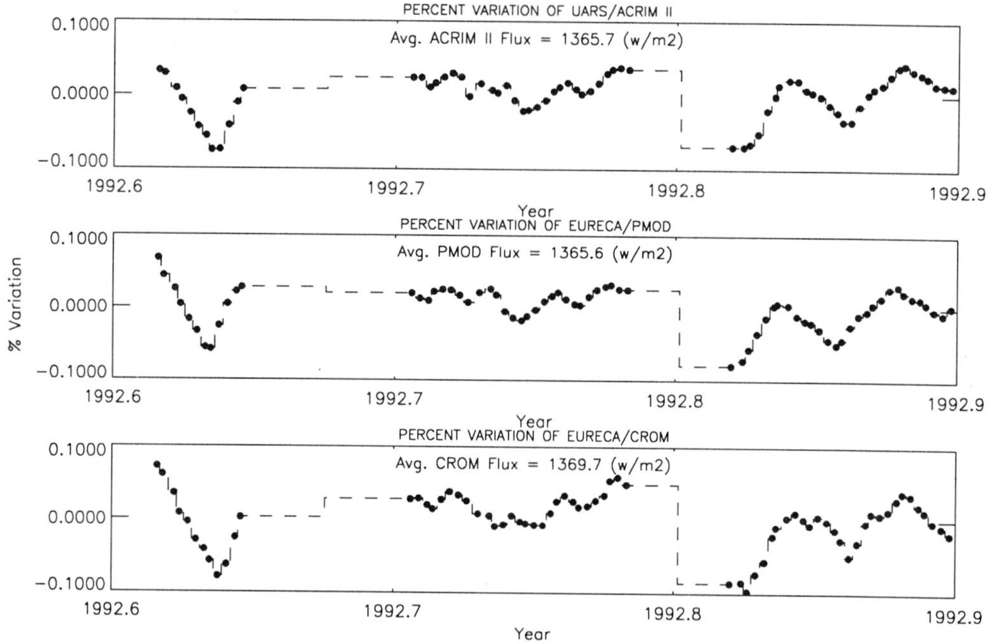

FIGURE 5. Comparisons of UARS/ACRIM II and the EURECA/PMOD and CROM results

lating with exposure to solar high energy fluxes and particles. A high initial rate of relative decline for the CROM may indicate an accelerated initial degradation similar to that found in early results from the ACRIM experiments.

4. Summary

The TSI variability detected thus far is directly proportional to solar activity, as would be required by a solar forcing function for climate change that might have been responsible for the 'Maunder or Spörer climate minima' of the 14th through the 17th centuries, and the 'Medieval Climate Maximum' of the 12th century. This implies the need to extend the high precision TSI database indefinitely with maximum available precision to detect subtle, long term TSI variation. A careful measurement strategy that relates successive solar monitoring experiments at a precision level of the instrumentation will be required to sustain the database since the uncertainty of TSI instrumentation on an absolute basis, even if operated at cryogenic temperatures, is inadequate for this purpose.

There are presently no U.S. experiments planned for the mid-to-late 1990's that have a high degree of probability for sustaining the TSI database until deployment of the next ACRIM experiment as part the Mission to Planet Earth (on the Earth Observation System 'Chemistry' satellite in late 2002). If the current experimental scenario persists, the first high precision TSI database will terminate in the late 1990's and the next will begin in 2002, with effectively no relationship between them of a sufficiently high precision to be useful for the long term study of either solar or climate variability.

Acknowledgments. The research described in this paper was completed by the Jet Propulsion Laboratory, California Institute of Technology, under a contract with the

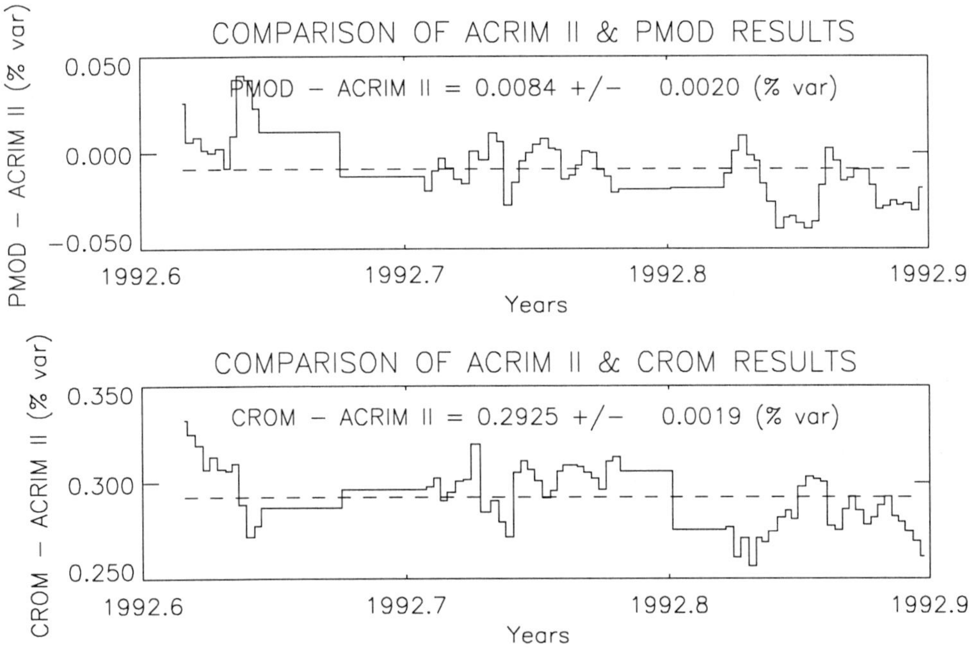

FIGURE 6. Comparison of UARS/ACRIM II with PMOD and CROM results as percentage differences in time series.

National Aeronautics and Space Administration. The author thanks C. Fröhlich and D. Crommelynck for preliminary EURECA/SOVA results.

REFERENCES

ANDERSON, R.Y. 1991 Solar variability captured in climatic and high-resolution paleoclimatic and high-resolution paleoclimatic records: A geological perspective. In *The Sun in Time* (ed. C.P. Sonett, M.S. Giampapa & M.W. Mathews). pp. 543–561. Univ. Arizona Press, Tucson, AZ, USA.

BRUSA, R.W. & FRÖHLICH, C. 1972 Entwicklung eines neuen Absolutradiometers. *Technical Note 1*. World Radiation Center, Davos, Switzerland.

CROMMELYNCK, D. 1981 Fundamentals of absolute pyrheliometry and objective characterization. *Technical Note* World Radiation Center, Davos, Switzerland.

CROMMELYNCK, D. 1993 Private communications.

CROMMELYNCK, D., DOMINGO, V., FICHOT, A. & LEE, R. 1994 Irradiance observations from the EURECA and ATLAS experiments. In *The Sun as a Variable Star: Solar and Stellar Irradiance Variations* (ed. J.M. Pap, C. Fröhlich, H.S. Hudson & S.K. Solanki). Cambridge University Press, in press.

EDDY, J.A. 1977 Historical evidence for the existence of the solar cycle. In *Solar Output and Its Variation* (ed. O.R. White). pp. 51–71, Univ. of Colorado Press, Boulder, CO, USA.

FRÖHLICH, C. 1993 Private communications.

FRÖHLICH, C. 1994 Reviews of space observations of total solar irradiance. In *The Sun as a Variable Star: Solar and Stellar Irradiance Variations*, (ed. J.M. Pap, C. Fröhlich, H.S. Hudson & S.K. Solanki). Cambridge Univ. Press, in press.

HOYT, D. & KYLE, L. 1990 An alternative derivation of the Nimbus-7 total solar irradiance variations. In *Climate Impact of Solar Variability* (ed. K.H. Schatten & A. Arking). NASA

CP-3086, 293–300.

ROMERO, J., WEHRLI, C. & FRÖHLICH, C. 1994 Solar total irradiance variability from SOVA2 on board EURECA. *Solar Phys.*, in press.

WILLSON, R.C. 1980 Solar irradiance observations from the SMM/ACRIM experiment. *American Geophysical Union*, Toronto, Canada, May 1980.

WILLSON, R.C. 1982 Solar irradiance variations and solar activity. *J. Geophys. Res.* **86**, 4319–4326.

WILLSON, R.C. 1984 Measurements of solar total irradiance and its variability. *Space Science Reviews* **38**, 203–242.

WILLSON, R.C. 1992 Solar total irradiance monitoring, past, present and future. *EOS Transactions, American Geophysical Union* **73**(14), pp. 242.

WILLSON, R.C. & HUDSON, H.S. 1981 Variations of solar irradiance. *Astroph. J. Lett.* **24**, 185–189.

WILLSON, R.C. & HUDSON, H.S. 1988 Solar luminosity variations in solar cycle 21. *Nature* **332**, 810–812.

WILLSON, R.C. & HUDSON, H.S. 1991 The Sun's luminosity over a complete solar cycle. *Nature* **351**, 42–44.

WILLSON, R.C., GULKIS, S., JANSSEN, M., HUDSON, H. S. & CHAPMAN, G. A. 1981 Observations of solar irradiance variability. *Science* **211**, 700–702.

Total Solar Irradiance Observations from the EURECA and ATLAS Experiments

By DOMINIQUE CROMMELYNCK[1], VICENTE DOMINGO[2], ALAIN FICHOT[1] AND ROBERT LEE III[3]

[1]Royal Meteorological Institute, Brussels, Belgium

[2]European Space Agency/ESTEC, Noordwijk, The Netherlands

[3]Langley Research Center, Hampton, VA 23681, USA

The total solar irradiance was observed by Differential Dual Absolute Radiometers (DIARAD) within the SOVA (*SO*lar *VA*riablity) experiment during the EURECA-I mission (*EU*ropean *RE*trievable *CA*rrier; August 1992 – June 1993) and within the SOLCON (*SOL*ar *CON*stant) experiment during the ATLAS-1 and 2 missions (*AT*mospheric *L*aboratory for *A*pplications and *S*ciences; March 1992 and April 1993). The results from these observations are presented and the metrology aspect addressed through comparison of the simultaneaous measurements with the two DIARADs on EURECA-I and ATLAS-2, respectively.

1. Introduction

The SOLCON (*SOL*ar *CON*stant) experiment of the ATLAS (*AT*mospheric *L*aboratory for *A*pplications and *S*ciences)program is the result of a collaboration between the Royal Meteorological Institute of Belgium (RMIB), the Space Science Department (SSD) of the European Space Agency (ESA), and the Langley Research Center (LRC) of the National Aeronautics and Space Adminiostration (NASA). The SOVA (*SO*lar *VA*riablity) experiment on EURECA (*EU*ropean *RE*trievable *CA*rrier) is lead by RMIB and in cooperation with the Physikalisch Meteorologisches Observatorium Davos (PMOD) and the SSD. Each experiment contains amoung other instruments a Differential Dual Absolute Radiometer (DIARAD) designed by Crommelynck (1973, 1975, 1982), the results of which are reported here. SOLCON contains also a 2 channel sunphotometer from RMIB and was flown on the ATLAS-1 and 2 missions launched by the Space Shuttles Atlantis and Endeavour on March 24, 1992 and April 7, 1993, respectively. SOVA contains further one 2-channel sunphotometer from RMIB, two PMO6 type absolute radiometers, two 3-channel sunphotometers, one high resolution radiometer and a high precision two axis sun sensor from PMOD. SOVA on EURECA was launched by Atlantis on July 31, 1992 and was brought back to Earth in June 1993 after a flight of more than nine months. The particular interest of this latter mission is that the exeriment is recovered and the performance of the instruments can be checked after the extended exposure to the space environment.

2. Objectives and instrumentation

The main scientific objective of the SOVA and SOLCON experiments is the observation of the solar constant and its variation in time. The SOVA experiment observed solar irradiance quasi-continuously for more than 9 months, whereas SOLCON measured during the four and five solar observation periods of ATLAS-1 and 2, respectively. The latter experiment is the continuation of the measurements on Spacelab 1 (Crommelynck & Domingo 1984). The data of the recent ATLAS missions are mainly aimed at comparisons between the different radiometers in space: ACRIM II on UARS (Upper Atmosphere Research Satellite of NASA), solar monitor on ERBS (Earth Radiation Budget Satellites of NASA), HF on NIMBUS 7 (NASA/NOAA), SOVA on EURECA and the other ACRIM on ATLAS. This will improve the metrology of total solar irradiance and aims eventually at the materialization of a World Space Radiometric Reference. Such a reference is needed to support the long-term accurate monitoring of the Sun (Crommelynck 1988).

The SOVA experiment is described in detail by Crommelynck *et al.* (1993). As the emphasis here is on DIARAD, some details of the data acquisition system and the instrument will be presented. The data acquisition is based on voltage-to-frequency converters which are calibrated against highly accurate reference voltages every 3 minutes. The measuring period extends over 8.2 s and yields for the measured voltages a resolution of 22 bits. The DIARAD has two cylindrical cavities with a flat bottom which are coated with a diffuse black paint and mounted on a common heat sink with a heat flux sensor in between. The receiver area is circular with a diameter of 10 mm. In contrast to most other solar radiometers, the two detectors are located side by side and both can be used for irradiance measurements which are identified as -L and -R (left and right). The symmetrical arrangement allows for true differential operation which automatically eliminates the effect of changing thermal environment by e.g. changing the temperature of the view-limiting muffler in front of the detectors. In addition, it allows for 16 different radiometric states, used for internal tests and redundancy in operation. During normal mode of operation one detector remains shaded and the other is alternatively irradiated and shaded; Both are heated electrically and the difference in output of the two heat flux sensors is used to drive a servo system which keeps the differential heat fields constant. The shutter is actuated every 99 seconds and the irradiance calculated from a function of the difference between electrical powers in the shaded and irradiated state. Both channels are completely and independently characterized in air and vacuum in order to determine the deviations from ideal behaviour. The uncertainty of the coefficients determined by the characterization provides an evaluation of the absolute accuracy of the detectors which is estimated for DIARAD to be $<0.1\%$.

3. Results from the SOVA-DIARAD on EURECA

On August 10, 1992 the SOVA experiment was switched on and, except for a few days, was pointed at the Sun continuously during nine months up to the time when EURECA was lowered to the rendez-vous orbit for the retrieval. In this orbit SOVA was powered up again and measurements were continued, but with reduced pointing accuracy. SOVA-DIARAD-L was run all the time and was compared to SOVA-DIARAD-R about every two months for a few orbits. An aging of SOVA-DIARAD-L was observed and was taken into account in the evaluation of the final data which are shown in Figure 1. The results of SOVA-PMO6 (Romero *et al.* 1994) are very similar and the relative difference at least one order of magnitude smaller than during comparison on the ground. This demonstrates the high quality of data which can be gained from space.

4. Results from SOLCON-DIARAD on ATLAS 1 and 2

During the ATLAS 1 and 2 missions both channels of SOLCON-DIARAD were operated and compared in such a way that the exposure of SOLCON-DIARAD-L was always longer than that of SOLCON-DIARAD-R. In the long run, after several ATLAS missions, this will enable us to identify a possible aging due to solar exposure. The comparative

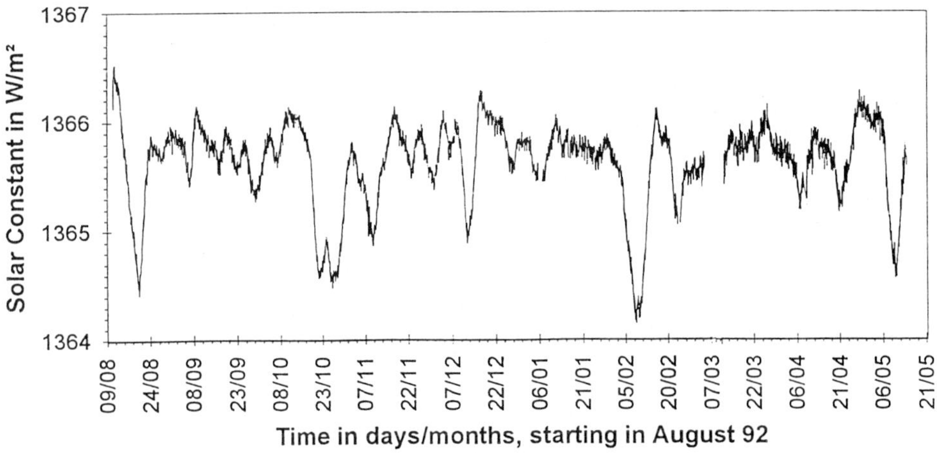

FIGURE 1. Total solar irradiance measured with SOVA-DIARAD-L on EURECA plotted as orbital means. These values shall be multiplied by 1.00091 to reflect the results of the ensemble average of the DIARADs determined by comparisons in space (Table 3).

Observation period	1	2	3	4
Day	085	089	091	092
Time (UT)	17:00	04:30	17:30	20:30
DIARAD-L	1367.65	1366.20	1366.56	1367.15
DIARAD-R	1366.92	1365.37	1365.80	-
Difference	0.73	0.83	0.76	-
Mean	1367.29	1365.79	1366.18	-
99% confid. int.	0.93	1.06	0.97	-

TABLE 1. Results of total solar irradiance determinations reduced to 1 AU obtained by DIARAD during 4 observation periods on the ATLAS 1 mission. All values are in Wm^{-2}.

Observation period	1	2	3	4	5
Day	099	101	103	105	106
Time (UT)	07:30 to 12:30	15:45	13:00	19:10	09:45
DIARAD-L	1367.15	1367.44	1367.39	1367.31	1367.27
DIARAD-R	1366.25	1366.65	1366.61	1366.53	1366.64
Difference	0.9	0.79	0.78	0.78	0.63
Mean	1366.70	1367.04	1367.00	1366.92	1366.96
99% confid. int.	1.15	1.01	1.00	1.00	0.81

TABLE 2. Results of total solar irradiance determinations reduced to 1 AU obtained by DIARAD during 5 observation periods on the ATLAS 2 mission. All values are in Wm^{-2}.

numbers are provided only at the moment that the switching is made between both channels.

The results are presented in Tables 1 and 2 and in Figures 2 and 3 together with the Preliminary Photometric Solar Index (PPSI from Sunspot Bulletin of the Sunspot Index Data Center, Bruxelles) to indicate the level of activity during the observations. Even during the short period of the ATLAS 1 mission one could observe the effect of spots crossing over the solar disk. During the ATLAS 2 mission the Sun was very quiet, not a single sunspot was visible during several days. The differences between the left and right

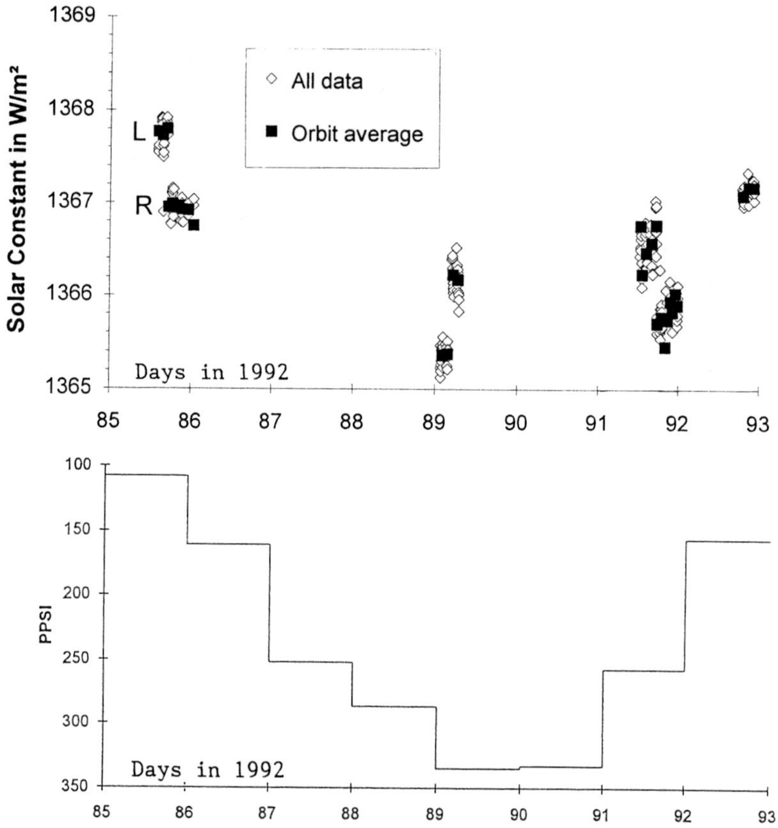

FIGURE 2. Total solar irradiance measured with SOLCON-DIARAD during the ATLAS 1 mission. The PPSI (Provisonal Photometric Sunspot Index, plotted upside down) is shown as indication of the level of activity during the observations.

channels did not show any significantly detectable aging. The ATLAS-1 results could be compared to the measurements obtained with ERBS and the mean solar constant value of SOLCON-DIARAD-L and R is about 1.5 Wm^{-2} higher than those obtained by ERBS at the same time.

5. Space absolute radiometric comparison

During the ATLAS 2 mission, observations of the solar constant was obtained quasi-simultaneously with DIARAD in SOLCON and SOVA. Fortunately the Sun was very quiet during these observation periods so that true simultaneity was not so important. The four values obtained during the different solar performances of the ATLAS-2 mission can thus be used each time to determine an average with its confidence interval of the population mean (at e.g. 99% confidence level). The aging of SOVA-DIARAD-L was corrected before the determination of the mean. The results presented in Table 3 provide in fact an objective state-of-the-art status of the accuracy that is obtained with DIARAD. If we consider only SOLCON-DIARAD-R, SOVA-DIARAD-R and SOLCON-DIARAD-L, the value of their mean would have a confidence interval of 0.7 Wm^{-2}. We believe that the value of the SOVA-DIARAD-L sensitive area is probably known with less accuracy than the three other channels. However SOVA-DIARAD-L can not be eliminated arbi-

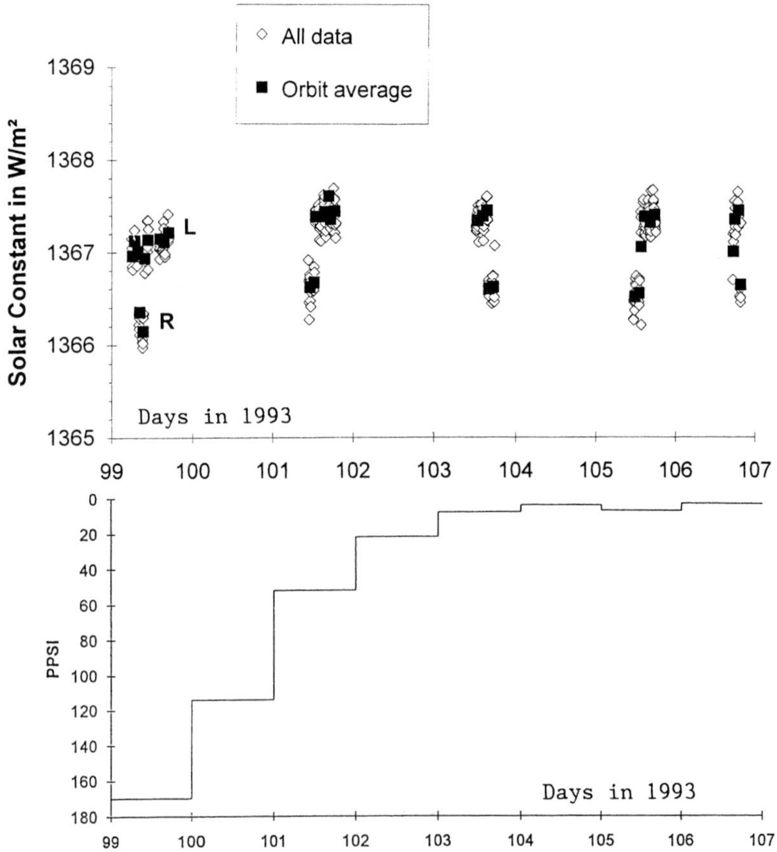

FIGURE 3. Total solar irradiance measured with SOLCON-DIARAD during the ATLAS 2 mission. The PPSI (Provisonal Photometric Sunspot Index, plotted upside down) is shown as indication of the level of activity during the observations.

trarily from the population of instruments. Note that the 99.9% confidence interval of about 0.15 Wm^{-2} on the measurements with one channel is partly due to instrumental noise and partly to solar noise. The relative contributions is under investigation.

Table 3 shows also that radiometric comparisons in space can be done with a much better precision than on the ground where the atmospheric turbidity and transmittance variability are the limiting effects. The space calibration of a stable high precision relative radiometer with an accurate and well defined absolute radiometer is limited only by the state–of–the–art absolute accuracy and the precision of both instruments. The results were obtained on April 11, 13 and 15, 1993 with very stable mean solar constant values respectively of 1366.49, 1366.45 and 1366.44 Wm^{-2}. The evaluated absolute error or uncertainty of these values is consistent with the measured differences at 99% confidence level.

It must be realised that the only way to improve further the accuracy and thus the knowledge of the *real solar constant* is to compare our values to other simultaneous observations also made with highly accurate absolute (in a strict sense) radiometers. It is then possible to calculate a new mean that would improve the actual confidence interval. However, if another set of measurements exists with a similar or better confidence

Date 1993 Time (UT)	9 April 08:00–17:30	11 April 09:00–18:00	13 April 11:00–19:00	15 April 07:30–20:00
SOVA-DIARAD-L	1364.92	1365.23	1365.21	1365.24
99.9% conf. int.	0.14	0.13	0.17	0.12
SOLCON-DIARAD-R	1366.25	1366.65	1366.61	1366.53
99.9% conf. int.	0.16	0.15	0.12	0.10
SOVA-DIARAD-R	1366.37	1366.64	1366.58	1366.68
99.9% conf. int.	0.14	0.12	0.16	0.40
SOLCON-DIARAD-L	1367.15	1367.44	1367.39	1367.31
99.9% conf. int.	0.12	0.12	0.14	0.16
Mean	1366.17	1366.49	1366.45	1366.44
99% conf. int.	1.42	1.47	1.46	1.43

TABLE 3. Total solar irradiance observations with SOLCON-DIARAD on ATLAS 2 and SOVA-DIARAD on EURECA I. All values are in Wm^{-2}. The aging of SOVA-DIARAD-L has been removed.

interval but with a mean outside our confidence interval this would be the indication of a metrological problem and/or of a wrong estimate of the uncertainty.

It is important and necessary that each design of an absolute radiometer has its uncertainty determinded by experimental characterization. A set of such instruments can then be compared by measuring the same stable source and the results be used to verify the uncertainty which can be achieved with this design (for DIARAD see Table 3).

6. Conclusions

The concept of DIARAD in SOLCON and SOVA has been proven to be efficient and has been operated succesfully on the ATLAS 1 and 2 missions and on EURECA. For the first time four detectors characterised independently have measured quasi-simultaneous the solar irradiance from space each with a precision of much better than 0.01% and a 99% confidence interval uncertainty of the mean of $\pm 0.1\%$ at a 99% confidence level.

Based on preliminary results ERBS, UARS/ACRIM II and PMO6-SOVA (referenced to the World Radiometric Reference on ground) are close to the mean of Table 3.

Acknowledgments. The RMIB was funded by the *Fonds de la Recherche Fondamentale Collective d'Initiative Ministérielle.* The quality of the engineering support of A. Chevalier, J. Donaldson, B. Penelle, U. Telljohann and their colleagues, the team members of the three collaborating institutions as well as the supporting teams of the space agencies resulted in an efficient and successful group achievement. C. Fröhlich is acknowledged for discussing and editing our manuscript.

REFERENCES

CROMMELYNCK, D. 1973 Théorie instrumentale en radiométrie absolue. *I.R.M. Publication série A* **81**, 1-50.

CROMMELYNCK, D. 1977 Calibration of radiation instruments for measurement of the radiant flux of an arbitrary Source. In Symposium on Optical Sensing and Probing of the Atmosphere (IS 23) - Assemblée Générale de l'UGGI, Grenoble (F), 1975. *Applied Optics* **16**, 302-305.

CROMMELYNCK, D. 1982 Fundamentals of absolute pyrheliometry and objective characterization. *Earth Radiation Science Seminars*, NASA Conference Publication 2239, 53-88.

CROMMELYNCK, D. 1985 The IRMB/SSD Absolute Radiometric Base, Objectives and Developments, Advances in Absolute Radiometry for Atmospheric and Environmental Research, Inc., Cambridge, Massachussetts, pp. 12-15.

CROMMELYNCK, D. 1988 Factors limiting the accuracy of absolute radiometry. In *New developments and applications in optical radiometry* Institute of Physics Conference series number **92**, NPL, London, 19-24.

CROMMELYNCK, D. & DOMINGO, V. 1984 Solar irradiance observations. *Sciences* **225**, 180-181.

CROMMELYNCK, D., DOMINGO, V., FICHOT, A., FRÖHLICH, C., PENELLE, B., ROMERO J. & WEHRLI, CH. 1993 Preliminary Results from the SOVA Experiment on Board the European Retrievable Carrier (EURECA). *Metrologia* **30**, 372–379.

ROMERO, J., WEHRLI, C. & FRÖHLICH, C. 1994 Solar total irradiance variability from SOVA 2 on board EURECA, *Solar Phys.* **152**, in press.

Irradiance Observations in Near-UV, Visible and Near Infrared Spectral Bands from Measurements Carried out during ATLAS-1 and EURECA-1 Missions

By GERARD THUILLIER[1], MICHEL HERSE[1], DIETRICH LABS[2], PAUL C. SIMON[3], DIDIER GILLOTAY[3] AND HOLGER MANDEL[2]

[1]Service d'Aeronomie du CNRS, BP3, F91371 Verrieres le Buisson, France

[2]Landessternwarte, Königstul, D-69117 Heidelberg 1, Germany

[3]Institut d'Aeronomie Spatiale de Belgique, 3 avenue Circulaire B-1180 Bruxelles, Belgium

1. Introduction

For the ATLAS and EURECA missions, we have used two identical instruments to measure the absolute solar spectral irradiance from 180 to 3200 nm. These instruments are calibrated by use of a blackbody and a set of lamp standards. The measurements are carried out with 1 nm bandpass up to 800 nm and 20 nm above. The instruments and calibration procedures are described by Thuillier *et al.* (1981). The platform capability of instruments retrieval after measurements allows a post-flight calibration which is essential for accurate measurements. The main results obtained up to now are:

- In the UV, the ATLAS-1 and EURECA-1 solar spectral irradiance are consistent with the SpaceLab 1 data obtained in 1983 (Labs *et al.* 1987). Figure 1 shows the ATLAS1 and SL 1 spectra. The origin of the existing differences is presently under investigation.
- In the visible domain, our measurements agree with the solar spectrum from Neckel & Labs (1984) within a few percent difference at certain wavelength.
- In the IR domain, the preliminary processing shows a spectrum close to the one obtained by Thekaekara (1974).

This instrument calibrated against the same standard and operated in space has measured the solar irradiance spectrum from 200 to 3000 nm. This instrument has participated in the ATLAS 2 mission and will continue to monitor the solar spectral irradiance as a function of time.

REFERENCES

LABS, D., NECKEL, H., SIMON, P. & THUILLIER, G. 1987 Ultraviolet solar irradiance measurement from 200 to 358 nm during the SpaceLab 1 mission. *Solar Phys.* **107**, 203–219.

NECKEL, H. & LABS, D. 1984 The solar radiation between 3300 and 12500 Å. *Solar Phys.* **90**, 205–258.

THEKAEKARA, M. P. 1974 Extraterrestrial solar spectrum, 3000-6100 Å at 1 Å intervals *Appl. Opt.* **13**, 518–521.

THUILLIER, G., SIMON, P.C., LABS, D., PASTIELS, R. & NECKEL, H. 1981 An instrument to measure the solar spectrum from 170 to 3200 nm. *Solar Phys.* **74**, 531–537.

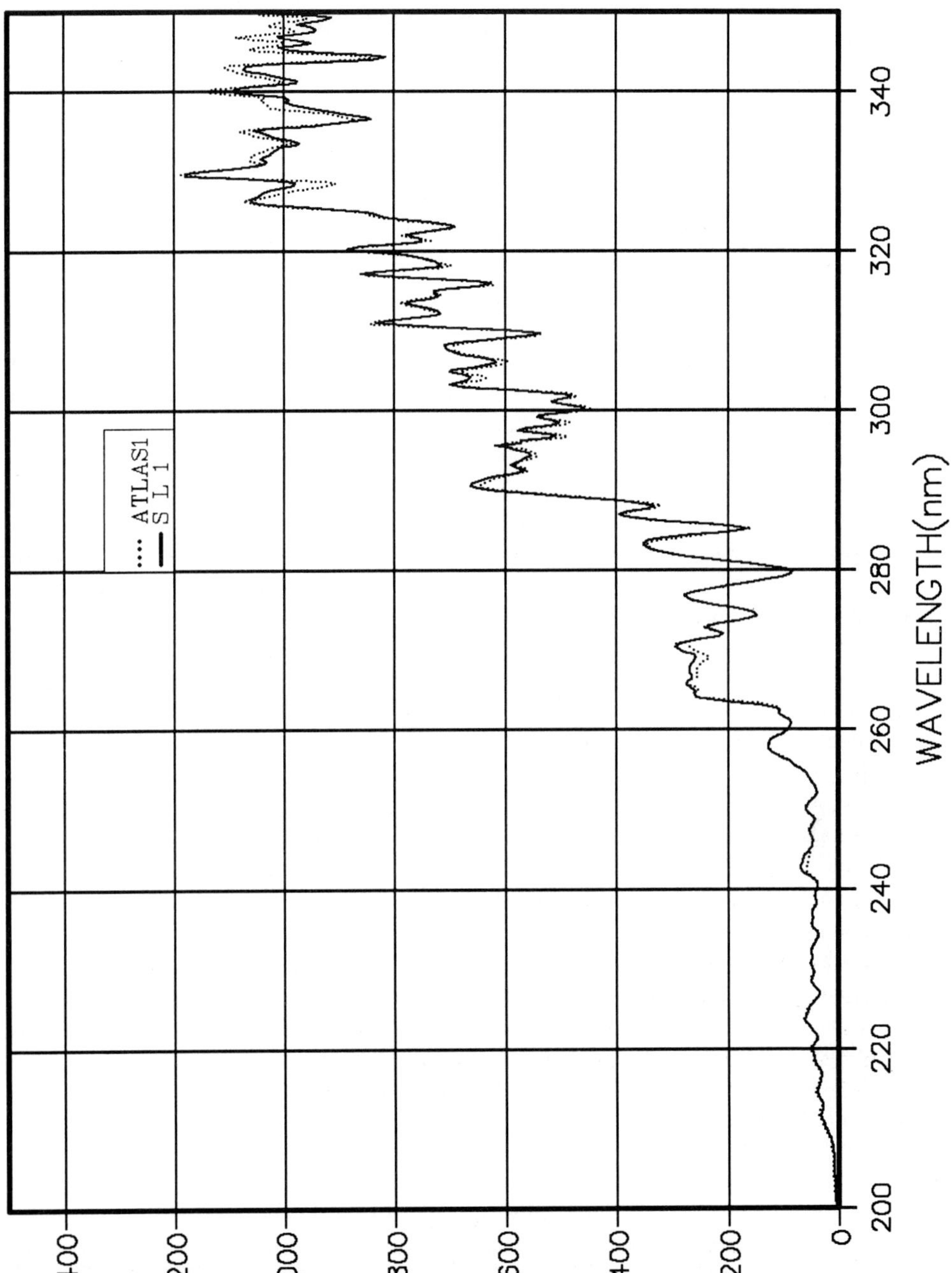

FIGURE 1. UV Solar irradiance from ATLAS1 mission.

Irradiance Observations from the UARS/SUSIM and ATLAS/SUSIM Experiments

By GUENTER BRUECKNER[1], LINTON E. FLOYD[2], PAUL A. LUND[3], DIANNE K. PRINZ[1] AND MICHAEL E. VANHOOSIER[1]

[1]E. O. Hulburt Center for Space Research, Naval Research Laboratory, Washington, DC 20375, USA

[2]Interferometrics Corporation, Vienna, VA 22180, USA

[3]Computation Physics Inc., Fairfax, VA 22031, USA

The SUSIM (Solar Ultraviolet Spectral Irradiance Monitor) on board the UARS (Upper Atmosphere Research Satellite) has measured the solar UV output from 120 nm to 400 nm on a daily basis since October 1991. A reference channel records a solar spectrum semi-annually only to reduce the instrument degradation of this channel and to provide long-term stability marks. Four deuterium lamps are used at monthly, semi-annual and annual intervals to provide long term calibration of the instrument. A preliminary analysis of the long term stability of SUSIM-UARS indicates that the precision of the instrument should be better than a few percent. The repeatability of two scans is better than 0.2%. A simplified SUSIM instrument is flying on NASA's ATLAS Spacelab missions anually to provide calibration points for the SUSIM-UARS.

The ultraviolet variability below 400 nm shows two distinguishable components, of chromospheric and photospheric origin. The chromospheric component is dominated by plage emission, and correlates well with other chromospheric indices, such as the 10830 Å index. The photospheric component is determined by the sunspot radiation deficit in umbras and penumbras. Sunspots, in specific cases, can modulate the continuum emission in the entire wavelength region from 120 nm to 400 nm. Below the Al edge at 210 nm the continuum emission can be reduced by up to 5% if the radiation deficit of sunspot is not being compensated by strong plage emission. This is caused by the increasing photosphere/penumbra contrast with decreasing wavelength and the large size of penumbras, which have a larger contrast in the ultraviolet than in the visible spectrum. However, strong plage emission dominates the UV variability below the Al edge. Above the Mg edge at 260 nm sunspots dominate the UV variability which correlates well with the variation of the total solar irradiance and anticorrelates with chromospheric indices. In the transition wavelength region between the Al edge at 210 nm and the Mg edge at 260 nm, the UV variability is caused by a complex mixture of the chromospheric and photospheric components. The wavelength dependence of the UV variability is complicated even further by the change from a chromospheric component in the center of strong Fraunhofer lines to a photospheric component in the wings of these lines. Emission line variability strictly follows the chromospheric component. One case of short-term variability over a period of several days shows an ultraviolet energy deficit of 0.5 W/cm^2 integrated over the wavelength region 120 nm to 400 nm, which is equal to the energy deficit of the total solar irradiance. It remains to be seen, whether this is a more general rule.

Irradiance Observations from the UARS/SOLSTICE Experiment

By GARY J. ROTTMAN[1], THOMAS N. WOODS[1], ORAN R. WHITE[1] AND JULIUS LONDON[2]

[1]High Altitude Observatory, National Center for Atmospheric Research, Boulder, CO 80307, USA

[2]University of Colorado, Boulder, CO 80309, USA

The Solar/Stellar Irradiance Comparison Experiment (SOLSTICE) is one of the two ultraviolet spectrometers on the Upper Atmosphere Research Satellite (UARS) measuring the full disk solar irradiance. This spectrometer covers the spectral range of 119 to 421 nm and typically makes ten full spectral scans per day. The instrument was designed to monitor changes in the solar radiation with a relative accuracy of better than 1%. To achieve this goal the spectrometer makes repeated observations of a number of bright blue stars; stars that individually should vary by only small fractions of a percent over time periods of thousands of years. The calibration stars are observed with the same optics and detectors used for the solar observations and any drift in the instrument response is directly determined by changes in the ensemble average flux from these calibration stars. The instrument performance and calibration are described and solar data from the first year of the UARS mission are presented. These UARS data show significant solar activity during late 1991 and early 1992 and then clearly mark a decrease as we move off the maximum of solar cycle 22. We are optimistic that an extended UARS mission, perhaps as long as ten years, is now possible.

1. Introduction

The Solar/Stellar Irradiance Comparison Experiment (SOLSTICE) is one of ten complementary instruments on the Upper Atmosphere Research Satellite (UARS). The UARS was launched in September of 1991 and its instruments provide the first systematic, detailed study of the Earth's stratosphere, mesosphere, and lower thermosphere. These satellite data are providing a new understanding of many atmospheric processes, including stratospheric ozone depletion. Ozone levels and other upper atmosphere properties are established by complex interactions of chemistry, dynamics and energy input. Four of the instruments (CLAES, ISAMS, MLS, and HALOE) use remote sensing of atmospheric emissions to measure temperature profiles and concentrations of ozone, water vapor, and other trace species. Two instruments (HRDI and WINDII) employ high-resolution interferometric techniques to measure Doppler shifts and infer the upper atmosphere wind fields. Four instruments (SOLSTICE, SUSIM, PEM, and ACRIM) measure the energy input which strongly influence both chemistry and dynamics. Of these four, three measure the radiation from the Sun, ACRIM measuring the total irradiance and SOLSTICE and SUSIM measuring the ultraviolet spectral irradiance. The PEM instrument, on the other hand, studies the particle influence on the atmosphere by making in situ measurements of charged particles and X-ray images of the atmosphere responding to incoming high energy electrons. This paper describes the observations now being made by SOLSTICE.

Ultraviolet radiation is strongly absorbed by the Earth's atmosphere, in fact shortward of 300 nm it is almost completely absorbed. The solar UV is the dominant energy input to the middle atmosphere and even small changes can have a significant influence.

Ultraviolet solar radiation can only be detected from space. Early attempts to measure it were sporadic and usually had inherent uncertainties exceeding the amount of solar variability. At UV wavelengths ($\lambda > 120$ nm) the Solar Mesosphere Satellite (SME) (1981 through 1989) and the NIMBUS-7 (1978 through 1989) carried instruments with adequate short-term stability to reliably monitor both short- and intermediate-term solar variations. In the context of this paper, short-term variations have periods of minutes to hours and refer to erruptive type phenomenon, for example flares. Intermediate-term variations have periods of days to weeks and refer to the more slowly varying evolution of magnetic structures on the Sun modulated by the 27-day rotation period. Since both SME and NIMBUS operated for time periods covering a significant portion of a solar cycle they also provided some plausible indications of solar cycle variability. At short wavelengths below 200 nm these data imply solar cycle variations of 10%, increasing to as much as a factor of two at Lyman-α (121.6 nm). At wavelengths longward of 200 nm the solar cycle variation decreases to a few percent or less and is quickly obscurred by the measurement noise and drift in the instrument response. These early instruments had no provision for accurately tracking their sensitivity and their data generally have a relative accuracy growing at a rate of 1-2% percent per year.

The accurate determination of long-term changes in the solar UV output is now coming of age. The SOLSTICE instrument on UARS and perhaps on future missions employs a technique that should adequately detect instrumental drift and provide photometric measurements of the Sun accurate to better than 1% over arbitrarily long time periods. The instrument has been designed to observe the Sun and, using the very same optics and detectors, to observe bright, blue stars. These stars of spectral classification O, B, and A are assumed to vary by only small fractions of one percent over time periods of thousands of years. The strength of the technique, however, is that many stars are used and the ensemble average flux from the stars provides that individual pathological stars can be easily identified and removed from the "calibration" set. The only differences between the solar and stellar observing configuration are: 1) the entrance aperture: 2 cm^2 for the stars and 10^{-4} cm^2 for the Sun, 2) the bandpass or exit slit: 1 to 6 nm for the stars and 0.1 nm for the Sun, and 3) the observing time : 100 to 1000 seconds for the stars and 1 second for the Sun. Figure 1 shows the SOLSTICE solar irradiance data together with the UV flux from the star α-Canis Majoris (amplified by eight orders of magnitude). For these bright blue stars we see that the ratio of solar to stellar flux levels is on the order of five to nine orders of magnitude, and SOLSTICE accommodates this large dynamic range by changing integration time and apertures. Details of the SOLSTICE instrument and its calibration are provided in two papers, Rottman *et al.* (1993) and Woods *et al.* (1993).

2. Observations

2.1. *Solar observations*

The UARS is a three-axis stabilized, Earth-oriented spacecraft with one side always facing the nadir (Reber 1990). From this spacecraft, observation of the Sun or other astronomical targets requires that the instruments be mounted on a pointing platform, decoupled from the satellite motion. In UARS this is accomplished by a two-axis gimbal platform, the Solar/Stellar Pointing Platform, SSPP.

The UARS orbit, at an inclination of 57 degrees and altitude near 585 km, provides a minimum of 60 minutes of daylight in each orbit, although the obscuration of the solar array and other "masks" constrain the effective solar viewing time as seen by the SSPP

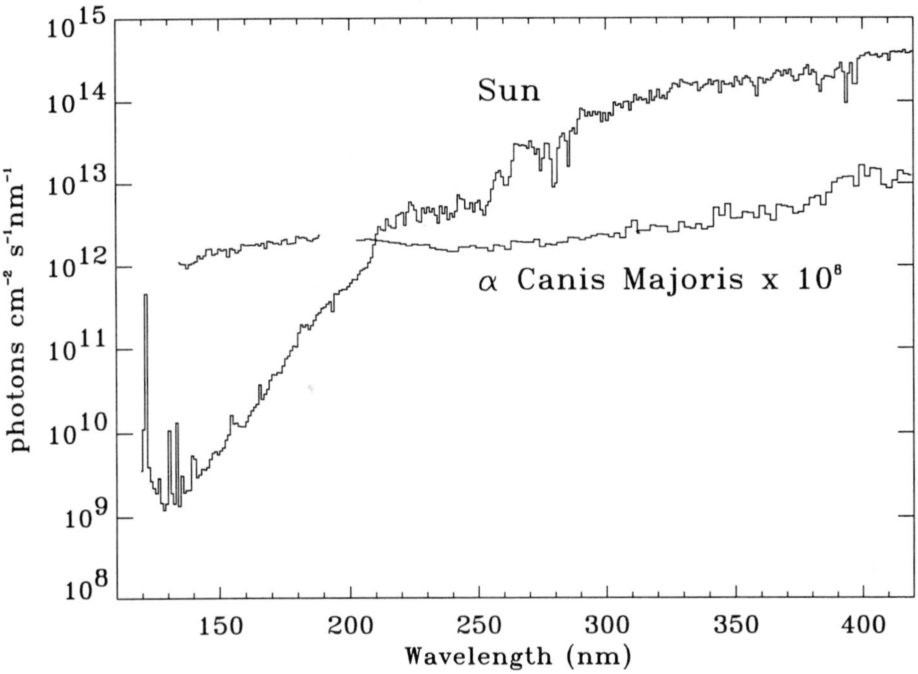

FIGURE 1. Comparison of the SOLSTICE Level 3AS solar spectrum with the UV flux ($\times 10^8$) from the star α-Canis Majoris.

instruments to approximately 35 minutes. Throughout this period the SSPP points the solar instruments at Sun center, and generally the alignment and accuracy of the tracking is better than a few minutes of arc. The solar data reported here were obtained by a scan mode of the spectrometer with a detector integration time at each grating position of either 1 or 0.1 second. A complete spectral scan covers 2048 grating steps while the three instrument channels operate simultaneously. Figure 2 displays a typical full spectral scan and, although compact, it illustrates the spectral coverage and spectral resolution (0.1 nm at the shortest wavelengths up to 0.2 nm at the long wavelength end).

Throughout a calendar day the SOLSTICE obtains a number of full or partial spectral scans. These data are transmitted by the spacecraft and collected at the UARS Central Data Handling Facility (CDHF) at the Goddard Space Flight Center (GSFC). There a processing code corrects the data for all known instrument and observing artifacts and applies a photometric calibration to convert the detector counts to geophysical units (S.I. units of W/m^2). Our primary science requirement is to provide one full spectrum per calendar day, and to achieve this, the data processing algorithm combines all individual observations to form a single spectrum. This daily average spectrum is also "binned" so that a value is reported for each 1.0 nm interval between 119 to 421 nm. These daily spectra are referred to as the SOLSTICE Level 3AS data product, and the solar spectrum provided in Figure 1 is an example.

The pre-flight calibration of the SOLSTICE was accurate to $\pm 3\%$ (Woods *et al.* 1993) but this uncertainty increased through ground testing and launch. We continue to cross-calibrate the SOLSTICE data with the data from the UARS SUSIM instrument (Brueckner *et al.* 1993) and the ATLAS SSBUV and SUSIM instruments. Our current validation

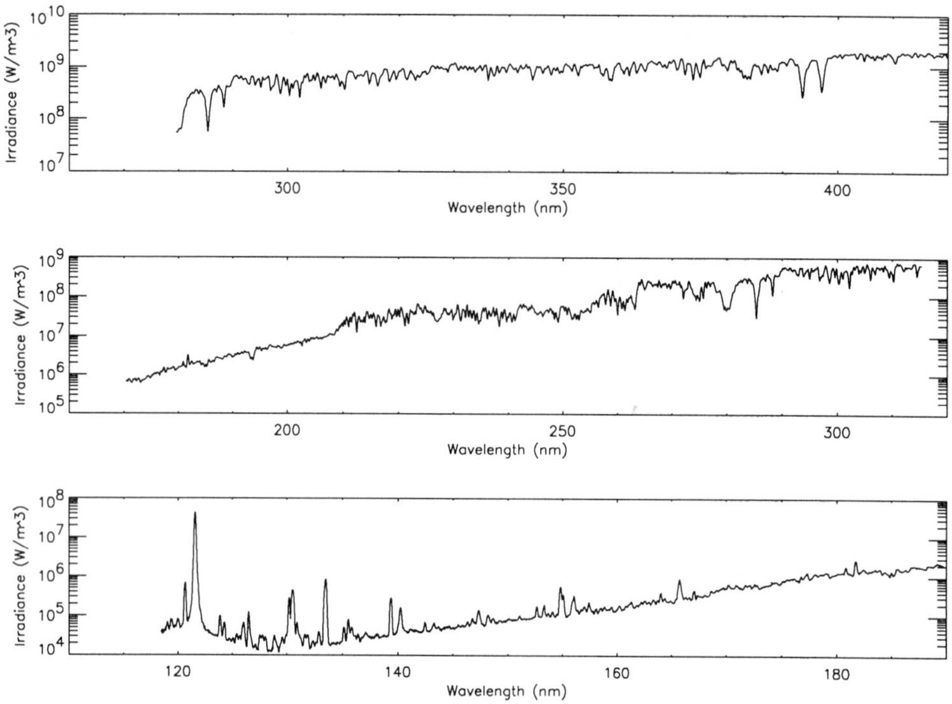

FIGURE 2. Bottom panel: Spectrum obtained from short wavelength channel of the SOLSTICE. Middle panel: Spectral data from middle wavelength channel. Top panel: Spectral data from the long wavelength channel.

efforts indicate that SOLSTICE absolute accuracy is ±2% for wavelengths longward of 200 nm and ±5% for shorter wavelengths.

2.2. *Stellar observations*

As described in the introduction, the SOLSTICE has the unique capability of tracking changes in instrument sensitivity by observing a set of bright, blue stars. The stars used for this in-flight calibration have been carefully selected on the basis of minimal variability (from all known sources of ground based and space observations), on the basis of brightness ($m_{vis} < 3$), and on the basis of their being isolated in the field of view of the instrument (no companion stars brighter than $m_{vis} < 8$ within the instrument field of view).

During each night portion of a UARS orbit, the SSPP is programmed to point SOLSTICE at a selected star and the instrument is configured for a stellar observation. The entrance slit is changed to a 2 cm^2 aperture (four orders of magnitude larger than the solar aperture) and the spectral bandpass, determined by the exit slit width, is increased to at least 1 nm (factor of 10). Usually these stellar experiments are at a fixed wavelength and have an integration or dwell time between 100 and 1000 seconds. The SSPP can only view to one side of the UARS; therefore at any given time, only half of our calibration stars can be observed. However as the UARS orbit precesses with a period of approximately 40 days, all of the calibration stars become available.

The stellar data are also processed at the CDHF with corrections for pointing, dark counts and stray light. As an example, the stellar flux provided in Figure 1 is from the star α-Canis Majoris. Figure 3 is the complete set of observations of this single star

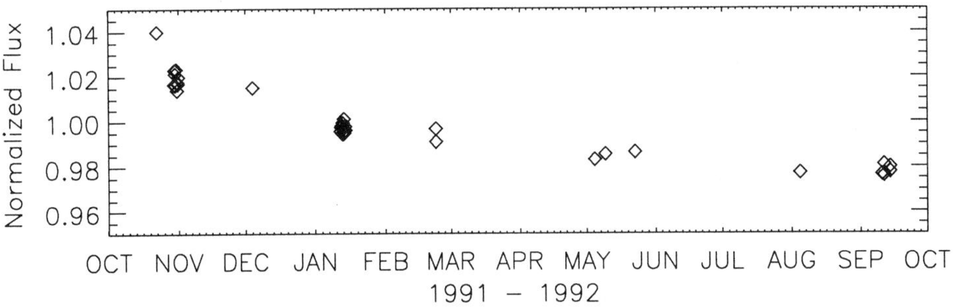

FIGURE 3. SOLSTICE observations of the star α-Canis Majoris and changes are presumed to represent degradation in the instrument sensitivity.

at 153.5 nm and illustrates an approximately exponential decrease throughout the year. In fact it is the combination of similar measurements from fifteen stars that defines the instrument degradation at 153.5 nm. Throughout the first year of SOLSTICE operations the amount of degradation at this wavelength is approximately 3%. Similar analysis is carried out at 54 different wavelengths spanning the interval 118 to 420 nm and the resulting correction factors are then applied to the solar data. The SOLSTICE data, corrected for instrument degradation, are referred to as the Level 3BS data product. These data are transferred from the UARS CDHF to the Distributed Active Archive Center (DAAC) also at the Goddard Space Flight Center, on a fixed schedule and from there they are available to the scientific community. The first year of SOLSTICE data, October 3, 1991 through September 12, 1992 will be available on the Goddard DAAC in the Fall of 1993.

3. Discussion

The Level 3BS SOLSTICE data provide accurate information on short, intermediate, and long-term solar variations. Short-term variations, perhaps associated with flare activity, are identified in the SOLSTICE data and set aside to a special "flare" file for further analysis. It is still premature to consider long-term or solar cycle variations since the UARS has only observed for a short, two-year period, perhaps covering only a portion of the downward trend of cycle 22. In this report we concentrate on the first year of SOLSTICE data and consider primarily the intermediate or 27-day variations.

Figure 4 provides time series for three different wavelengths, Lyman-α, 155 nm, and 240 nm where the ordinates represent a range of $\pm 25\%$, $\pm 20\%$, and $\pm 2.5\%$, respectively. From a solar perspective these three wavelengths represent emission originating predominantly in the chromosphere (Lyman-α), chromosphere/transition region (155 nm), and photosphere (240 nm). From the perspective of atmospheric influences, the Lyman-α is extremely important to mesospheric photochemistry, especially through photodissociation of water vapor; the 155 nm radiation is important to the dissociation of molecular oxygen in the lower thermosphere; and the 240 nm radiation is important to the photodissociation of ozone in the upper stratosphere. All three data sets provide striking evidence of the 27-day variability of the Sun, especially through the spring of 1992. Late in 1991 there appear to be two active longitudes and the time signal has an approximate 13 day periodicity. In mid-summer of 1992, the UARS instruments were turned off in

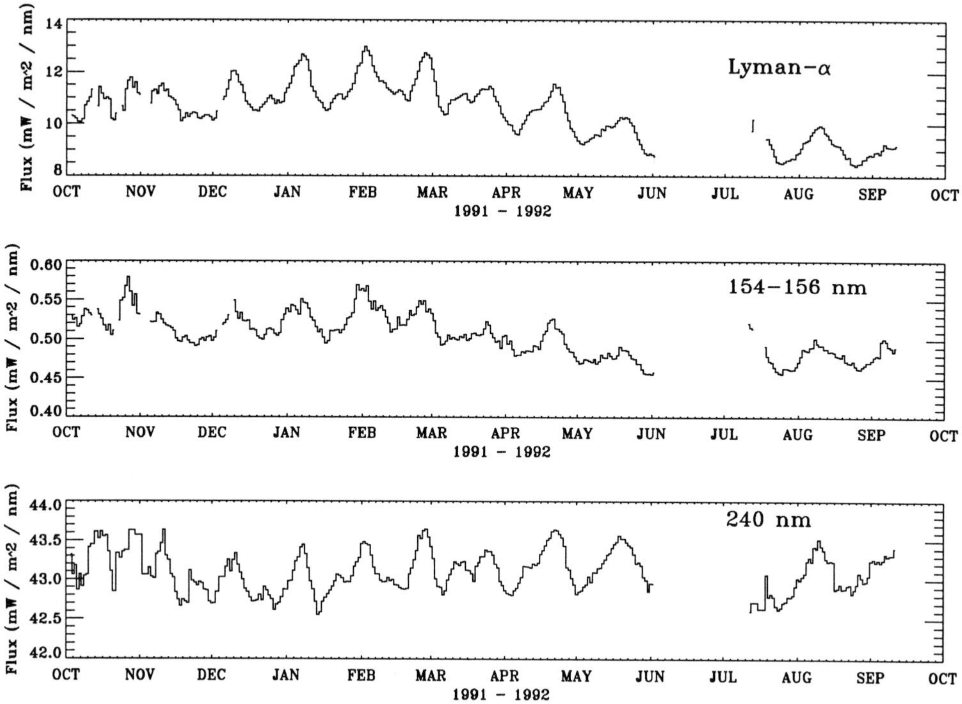

FIGURE 4. One year of SOLSTICE solar data at three separate wavelengths. (The enhanced tick marks denote the beginning of each month.)

order to trouble shoot an anomaly in the solar array drive mechanism and there is a resulting break in the time series.

The magnitude of intermediate term variability is a strong function of wavelength (Rottman 1988; London *et al.* 1993). For example early in 1992 when the 27-day signal is most evident in the SOLSTICE data, a ratio of the irradiance for February 24th (moderately high activity) to March 4th (lower activity) provides a magnitude and wavelength dependence as illustrated in Figure 5. This figure attests to the precision and short-term relative accuracy of the SOLSTICE data. At the shortest wavelengths the emission lines and underlying continuum display strong variability, for example 17% at Lyman-α, and this decreases to only 3% near 200 nm. Near 208 nm, corresponding to the absorption edge for neutral aluminum in the solar atmosphere, the variation abruptly decreases to approximately 1%, and maintains this low level out to 300 nm. Beyond 300 nm the variation is less than 0.5% and approaches the instrumental noise and uncertainty in the two individual measurements. The strong Fraunhofer absorption lines of CaII at 393 nm and MgII at 280 nm stand out in Figure 5 with approximately 2.5% and 3.5% variability, respectively. We emphasize that in spectral regions with both lines and continuum, for example the 1 nm bins including these strong Fraunhofer lines, the reported variations are a function of instrument resolution, and extreme care must be exercised when comparing data sets from two different instruments.

A major scientific goal of the UARS SOLSTICE is to establish the change of the UV irradiance over a solar cycle. The UARS measurements began in late 1991 and we presume that this time period may be representative of solar maximum. Early in 1992 the solar activity decreased precipitously (White *et al.* 1993) and this decrease in

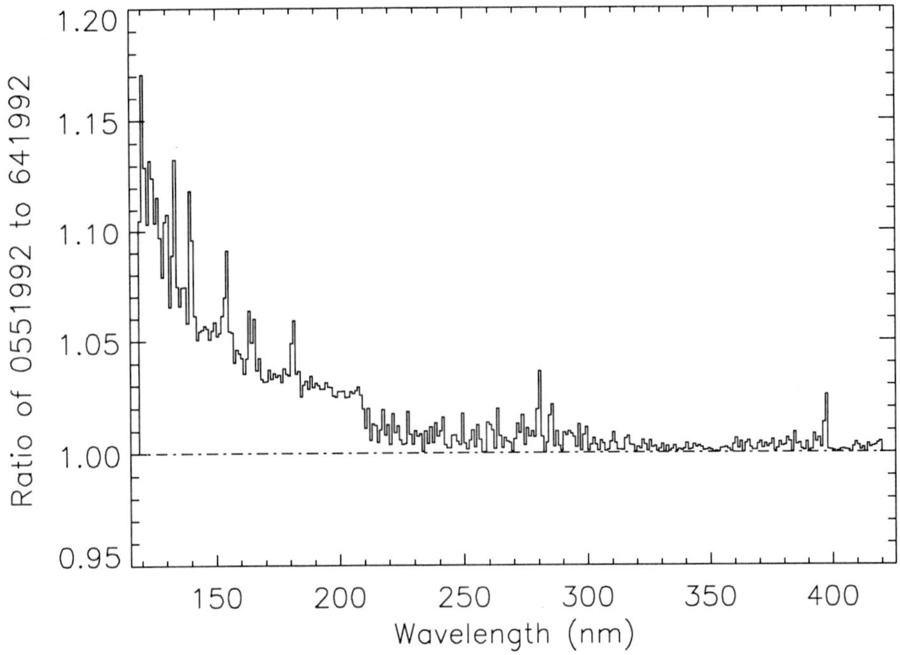

FIGURE 5. Ratio of the solar spectrum for February 24, 1992 to March 4, 1992 as an illustration of typical intermediate term (27-day) variation.

solar activity is attributed to the disappearance of large active regions, especially in the southern hemisphere of the Sun. We are encouraged that the stellar calibration technique of SOLSTICE is working well and that it will provide a final solar data set accurate to 1% over arbitrarily long time periods, in particular over time periods on the order of a solar cycle. Recording the solar cycle variability will require that the UARS mission be continued through solar minimum, now estimated to occur as early as 1995.

If the UARS solar data set extends from solar maximum until solar minimum we will be able to provide a measurement of solar cycle variability. The SOLSTICE data set will have a relative accuracy of better than one percent thereby providing a measurement of solar-cycle variability valid to at least this level. The wavelength dependence of the flux ratio, solar maximum to solar minimum, could be displayed similar to the ratio of intermediate term variability shown in Figure 5. Would the solar cycle variability be a simple amplification of the rotational variability shown in the figure? If so, two-component models with a quiet Sun and active region component may suffice to describe the solar cycle (Lean 1987). A more complicated relation between intermediate and long-term variability requires additional terms, for example an active network component, to fully describe the solar variations.

4. Summary

The UARS SOLSTICE has been in operation since October 3, 1991 and the data reported here cover only the first year. These solar data provide detailed time variations and a precise measurement of intermediate-term solar variability. This information is essential to many of the atmospheric investigations being conducted by the UARS, for it allows an accurate evaluation of the solar effect in the highly variable atmospheric signals. Moreover, these solar data provide important new information on the Sun's variability. Experimental techniques have improved to the point that measurement of intermediate-term solar variability is now straightforward. With proper attention to instrument design, calibration and especially contamination control, spectral photometry at the 0.1% level (relative accuracy) for time periods of one or two months are now routine.

The measurement of long-term solar variations, on the other hand, requires a special technique to accurately and precisely track changes in instrument sensitivity. The SOLSTICE achieves this in a unique way by directly comparing the solar irradiance to the UV flux from a number of bright blue stars. The assumption that the ensemble average flux from these stars is constant now appears valid, certainly to the 1% level and most likely to even a higher degree of relative accuracy. We are optimistic that as UARS continues to operate for several more years, the solar data set will extend from late 1991, near solar maximum of cycle 22, through 1995, the next solar minimum. The resulting time series will provide a first accurate and reliable estimate of solar cycle variability in the ultraviolet spectral range between 120 and 400 nm. Furthermore, the SOLSTICE technique allows that future observations of the Sun, made by comparing solar irradiance to the flux of the same bright, blue stars used by UARS, can be directly related back to today's UARS observations. A second version of the SOLSTICE has been proposed and accepted for the Earth Observing System (EOS). This second version of SOLSTICE is now included as one of the five scientific instruments on the EOS Chemistry Mission, scheduled for launch in the year 2002.

Acknowledgments. The success of SOLSTICE is due to the dedication, expertise and hard work of many individuals. We recognize the contributions of the engineers at the University of Colorado who designed, built, and tested this fine instrument; Tom Sparn, Dan Gablehouse, our staff and students who operate the instrument and monitor its performance; and Barry Knapp, Chris Pankratz, Paul Willis and their assistants who process and validate these data. NCAR is sponsored by the National Science Foundation and the SOLSTICE program is supported by the NASA / NSF Interagency Agreement No. S-87289-E.

REFERENCES

Brueckner, G. E., Edlow, K. L., Floyd, L. E., Lean, J. & Van Hoosier, M. E. 1993 The Solar Ultraviolet Spectral Irradiance Monitor (SUSIM) Experiment on board the Upper Atmosphere Research Satellite (UARS). *J. Geophys. Res.* **98**, 10,695–10,711.

Lean, J. 1987 Solar ultraviolet irradiance variations: A review. *J. Geophys. Res.* **92**, 839–868.

London, J., Rottman, G. J., Woods, T. N. & Wu, F. 1993 Time Variations of the solar UV irradiance as measured by the SOLSTICE (UARS) instrument. *Geophys. Res. Letts.* **20**, 1315–1318.

Reber, C. 1990 The Upper Atmosphere Research Satellite. *Eos Trans. AGU* **71**, 1867–1873.

Rottman, G. J. 1988, Observations of solar UV and EUV variability. *Adv. Space Res.* **8, 7**, 53–66.

Rottman, G. J., Woods, T. N. & Sparn, T. P. 1993 Solar stellar irradiance comparison experiment I: 1 Instrument design and operation. *J. Geophys. Res.* **98**, 10,667–10,677.

White, O. R., Rottman, G.J. Woods, T. N., Keil, S. I. Livingston, W. C., Tapping, K. F., Donnelly, R. F. & Puga, L. 1993 Change in the UV output of the Sun in 1992 and its effect in the thermosphere. submitted to *Geophys. Res. Letts.*

Woods, T. N., Rottman, G. J. & Ucker, G. 1993 Solar stellar irradiance comparison experiment I: 2 Instrument calibration. *J. Geophys. Res.* **98**, 10,679–10,694.

Middle Ultraviolet Solar Spectral Irradiance Measurements, 1985–1992, from the SBUV/2 and SSBUV Instruments

By RICHARD P. CEBULA[1], ERNEST HILSENRATH[2] AND MATTHEW T. DELAND[1]

[1]Hughes STX Corporation, Greenbelt, MD 20770, USA

[2]NASA/Goddard Space Flight Center, Code 916, Greenbelt, MD 20771, USA

Solar spectral irradiance measurements from the NOAA-9 and NOAA-11 SBUV/2 instruments and SSBUV are presented and compared. Inter-instrument solar spectral irradiance differences on the order of 5-10% are observed. Internal comparisons of the SBUV/2 solar data suggest instrument sensitivity changes ranging from less than 1% per year near 400 nm to approximately 5% per year near 200 nm.

The mission-to-mission accuracy of the SSBUV solar spectral irradiance measurements is better than ±2%. Initial results of our investigation into the feasibility of using SSBUV data to maintain the long-term calibration of the SBUV/2 instruments for solar spectral irradiance monitoring are reviewed.

1. Introduction

A data set whose length is longer than a single solar cycle and which accurately characterizes both short-term and long-term ultraviolet (UV) solar change is needed in order to understand the role of solar UV variations on observed stratospheric ozone trends. Daily solar spectral irradiance measurements have been made in the region 160 to 405 nm at 1 nm resolution by the Solar Backscatter Ultraviolet model 2 (SBUV/2) instruments on the NOAA-9 and NOAA-11 operational satellites since March 1985. These instruments, whose primary objective is to monitor long-term changes in stratospheric ozone, continue the solar irradiance measurements made from 1978 through 1987 by the Nimbus-7 SBUV instrument (Schlesinger & Cebula 1992). The first SBUV/2 instrument was launched onboard the NOAA-9 satellite in December 1984 and began solar measurements in March 1985. The initial, or "Day 1", NOAA-9 SBUV/2 solar spectrum is shown in Figure 1. NOAA-9 data from March 1985 through January 1993 were processed using a simplified algorithm which assumes constant instrument sensitivity. Annual comparisons of NOAA-9 solar measurements to the Day 1 measurement are shown in Figure 2. These data were not corrected for long-term solar irradiance change. The Mg II core-to-wing proxy index of solar activity indicates that solar activity was at a minimum and roughly constant between March 1985 and March 1987, began to increase in spring 1987, and increased more rapidly starting in early 1988 (Donnelly 1990; Cebula *et al.* 1992). As measured by the Mg II index, solar activity was approximately 8% higher in 1989–1990 than in spring 1985. During the first two years operation, when the mean solar UV output was nearly constant, the observed instrument solar irradiance output changes arise solely due to instrument change. The data indicate that the instrument's sensitivity decreased by roughly 1.5% at 400 nm to as much as approximately 11% near 200 nm over this period. The rate of sensitivity decrease apparently slowed in the third year of operation, especially shortward of roughly 260 nm. This apparent decline in the instrument degradation rate is attributable to an increase in the solar irradiance offsetting the continued instrument throughput decrease. DeLand & Cebula (1993) have

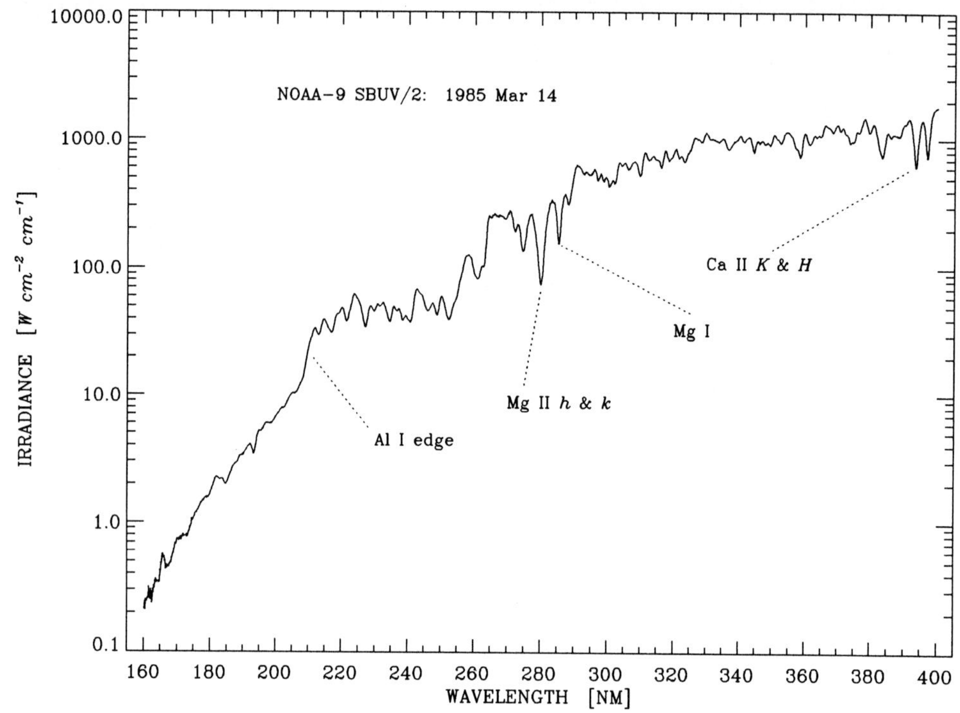

FIGURE 1. NOAA-9 SBUV/2 "Day 1" solar spectral irradiance.

estimated long-term solar spectral irradiance variations by scaling the Mg II index to other UV wavelengths. Using their methodology, we derive an approximate 9% increase in the solar irradiance at 200 nm. Correcting for solar change, we conclude that the instrument's sensitivity continued to decline at a mean rate of 4-5%/yr near 200 nm.

The SBUV/2 instruments include an onboard calibration system in order to assess time-dependent diffuser reflectance changes in order to correctly derive long-term ozone amounts. Unfortunately, the NOAA-9 SBUV/2's system failed shortly after launch (Frederick *et al.* 1986). The second SBUV/2 instrument was launched onboard the NOAA-11 satellite in September 1988 and has been making daily solar spectral irradiance measurements from February 1989 through the present. The onboard calibration system was redesigned for the NOAA-11 instrument, and the resulting data have proven highly useful in monitoring long-term diffuser plate degradation (Weiss *et al.* 1993). The NOAA-11 SBUV/2 solar data presented here incorporate a correction for diffuser degradation, a recently derived inflight goniometry that is a 1-2% improvement on the previously used preflight goniometry, and an updated correction for photomultiplier tube change. The largest of these changes is the diffuser degradation correction, which ranges from roughly 0.4%/yr near 400 nm to 1.7%/yr near 160 nm. In addition, the absolute radiometric calibration of the NOAA-11 instrument has recently been revised to correct for the effects of an error in the current setting of the spectral irradiance standard used during the calibration. This adjustment ranges from approximately 2.7% at 400 nm to 5.5% near 160 nm. The calibration of the NOAA-9 instrument contains a corresponding error, however a revised calibration is not yet available. As a first approximation the NOAA-9 data presented here were corrected using the NOAA-11 wavelength dependent adjustment factor.

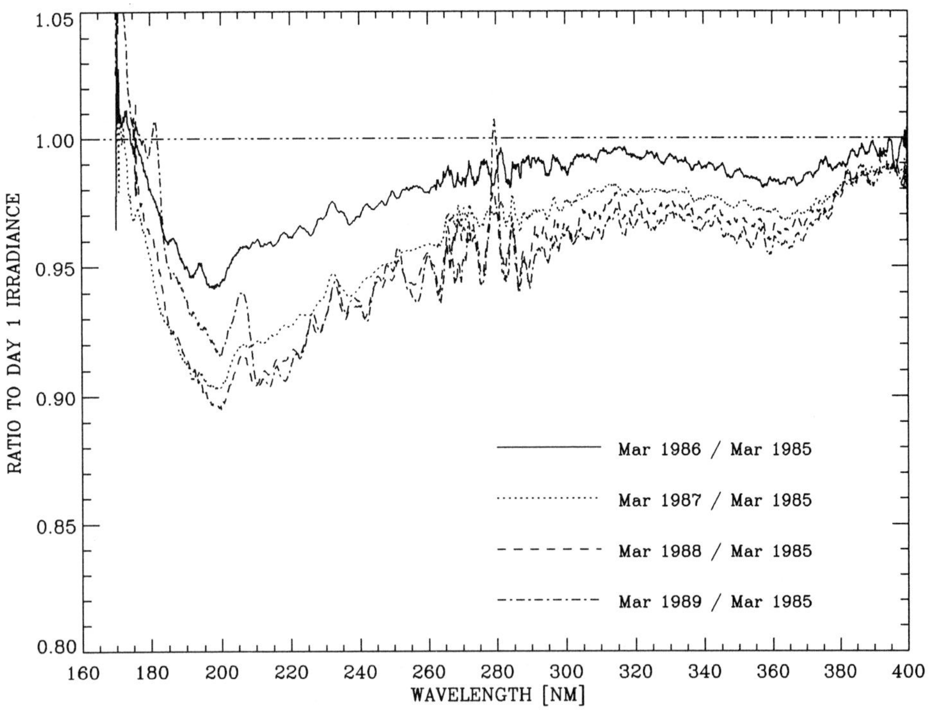

FIGURE 2. Annual comparisons of NOAA-9 SBUV/2 solar spectral irradiance measurements to the initial measurement.

The SBUV/2 irradiances presented herein replace the data previously presented (Cebula *et al.* 1991; Cebula & DeLand 1992).

Figure 3 presents annual comparisons of the NOAA-11 SBUV/2 solar spectral irradiance output to the Day 1 irradiance. Using the technique of DeLand & Cebula (1993), we estimate that from February 1989 through February 1992 the average solar output varied by 2% or less between 160 and 200 nm, approximately 1% near 200 nm, roughly 0.5% between 210 and 260 nm, and less than 0.2% longward of 260 nm. Consequently, Figure 3 provides a measure of long-term NOAA-11 instrument change, which is seen to range from less than 0.5%/yr near 400 nm to roughly 5%/yr near 200 nm. Thus, the rate of sensitivity decrease of the NOAA-11 instrument optics *alone* is comparable to the rate of NOAA-9 instrument optics plus diffuser reflectivity change.

The SBUV/2 onboard calibration system only provides data on one element of the optical system, with no provision for the end-to-end calibration required for accurate long-term solar irradiance observations. Hence, alternate methods must be used to maintain the calibration of the SBUV/2 instruments. The Shuttle SBUV (SSBUV) experiment provides one such method. SSBUV is making middle UV solar irradiance measurements on a decade-long time scale with an instrument that is calibrated preflight, in-flight, and post-flight (Hilsenrath *et al.* 1988). The SSBUV instrument was the engineering model of the SBUV/2-series and, except for replacement of the reflection diffuser with a transmission diffuser to facilitate solar measurements from the Space Shuttle, is optically identical to the satellite instruments.

SSBUV has flown five times during the period October 1989 through April 1993, and four additional SSBUV missions are scheduled through at least 1997.

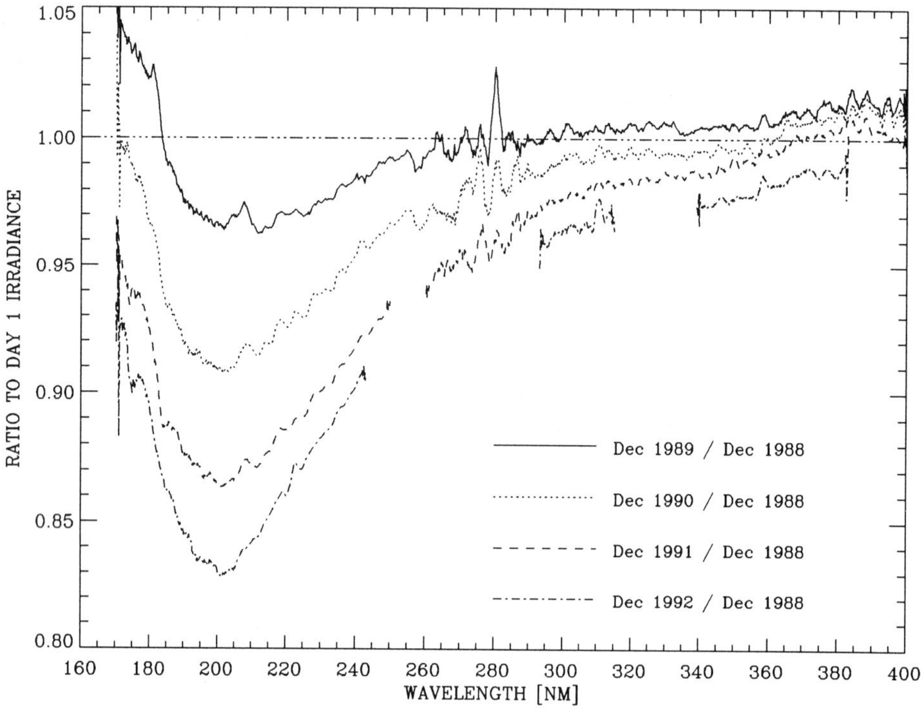

FIGURE 3. Annual comparisons of NOAA-11 SBUV/2 solar spectral irradiance measurements to the initial measurement.

SSBUV solar spectral irradiance data from the first four missions are compared in Figure 4. Each mission's data were processed using that mission's postflight calibration, since the postflight calibrations most accurately represent the instrument sensitivity on orbit (Cebula & Hilsenrath 1992). Shortward of 260 nm the SSBUV-4 calibration is provisional. Middle UV solar activity, evaluated using 27-day averages of the Mg II index, varied by only 3% amongst the first four SSBUV missions. Therefore, we have not corrected for UV solar activity variations in this comparison. Longward of 300 nm, where long-term solar changes are negligible, the peak-to-peak reproducibility of the SSBUV solar spectral irradiance is better than 1.5%. The ±1% spectral structure seen in the 220 to 290 nm region is indicative of a small error in the SSBUV-1 calibration, and closely parallels, although is an order of magnitude smaller than that mission's calibration changes.

The mission-to-mission spectral differences seen just shortward of the Al edge near 210 nm mimic the spectral shape of the Mg II proxy index scale factors (DeLand & Cebula 1993), which is distinct from the spectral shape of SSBUV calibration changes (Cebula & Hilsenrath 1992), and therefore are indicative of true solar spectral irradiance variations from October 1989 through March 1992. Cebula & Hilsenrath (1993) showed a 1% agreement between the solar spectral irradiance changes measured by SSBUV between October 1990 and August 1991 and the changes predicted by the Mg II index and scale factors. The Mg II indexes measured by SSBUV and NOAA-11 SBUV/2 show excellent agreement.

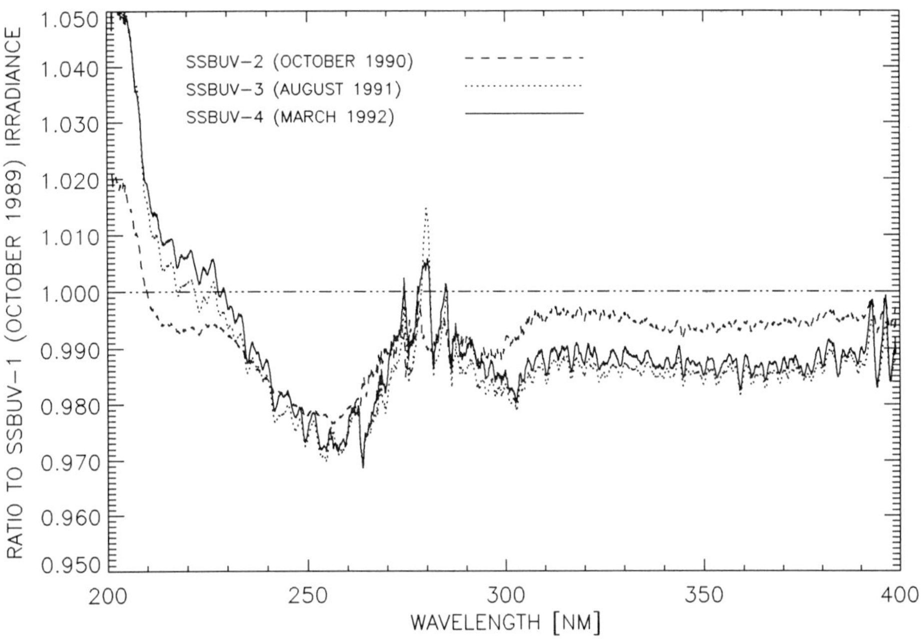

FIGURE 4. Ratios of the SSBUV-2, -3, and -4 solar spectral irradiance measurements to the SSBUV-1 measurement. No correction for solar irradiance variation was made in these comparisons.

2. Inter-instrument comparisons

The initial solar spectral irradiances measured by the two satellite instruments and SSBUV are compared in Figure 5. Whereas the first NOAA-11 and the SSBUV-1 measurements were made near the peak of solar cycle 22, NOAA-9 began solar observations at solar minimum. In making this comparison we have corrected for long-term solar variation using the composite Mg II index and associated scale factors. As discussed in DeLand & Cebula (1993), noise in the raw SBUV/2 data gives rise to an approximate 1% noise in the Mg II index, and caution must be exercised in using individual Mg II index measurements. The solar change corrections used here, 5.5% for March 1985 to October 1989 and 1.5% for February 1988 to October 1989, are 1.5% less than the corrections derived using the Mg II index. The solar spectral irradiances measured by these instruments agree to within -5 to +10% between 200 and 400 nm. The NOAA-9 irradiance has a calibration-induced anomaly near 232 nm relative to the other instruments; this feature will be reduced once the instrument's calibration is revised. Longward of 250 nm the NOAA-9 and 11 and irradiances are larger than the SSBUV irradiance. The NOAA-9 irradiance shows a clear bias with respect to the other instruments above 360 nm. The commonality of the two curves between 200 and 250 nm in Figure 5 signifies that the SSBUV irradiances are biased with respect to the two satellite instruments in this region. We plan to further investigate these differences via comparisons of these data with SOLSPEC, SOLSTICE, and SUSIM solar irradiance data.

Figure 6 presents the ratio of the NOAA-11 and SSBUV solar spectral irradiances for each of the first four SSBUV missions, relative to the Day 1 comparison. This pre-

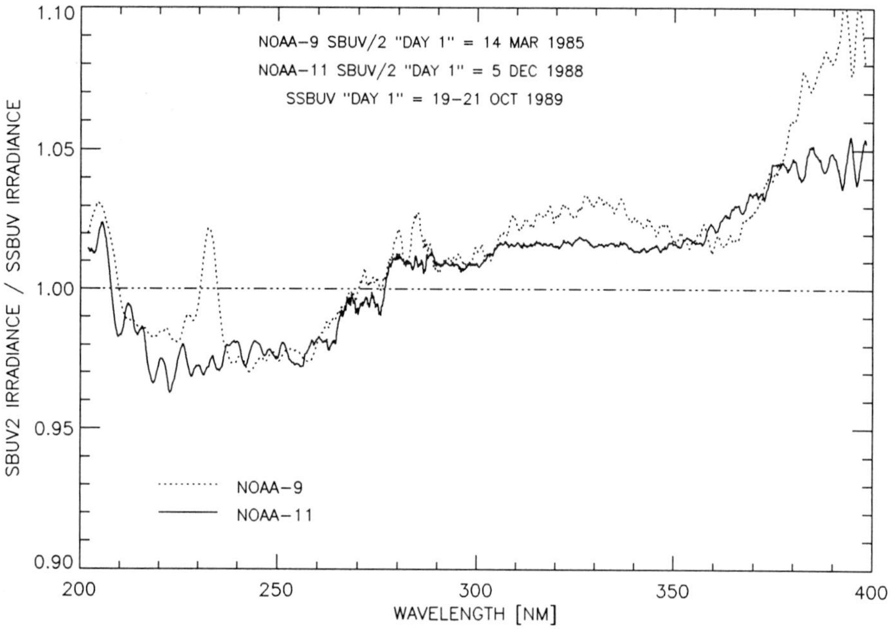

FIGURE 5. Ratios of the "Day 1" NOAA-9 SBUV/2 (dotted curve) and "Day 1" NOAA-11 SBUV/2 (solid curve) solar irradiances to the SSBUV-1 solar irradiance. These data were adjusted for solar variation.

sentation removes irradiance differences due to absolute calibration bias. As discussed previously, the satellite data were corrected for diffuser plate degradation but not for changes in the throughput of the monochromator. These comparisons indicate a significant spectrally dependent drift in the NOAA-11 instrument's sensitivity over four years, ranging from less than 1% near 400 nm to approximately 20% near 200 nm. The SBUV/2 drift assessment presented in Figure 6 and the internal assessment presented in Figure 3 exhibit excellent agreement and indicate that SSBUV solar data can be used to monitor and correct for changes in the radiometric sensitivity of the SBUV/2 instruments, particularly at wavelengths shortward of 300 nm, where long-term solar spectral irradiance changes are significant.

3. Conclusion

The SBUV/2 instruments onboard the NOAA operational satellites provide high quality daily measurements of the middle UV solar spectral irradiance, however protracted instrument changes prevent these data from being directly used to assess long-term solar change. The SSBUV experiment provides solar spectral irradiance measurements which are accurate with respect to one another to within ±1-2% from 200 to 400 nm. Although each SSBUV mission provides only a snapshot of the solar output, the work summarized in this paper shows that SSBUV data can be used in conjunction with the simultaneous SBUV/2 data to assess both short and long-term solar spectral irradiance variations. Our ultimate goal is to combine the SSBUV-corrected solar data from NOAA-9 and

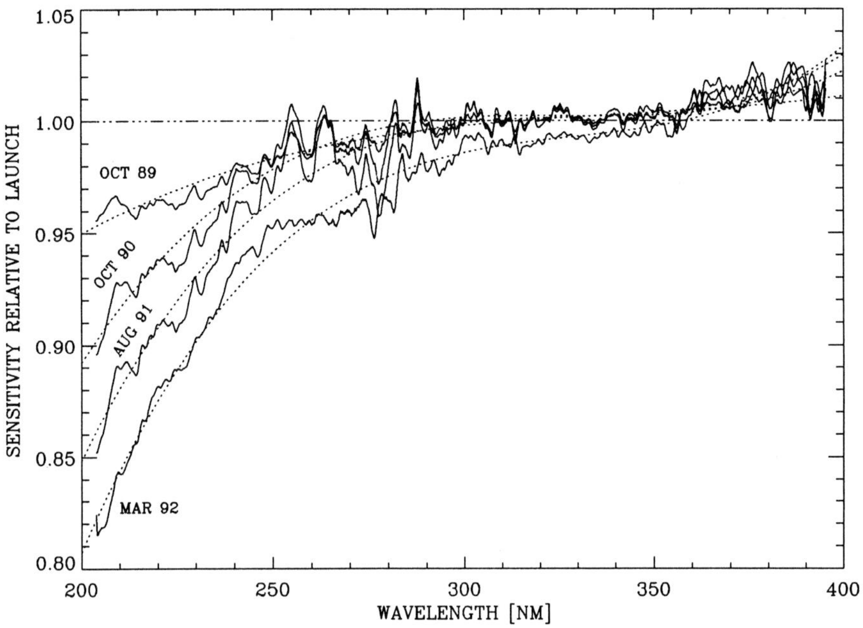

FIGURE 6. Ratios of the NOAA-11 SBUV/2 and the SSBUV solar spectral irradiances, October 1989 through March 1992, relative to the "Day 1" comparison. These curves provide a measure of the long-term drift in the satellite instrument. The smooth curves represent third order polynomial fits to the data.

NOAA-11 SBUV/2 with Nimbus-7 SBUV solar data to produce a single, self-consistent middle UV solar spectral irradiance data set covering the period 1978 through at least 1992.

Acknowledgments. This work was supported by NASA contract NAS5-31755. The SBUV/2 data were provided by H. D. Bowman, J. H. Lienesch and W. G. Planet of NOAA/NESDIS.

REFERENCES

CEBULA, R. P. & DELAND, M. T. 1992 The SBUV/2 monitors on the NOAA-9 and NOAA-11 satellites. In *Proceedings of the Workshop on the Solar Electromagnetic Radiation Study for Solar Cycle 22* (ed. R. F. Donnelly). pp. 239–249. NOAA ERL SEL, Boulder, CO, USA.

CEBULA, R. P., DELAND, M. T., HILSENRATH, E., SCHLESINGER, B. M., HUDSON, R. D. & HEATH D. F. 1991 Intercomparisons of the solar irraddiance measurements from the Nimbus-7 SBUV, the NOAA-9 and NOAA-11 SBUV/2, and the STS-34 SSBUV instruments: A preliminary study. *J. Atmos. Terr. Physics* **53**, 993–997.

CEBULA, R. P., DELAND, M. T. & SCHLESINGER, B. M. 1992 Estimates of solar variability using the Solar Backscatter Ultraviolet (SBUV) 2 Mg II index from the NOAA-9 satellite. *J. Geophys. Res.* **97**, 11613–11620.

CEBULA, R. P. & HILSENRATH, E. 1992 Ultraviolet solar irradiance measurements from the SSBUV-1 and SSBUV-2 missions. In *Proceedings of the Workshop on the Solar Electro-*

magnetic Radiation Study for Solar Cycle 22 (ed. R. F. Donnelly). pp. 250–264. NOAA ERL SEL, Boulder, CO, USA.

CEBULA, R. P. & HILSENRATH E. 1993 SSBUV middle ultraviolet solar spectral irradiance measurements. In *Proc. of the Quadrennial Ozone Symposium*, in press.

DELAND, M. T. & CEBULA, R. P. 1993 Composite Mg II solar activity index for solar cycles 21 and 22. *J. Geophys. Res.* **98**, 12809–12823.

DONNELLY, R. F. 1990 Solar UV temporal variation during solar cycle 22 and the twentieth century. In *Climate Impact of Solar Variability* (ed. K. H. Schatten & A. Arking). NASA CP-3086, 328–335.

FREDERICK, J. E., CEBULA, R. P. & HEATH D. F. 1986 Instrument characterization for the detection of long-term changes in stratospheric ozone: An analysis of the SBUV/2 radiometer. *J. Atm. Oceanic Tech.* **3**, 472–480.

HILSENRATH, E., WILLIAMS, D. & FREDERICK, J. 1988 Calibration of long term data sets from operational satellites using the Space Shuttle. In *Proc. SPIE* **924**, 215–222.

SCHLESINGER, B. M & CEBULA, R. P. 1992 Solar variation 1979-1987 estimated from an empirical model for changes with time in the sensitivity of the Solar Backscatter Ultraviolet Instrument. *J. Geophys. Res.* **97**, 10119–10134.

WEISS, H., CEBULA, R. P. LAAMANN, K. & MCPETERS R. D. 1993 Performance evaluation of the Solar Backscatter Ultraviolet Radiometer, Model 2 (SBUV/2) inflight calibration system. In *Proc. of the Quadrennial Ozone Symposium*, in press.

The Variable Structure of the X-ray Corona as Viewed by Yohkoh

By KAZUNARI SHIBATA

Solar Physics Division, National Astronomical Observatory, Mitaka, Tokyo 181, Japan

Recent findings on the variable structure of the corona by the soft X-ray telescope on board *Yohkoh* are reviewed, with emphasis upon the structure and evolution of active regions, and the long term evolution of the global structure of the corona.

1. Introduction

The Solar-A spacecraft (Ogawara *et al.* 1991) was successfully launched on August 30, 1991 (Ogawara *et al.* 1992), and was named *Yohkoh* which means "sunbeam" in Japanese. *Yohkoh* carries a soft X-ray telescope (SXT) (Tsuneta *et al.* 1991) that has been producing an unprecedented sequence of whole-Sun images at 3 - 60 Å (Acton *et al.* 1992). The routine data typically consist of some tens of whole-Sun images (FFI = full frame images; see Figure 1) per day, with much higher time resolution for specific observations of restricted areas (PFI = partial frame images), such as active regions or flares. The X-ray emission reveals the connectivity of the coronal magnetic field, and also shows where strong coronal heating is occurring. In particular, the movie of the whole-Sun SXT images has revealed that the solar corona is much more dynamic than had been thought, *i.e.*, the corona is full of transient loop brightenings (Shimizu *et al.* 1992), jets (Shibata *et al.* 1992b), expanding magnetic loops (Uchida *et al.* 1992), global restructuring of coronal fields (Tsuneta *et al.* 1992a), *etc.*

In this article, I shall review recent findings on variable structure of corona by *Yohkoh*/SXT, mainly on the following subjects: (1) structure and evolution of active regions (from emerging flux regions to global restructuring events) and (2) long term evolution of global structure of the corona (mainly based on X-ray synoptic charts). A preliminary report on the white light observations by the aspect sensor of *Yohkoh*/SXT is given by Hudson (1994). The reader should refer to the *Yohkoh* special issue of *Publ. Astr. Soc. Japan* (1992, vol. 44, No. 5), Acton *et al.* (1992), and Tsuneta & Lemen (1993) for reviews of initial discoveries by *Yohkoh*/SXT, and to Hudson (1993), Uchida (1993), and Tsuneta (1993) for reviews of flare observations by *Yohkoh*/SXT. See also Antonucci (1994) for a review of earlier soft X-ray observation.

2. Structure and evolution of active regions

Structure and Evolution of EFRs: Emerging flux regions are observed as very bright compact regions in SXT images, and hence are easily recognized in FFIs. They rapidly grow in size and in brightness (Figure 2), with rapid fluctuations in shape and brightness (Kawai *et al.* 1992), often accompanied by small flares and jets (Shibata *et al.* 1992b). In general, the SXT bright features coincide well in space with Hα arch filament systems (Kawai *et al.* 1992). Ishido *et al.* (1992) studied the size evolution of several EFRs and found that the apparent velocity of the increase in size of EFRs is large ($\sim$3 - 5 km/s) in the very early phase (less than 5 hours after birth), but is small ($\sim$1 - 2 km/s) in the later phase. This evolution is similar to that of the separation velocity of the two footpoints of the emerging flux (Harvey & Martin 1973; Chou 1993), although the expansion speed of

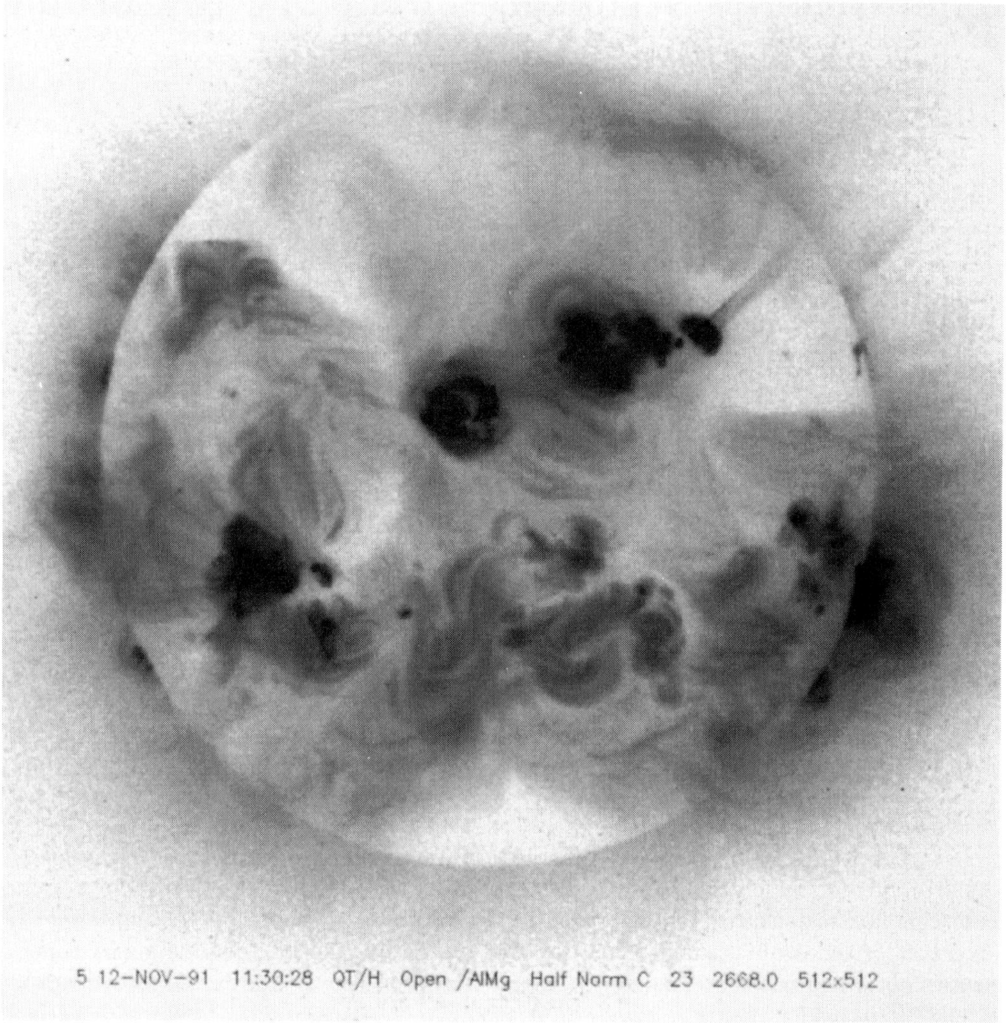

FIGURE 1. A full Sun soft X-ray image (negative) taken by *Yohkoh*/SXT on Nov. 12, 1991. The image was taken at an effective energy of ~ 1 keV with 5" pixels and 2.67 sec exposure time (a compressed image is shown here). Note that an X-ray jet is ejected from a small active region (NOAA 6918) in the northwest (upper right) quadrant of the solar disk (Shibata *et al.* 1992b).

coronal X-ray loops is larger than the separation velocity (< 1.2 km/s). Large expansion speeds of the coronal parts of emerging flux loops has also been found theoretically by Shibata *et al.* (1989) using 2D MHD numerical simulations, and the model has been extended to 3D by Matsumoto *et al.* (1993).

A preliminary study of the thermal evolution of an EFR (NOAA7150) by Su *et al.* (1993) using the filter ratio method (Hara *et al.* 1992) has shown that the average temperature of the EFR increases with age from 2×10^6 K to 3.5×10^6 K in 3 days. Considering the very rapid increase in size of EFRs, this seems to be consistent with the well-known scaling law of Rosner *et al.* (1978), but may not be necessarily in agreement with the analysis of Skylab data by Little and Krieger cited by Webb (1981).

Interaction of EFRs with Pre-existing Magnetic Fields: Shibata *et al.* (1993a) have

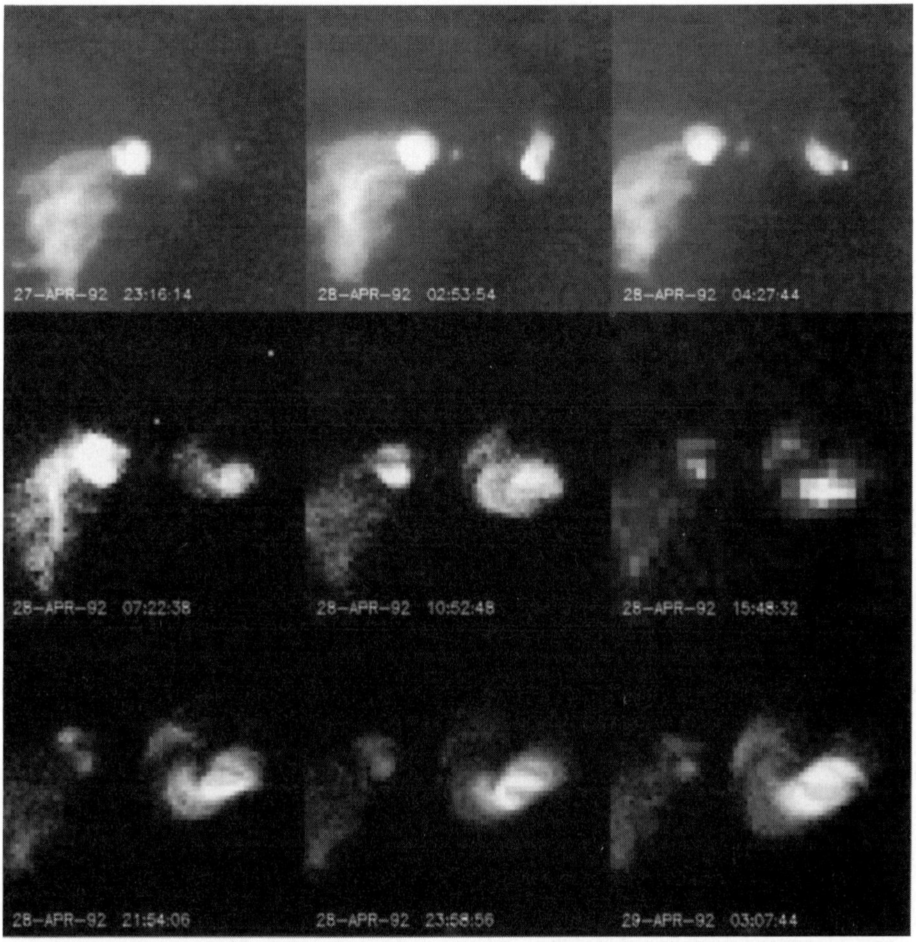

FIGURE 2. The rapid growth of emerging flux region, NOAA7150 (courtesy of G. Kawai). The size of each frame is 640" x 640".

studied the interaction between EFRs and overlying coronal magnetic fields, and found that there are fundamentally two types of interaction (reconnection) (see Figure 3):

(1) *Anemone-jet type*: This type occurs when emerging flux appears in a coronal hole (Figure 4 for a typical example). The EFR (or AR) looks like a "sea anemone", and an α-type sunspot is sometimes seen at the center of the "anemone". In many regions, jets (Shibata *et al.* 1992b) are often ejected from the AR in the vertical direction, which suggests reconnection between emerging flux and the (nearly vertical) coronal fields of the surrounding unipolar region.

(2) *Double-loops (Two-sided-loops) type*: This type is seen when emerging flux appears in a quiet region. Large-scale loop brightenings occur on both sides of the emerging flux, suggesting reconnection between the emerging flux and an overlying (nearly horizontal) coronal magnetic field. The loop brightenings seem to correspond to the jets occurring in closed loop systems. A reconnection model of two-sided loops has been developed by Yokoyama & Shibata (1993) by using 2D resistive MHD numerical simulations (Shibata *et al.* 1992a), extending their previous emerging flux model (Shibata *et al.* 1989).

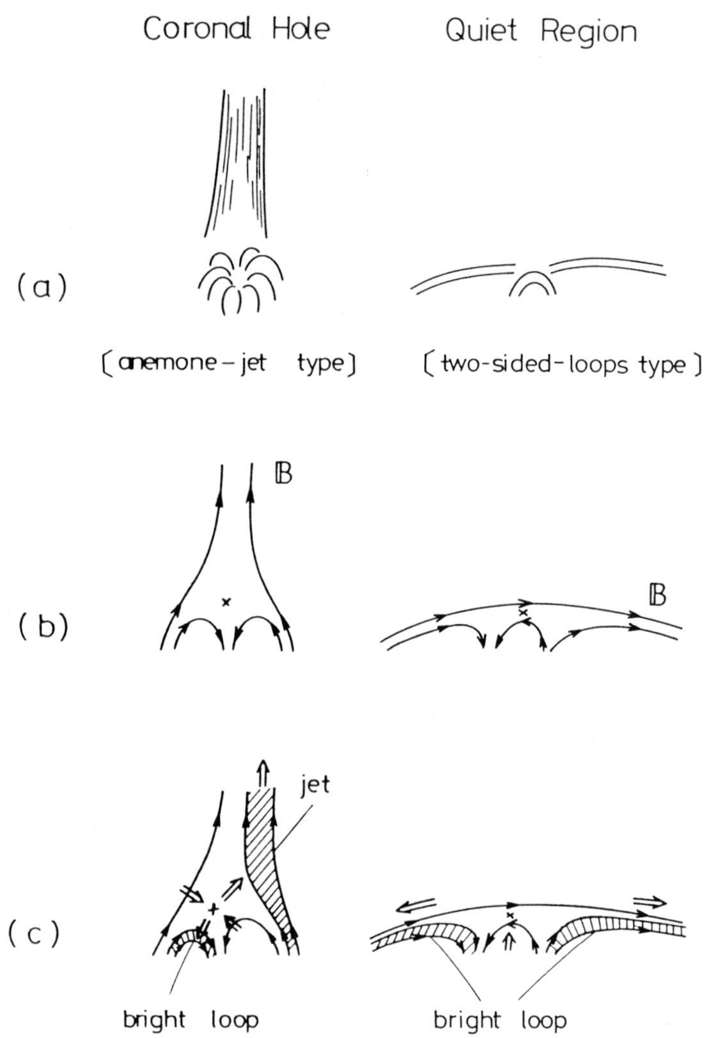

FIGURE 3. Schematic illustration of two types of interaction between emerging flux regions and overlying coronal fields (Shibata *et al.* 1993a).

Jets: The *Yohkoh*/SXT has revealed that there are many jet-like features in the solar corona, *i.e.*, transitory X-ray enhancements with an apparent collimated motion (Shibata *et al.* 1992b; Strong *et al.* 1992). According to preliminary studies, the length of jets found in FFI ranges from 10^4 km to more than 4×10^5 km, the apparent projection velocity is within a few 10 and 1000 km/s, though most of jets have velocities less than a few 100 km/s, and the total kinetic energy of these jets is estimated to be $10^{25} - 10^{28}$ erg. Almost all jets except for limb events are associated with flares in X-ray bright points (see Strong *et al.* 1992; Nitta *et al.* 1992; Harvey *et al.* 1993, for study of XBPs found by *Yohkoh*) emerging flux regions, and active regions. They sometimes occur several times from the same X-ray feature. It is likely that some flare-like loop brightenings (e.g., Shimizu *et al.* 1992) correspond to jets occurring in closed loop systems. As discussed above, various types of magnetic reconnection especially interactions between emerging

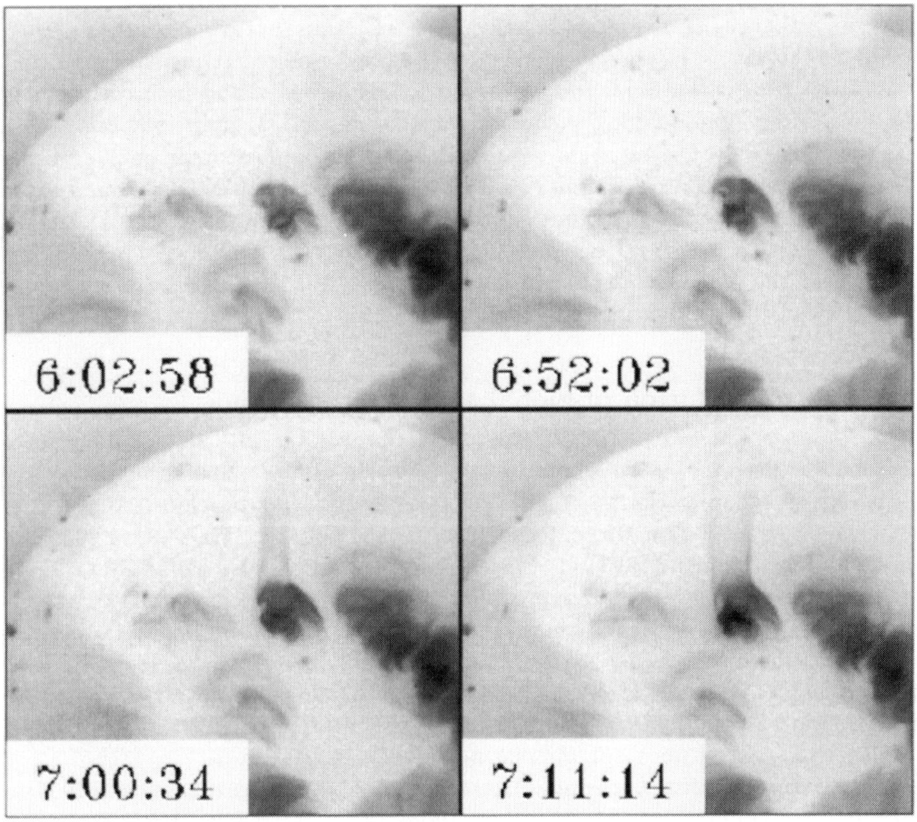

FIGURE 4. A typical example of an "anemone-jet" that occurred in NOAA 7001 on 1992 Jan. 11 (Shibata *et al.* 1993b). The shape of the active region at the footpoint of the jet looks like sea anemone, and hence this AR is called *anemone-AR*. The surrounding region is a coronal hole. The maximum length is probably greater than 3 x 10^5 km, and the apparent velocity is within the range of 90 – 240 km/s.

flux (or expanding magnetic flux) and pre-existing fields have been suggested to play a central role in generating these X-ray jets (Shibata *et al.* 1992b).

AR Expansion and CMEs: The SXT movie revealed that almost all active regions show continual or intermittent expansion with a velocity from a few to a few 10 km/s (Uchida *et al.* 1992). (It is interesting to note here that the opposite case also occurs; *shrinking loops* were clearly observed in an active region NOAA7420 during the week of Feb. 7 – 14, 1993.) The slower events (at a few km/s) seem to be a result of growth of ARs due to succesive emergence of magnetic flux as discussed above, whereas the faster expansion (at a few 10 km/s) may be physically more similar to a CME (Watanabe *et al.* 1992; Klimchuk *et al.* 1993). A typical example of the faster events occurred at the east limb of the Sun on Dec. 9, 1991, as reported by Uchida *et al.* (1992).

Global Restructuring of Coronal Magnetic Fields Associated with Filament Eruption: It has been found that large arcade structures suddenly (in a time scale of 30 min to 1 day) appear in SXT images after filament eruption (or filament disappearance) (Watanabe *et al.* 1992; McAllister *et al.* 1992; Tsuneta *et al.* 1992a; Hanaoka *et al.* 1993). The size of arcade structure ranges from 10^5 km to $\sim 10^6$ km, making them almost global in scope in extreme cases. Such a global restructuring event occurred on Nov. 12, 1991, near the polar region (Tsuneta *et al.* 1992a) in association with a disappearance of the polar crown filament. In this event, the height of the arcade (a few hundred thousand km) gradually increases with time with a velocity of a few km/s. A cusp-like structure is seen sometimes at the tops of the loops forming such an arcade, suggesting magnetic reconnection as in a classical CSHKP (= Carmichael-Sturrock-Hirayama-Kopp-Pneuman) model (Sturrock 1992) of CME-related flares (Shibata 1991). Note that *Yohkoh*/SXT has found excellent examples of this class of flares, which show beautiful cusp shapes, e.g., an LDE flare that occurred on Feb. 21, 1992 at the west limb (Tsuneta *et al.* 1992b). A helmet streamer observed near the pole, such as the event on Jan. 24, 1992, (Hiei *et al.* 1993), may be the side view of this kind of global restructuring event.

3. Long term evolution of global structure of corona

Rigid Rotation of Coronal Holes: The long-term evolution of the global structure of the corona has been extensively studied by Takahashi (1993) and Kozuka *et al.* (1993) by using X-ray synoptic charts. Figure 5 shows X-ray synoptic charts for Carrington rotations 1851 (from 5 Jan. 92 to 1 Feb. 92) to 1854 (from 27 Mar. 92 to 23 Apr. 92) made by Slater & Lemen (1993). It is seen that the dark coronal hole (CH) channel at 45° (Carrington longitude) keeps approximately the same shape for more than 4 months (from Feb. 92 to May 92), *i.e.*, shows rigid rotation (Takahashi 1993, Hayashi 1992). This confirms the discovery by Skylab (Timothy *et al.* 1975). From systematic analysis of X-ray synoptic charts, Takahashi (1993) found that low-latitude CHs rotate nearly rigidly but high latitude CHs do not, although the rotation rates of high latitude CHs are higher than that of photospheric plasma.

Active Nests and Active Longitudes: Takahashi (1993) and Kozuka *et al.* (1993) have noted the X-ray counterparts of active longitudes and nests (Zwaan 1987) seen in other ways. It is seen from Figure 5 that there are fundamentally two active longitudes, around 0° and 180° in the *Yohkoh* time interval. The δ-type sunspots producing many flares often appear in active longitudes. It is interesting to note here that these active longitudes correspond to giant bipolar magnetic regions (GBMR) situated on the global neutral line (Saito *et al.* 1993). On the basis of the 'triple dipole model' (Saito *et al.* 1989) of global solar magnetic field distribution, Saito predicted the existence of very large coronal loop structures above GBMRs in active longitudes. Such very large coronal loop structures have actually been observed on the limb in FFI images when the GBMR crosses the limb (Saito *et al.* 1993).

Anemone-ARs: The *Yohkoh*/SXT has revealed that many of ARs appearing in coronal holes show structure similar to "sea-anemone" as described above. Such ARs are called *anemone-ARs* (see Figure 4). The *anemone-ARs* tend to appear in coronal holes (far from the active longitudes) in the northern hemisphere during January 1992 – March 1993 (Shibata *et al.* 1993c). This tendency is opposite to the number of sunspots, which are more in the southern hemisphere during the same period. This as well as the apparently less-sheared field structure of the anemone-ARs suggest that the anemone-ARs are a signature of less-active ARs (*i.e.*, α-type spots) in spite of much magnetic flux, and that the origin of activity (or anti-activity) may be deeply rooted in the convection zone since the origin of the anemone-ARs seems to be closely related to the origin of coronal holes.

Topology of Global Coronal Magnetic Fields: The SXT movie has clearly shown that the topological pattern of global coronal magnetic fields changes with time as ARs emerge (as discussed above), evolve, and decay. It is not a rare phenomenon that the magnetic

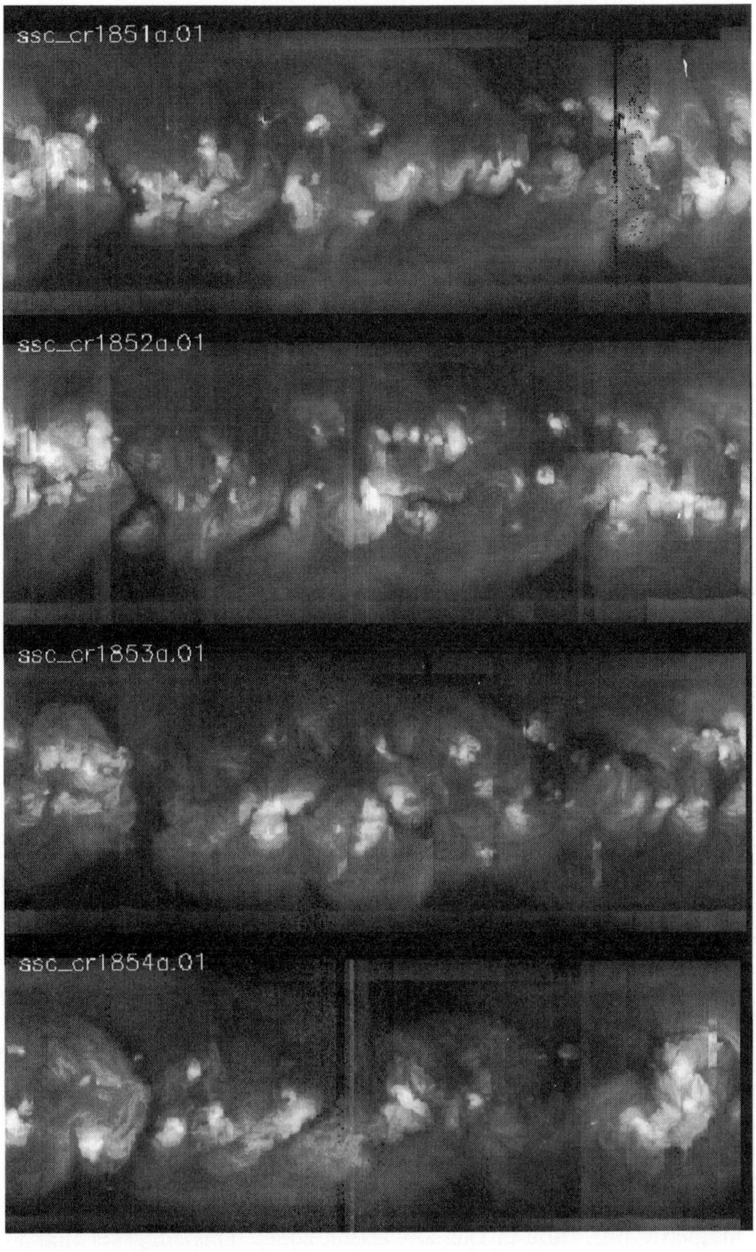

FIGURE 5. X-ray synoptic charts during Carrington rotation 1851–1854 (Jan. - Apr. 1992) made by Slater & Lemen (1993).

field in an AR in the northern hemisphere connects to an AR in the southern hemisphere (*i.e.*, trans-equatorial loops), as observed by Svestka & Howard (1979). The boundary of a coronal hole (CH) depends strongly on the evolution of the neighboring AR. In some cases, a small CH appeared after the filament eruption (Watanabe *et al.* 1992). It is also found that the shape of loops in the ARs in active longitudes are often sheared and twisted; the direction of the shear is opposite in northern and sourthen hemispheres, and

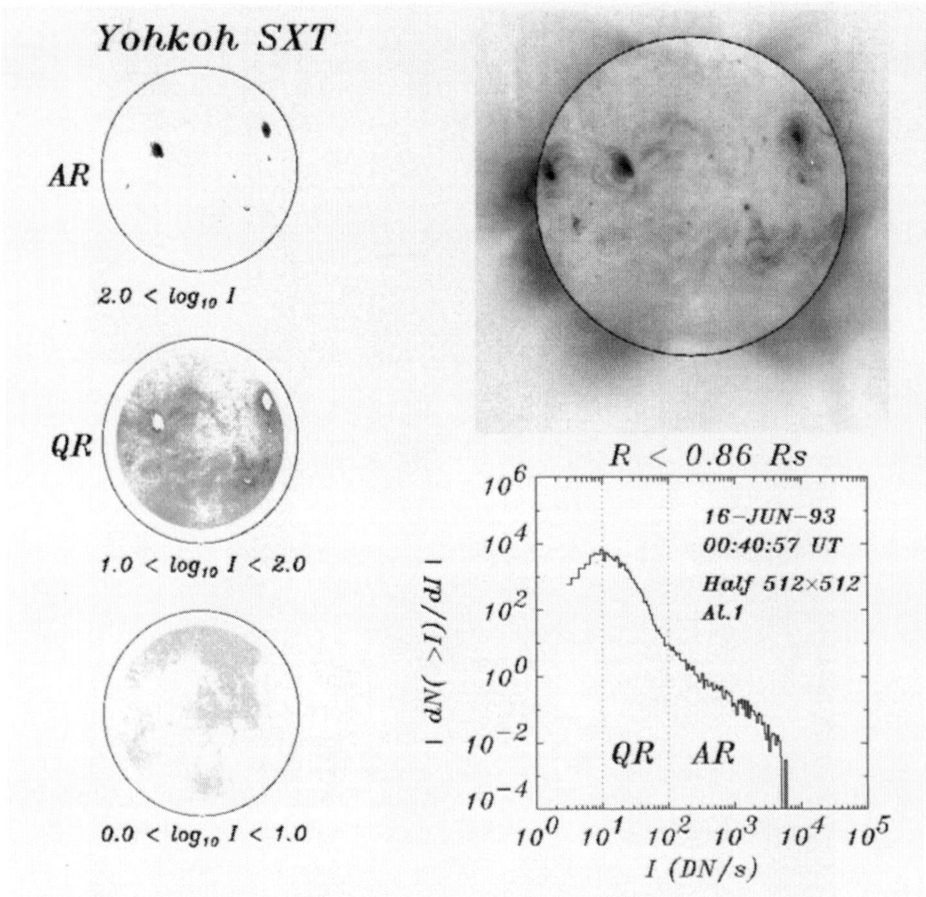

FIGURE 6. The number-distribution of soft X-ray intensity per pixel in an SXT/FFI image taken with thin aluminum filter on 1993 June 16 (shown in right bottom), *i.e.*, the number of pixels with intensity between I and I + dI as a function of I. The FFI image is shown in the right top, and the three images in the left show the spatial distributions of AR-pixels (top), QR-pixels (middle), and darker pixels (including coronal hole). The pixels near the limb and outside the Sun are excluded from statistics. All these figures are from Hara (1993).

is consistent with the direction expected from the Coriolis force acting on a rising (and expanding) magnetic flux tube in the convection zone. This suggests that not only the formation of ARs but also the generation of the twist and shear (*i.e.*, origin of activity) might be deeply rooted in the convection zone.

Long-Term Variation of the Soft X-ray Intensity: According to Hara (1993), the fraction of total soft X-ray intensity (observed by SXT) emitted from ARs is about 90 percent in an active phase (e.g., in Oct. 1991), while it is about 30 percent in a less active phase (e.g., in June, 1992). Hara (1993) further studied the distribution of soft X-ray intensity per pixel in SXT/FFI images (Figure 6), and found that the distribution pattern is different between quiet regions (QR) and active regions (AR). Figure 6 shows that the distribution for AR-pixels is nearly power-law while that for QR-pixels is not. This power-law distribution for AR-pixels reminds us of the power-law distribution of the solar flare occurrence against total flare energy (e.g., see Hudson 1991) and the idea of episodic heating (Sturrock *et al.* 1984; Parker 1988). Hara (1993)'s result suggests that the different heating mechanisms might work in QR and AR.

4. Summary

The *Yohkoh*/SXT has revealed that *the solar corona is much more dynamic than has been thought*, *i.e.*, the non-flaring corona is full of loop brightenings, jets, AR expansions from continual growth of EFRs to transient CMEs, global restructuring of coronal magnetic field beyond AR size, and so on. These are controlled by global pattern of magnetic field, such as coronal holes, active longitudes, and active nests. The total X-ray intensity is dominated by ARs (especially in active longitudes) in active phase, but the detailed spectral characteristics of X-ray emission seem to depend on the evolutionary phase of ARs.

Acknowledgments. The video movie shown during the conference was made with the help of the following people; L. Acton, H. Ando, M. Bruner, S. Freeland, J. Hirayama, H. Hudson, J. Khan, J. Lemen, G. Linford, R. Matsumoto, M. Morrison, N. Nitta, Y. Ogawara, G. Slater, K. T. Strong, S. Tsuneta, and Y. Uchida. I would like to thank these people for their excellent work. I would also like to sincerely thank Dr. H. S. Hudson for his many helpful comments on this manuscript, and Prof. T. Hirayama for encouragement and useful discussion. Thanks are also due to Y. Ishido, H. Hara, G. Kawai, T. Takahashi, Y. Kozuka, and Prof. T. Saito for allowing me to use their results before publication in the conference talk and/or in this manuscript. Finally, I greatly acknowledge all *Yohkoh* team members for their tremendous contribution on developing both hardware and software, and on doing the daily work of operating *Yohkoh* successfully.

REFERENCES

Acton, L., Tsuneta, S., Ogawara, Y., Bentley, R., Bruner, M., Canfield, R., Culhane, L., Doschek, G., Hiei, E., Hirayama, T., Hudson, H., Kosugi, T., Lang, J., Lemen, J., Nishimura, J., Makishima, K., Uchida, Y. & Watanabe, T. 1992 The *Yohkoh* Mission for high-energy solar physics. *Science* **258**, 618–625.

Antonucci, E. 1994 Variations of coronal radiations in X-ray related to coronal holes, active region loop systems, bright points. In *The Sun as a Variable Star: Solar and Stellar Irradiance Variations* (ed. J.M. Pap, C. Fröhlich, H.S. Hudson & S.K. Solanki). Cambridge University Press, in press.

Chou, D. Y. 1993 Structure of emerging flux regions. In *The Magnetic and Velocity Fields of Solar Active Regions* (ed. H. Zirin, G. Ai & H. Wang). Astronomical Society of the Pacific Conference Series **Vol. 46**, pp. 471–478.

Hanaoka, Y. Kurokawa, H., Enome, S., Nakajima, H., Shibasaki, K., Nishio, M., Takano, T., Torii, C., Sekiguchi, H., Kawashima, S., Bushimata, T., Shinohara, N., Irimajiri, Y., Koshiishi, H., Shiomi, Y., Nakai, Y., Funakoshi, Y., Kitai, R., Ishiura, K. & Kimura, G. 1993 Simultaneous observations of a prominence eruption followd by a coronal arcade formation in radio, soft X-rays and Hα. *Publ. Astron. Soc. Japan*, in press.

Hara, H., Tsuneta, S., Lemen, J. R., Acton, L. W. & McTiernan, J. M. 1992 High-temperature plasmas in active regions observed with the Soft X-ray Telescope aboard *Yohkoh*. *Publ. Astr. Soc. Japan* **44**, L135–L140.

Hara, H. 1993 The X-ray intensity distribution of the corona and its variability associated with solar activity. In *New Look at the Sun with Emphasis on Advanced Observations of Coronal*

Dynamics and Flares – What Do We See with YOHKOH and Nobeyama Radioheliograph –. Univ. Academy Press, in press.

HARVEY, K. L. & MARTIN, S. F. 1973 Ephemeral active regions. *Solar Phys.* **32**, 389–402.

HARVEY, K. L., NITTA, S., STRONG, K. T. & TSUNETA, S. 1993 The relationship of X-ray bright points to the photospheric magnetic fields. In *X-ray Solar Physics from Yohkoh* (ed. Y. Uchida, H. Hudson, T. Watanabe & K. Shibata). Univ. Academy Press, in press.

HAYASHI, K. 1992 Rotation of coronal holes. Undergraduate thesis (in Japanese), Dept. Astronomy, Univ. Tokyo.

HIEI, E., HUNDHAUSEN, A. J. & SIME, D. G. 1993 Reformation of a coronal helmet streamer by magnetic reconnection after a coronal mass ejection. *J. Geophys. Res.*, in press.

HUDSON, H. S. 1991 Solar flares, microflares, nanoflares, and coronal heating. *Solar Phys.* **133**, 357–369.

HUDSON, H. S. 1994 Spatial resolution of solar total irradiance variability: The *Yohkoh* white-light observations. In *The Sun as a Variable Star: Solar and Stellar Irradiance Variations* (ed. J.M. Pap, C. Fröhlich, H.S. Hudson & S.K. Solanki). Cambridge University Press, in press.

HUDSON, H. S. 1993 The *Yohkoh* context for high-energy particles in solar flares. In *Proc. Waterville Workshop on Solar Flares*, in press.

ISHIDO, Y., SHIBATA, K., TANAKA, N. & KUROKAWA, H. 1992 Observation of EFRs using SXT. In *Proc. of "Study of Solar Active Phenomena induced by Magnetic Shear – New Results of Yohkoh Observations –" held in Mitaka, Tokyo, Japan at Feb. 12 – 14 in 1992*, pp. 145–149.

KAWAI, G., KUROKAWA, H., TSUNETA, S., SHIMIZU, T., SHIBATA, K., ACTON, L. W., STRONG, K. T. & NITTA, N. 1992 Comparison between Hα and *Yohkoh* soft X-ray images of emerging flux regions. *Publ. Astr. Soc. Japan*, **44**, L193–L198.

KAWAI, G. 1993 Emerging flux regions observed in soft X-ray and H-alpha. In *New Look at the Sun with Emphasis on Advanced Observations of Coronal Dynamics and Flares – What Do We See with YOHKOH and Nobeyama Radioheliograph –*, in press.

KLIMCHUK, J. A., ACTON, L. W., HARVEY, K. L., HUDSON, H. S., KLUGE, K. L., SIME, D. G., STRONG, K. T. & WATANABE, TA. 1993 Coronal eruptions observed by *Yohkoh*. In *X-ray Solar Physics from Yohkoh* (ed. Y. Uchida, H. Hudson, T. Watanabe & K. Shibata). Univ. Academy Press, in press.

KOZUKA, Y., WATANABE, TA., KOJIMA, M., OHYAMA, M., TSUNETA, S., SAITO, T. & WATARI, S. 1993 Large scale coronal and solar wind structures. In *X-ray Solar Physics from Yohkoh* (ed. Y. Uchida, H. Hudson, T. Watanabe & K. Shibata). Univ. Academy Press, in press.

MATSUMOTO, R., TAJIMA, T., SHIBATA, K. & KAISIG, M. 1993 Three-dimensional magnetohydrodynamics of the emerging magnetic flux in the solar atmosphere. *Astrophys. J.* **414**, 357–371.

MCALLISTER, A., UCHIDA, Y., TSUNETA, S., STRONG, K. T., ACTON, L. W., HIEI, E., BRUNER, M. E., WATANABE, TA. & SHIBATA, K. 1992 The structure of the coronal soft X-ray source associated with the dark filament disappearance of 1991 September 28 using the *Yohkoh* Soft X-ray Telescope. *Publ. Astr. Soc. Japan* **44**, L205–L210.

NITTA, N., BASTIAN, T. S., ASCHWANDEN, M. J., HARVEY, K. L. & STRONG, K. T. 1992 Simultaneous observations of coronal bright points in X-ray and radio wavelengths. *Publ. Astr. Soc. Japan* **44**, L167–L172.

OGAWARA, Y., TAKANO, T., KATO, T., KOSUGI, T., TSUNETA, S., WATANABE, T., KONDO, I. & UCHIDA, Y. 1991 The Solar-A Mission: An overview. *Solar Phys.* **136**, 1–16.

OGAWARA, Y., ACTON, L. W., BENTLEY, R. D., BRUNER, M. E., CULHANE, J. L., HIEI, E., HIRAYAMA, T., HUDSON, H. S., KOSUGI, T., LEMEN, J. R., STRONG, K. T., TSUNETA, S., UCHIDA, Y., WATANABE, T. & YOSHIMORI, M. 1992 The status of *Yohkoh* in orbit: An introduction to the initial scientific results. *Publ. Astr. Soc. Japan* **44**, L41–L44.

PARKER, E. N. 1988 Nanoflares and the solar X-ray corona. *Astrophys. J.* **330**, 474–479.

ROSNER, R., TUCKER, W. H. & VAIANA, G. S. 1978 Dynamics of the quiescent solar corona. *Astrophys. J.* **220**, 643–665.

SAITO, T., OHKI, T., OLMSTED, C. & AKASOFU, S. I. 1989 A representation of the magnetic neutral line on the solar source surface in terms of the Sun's axial dipole at the center and two equatorial dipoles in the photosphere. *J. Geophys. Res.* **94**, 14993–14999.

SAITO, T., MINAMI, S., KOZUKA, Y., TAKAHASHI, T., HUDSON, H., TSUNETA, S. & WATANABE, T. 1993 Large-scale structure of the solar plasma corona – An analysis of it Yohkoh SXT images –. In *Proc. of the Fourth International Conference on Plasma Physics and Controlled Nuclear Fusion* (ESA SP-351). pp. 215–218.

SHIMIZU, T., TSUNETA, S., ACTON, L. W., LEMEN, J. R. & UCHIDA, Y. 1992 Transient brightenings in active regions observed by the Soft X-ray Telescope on *Yohkoh. Publ. Astr. Soc. Japan* **44**, L147–L153.

SHIBATA, K., TAJIMA, T., STEINOLFSON. R. & MATSUMOTO, R. 1989 Two-dimensional magnetohydrodynamic model of emerging magnetic flux in the solar atmosphere. *Astrophys. J.* **345**, 584–596.

SHIBATA, K. 1991 Theoretical models of solar flares. In *Flare Physics in Solar Activity Maximum 22* (ed. Y. Uchida, R. Canfield, T. Watanabe & E. Hiei). Lecture Note in Physics No. 387, Springer Verlag, pp. 205–218.

SHIBATA, K., NOZAWA, S. & MATSUMOTO, R. 1992a Magnetic reconnection associated with emerging magnetic flux. *Publ. Astr. Soc. Japan* **44**, 265–272.

SHIBATA, K., ISHIDO, Y., ACTON, L. W., STRONG, K. T., HIRAYAMA, T., UCHIDA, Y., MCALLISTER, A. H., MATSUMOTO, R., TSUNETA, S., SHIMIZU, T., HARA, H., SAKURAI, T., ICHIMOTO, K., NISHINO, Y. & OGAWARA, Y. 1992b Observation of X-ray jets with the *Yohkoh* Soft X-ray Telescope. *Publ. Astr. Soc. Japan* **44**, L173–L179.

SHIBATA, K., NITTA, N., MATSUMOTO, R., TAJIMA, T., YOKOYAMA, T., HIRAYAMA, T. & HUDSON, H. 1993a Two types of interaction between emerging flux and coronal magnetic field. In *X-ray Solar Physics from Yohkoh* (ed. Y. Uchida, H. Hudson, T. Watanabe & K. Shibata). Univ. Academy Press, in press.

SHIBATA, K., NITTA, N., STRONG, K. T., MATSUMOTO, R., YOKOYAMA, T., HIRAYAMA, T., HUDSON, H. & OGAWARA, Y. 1993b A gigantic coronal jet ejected from a compact active region in a coronal hole. *Nature*, submitted.

SHIBATA, K., NITTA, N. & HARA, H. 1993c Characteristics of active regions appearing in coronal holes, to be submitted.

SLATER, G. & LEMEN, J. 1993, private communication.

STRONG, K. T., HARVEY, K. L., HIRAYAMA, T., NITTA, N., SHIMIZU, T. & TSUNETA, S. 1992 Observations of the variability of coronal bright points by the Soft X-ray telescope on *Yohkoh. Publ. Astr. Soc. Japan* **44**, L161–L166.

STURROCK, P. A., KAUFMAN, P., MOORE, R. L. & SMITH, D. F. 1984 Energy release in solar flares. *Astrophys. J.* **94**, 341–357.

STURROCK, P. A. 1992 Emerging picture of eruptive solar flares. In *Proc. IAU Colloq. No. 133, "Eruptive Solar Flares"* (ed. Svestka, Z., Jackson, B. L. & Machado, M. E.). Spring-Verlag, pp. 397–409.

SU, Q., HIRAYAMA, T., SHIBATA, K., ICHIMOTO, K., TSUNETA, S., HARA, T., ACTON, L. & LEMEN, J. 1993 Soft X-ray loops at the beginning of a rapidly growing sunspot group, in preparation.

SVESTKA, Z. & HOWARD, R. 1979 Interconnecting loops: morphology of the sudden brightenings. *Solar Phys.* **63**, 297–318.

TAKAHASHI, T. 1993 Long-term variation of large scale solar X-ray structures with the Soft X-ray Telescope aboard *Yohkoh* satellite. Master thesis (in Japanese), Dept. of Astronomy, Univ. of Tokyo.

TIMOTHY, A. F., KRIEGER, A. S. & VAIANA, G. S. 1975 The structure and evolution of coronal holes. *Solar Phys.* **42**, 135–156.

TSUNETA, S., ACTON, L. W., BRUNER, M., LEMEN, J., BROWN, W., CARAVALHO, R., CATURA, R., FREELAND, S., JURCEVICH, B., MORRISON, M., OGAWARA, Y., HIRAYAMA, T. & OWENS, J. 1991 The Soft X-ray Telescope for the Solar A Mission. *Solar Phys.* **136**, 37–67.

TSUNETA, S., TAKAHASHI, T., ACTON, L. W., BRUNER, M. E., HARVEY, K. L. & OGAWARA, Y. 1992 Global restructuring of the coronal magnetic fields observed with the *Yohkoh* Soft X-ray Telescope. *Publ. Astr. Soc. Japan* **44**, L211–L214.

TSUNETA, S., HARA, H., SHIMIZU, T., ACTON, L. W., STRONG, K. T., HUDSON, H. & OGAWARA, Y. 1992 Observation of solar flare at the limb with the *Yohkoh* Soft X-ray Telescope. *Publ. Astr. Soc. Japan* **44**, L63–L69.

TSUNETA, S. & LEMEN, J. R. 1993 Dynamics of the solar corona observed with the *Yohkoh* Soft X-ray Telescope. In *Proc. "Advances in Stellar and Solar Coronal Physics"* (ed. J. Linsky & S. Serio). Kluwer, in press.

TSUNETA, S. 1993 Highlights of *Yohkoh* results. In *Proc. of the Fourth International Conference on Plasma Physics and Controlled Nuclear Fusion* (ESA SP-351). pp. 75–83.

UCHIDA, Y., MCALLISTER, A., STRONG, K. T., OGAWARA, Y., SHIMIZU, T., MATSUMOTO, R. & HUDSON, H. 1992 Continual expansion of the active-region corona observed by the Soft X-ray Telescope on *Yohkoh. Publ. Astr. Soc. Japan* **44**, L155–L160.

UCHIDA, Y. 1993 Some initial results from the *Yohkoh* observations of solar high-temperature and high-energy phenomena. In *Proc. of COSPAR Symp.*, in press.

WATANABE, TA., KOZUKA, Y., OHYAMA, M., KOJIMA, M., YAMAGUCHI, K., WATARI, S., TSUNETA, S., JOSELYN, J. A., HARVEY, K. L., ACTON, L. W. & KLIMCHUK, J. A. 1992 Coronal/interplanetary disturbances associated with disappearing solar filaments. *Publ. Astr. Soc. Japan* **44**, L199–L204.

WEBB, D. F. 1981 Active region structures in the transition region and corona. In *Proc. Skylab Workshop "Solar Active Region"* (ed. F. Q. Orrall). U. Colorado Press, pp. 165–198.

YOKOYAMA, T. & SHIBATA, K. 1993 Numerical simulation of magnetic reconnection associated with emerging flux in the solar atmosphere. In *Proc. of the Fourth International Conference on Plasma Physics and Controlled Nuclear Fusion* (ESA SP-351). pp. 203–206.

ZWAAN, C. 1987 Elements and patterns in the solar magnetic field. *Ann. Rev. Astr. Ap.* **25**, 83–111.

Summary of Future Space Observations of Solar Irradiances at Different Wavelengths

By VICENTE DOMINGO

Space Science Department, European Space Agency, ESTEC, Noordwijk, The Netherlands

The variations of solar irradiance can only be measured with a reasonable precision from space for almost all wavelengths. Yet the history of solar irradiance determinations is, with few exceptions, the history of efforts to patch the infrequent or too discontinuous space-based measurements by devising mechanisms for proxy determination of solar irradiance fluxes from ground-based obtained parameters. Although it appears that in the near future the situation will not be too different, there is some increase in the coverage of the spectral range that will be monitored. From gamma-rays to infrared light there are instruments that are now operating (e.g. GOES, *Yohkoh*, UARS) or are planned to operate in approved space missions (e.g. SOHO, ATLAS), or form part of proposed, not yet approved, missions. New techniques are being developed specifically for the measurement of solar irradiance variations with increased precision (e.g. cryogenic radiometer) in the next generation of space-borne measurements.

1. Introduction

The observation of solar irradiance variations in the past is the object of several papers in these proceedings. Therefore we refer to those papers for a review of the status of the subject.

This paper summarizes what may be expected to be the coverage of solar irradiance measurements in the near future. To do so we list the spacecraft that are now in operation and carry solar irradiance measuring instruments, the spacecraft that will be launched in the near future because they are being built now and those that probably will be launched because, either they have been approved and not yet financed or are likely to be approved.

Although we are primarily interested in irradiance measuring instruments, we also list the imaging instruments because they can be used as an alternative to approximate the measurements of solar irradiance. Even though the irradiance obtained in this way will have a larger error than if obtained with a radiometer, but in the absence of direct irradiance observations it will be the best one that we can get.

2. Spacecraft now in operation

The following spacecraft carry instruments that measure the solar irradiance. These spacecraft have limited nominal lifetimes but it is expected that they will be maintained in operation as long as their instruments are operational (Reinhard & Battrick 1990; RISE 1990).

• The Upper Atmosphere Research Satellite (UARS) is a spacecraft designed to provide the global data base necessary for understanding the coupled chemistry and dynamics of the stratosphere and mesosphere, the role of solar radiation in driving the chemistry and dynamics, and the susceptibility of the upper atmosphere to long-term changes in the concentration and distribution of key atmosphere constituents, particularly ozone. It carries instruments to measure the solar input: Active Cavity Radiometer Irradiance Monitor II (ACRIM II), Solar-Stellar Irradiance Monitor (SOLSTICE) and Solar Ultraviolet Spectral Irradiance Monitor (SUSIM). Details of these experiments are

Spacecraft & launch	Lifetime	Experiment	Wavelength	Spectral resolution
UARS 1991	design: 3 yrs expectation: a few years more	ACRIM II	total irradiance	
		SUSIM	120-400nm	0.15 or 5nm step
		SOLSTICE	115-430nm	0.12nm step
NOAA-9, -11	Continuous series of S/C	SBUV/2	160-405nm	1nm step
GOES	Continuous series of S/C	1-8Å		
ATLAS once per year, one solar cycle (to about 2002)	7-9 days	ACRIM	total irradiance	
		SOLCON	total irradiance	
		SOLSPEC	170-320nm	step: 1nm visible 20nm infrared
		SUSIM	120-400nm	0.15 or 5nm step
Yohkoh 1991	design life will expire in August 1994, expected several years more	BCS	S XV line (5.0385Å) CA XIX (3.1769) FE XXV (1.8509) Fe XXVI (1.7780)	
		WBS	2-30 keV, 20-400 keV, 0.2-100 MeV	

TABLE 1. Now operating spacecraft: Solar irradiance (see explanation of experiment acronyms in the main text).

given in this volume by Willson (1994), Rottman *et al.* (1994), and Brueckner *et al.* (1994). It orbits the Earth at 600 km altitude, at an inclination of 57 degrees. Launched in September 1991, its design lifetime ends in 1994, but it is expected to be operational for several more years.

• The Atmospheric LAboratory Spacecraft (ATLAS) is a platform that is brought into space within the cargo bay of the Space Shuttle, about once a year. The duration of the observations is of the order of 10 to 20 hours in a span of about 7 to 9 days. The last flight took place in April 1993. The next one is planned for 1994, and thereafter once a year until about 2001. It carries two total solar irradiance (solar constant) measuring instruments, the Active Cavity Radiometer Irradiance Monitor (ACRIM) and the Solar CONstant experiment (SOLCON), both of the electrical substitution radiometer type (Willson 1994; Crommelynck *et al.* 1994), SOLar SPECtrum (SOLSPEC) (Thuillier *et al.* 1994) and the SUSIM, already mentioned in the description of UARS, instruments measuring the spectral solar irradiance.

• The National Oceanographic and Atmospheric Administration satellites, NOAA-9 and NOAA-11 form part of a continuing program to monitor the Earth's atmosphere. They carry an instrument (SBUV/2) that monitors the solar ultraviolet spectrum (Cebula *et al.* 1994).

• The Geostationary Operational Environment Satellite (GOES) series continues monitoring the high energy region of the spectrum, sensitive to energetic solar activity (Aschwanden 1994)

Spacecraft & launch	Lifetime	Experiment	Wavelength	Spatial resolution
Yohkoh 1991	design life will expire in August 1994, expected several years more	SXT	0.3-6nm	2.5"
		HXT	15-100 keV (4 bands)	5"

TABLE 2. Now operating spacecraft: Solar imaging experiments (see explanation of experiment acronyms in main text).

Spacecraft & launch	Lifetime	Experiment	Wavelength	Spectral resolution
Coronas early 1994	1 year, more expected	SUVR	0.3-2.5, 0.3-12 121.6(HLy-α), <130nm	
		VUSS	20-58nm	L/DL=400
		IRIS	2-300 keV, diverse bands	
		DIFOS	520, 710, 400-1000nm	bandpass 100nm
SOHO July 1995	2 years (+4 possible)	VIRGO	total irradiance 402, 500, 862nm	bandpass 5nm
		SEM	30.5nm	
TIMED 2000-2001	2 years	SEE	0.1-200, 0.1-35, 0.1-90nm	step 1nm, ? ?

TABLE 3. Future spacecraft: Solar irradiance (see explanation of experiment acronyms in the main text).

• The *Yohkoh* satellite is a mission to investigate high energy phenomena on the Sun through X-ray and gamma-ray observations. The Hard X-ray Telescope (HXT) and the Soft X-ray Telescope (SXT) produce images of the solar corona, and the Wide Band Spectrometer (WBS) and the Bragg Crystal Spectrometer (BCS) monitor several spectral lines and bands both in the soft X-rays and in the hard X- and gamma-rays regions of the spectrum (see details by Shibata 1994). Launched in 1991, it has completed its design lifetime but is expected to remain operational for several years. It orbits the Earth at about 600 km altitude in a low inclination orbit.

Tables 1 and 2 summarize the main characteristics of the solar monitoring experiments now in operation.

3. Spacecraft now being built and planned spacecraft

We include here spacecraft that are now under development, spacecraft that have been approved, but have not been funded yet, and one spacecraft that has been proposed for funding, but is not yet approved.

• *Coronas-I* is a mission to investigate solar flares in a broad range of the electromagnetic spectrum complemented by an instrument to measure solar luminosity oscillations for helioseismological research. *Coronas* will be placed in a quasi sun-synchronous orbit at 500 km altitude. It will be launched in early 1994, with a nominal lifetime of 1

Spacecraft & launch	Lifetime	Experiment	Wavelength	Spatial resolution
Coronas 1994	1 year +	RES-C	190-205, 8.41-8.43, 1.85-1.87Å	6"×90" 6"×6" 2'
SOHO 1995	2 years (+4)	VIRGO/LOI	500nm	8'(12 pixels)
		MDI	6768Å	4" and 1"
		SUMER	50-160nm	2"
		CDS	15-80nm	3"
		EIT	171Å FeIX, 195 FeXII 284 FeXV, 304 HeII	5"
GOES-7 1997	continuous series	X-ray imager	1-8Å	1-2"
SWATH 1996(?)	1 year	XUV telescope	193Å	1"
TRACE 1997	1 year		3500, 1700, 1570, 1550, 1216, 285, 195, 171Å	1"

TABLE 4. Future spacecraft: Solar imaging experiments (see explanation of experiment acronyms in the main text).

year. *Coronas* carries instruments that monitor the electromagnetic spectrum in many wavelength bands from X-rays to ultraviolet. Particularly interesting, for the subject of this conference, are the Solar Ultraviolet Radiometer (SUVR) and the Vacuum Ultraviolet Solar Spectrometer (VUSS). The 3-channel solar white light photometer provides continuous measurement of 3 spectral bands in the visible. The Solar X-ray Spectrometer (RES-C) produces monochromatic images of the Sun in three spectral ranges in the EUV and soft X-ray ranges.

• The Solar and Heliospheric Observatory (SOHO) has been designed to study the physics of the solar corona, the generation of the solar wind and the internal structure of the Sun. For that it carries instruments that will perform helioseismic measurements, spectroscopic plasma diagnostics and *in situ* solar wind particle analysis measurements. SOHO will be in a heliocentric orbit around the L1 Sun-Earth Lagrangian point. It will be launched in July 1995, with a nominal lifetime ending in December 1997, but with consumables that will allow it to stay at least another 4 years. The Variations of Solar Irradiance and Global Oscillations (VIRGO) experiment will perform total solar irradiance (solar constant) measurements and spectral irradiance measurements in the visible, in 3 wavelength bands, each about 5 nm wide. The Solar Extreme-ultraviolet Monitor (SEM) instrument monitors the EUV radiation around 30.5 nm. Several instruments will produce images of the full solar disk in the EUV wavelength range (Solar Ultraviolet Measurements of Emitted Radiation – SUMER, Coronal Diagnostic Spectrometer – CDS, and Extreme-ultraviolet Imaging Telescope – EIT) and one, the Michelson Doppler Imager (MDI) will produce images in the visible.

• The Space Weather and Terrestrial Hazards (SWATH) experiment is designed to place into low-Earth orbit a set of instruments for measurement and study of local space hazards. The mission has three major scientific objectives: 1.) to monitor, track and study large disturbances originating at the Sun and propagating to the near-Earth environment; 2.) to monitor and study the activity of the solar corona; 3.) to track and measure space debris in low-earth orbit. It carries a small XUV telescope operating at 193 Å wavelength, which images the solar corona at two million degrees temperature. A second telescope contains a large multilayer-coated mirror for imaging of the solar corona at arcsecond spatial resolution; a moveable filter also provides imaging of the solar photosphere at high resolution. The satellite will be placed in a polar, sun-synchronous orbit for nearly continuous monitoring of the Sun and for pole-to-pole mapping of the space debris populations. If funding is confirmed it could be launched in 1996.

- A new GOES satellite, to be launched around 1997 (?) will carry an X-ray imager operating in the 1-8 Å band.
- The Thermosphere–Ionosphere–Mesosphere Energetics and Dynamics (TIMED) mission is intended to carry out the first detailed, space-based study of physical and chemical processes in the terrestrial atmosphere between about 60 and 180 km altitude. It has not yet been funded, but it is planned to be launched around 2000 or 2001 with a nominal lifetime of 2 years. It will carry a Solar EUV Experiment (SEE) measuring spectral irradiance between 0.1 and 200 nm with a combination of spectrometers and filtered photodiods.
- TRACE – the Transition Region And Coronal Explorer is a single-instrument solar mission to be put into a sun-synchronous polar orbit and will obtain continuous solar observations for about 8 months per year. It will collect images of the Sun at 8 wavelength bands, between 17 and 250 nm, with 1-arcsec spatial resolution and excellent temporal resolution and continuity. At the moment TRACE is a proposed mission that, if approved, could be launched around 1997–1998.

Tables 3 and 4 list the main characteristics of the experiments being developed and expected to be in space in the immediate future. Figures 1 and 2 summarize the spectral coverage versus the expected solar radiometric observations.

4. Conclusion

4.1. *Solar total irradiance (solar constant)*

Given the accuracy of the absolute radiometers (Fröhlich 1994), the discovery of the variation of the solar constant during a solar cycle was possible because the solar irradiance measuring instruments aboard NIMBUS-7, the Solar Maximum Mission (SMM), and the Earth Radiation Budget Satellite (ERBS) were operating long enough in a continuous manner.

The variation of the solar constant from cycle to cycle is expected to be well below one part per thousand. Since the present generation of radiometers can only measure the solar irradiance with an absolute precision of the same order, one can expect to observe these long-term variations only if one observes with one radiometer continuously, or if one is able to offset the relative instrumental error between two or more radiometers by performing simultaneous measurements of the solar constant with them and comparing the results.

The successful series of measurements started with NIMBUS-7, the Solar Maximum Mission and ERBS and continued by the ACRIM II instrument aboard UARS may, if UARS operates long enough, be connected with the measurements that will start to be made from SOHO in the second half of 1995 (Figure 1). If this is so the solar constant variation along one solar cycle detected by the ACRIM instrument may be continued to be observed during a second solar cycle. The chart (Figure 1) also shows that, unless

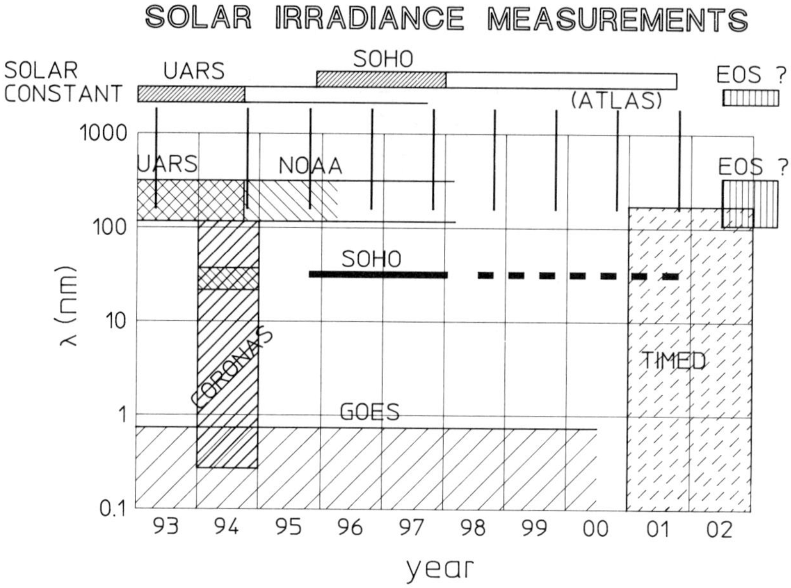

FIGURE 1. Time history of the expected solar irradiance measurements in the present decade at different wavelengths (the acronyms of the spacecraft are explained in the main text).

some action is undertaken by the space agencies, there is a large probability that the solar constant measurement series will be interrupted after SOHO. To be able to observe longer time trends in solar variations (those that may be of significance for the Earth's climate) the space agencies should make sure that continued measurements of well intercompared instruments are continued for the next solar cycles.

A complementary action that should be taken is to support the development and operation in space of a new generation of radiometers with higher absolute precision such as the cryogenic radiometers that are being proposed for future observations (Martin & Fox 1992). Increasing the current absolute precision (about ±0.15%) of the measurements by about an order of magnitude may eliminate the need to obtain continuous measurements made with the same instrument or with instruments that have been compared in space with the previous ones.

4.2. *Spectral irradiance measurements*

The two main regions of the electromagnetic spectrum of interest for aeronomy are the Ultraviolet (UV), from Lyman-α to 400 nm, and the Extreme Ultraviolet (EUV), 5 nm to Lyman-α.

There is a continuous monitoring of the solar UV from the NOAA satellites that will last at least until 2000, and in addition now UARS is producing measurements from its two UV radiometers. Some time after 2000, the Earth Observation System (EOS) will carry a UV instrument as well, but the launch dates are not yet fixed.

More unclear are the prospects of EUV continuous monitoring. From now to the indefinite future (> 2000), the only programs that include EUV monitoring are *Coronas*, to be launched in 1994, SOHO, starting late 1995, and TIMED, possibly around 2000-2001. Such measurements are not easy, but they are needed for upper atmosphere research.

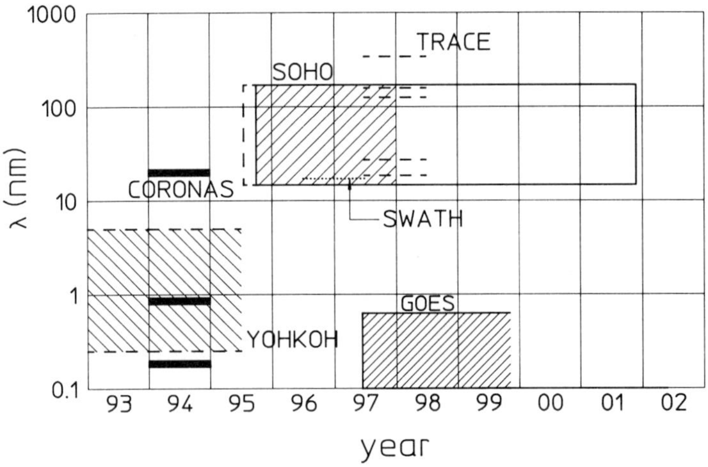

FIGURE 2. Time history of the expected solar imaging observations from space, in the present decade, that may be useful for radiometric measurements (the acronyms of the spacecraft are explained in the main text).

4.3. *Solar imaging experiments*

The solar imaging instruments have not been conceived to obtain radiometric results, yet when there are no better measurements they can be used to determine the solar irradiance by careful integration of the images. This will be the case of the EUV instruments aboard SOHO, that will be able to produce radiometric measurements of the EUV irradiance. These measurements will have, of course, limited accuracy, but they will be an improvement over the proxy indices (Tobiska & Barth 1990) used now to compute the solar EUV fluxes. Figure 2 represents the expected supplementary cover that the solar imaging instruments will provide over the following years.

REFERENCES

ASCHWANDEN, M. 1994 Irradiance observations from the GOES satellite. *Solar Phys.*, in press.

BRUECKNER, G., FLOYD, L.E., LUND, P.A., PRINZ, D.K. & VANHOOSIER, M.E. 1994 Irradiance observations from the UARS/SUSIM and ATLAS/SUSIM Experiments. In *The Sun as a Variable Star: Solar and Stellar Irradianec Variations* (ed. J.M. Pap, C. Fröhlich, H.S. Hudson & S.K. Solanki). Cambridge Univ. Press, in press.

CEBULA, R.P., HILSENRATH, E. & DELAND, M.T. 1994 Middle ultraviolet solar spectral irradiance measurements, 1985-1992, from the SBUV/2 and SSBUV instruments. In *The Sun as a Variable Star: Solar and Stellar Irradiance Variations* (ed. J.M. Pap, C. Fröhlich, H.S. Hudson & S.K. Solanki). Cambridge Univ. Press, in press.

CROMMELYNCK, D., DOMINGO, V., FICHOT, A. & LEE, R. 1994 Solar irradiance observations from the EURECA and ATLAS experiments. In *The Sun as a Variable Star: Solar and Stellar Irradiance Variations* (ed. J.M. Pap, C. Fröhlich, H.S. Hudson & S.K. Solanki). Cambridge Univ. Press, in press.

FRÖHLICH, C. 1994 Irradiance observations of the Sun. In *The Sun as a Variable Star: Solar and Stellar Irradiance Variations* (ed. J.M. Pap, C. Fröhlich, H.S. Hudson & S.K. Solanki). Cambridge Univ. Press, in press.

REINHARD, R. & BATTRICK, B. 1990 Inter-Agency Consultative Group for Space Science (IACG). *Key Data on Participating Projects* ESA-SP- 1107. Paris, November 1990.

RISE 1990 Radiative Inputs of the Sun to Earth. A Research Plan for the 1990's on Solar

Irradiance Variation. National Science Foundation Report. February 1990.

MARTIN, J. E. & FOX, N. P. 1992 Absolute radiometric measurements in space. In *Proceedings of an ESA Syposium on Photon Detectors for Space Instrumentation* ESA SP-356, December 1992.

ROTTMAN, G.J., WOODS, T.N, WHITE, O.R. & LONDON, J. 1994 Irradiance observations from the UARS/SOLSTICE experiment. In *The Sun as a Variable Star: Solar and Stellar Irradiance Variations* (ed. J.M. Pap, C. Fröhlich, H.S. Hudson & S.K. Solanki). Cambridge Univ. Press, in press.

SHIBATA, K. 1994 The variable structure of the X-ray corona as viewed by *Yohkoh.* In *The Sun as a Variable Star: Solar and Stellar Irradiance Variations* (ed. J.M. Pap, C. Fröhlich, H.S. Hudson & S.K. Solanki). Cambridge Univ. Press, in press.

TOBISKA, W.K. & BARTH, C.A. 1990 A solar EUV Flux model. *J. Geophys. Res.* **95**, 8243-8251.

WILLSON, R.C. 1994 Irradiance observations of SMM, Spacelab 1, UARS, and ATLAS experiments. In *The Sun as a Variable Star: Solar and Stellar Irradiance Variations* (ed. J.M. Pap, C. Fröhlich, H.S. Hudson & S.K. Solanki). Cambridge Univ. Press, in press.

Photometric Variations of Solar Type Stars

By RICHARD R. RADICK

Air Force Phillips Laboratory, Solar Research Branch, Sunspot, NM 88349, USA

High precision measurements of photometric variability among solar type stars have now been made since 1980. These observations clearly show that year-to-year brightness variations connected with magnetic activity are a widespread phenomenon among such stars. They also suggest that the Sun's potential for long-term white light variability may be significantly understated by measurements of solar total irradiance during the 1980s.

1. Introduction: solar activity and variability

Because our experience with the Sun so strongly influences our thinking about stellar behavior, it seems appropriate to begin this discussion of variability among solar type stars with a brief review of its solar paradigm. Our immediate attention will be focused on global properties of solar variability that can be (more or less) directly compared to analogous stellar behavior and used to help interpret it. In turn, our maturing stellar perspective may eventually provide a broader context for understanding and even predicting solar variability itself.

It is generally agreed that solar variability arises from the presence and evolution of magnetic fields on the Sun's surface. On the global scale, these magnetic fields are involved in the nonradiative heating of the Sun's chromosphere and corona. This nonradiative heating is designated by the term 'activity' (the term 'variability' refers more narrowly to the temporal behavior and, indirectly, to the spatial inhomogeneity of the activity). The chromosphere and corona get rid of their excess heat by radiating it away in specific spectral lines, such as the H- and K-lines of singly ionized calcium. Monochromatic pictures of the Sun in such lines (spectroheliograms) map the instantaneous distribution of the nonradiative heating, and presumably also faithfully map the underlying magnetic fields.

Clearly, we would like to have time series of magnetic flux measurements for both the Sun and its stellar analogues. Although such time series are available for the Sun, they are extremely difficult to generate for stars. Therefore, we are forced to resort to proxies, such as the chromospheric Ca emission index, which are accessible to stellar observations. It is important, however, to recognize that any proxy is likely to be imperfect. For example, the chromospheric Ca emission index suffers from two distinct defects: (1) It is sensitive to non-magnetic effects, such as acoustic heating and abundance differences, but (2) It is relatively insensitive to the magnetic flux contained in sunspots.

The global solar Ca K-line emission comprises the sum of two components — emission from the network, and emission from active region plages. The plage component increases dramatically from solar minimum to solar maximum, whereas the variation of the network component is less pronounced. The Sun's global white light variations are more complex. The conventional picture invokes three components; network faculae, active region faculae, and dark sunspots, with the net variation representing the competitive balance between the bright faculae and the dark spots (e.g., Foukal & Lean 1988). On the time scale of the 11-yr solar activity cycle, roughly half of the total positive contribution from the faculae is cancelled by the opposing flux deficit from the sunspots, leaving a net white light variation that approximately tracks the cyclic variation of solar activity as measured, for example, by the global Ca K-line emission.

By appropriately suppressing spatial resolution, a time series of disk integrated solar fluxes may be created. Such time series are available for both Ca K-line emission (from the National Solar Observatory: Keil 1992; Livingston 1992) and white light total irradiance (from SMM/ACRIM or NIMBUS-7/ERB), and provide solar data that may be compared to stellar observations with relative ease.

2. Stellar activity and variability

2.1. *The co-ordinated Lowell and Mount Wilson programs*

Modern synoptic measurements of photometric variability among solar type stars have been made since 1980, when observations of lower main sequence stars in the Hyades, a nearby cluster 10-15% the age of the Sun, were inaugurated at the Cloudcroft Observatory (Radick 1991 and references therein). This program was moved to the Lowell Observatory in 1982, where photometric monitoring of a few Hyades stars continues up to the present time. Between 1984 and 1987, stars in the Hyades-age Coma Berenices cluster were added. The heart of the Lowell photometric program, however, traces back to 1984, when regular observations of 33 field stars, selected from among those of the Mount Wilson HK Project, were begun (Lockwood 1994). The HK Project itself can be traced back to 1966, when regular spectrophotometric observations of chromospheric Ca H+K emission from about 100 stars began at the Mount Wilson Observatory (Wilson 1978). All but two of the 33 stars of the Lowell program are among these HK Project 'veterans.'

The Lowell and Mount Wilson observations demonstrate that both broadband and chromospheric Ca H+K emission variations are common among solar type stars. The time scale of variation ranges from the night-to-night interval between the individual observations to the present 27 year duration of the HK Project. The amplitude of variation shows the dependences on stellar mass (or temperature) and rotation rate (or age) predicted by stellar dynamo theory (e.g., Radick 1992). Cyclic variability reminiscent of the 11-yr solar activity cycle, however, appears to be confined largely, if not exclusively, to the older, solar-age stars.

Although photometric brightness and Ca H+K emission variations are clearly connected phenomena, their relationship is neither simple nor accurately predictable. Among young stars, the year-to-year brightness and H+K emission variations are anti-correlated, whereas for the older, solar-age stars, they tend to correlate directly, as they do for the Sun. Although the Sun's chromospheric K-line emission variation appeared to be normal during the 1980's, the amplitude of its white light variation seems to have been distinctly less than the variation of otherwise similar stars (Radick *et al.* 1990; Lockwood *et al.* 1992).

2.2. *Some specific solar-stellar comparisons*

Figure 1 displays the time series data for the Sun and two stars, HD10476 and HD1835. Although it is somewhat cooler and less massive than the Sun, HD10476 (= HR493, $B--V = 0.84$, $Sp = K1V$) has an average level of chromospheric activity, as measured by Ca H+K emission, that is very close to that of the Sun. Like the Sun, it is a slow rotator, with a period of about 35 days. These circumstances imply that it is comparable in age to the Sun. In contrast, HD1835 (= HR88, $B--V = 0.66$, $Sp = G2V$) is a typical young star. It is a member of the Hyades group, which suggests that its age is about 10-15% that of the Sun. This is confirmed by the star's relatively rapid rotation (its period is about 8 days) and high average level of chromospheric activity (about three times the solar level). Its temperature and mass closely match those of the Sun.

The upper panels in Figure 1, which are all plotted with the same ordinate scale,

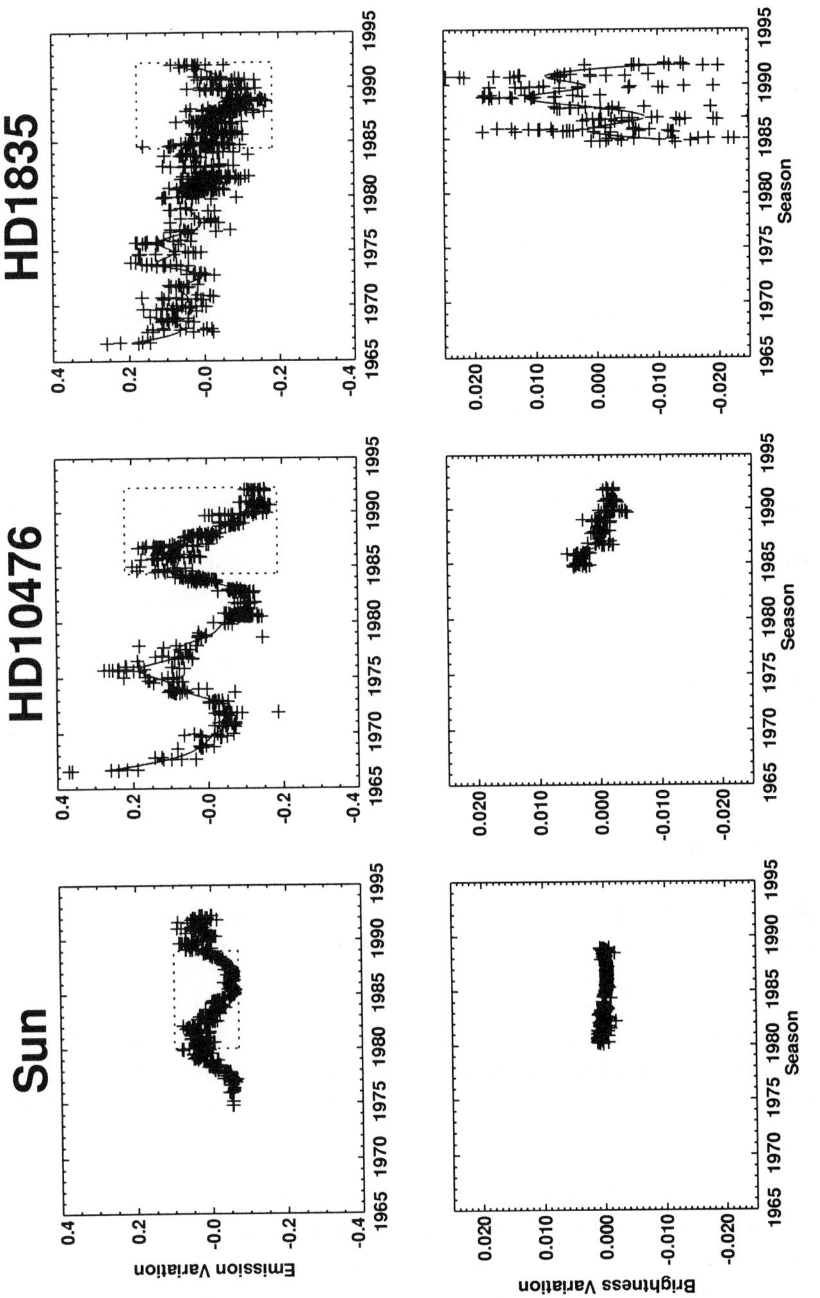

FIGURE 1. Time series data for the Sun and two stars

show the (fractional) variation in chromospheric emission. The lower panels, also plotted using a single ordinate scale, show the corresponding photometric variations. The box within each of the upper panels encloses the interval of the contemporaneous broadband observations shown below. Embedded in each time series is a cubic spline fitted to the annual mean values, which can be helpful for tracing out year-to-year fluctuations, especially for cases like the one shown in the lower right panel where the annual changes are masked by strong night-to-night variations. The solar data are the NSO K-index measurements (Livingston 1992) and the SMM/ACRIM radiometry (Willson 1992). The stellar data are from the Mount Wilson and Lowell Observatories.

Both the Sun and HD10476 show clear activity cycles approximately a decade in length. The amplitude of the solar cycle, however, is less than half that of HD10476. Like many young stars, HD1835 shows strong, random-looking variability, superimposed on a long-term secular trend (in this case, downward), but no obvious activity cycle.

The Sun's white light cyclic variation (lower left panel) is almost imperceptible on the plotted scale. In contrast, the photometric variation of HD10476, which is 5-6 times larger than that of the Sun, can clearly be seen to track the downward trend of its Ca H+K emission. As mentioned previously, this correlated behavior is a hallmark of older, less active stars, including the Sun. This suggests that the cyclic white light variability of solar-age stars as a class is governed by bright features, as it is for the Sun. In contrast to either the Sun or HD10476, the photometric variations of HD1835 are very pronounced, on both the night-to-night and the year-to-year time scales. The variability amplitude amounts to 2-3%, on both time scales. Furthermore, the photometric variations do not track the H+K variations on the year-to-year time scale, but rather mirror them in an anticorrelated pattern that is characteristic of young stars. This suggests that, in contrast to solar-age stars, the year-to-year photometric variations of young stars are controlled by dark features.

When the year-to-year variation is suppressed, the relation between the Ca H+K and broadband residuals (whenever it is statistically significant) is almost always found to be an anticorrelation, for solar-age stars as well as young stars. This suggests that (1) The surface features of solar type stars, both young and old, are consistent with active regions: dark spots temporally and spatially associated with bright plages, and (2) spots dominate short-term variation, for all stars.

Figure 1 illustrates two points: (1) It is not hard to find sunlike, solar-age stars (like HD10476) that vary more than the Sun, and (2) Chromospheric and broadband variations are not simply proportional, which raises a warning flag about the common practice of using one activity indicator as a proxy for another.

2.3. *Characteristics of the entire sample*

In Figure 2, we turn our attention to the entire sample of 33 stars. In the two panels of this diagram, chromospheric and photometric variability are plotted against average chromospheric activity. In both cases, variability is measured by the rms dispersion of annual mean values (in fractional units), and average activity ('chromospheric emission ratio') is measured relative to bolometric luminosity and plotted on a logarithmic scale. A point representing the Sun has also been plotted in each panel. In the lower panel, stars which are judged to be variable on the basis of the internal evidence of the photometry alone are indicated by asterisks, and stars whose variability cannot be clearly discerned relative to the measurement noise are designated by crosses.

The upper panel shows that the amplitude of Ca H+K emission variation increases with average activity, and that the Sun is near the center of the range in variability for stars of comparable average activity. The amplitude of photometric variation also increases with average activity, but more abruptly. Young, active stars almost always show strong photometric variability. Indeed, the lower panel of Figure 2 suggests that the young Sun probably varied by more than a percent from year to year, like HD1835 or even the star in the upper right corner of the diagram (HD129333, whose age is probably only a few percent that of the present Sun). Among stars comparable to the Sun in

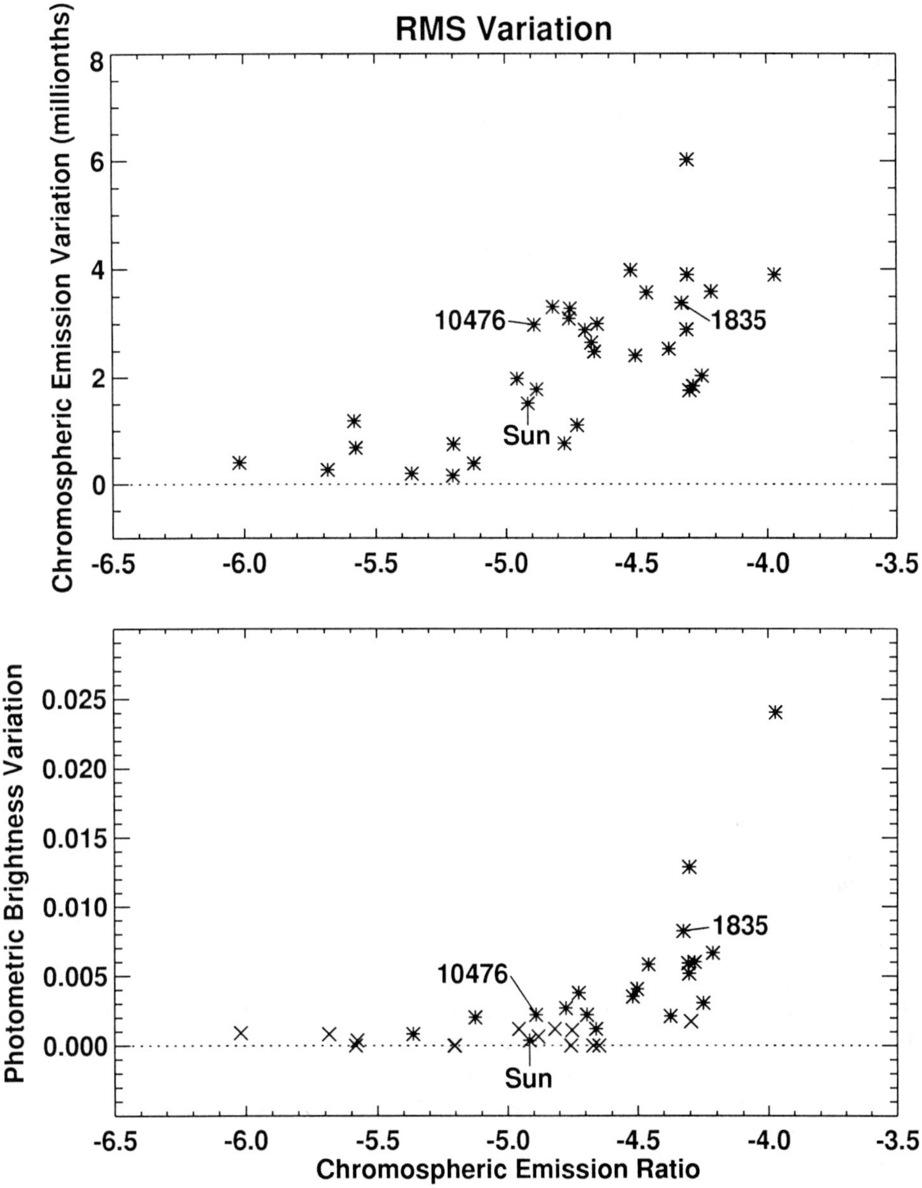

FIGURE 2. Cyclic chromospheric and photometric variability as a function of mean chromospheric activity.

average activity, about half vary, some by as much as about 0.5%. Stars even less active than the Sun are only occasionally found to vary, and at relatively low amplitudes. It is interesting that the Sun appears to be near the bottom of the photometric distribution, even though its amplitude of Ca H+K variation seems normal. The Sun is not the only star for which the amplitudes of photometric and chromospheric variation are not very tightly coupled. In other words, chromospheric variation is not a particularly precise predictor for photometric variability. Likewise, average activity predicts photometric variability to only a factor of two or three.

3. Analysis: calibration issues and other matters

The surprisingly low white light variability shown by the Sun evokes concern that it might be an artifact. This, in turn, has led to a detailed and continuing examination of a long list of selection and calibration issues aimed at reducing this possibility. These tend to involve either (1) relating the solar to the stellar measurements, or (2) identifying and correcting systematic effects in the observations. The flavor of this process can perhaps be appreciated from a few examples.

The two NSO K-indices and the Mount Wilson S-index, all of which measure chromospheric Ca emission, are surprisingly difficult to relate in a satisfactory manner. Except for a handful of measurements made between late 1975 and early 1977, there are no contemporaneous data presently available that directly link the K-indices with the (H+K) S-index, and the available overlapping observations are inadequate to properly constrain even a linear (i.e., two-parameter) regression. In order to sample a wider range of activity, Duncan *et al.* (1991) resorted to high dispersion spectrograms of the K-line region from Lick Observatory for 15 stars from the Mount Wilson HK Program. Seven of these 15 stars have only one or two measurements of the S-index, and none have more than 30. The K-line spectrograms are all single plates. Because the chromospheric emission for several of the 15 stars is known to vary strongly and rapidly, and because none of the observations are simultaneous, it is virtually assured that the corresponding measurements for a good number of these stars sample different activity levels. Juggling difficulties such as these, a recent re-evaluation of the calibration led to the relation: $S = 1.48 * K + 0.041$, with standard errors of 0.07 and 0.013 in the slope and intercept, respectively (the original calibration yielded $S = 1.54 * K + 0.040$, with no error estimates). The bottom line of all this is that the uncertainty in the value for the average chromospheric emission ratio (the abscissa of Figure 2) for the Sun amounts to about 20%.

If anything, the issues relating to the calibration and correction of the photometric observations can be even more vexing (Lockwood 1993). Besides the noise arising from sources such as scintillation, counting statistics, sky transparency fluctuations, and centering and guiding errors, attention must be given to effects such as variability among the photometric reference stars and long-term, secular drifts in the instrumental response.

The variability of the reference stars must be handled on a case by case basis. Fortunately, it is often relatively easy (at least, after a few years have elapsed!) to determine that a particular reference star is not stable, and even to estimate with some accuracy its contribution to the rms dispersions that we have adopted to measure variability. Typically, the contribution to the composite dispersion (calculated from annual mean values) from reference star variability is about 0.0006, and ranges from 0.0003 to 0.0029.

Like reference star variability, secular drifts become apparent only after the fact. For the Lowell photometry, the principal effect appears to be a gradual drift in the color response of the photomultiplier tube that aliases brightness changes whenever there are color differences among the program and reference stars. Because a stable configuration for the Lowell instrumentation has been deliberately maintained, it has been possible to use observations dating back as far as 1973 to trace and calibrate this effect. For the present measurements, its contribution to the composite relative dispersion is typically 0.0002, and ranges from 0.0000 to 0.0010.

Another issue involves the relation between variability measured in a finite bandpass (the Lowell photometry, here the sum of measurements made in Strömgen b and y filters centered at 472 and 551 nm, respectively) and variability in total irradiance (the SMM/ACRIM solar radiometry, which is effectively bolometric). Were the Sun to radiate as a black body, it is not hard to estimate that its Strömgen $b+y$ variation would be 25-30% larger than its bolometric variation. When scaled by this factor, the rms variability of the Sun becomes 0.0004 (on the Lowell system). However, the solar total irradiance is known to be characterized by substantial excess variation in the extreme UV, relative to the amount predicted by the black body model. Because this EUV excess must be counterbalanced elsewhere, the true correction is probably smaller than the factor of 25-30% stipulated by the black body model, but by how much is not clear.

4. Summary

Available time series data now permit meaningful comparisons of the variability of the Sun and its stellar analogues on time scales approximating that of the solar activity cycle. When this is done, some reassuring patterns emerge. Among the more notable of these is the indisputable tendency for solar-age stars, including the Sun, to show cyclic white-light variations that approximately track the corresponding activity fluctuations. However, the amplitude of the Sun's white light variation, as measured during the 1980s, seemed to be significantly less than that of many of its stellar analogues, despite that fact that the amplitude of the Sun's chromospheric emission variation was entirely typical for a star of its type. Although the relation between the broadband observations for the Sun and its stellar analogues is subject to several adjustments which tend to narrow this discrepancy (from perhaps a factor of eight, for the uncorrected data, to about a factor of four), they do not eliminate it. Thus, if this result is accepted at face value, it implies that the coupling between photometric and chromospheric variations is not very tight, raising a warning about the casual use of one variability indicator as a proxy for another.

Acknowledgments. This work was supported by the Air Force Office of Scientific Research under Task 2311G3.

REFERENCES

Duncan, D.K., Vaughan, A.H., Wilson, O.C., Preston, G.W., Frazer, J., Lanning, H., Misch, A., Mueller, J., Soyumer, D., Woodard, L., Baliunas, S.L., Noyes, R.W., Hartmann, L.W., Porter, A., Zwaan, C., Middelkoop, F., Rutten, R.G.M. & Mihalas, D. 1991 Ca II H and K measurements made at Mount Wilson Observatory, 1966–1983. *Astrophys. J. Suppl.* **76**, 383–430.

Foukal, P.V. & Lean, J. 1988 Magnetic modulation of solar luminosity by photospheric activity. *Astrophys. J.* **328** 347–357.

Keil, S.L. 1992 Private communication.

Livingston, W.C. 1992 Private communication.

Lockwood, G.W. 1994 Irradiance Observations of Stars. In *The Sun as a Variable Star: Solar and Stellar Irradiance Variations* (ed. J.M. Pap, C. Fröhlich, H.S. Hudson, & S.K. Solanki). Cambridge University Press, in press.

Lockwood, G.W., Skiff, B.A., Baliunas, S.L. & Radick, R.R. 1992 Long-term solar brightness changes estimated from a survey of Sun-like stars. *Nature* **360**, 653–655.

Radick, R.R. 1991 The luminosity variability of solar-type stars. In *The Sun in Time* (ed. C.P. Sonett, M.S. Giampapa, & M.S. Matthews). pp. 787–808. Univ. of Arizona Press, Tucson, AZ, USA.

Radick, R.R. 1992 Luminosity variability of lower main-sequence stars. In *The Solar Cycle* (ed. K.L. Harvey). Astronomical Society of the Pacific Conference Series **27**, pp. 450–464.

RADICK, R.R., LOCKWOOD, G.W. & BALIUNAS, S.L. 1990 Stellar activity and brightness variations: a glimpse at the Sun's history. *Science* **247**, 39–44.

WILSON, O.C. 1978 Chromospheric variations in main-sequence stars. *Astrophys. J.* **226**, 379–396.

WILLSON, R.C. 1992 Private communication.

Photometric Observations of the Sun

By GARY A. CHAPMAN

San Fernando Observatory, Department of Physics and Astronomy,
California State University, Northridge, CA 91330, USA

Ground-based calorimetry and photometry of the Sun have been carried out for many years. Following the early years, ground-based photometry has largely replaced ground-based calorimetry, in part due to the advent of airborne and spaceborne detector systems for the broad-band measurement of the solar irradiance and the realization of the difficulty of correcting calorimetry measurements for the effects of the terrestrial atmosphere. Even from spacecraft, recent measurements of the total solar irradiance range from about 1367 to 1374 W/m^2. Most of this difference can be ascribed to differences in instrumental scales, while a variation of about 1 to 2 W/m^2 appears to be due to solar variability. The quiet Sun may also change, globally, over longer time scales. Using disparate data to understand solar variability will require cooperation between a number of current groups, supported by various governments, covering several zones of longitude.

1. Introduction

Studying the Sun's output of energy (luminosity) has been a long-standing endeavor. Within the past decade or so, satellites have measured the solar output and have very clearly revealed variability. However, while the satellite measurements have agreed on the actual solar luminosity, agreeing for the most part on the short- and long-term temporal changes. Ground-based measurements, on the other hand, are hampered by the uncertainties in the opacity of the terrestrial atmosphere and its variations and cannot accurately determine the solar luminosity but might be able to measure relative changes in it. This review will attempt to place ground-based efforts into context with space-based experiments. An interesting, broad review of ground-based photometric measurements is that of LaBonte (1988). A general review of solar luminosity variability can be found in (Hudson 1988), Fröhlich *et al.* (1991) and Fröhlich (1992). We will not review non-imaged programs here. For a good review of these see White (1988) and Livingston (1992).

One of the reasons for carrying out ground-based programs of relative photometry is to help understand the cause of the solar variations seen by satellites. Most of the satellite data have been obtained without any spatial resolution on the solar disk. By having photometric images obtained at selected wavelengths, the satellite measurements can be modeled using selected contrast features seen in ground-based images. We assume that there is a non-changing quiet Sun output and that photometrically detectable features add to or subtract from this quiet Sun output. For relative ground-based photometry, the (assumed) quiet Sun is our photometric standard. If this assumption is not true, we can only find out by doing our very best in removing the effects of features on the solar output.

It has been shown that most of the variations can be attributed to magnetic features (Foukal & Lean 1988; Chapman *et al.* 1992). By modeling their effects on the solar output, we hope to gain a better understanding of the flow of energy in and around a solar active region and of the solar dynamo itself. A few questions still unanswered are: (1) where does the thermal energy go that is diverted by sunspots, (2) how much of this energy, if any, is stored in the convection zone, (3) what fraction of a sunspot's missing energy is radiated by faculae and network and over what time scale?

2. Past and recent work

2.1. *The effect of sunspots on solar irradiance*

Attempts in the 1980's to understand the cause of variations in total irradiance measurements relied on synoptic observations of sunspot area and positions. An early example is that of Willson *et al.* (1981) in which about half of the variance could be removed from the ACRIM I signal (Hudson 1988, and references therein). The sunspot effect was modeled by a Photometric Sunspot Index (PSI) in which it was assumed that the temperature structure of all sunspots is the same and that their contrast does not depend on position on the disk. The quiet Sun limb darkening was assumed to be that of the Eddington gray atmosphere. A similar index was derived in Foukal (1981) (see also Hudson *et al.* 1982). Because of its widespread use we give the PSI (Ψ):

$$PSI = \alpha \sum_i A_s^i \mu_i (3\mu_i + 2)/2 \tag{2.1}$$

where the sum is taken over all sunspots, A_s is the corrected sunspot area in micro-hemispheres and μ_i is the cosine of the heliocentric angle for the i'th sunspot. The value of α was found to be 0.33, but more about that later. This simple form has no free parameters. On the other hand, it may be overly simple. Although several authors quote a mean umbral to spot area of 0.17, this ratio can vary over a wide range. A more detailed model is the sunspot blocking of Hoyt & Eddy (1982) where they no longer assume the Eddington gray limb darkening law and they include the umbral and penumbral areas separately. Donnelly *et al.* (1992) found for the period 1980-1982 that the PSI and the Hoyt-Eddy sunspot blocking function were correlated with a correlation coefficient of 0.92. However, the accuracy of this source of umbral areas is not known. Lean & Foukal (1990) found a linear regression between their sunspot blocking, P_s, (identical to 2.1) and the Hoyt-Eddy relation with a correlation coefficient of 0.997 but where the P_s values were systematically higher by about 20%. Also, the contrast of individual sunspots can be quite variable (Brandt *et al.* 1990; Steinegger *et al.* 1990; Beck & Chapman 1993). It has been known for some time that individual sunspots vary and that umbrae can have darker or brighter parts.

2.2. *The effect of faculae on solar irradiance*

Bright features are associated with magnetic fields. However, the intensity-magnetic field relationship is more complicated than for sunspots. Common terms are faculae, network (active and quiet), plage, and filigree. Faculae have long been seen in white light beyond a radius on the solar disk of about 0.7 to 0.8 ($\mu \approx 2/3$). The network, which at high resolution has the appearance of filigree, is seen most readily using narrow-band filters or spectroheliographs tuned into the core or near-wings of absorption lines. We understand that these features are connected to flux tubes or collections of flux tubes and that their appearance is determined partly by their magnetic flux distribution and partly by their position on the solar disk Wilson (1981), Zwaan (1987), and references therein. We will refer to these non-spot magnetic features as faculae or flux tubes for the time being. As with sunspots, a simple model has been developed to describe the irradiance fluctuation due to faculae, called the Photometric Facular Index (PFI) (Foukal 1981; Hirayama *et al.* 1984). As with the PSI, the quiet Sun limb darkening is assumed gray. We show here the simple expression of Chapman *et al.* (1992):

$$PFI = C_p \sum_i A_p^i \mu_i (3\mu_i + 2)(\frac{1}{\mu_i} - a) \tag{2.2}$$

where A_p^i refers to the corrected area in millionths of the solar hemisphere of an active region plage as seen on spectroheliograms obtained in the central Åor so of the Ca II K-line. The parameters C_p and a are empirically determined. The parameter a determines whether the faculae are dark, bright, or invisible at disk center. An early attempt to combine parameters for sunspots and faculae by Sofia *et al.* (1982) used an assumed ratio of facular to sunspot areas rather than relying on published areas of Ca II plage.

Plage areas may be subject to significant measurement uncertainties. When the McMath-Hulburt Observatory closed, the Ca II plage observations, which were used to determine the published Ca II plage area, were obtained from Sacramento Peak Observatory, Mt. Wilson Observatory, and then Big Bear Solar Observatory. During this time interval there was a change in the plage brightness index which was anticorrelated with plage area. The plage brightness increased by about 50% during this time interval (Schatten *et al.* 1985).

An example of the use of sunspot and plage data in modeling solar irradiance variations over long time periods is shown by Foukal & Lean (1988). The sunspot irradiance deficits were derived from a sunspot "blocking" function virtually the same as 2.1. The facular irradiance excess was derived from a model that used full disk images obtained in the HeI 1083 nm line. HeI may be a more objective measure of magnetic activity than Ca II plage areas, which were often estimated by eye.

In Chapman & Meyer (1986), C_p was found to be 0.019 when a was assumed to be 0.0. However, Vršnak *et al.* (1991) found that C_p correlates with the plage intensity index ($r > 0.67$) and has a mean value of about 0.0166 ± 0.01. This is fairly close to the value of 0.0185 mentioned in Chapman (1987) and probably well within the limits of accuracy for plage areas themselves.

One problem with using plage areas to define a facular index is that the Ca II plage covers a larger area than the faculae because of fringing of the magnetic field lines in the chromosphere and more extensive heating. Another problem is that the network elements that lie outside the boundaries determined by the Ca II plage are not included. If we are attempting to model satellite measurements in the UV we may need a three-component model (Lean *et al.* 1982). Pap *et al.* (1992) studied the variation in the ACRIM I signal, smoothed with a 27-day running mean and fitted with a two-component model. The irradiance deficits were based on sunspot blocking and the excesses were based on the Ca II plages. There was a significant discrepancy in 1980 and at other times suggesting that two parameters may be insufficient for a good fit to total irradiance data.

Ground-based, full-disk photometry can model the variations in total irradiance from satellites fairly well. For example, in a recent study (Chapman *et al.* 1992) fluctuations in the ACRIM I data were fit to about 100 ppm over a 21-day interval using sunspot and facular terms based on 2.1 and 2.2. In this study, the quantity A_p represented elements of faculae rather than plage. The facular elements were defined by a 10 Å bandpass filter in the center of the Ca II K-line as detected by the Cartesian Full Disk Telescope (CFDT1) at the SFO, using a pixel size of $5'' \times 5''$. Each detected facular pixel had a contrast at disk center of 4.8% increasing with $1/\mu$ (in order to compensate for selection effects caused by the increasing contrast towards the limb). The values of C_p and a from this study were found to be 0.030 and 1.1, respectively, where $a = 1.1$ means that the faculae are dark at disk center, having a contrast of -0.93%. (This value is not inconsistent with values mentioned by Topka *et al.* (1992) using very high spatial resolution images.) Sunspots were determined from pixels having a contrast of -8.5% in photometric images,

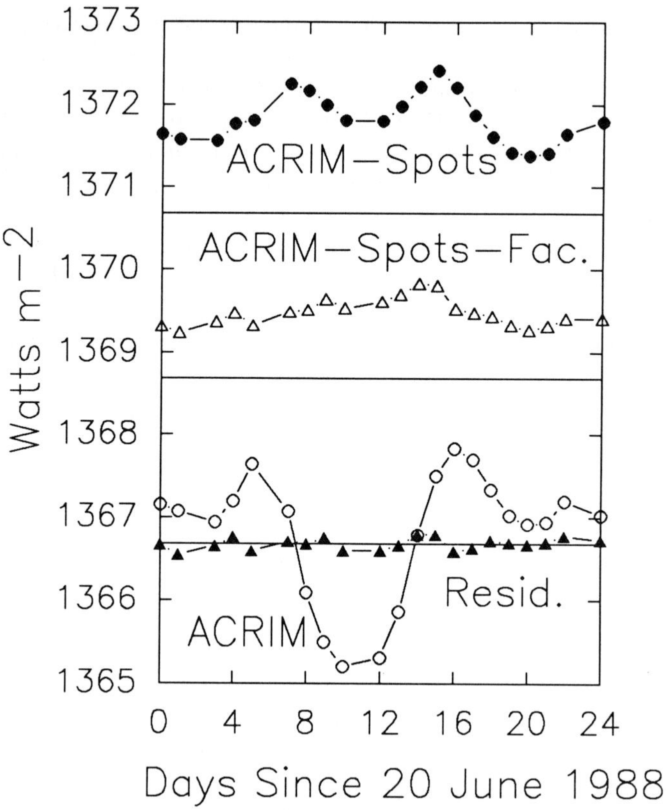

FIGURE 1. Plot of the ACRIM I signal, ACRIM I minus modeled spot contribution, ACRIM I minus both modeled spot and facular contributions, and residuals from fitting spot, facular, and NOAA-9 (SBUV/2) contributions to ACRIM I. The ACRIM I minus spot and faculae has been shifted upward by 2 W/m^2, and the ACRIM I minus spots by 4 W/m^2, for clarity (from Chapman *et al.* 1992).

(6723 Å). A photometric deficit (DEF) was calculated according to

$$DEF = \sum_i \frac{\Delta I_i}{I} \phi(\mu_i) \tag{2.3}$$

where $\Delta I_i/I$ is the the contrast of a pixel, and ϕ is the quiet Sun limb darkening at that same value of μ. During this brief interval in 1988 from 20 June to 14 July, for 21 days the ACRIM I total irradiance could be fit to the sunspot deficit and a modeled facular excess with a coefficient of multiple determination, $R^2 = 0.974$ (see Figure 1). For 109 days of data from June through December in 1988, the total irradiance from Nimbus-7 was fit to sunspot deficits and the same facular excess model as above. The value of R^2 was 0.9018, with a standard deviation of approximately 140 ppm (Chapman *et al.* 1994a). Use of the Mg II core-wing ratio from NOAA-9 (Donnelly 1992) to account for faint network elements, improved the correlations in both of these studies by about 0.01 in the value of R^2.

Other full-disk photometric measurements have been recently reported by Nishikawa (1990a, 1990b), Nishikawa & Hirayama (1990), and Lin & Kuhn (1992). Nishikawa

(1990b) finds that the active region network has a contribution to fluctuations of the total irradiance that is approximately half that of active region faculae.

Other photometric observations of the full solar disk, restricted to the limb regions, can also be fit to variations in satellite measurements of the total irradiance. Kroll *et al.* (1990) and Kroll & Hill (1991) used the Santa Catalina Laboratory for Experimental Relativity by Astrometry (SCLERA) telescope to fit changes in limb darkening, associated with solar activity, to variations in the solar irradiance. Kuhn & Libbrecht (1985) and Kuhn *et al.* (1988) have fit measurements of latitudinal fluctuations in photospheric temperature near the solar limb, obtained from the Princeton Distortion Telescope (PDT), to variations in the ACRIM I total irradiance. Kuhn *et al.* (1988) have been able to obtain fits to total irradiance measurements from satellites with high correlations. Kroll *et al.* (1990), on the other hand, have found evidence for a change in the limb darkening that is not like the solar cycle. Observations of just the limb may make the effects of active regions across the solar disk difficult to identify due to the time lag for regions to travel from central parts of the solar disk to the limbs.

The contrast of flux tubes and faculae as a function of center-to-limb distance is poorly known. Even the sign of the contrast with μ is in some dispute (Chapman & Klabunde 1982; Libbrecht & Kuhn 1984, 1985; Wang & Zirin 1987; Lawrence & Chapman 1988; Foukal *et al.* 1993; Chapman & Zielger 1993). Figure 2 shows several different measurements of the center-to-limb variation (CLV) in the contrast of faculae in the mid-visible. The different amplitudes can be explained partly by the differing spatial resolution of the observations. However, part of the discrepancy may have to do with different definitions of contrast. Clearly, the CLV of faculae needs more observational work, preferably photometric, because of the ability to digitize data at high speed and sum them in real time in order to overcome the low signal to noise ratio and the large intensity gradient at the limb. We must be careful to remember that whatever the contrast of flux tubes is in the visible, their contrasts in the UV and IR are quite different (Bonnet *et al.* 1982; Foukal *et al.* 1990, 1993).

For modeling total solar irradiance changes, a central question concerns the bolometric contrast of faculae. The contrast of flux tubes in the mid-visible may systematically vary according to magnetic flux (cf. Stenflo & Harvey 1985). There is evidence that smaller magnetic elements have a positive contrast near disk center, whereas larger magnetic elements have zero or negative contrast near disk center (Lawrence *et al.* 1991, 1993; Topka *et al.* 1992). Some of these features are reported to have contrasts of about 30% when viewed under good seeing and when corrected for losses in contrast due to diffraction effects (Muller & Keil 1983; Solanki & Brigljević 1992).

These features would all have strong positive contrasts if viewed with a spectroheliograph (Lawrence *et al.* 1991), in an absorption line with a narrow filter (Topka *et al.* 1992), or in the 1600 Å region corresponding to the temperature minimum (Bonnet *et al* 1982). What is needed is the bolometric contrast as a function of disk position and magnetic flux. For example, the UV spectral irradiance of non-spot, magnetic flux elements can be important, perhaps contributing as much as one-fifth to one-third of the long-term variations in the total irradiance variation (Lean 1989; London *et al.* 1989).

3. Recent work

3.1. *Sunspots*

Several recent studies have shown that the average contrast of sunspots is not constant (Brandt *et al.* 1990; Steinegger *et al.* 1990; Beck & Chapman 1993; Chapman *et al.*

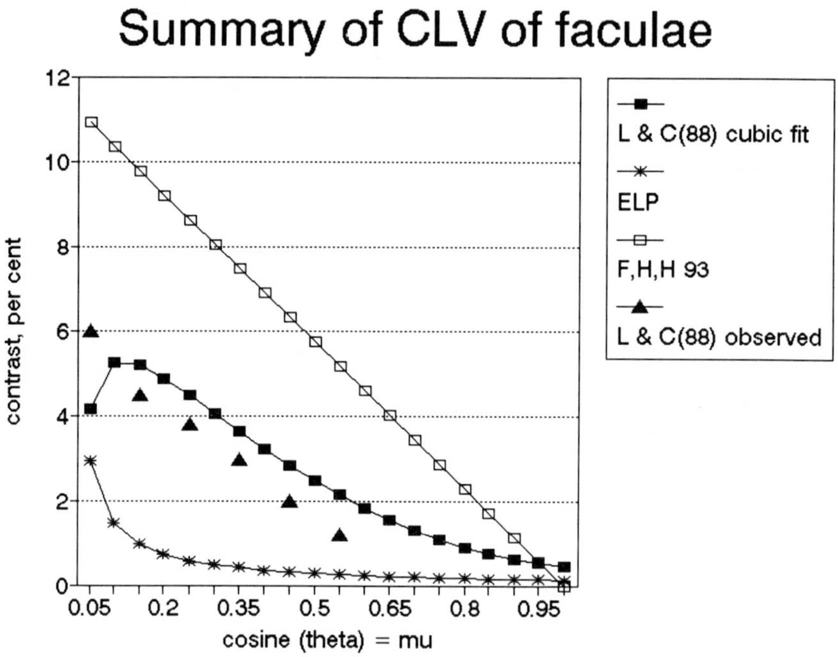

FIGURE 2. Facular contrast from several sources. The filled squares are from a cubic fit to S-series diode array observations (Lawrence & Chapman 1988), the asterisks are from ELP observations (Chapman & Ziegler 1993), the open squares are from observations at the NSO (Foukal *et al.* 1993), and the filled pyramids are based on S-series diode array observations (Lawrence & Chapman 1988).

1994b). First, there is evidence that the value of α used in 2.1 is too large; the mean value should be near 0.29 rather than 0.33 or 0.35. There are two points involved here. First, the observed contrast of a sunspot must be properly corrected for stray light. Second, the mean bolometric contrast will determine the correct contribution to the fluctuation in the total irradiance. For these purposes, it is not particularly useful to measure the darkest part of the umbra as has often been done in the past. Furthermore, without imaging the sunspot it is difficult to know much about the distribution of intensity within the umbra. What is needed is a properly weighted, bolometric, mean sunspot contrast.

From an inspection of 2.1, we can see that α is the mean bolometric contrast of a sunspot, which can be determined from the effective temperature of the entire sunspot, using the Stefan-Boltzmann law. Recent high resolution measurements of sunspots have been made by Brandt *et al.* (1990) and Steinegger *et al.* (1990). They showed a wide variation in α between different sunspot groups, from about 0.15 to approximately 0.4. Furthermore, they showed that α is correlated with the area of the sunspot group. For 130 spots and spot groups, they found that $\alpha = 0.2231 + 0.0244 * \log(A_s)$. The mean for all groups was about 0.27 (Brandt *et al.* 1992). Beck & Chapman (1993) found that $\alpha = 0.220 \pm 0.016 + (0.340 \pm 0.058) * A_u/A_p$ with a correlation coefficient of 0.602 ($p < 0.005$), where A_u and A_p are the umbral and penumbral areas, respectively. The mean value of α for 63 sunspot groups was 0.304.

Another important question concerning α is whether or not it is independent of position

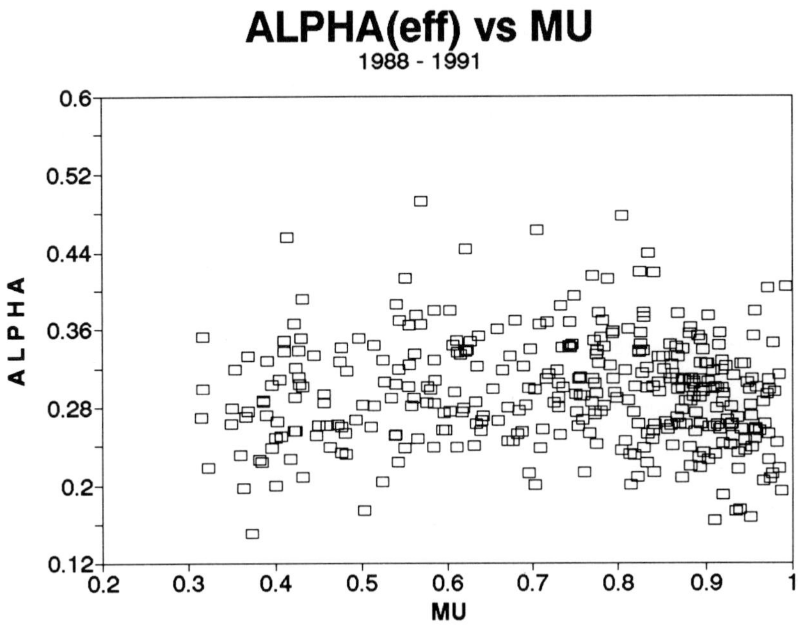

FIGURE 3. Bolometric contrast, α_{eff}, versus μ for 377 observations of sunspot groups from 1988 through 1992 (from Chapman *et al.* 1994b).

on the disk. From a study of 377 measurements of sunspot groups at SFO, during their disk transit, an effective α (α_{eff}) has been determined (Chapman *et al.* 1994b). The value of α_{eff} is given by

$$\alpha_{eff} = DEF/2\Psi \tag{3.4}$$

where DEF (2.3) is the photometrically measured sunspot deficit and Ψ is given by 2.1 with the sunspot areas being determined from the same pixels from the same photometric full disk images as for the quantity DEF. The wavelength of these images is 6732 Å. They found that $\alpha_{eff} = 0.218 \pm 0.018 + (0.683 \pm 0.028)A_u/A_s$ with $r^2 = 0.878 (N = 86)$. The mean value of α_{eff} for these measurements is 0.293, very close to values of α from other studies, mentioned above. The value of α_{eff} appears to be relatively free from the effects of stray light; its variation shows no correlation with sky transparency measured at 530 nm ($r^2 = 0.00065$). Finally, Chapman *et al.* (1994b) found no dependence of α_{eff} on μ, the cosine of the heliocentric angle ($r = 0.0005; N = 377$). Figure 3 shows the quantity α_{eff} plotted as a function of μ.

3.2. *Non-sunspot magnetic features*

3.2.1. *Brightness related to magnetic flux*

As we are finding, non-spot magnetic features such as faculae or photospheric network can be bright or dark, depending on their location on the disk, their wavelength, and their magnetic flux. All non-spot magnetic features are bright when seen at a wavelength corresponding to the upper photosphere. The wings of the Ca II K line are formed there and are very useful in mapping out the location of non-spot magnetic features. However, knowing the bolometric contrast requires knowing something about the magnetic flux in each feature. This in turn requires high spatial resolution, and a magnetogram signal that is linear with magnetic flux. For example, if two magnetic features of equal flux but opposite polarity are not spatially resolved their flux will cancel in the magnetogram but their positive contrast in the wing of the K line will not cancel.

An interesting way of modeling irradiance variations is to convert the non-spot magnetic signal in a magnetogram into an irradiance fluctuation. To the extent that irradiance variations are caused by magnetic features (and there is yet no convincing evidence that this is not the case) we want to detect only magnetic features. Furthermore, since a magnetogram is a differential signal, the limb darkening is removed. Such an analysis was carried out with Mt. Wilson magnetograms (Chapman & Boyden 1986) and should be carried out with KPNO magnetograms. Harvey (1992) has used the total magnetic flux from KPNO magnetograms as an indicator of the level of solar activity in comparison to other solar measurements. Data from the newly installed Spectro-Magnetograph (Jones *et al.* 1992) should be even more useful for irradiance studies because it will have full disk information on both the continuum and the line core.

Detailed analyses on the behavior of flux tubes in the continuum as a function of magnetic flux and disk position have been carried out by a group using Lockheed's Solar Optical Universal Polarimeter (SOUP) in the Canary Islands (Topka *et al.* 1992) and a group using the Video Spectra-Spectroheliograph (VSSHG) at the San Fernando Observatory (Lawrence *et al.* 1991, 1993).

Title *et al.* (1992), from filtergrams having a spatial resolution approaching $0.35''$, find that all bright continuum features have a magnetic signal, but that the reverse is not true, some magnetic features do not have associated bright points either in the continuum or in the line center of Ni I, at 6768 Å. Topka *et al.* (1992), from more extensive data using SOUP for four active regions at various locations on the solar disk, found that the continuum contrast for all magnetic fluxes was negative at $\theta = 5°$, but became zero at 20° for a magnetic signal of about 400 Gauss.

Lawrence *et al.* (1991) using data from the VSSHG, show much the same results, but due to the lower spatial resolution the contrasts of the flux tubes are lower and the magnetic flux where the maximum positive continuum contrasts are seen, away from disk center, are 180 Gauss compared to about 400 Gauss for the SOUP data. Lawrence *et al.* (1991) calculated the irradiance effects of collections of magnetic flux tubes, based on their contrast and their limb position, taking into account the weighting of the quiet Sun. Depending on the method of defining the quiet Sun, they found an irradiance variation of from 0.3 ± 0.1 ppm to about 2.1 ± 0.7 ppm near disk center ($0.79 < \mu < 0.97$). If the zero contrast is defined by the line core contrast being less than 4% then the average quiet Sun background is shifted an average of $(-0.06 \pm 0.02)\%$ relative to defining the zero contrast by those pixels having a magnetic flux less than 80 Gauss. For low and high spatial resolution, determining the quiet Sun intensity is a difficult problem.

3.2.2. *Synoptic measurements of bright features*

While it is desirable to have very high resolution magnetic and brightness maps of the Sun at several wavelengths, this not yet generally done. We do have the daily NSO/KPNO spectra-magnetograph program at 6302.5 Å with $1''$ pixels. However, the core of 6302.5 does not sample as high in the solar atmosphere as the wings of the Ca II K-line. On the other hand, the several synoptic programs that obtain full disk Ca II K-line images do not obtain full-disk magnetograms with a possible exception being the NSO/KPNO. Presently, the NSO/SPO Ca II K-line images are obtained photographically with very high spectral resolution. Of particular interest are images obtained in the wing of the K-line, between 1 and 3 Å from line center. At this spectral position, the complicating effects of the chromosphere are mostly avoided. At San Fernando Observatory (SFO), K-line images are obtained with filters having a 10 Å and 3 Å bandpass centered on the K-line. The effective passband is split into two passbands on either side of the center of the K-line in the region 1 to 4 Å from line center. Other filter-defined monitors are at Big Bear Solar Observatory (BBSO), the National Solar Observatory's Kitt Peak National Observatory (NSO/KPNO), and Mees Solar Observatory (MSO). These last three use two-dimensional CCD detectors, whereas at SFO linear diode arrays are used. At BBSO the filter has a bandpass of 1.2 Å and a pixel size of 4″ (Marquette 1992). The NSO/KPNO filter has a bandpass of 1 nm and a pixel size of about 2.4″ (Harvey 1994). The MSO has two K-line systems with one designed for daily full disk images. It has a bandpass of 0.3 Å and a pixel size of 5″ × 5″ (Labonte, private communication). At SFO, beginning in July 1992, another photometric telescope (CFDT2) with a pixel size of 2.5″ X 2.5″ went into operation. The new telescope, in addition to other filters, has a 10 Å and a 3 Å filter at the K-line. The Freiburg group is beginning a study of faculae as well as carrying on with work on sunspots. Details of their, and other, photometric programs are lacking at this time. Apparently a new program of full disk photometry at several wavelengths is being started at BBSO.

A simple global quantity is the irradiance excess of bright features obtained by adding up the contrast of each bright pixel weighted by the quiet Sun limb darkening. Changes in this quantity from day to day can then be compared with changes in spectral irradiance from satellite measurements. For 125 days during the interval from 3 October 1991 to 1 April 1992, the SFO 10 Å K-line spectral irradiance excess data (5″ pixels) are highly correlated with the ratio of the SOLSTICE core-to-wing intensity of the Mg II h and k line (R_{MgII}, White, private communication). In a linear regression (Wiener 1994) the following relation was found; for the SOLSTICE core-to-wing ratio $R_{MgII} = 0.1537 \pm 0.0021 + (3.98 \pm 0.16) \times 10^{-6}$ times the CFDT1 10 Å K-line irradiance in ppm relative to the quiet Sun ($r^2 = 0.842$). The facular area from this instrument, for 27 January 1991, interpolated using data from 25 to 29 January 1991, was 44,000 micro-hemispheres, which was 17,000 micro-hemispheres larger (about 60%) than the area determined by BBSO from their Ca II plage data for the same date (Marquette 1992). Considering the different criteria for a "plage element", the agreement is not bad.

4. Future work

4.1. *Improvements in spatial and temporal resolution*

With improved temporal coverage of the solar irradiance by spacecraft, it will become more important to increase the tempo of ground-based measurements by having more stations around the world. Improvements in temporal resolution are needed for those times when significant changes in the solar irradiance occur over a short time span. How much of this change is due to sunspots growing or shrinking, how much is due to "bright" magnetic features growing or shrinking, and how much is due to the changing aspect of these features? An examination of the ingress of two sunspot groups around 27 June 1988 show a change of about 1 W/m^2 (about 730 ppm) per day in the ACRIM I irradiance. Such a rapid change implies that in order to model total irradiance data from a precision satellite experiment with a precision of order 100 ppm, ground-based observations should be taken with intervals of not more than a few hours. This will require irradiance monitors at a number of longitudes, similar to those of the GONG.

The spatial resolution needed to better understand satellite measurements, or to independently determine solar irradiance variations, is not clear. At SFO, measurements of the facular irradiance excess in the K-line with a 10 Å bandpass filter are in very close agreement whether the pixel size is 5″ or 2.5″. On the other hand, the area and irradiance fluctuations seen with the 3 Å filter and 2.5″ pixels are about twice those obtained with the 10 Å filter.

Presumably, with smaller pixel size and active stabilization the contrast with a 10 Å filter will be higher and the detected number of bright pixels will also go up. A narrower bandpass should help too. These are things to be considered in the SunRISE program, particularly the Precision Solar Photometric Telescopes (PSPT's) undergoing design at the NSO (Foukal *et al.* 1990; Kuhn 1994).

5. Summary

This is an exciting time for ground-based photometric observations of the Sun. It is important, over the next decade, to sort out solar from terrestrial effects on possible climate changes. Yet there are fundamental problems in understanding the cause of solar variability. Their solution will require, among other things, the acquisition of high precision photometric images across the visible spectrum. The solution to these problems will require high quality data and careful data analysis. The most useful data will be from full disk photometric instruments and instruments with polarimetric capabilities, in order to sort out the influence of magnetic flux. Progress is being made, in spite of great difficulties. In closing, it is argued that ground-based photometry, using photoeletric detector systems, can make a significant contribution to our understanding of variations in the solar output. This is a challenging task for the solar community.

Acknowledgments. I appreciate direct and indirect help in preparing this review, and in some of the research presented here, from the following: A. M. Cookson, J. J. Dobias, J. Lawrence, C. Mach, and S. R. Walton, and financial support from NASA Grant NAGW-3017 and NSF Grant ATM-9115111.

REFERENCES

BRANDT, P. N., SCHMIDT, W. & STEINEGGER, M. 1990 On the area ratio of sunspots. *Solar Phys. (Lett.)* **129**, 191–194.

BRANDT, P. N., SCHMIDT, W. & STEINEGGER, M. 1992 Photometry of sunspots observed at Tenerife. In *Proceedings of the Workshop on the Solar Electromagnetic Radiation Study for Solar Cycle 22* (ed. R. F. Donnelly). pp. 130–141. NOAA ERL SEL, Boulder, CO, USA.

BECK, J. G. & CHAPMAN, G. A. 1993 A study of the contrast of sunspots from photometric images. *Solar Phys.* **146**, 49–60.

BONNET, R. M., BRUNER, M., ACTON, L. W., BROWN, W. A., DECAUDIN, M. & FOING, B. 1982 Rocket photographs of fine structure and wave patterns in the solar temperature minimum. *Astron. Astrophys.* **111**, 125–129.

BRUNING, D. H. & LABONTE, B. J. 1983 Interpretation of solar irradiance variations using ground-based observations. *Ap. J.* **271**, 853–858.

CHAPMAN, G. A. & KLABUNDE, D. P. 1982 Measurements of the limb darkening of faculae near the solar limb. *Ap. J.* **261**, 387–395.

CHAPMAN, G. A., HERZOG, A. D. & LAWRENCE, J. K. 1986 Time-integrated energy budget of a solar activity complex. *Nature* **319**, 654–655.

CHAPMAN, G. A. & BOYDEN J. E. 1986 Solar irradiance variations derived from magnetograms. *Ap. J.* **302**, L71–L73.

CHAPMAN, G. A. & MEYER, A. D. 1986 Solar irradiance variations from photometry of active regions. *Solar Phys.* **103**, 21–31.

CHAPMAN, G. A. 1987 Variations of solar irradiance due to magnetic activity. *Ann. Revs. Astron. Astrophys.* **25**, 633–667.

CHAPMAN, G. A., HERZOG, A. D., LAWRENCE, J. K., WALTON, S. R., HUDSON, H. S. & FISHER, B. M. 1992 Precise ground-based solar photometry and variations of total irradiance. *J. Geophys. Res.* **97**, 8,211–8,219.

CHAPMAN, G. A., COOKSON, A. M. & HOYT, D. V. 1994a Solar irradiance from Nimbus-7 compared with ground-based photometry. *Solar Phys.*, in press.

CHAPMAN, G. A., COOKSON, A. M. & DOBIAS, J. J. 1994b Observations of changes in the bolometric contrast of sunspots. *Ap. J.*, submitted.

CHAPMAN, G. A. & ZIEGLER, B. 1993 Analysis of 1983 observations with an extreme limb photometer. *Solar Phys.*, submitted.

DONNELLY, R. F., HUDSON, H. S., PAP, J. & WILLSON, R. C. 1992 Estimating solar chromospheric UV fluxes from sunspot and solar radio data. In *Proceedings of the Workshop on Electromagnetic Radiation Study for Solar Cycle 22* (ed. R.F. Donnelly). pp. 275–285. NOAA ERL SEL, Boulder, CO, USA.

DONNELLY, R. F., ED. 1992 *Proceedings of the Workshop on Electromagnetic Radiation Study for Solar Cycle 22* NOAA ERL SEL, Boulder, CO, USA.

FOUKAL, P. V. 1981. Sunspots and changes in the global output of the Sun. In *The Physics of Sunspots* (ed. L.E. Cram & J.E. Thomas). pp. 391–423. Sacramento Peak Observatory, Sunspot, NM, USA.

FOUKAL. P. & LEAN, J. 1986 The influence of faculae on total solar irradiance and luminosity. *Ap. J.* **302**, 826–835.

FOUKAL, P. & LEAN, J. 1988 Magnetic modulation of solar luminosity by photospheric activity. *Ap. J.* **328**, 347–357.

FOUKAL, P. V., BRUECKNER, G. E., HUDSON, H. S., LIVINGSTON, W. C., NOYES, R. W. & ROBLE, R. G. 1990 *RISE, Radiative Inputs of the Sun to Earth: A Research Plan for the 1990s on Solar Irradiance Variations.* Prepared for The Division of Atmospheric Sciences, National Science Foundation.

FOUKAL, P., LITTLE, R., GRAVES, J., RABIN, D. & LYNCH, D. 1990 Infrared imaging of faculae at the deepest photospheric layers. *Ap. J.* **353**, 712–715.

FOUKAL, P., HARVEY, K. & HILL, F. 1993 Do changes in the photospheric magnetic network cause the 11-year variation of total solar irradiance? *Ap. J. (Lett.)*, submitted.

FRÖHLICH, C. & PAP, J. 1989 Multi-spectral analysis of total solar irradiance variations. *Astron. Astrophys.* **220**, 272–280.

FRÖHLICH, C., FOUKAL, P. V., HICKEY, J. R., HUDSON, H. S. & WILLSON, R. C. 1991 Solar irradiance variability from modern measurements. In *The Sun in Time* (ed. C.P. Sonett, M.S. Giampapa & M.S. Matthews). pp. 11–29. Univ. Arizona Press, Tucson, AZ, USA.

FRÖHLICH, C. 1992 Solar irradiance variability. In *Proceedings of the Workshop on Electromagnetic Radiation Study for Solar Cycle 22* (ed R.F. Donnelley). pp. 1–10. NOAA ERL SEL, Boulder, CO, USA.

HARVEY, K. L. 1992 Measurements of solar magnetic fields as an indicator of solar activity evolution. In *Proceedings of the Workshop on the Solar Electromagnetic Radiation Study for Solar Cycle 22* (ed. R. F. Donnelly). pp. 113–129. NOAA ERL SEL, Boulder, CO, USA.

HARVEY, K. 1994 Irradiance models based on solar magnetic fields. In *The Sun as a Variable Star: Solar and Stellar Irradiance Variations* (ed. J.M. Pap, C. Fröhlich, H.S. Hudson & S.K. Solanki). Cambridge Univ. Press, in press.

HIRAYAMA. T., OKAMOTO, T. & HUDSON, H. S. 1984 Facular limb darkening functions for irradiance modeling. In *Solar Irradiance Variations on Active Region Time Scales* (ed. B. J. LaBonte, G. A. Chapman, H. S. Hudson & R. C. Willson). NASA CP-2310, 59–71.

HOYT, D. V. & EDDY, J. A. 1982 *An Atlas of Variations in the Solar Constant caused by Sunspot Blocking and Facular Emissions from 1874 to 1981.* NCAR Tech. Note TN-194, Boulder, CO, USA.

HUDSON, H. S. 1988 Observed variability of the solar luminosity. *Ann. Revs. Astron. Astrophys.* **26**, 473–507.

HUDSON, H. S. SILVA, S., WOODARD, M. & WILLSON, R. C. 1982 The effects of sunspots on solar irradiance. *Solar Phys.* **76**, 211–219.

Jones, H. P., Duvall, Jr., T. L., Harvey, J. W., Mahaffey, C. T., Schwitters, J. D. & Simmons, J. E. 1992 The NASA/NSO spectromagnetograph. *Solar Phys.* **139**, 211–232.

Kroll, R. J., Hill, H. A. & Beardsley, B. J. 1990 A correlation between changes in solar luminosity and differential radius measurements. In *Climate Impact of Solar Variability* (ed. K. H. Schatten & A. Arking). NASA CP-3086, pp. 65–72.

Kroll, R. J. & Hill, H. A. 1991 The SCLERA program for monitoring solar limb darkening and diameter. Poster paper presented at the *NSF/NCAR Workshop on Solar-Terrestrial Impacts on Global Change*, Boulder, CO, USA, May 8-10, 1991.

Kuhn, J. R., Libbrecht, K. G. & Dicke, R. H. 1988 The surface temperature of the Sun and changes in the solar constant. *Science* **242**, 908–911.

Kuhn, J. R. & Libbrecht, K. G. 1985 Observations of a solar latitude dependent limb brightness variation. *Ap. J.* **290**, 758–764.

Kuhn, J. R 1994 Brightness observations of the Sun. In *The Sun as a Variable Star: Solar and Stellar Irradiance Observations* (ed. J.M. Pap, C. Fröhlich, H.S. Hudson & S.K. Solanki). Cambridge Univ. Press, in press.

LaBonte, B. J. 1988 Ground-based photometric measurements. In *Solar Radiative Output Variations* (ed. P. V. Foukal). pp. 156–175. NCAR, Boulder, CO, USA.

Lawrence, J. K. & Chapman, G. A. 1988 Photometric observations of facular contrasts near the solar limb. *Ap. J.* **335**, 996–1004.

Lawrence, J. K., Chapman, G. A. & Walton, S. R. 1991 Weak magnetic fields and solar irradiance variations. *Ap. J.* **375**, 771–774.

Lawrence, J. K., Topka, K. P. & Jones, H. P. 1993 Contrast of faculae near the disk center and solar variability. em J. Geophys. Res. **98**, 18,911–18,918.

Lean, J. 1988 Empirical modeling of UV and EUV flux variations. In *Solar Radiative Output Variations* (ed. P. Foukal). pp. 113–124. NCAR, Boulder, CO, USA.

Lean, J. & Foukal, P. 1988 A model of solar luminosity modulation by magnetic activity between 1954 and 1984. *Science* **240**, 906–908.

Lean, J. & Foukal, P. 1990 Solar cycle modulation of total irradiance: an empicrical model from 1874 to 1988. In *Climate Impact of Solar Variability* (ed. K. H. Schatten & A. Arking). NASA-3086, pp. 197–204.

Lean, J. L., White, O. R., Livingston, W. C., Heath, D. F., Donnelly, R. F. & Skumanich, A. 1982 A three-component model of the variability of the solar ultraviolet flux. *J. Geophys. Res.* **87**, 10,307–10,317.

Lean, J. 1989 Contribution of ultraviolet irradiance variations to changes in the Sun's total irradiance. *Science* **244**, 197–200.

Libbrecht, K. G. & Kuhn, J. R. 1984 A new measurement of the facular contrast near the solar limb. *Ap. J.* **277**, 889–896.

Libbrecht, K. G. & Kuhn, J. R. 1985 On the facular contrast near the solar limb. *Ap. J.* **299**, 1047–1050.

Lin, H. & Kuhn, J.R. 1992 Precision IR and visible solar photometry, I. *Solar Phys.* **141**, 1–26.

Livingston, W. C. 1992 Observations of solar spectral irradiance variations at visible wavelengths. In *Proceedings of the Workshop on Solar Electromagnetic Radiation Study for Solar Cycle 22* (ed. R.F. Donnelly). pp. 11–19. NOAA ERL SEL, Boulder, CO, USA.

London, J., J. Pap & G.J. Rottman 1989 Observed solar near UV variability: A contribution to variation of the solar constant. In *Handbook for Middle Atmosphere Program* (ed. J. Lastovićka, T. Miles & A. O'Neill). Vol. **29**, pp. 9.

Marquette, W. H. 1992 In *Proceedings of the Workshop on Solar Electromagnetic Radiation Study for Solar Cycle 22* (ed. R.F. Donnelly). pp. 154–159. NOAA ERL SEL, Boulder, CO, USA.

Muller, R. A. & Keil, S. L. 1983 The characteristic size and brightness of facular points in the quiet photosphere. *Solar Phys.* **87**, 243–250.

NISHIKAWA, J. 1990 A new system of precise full-disk solar surface photometry. *Ap. J. Suppl.* **74**, 315–323.

NISHIKAWA, J. 1990 Estimation of total irradiance variations with the CCD solar surface photometer. *Ap. J.* **359**, 235–245.

NISHIKAWA, J. & HIRAYAMA, T. 1990 Facular structures derived from precise two-color contrast observations. *Solar Phys.* **127**, 211–228.

PAP, J., WILLSON, R.C. & DONNELLY, R.F. 1992 Two–parameter model of total solar irradiance variation over the solar cycle. In *The Solar Cycle* (ed. K. Harvey). ASP **27**, 491–502.

SCHATTEN, K. H., MILLER, N., SOFIA, S., ENDAL, A. S., CHAPMAN, G., & HICKEY, J. 1985 The importance of improved facular observations in understanding solar constant variations. *Ap. J.* **294**, 689–696.

SOLANKI, S. K. & BRIGLJEVIĆ, V. 1992 Continuum brightness of solar magnetic elements. *Astron. Astrophys.* **262**, L29–L32.

SOFIA, S., OSTER, L. & SCHATTEN, K. 1982 Solar irradiance modulations by active regions during 1980. *Solar Phys.* **80**, 87–98.

STEINEGGER, M., BRANDT, P. N., PAP, J. & SCHMIDT, W. 1990 Sunspot photometry and the total solar irradiance deficit measured in 1980 by ACRIM. *Astrophysics. Space Sci.* **170**, 127–133.

STENFLO, J., O. & HARVEY, J. W. 1985 Dependence of the properties of magnetic fluxtubes on area factor or amount of flux. *Solar Phys.* **95**, 99–118.

TITLE, A. M., TOPKA, K. P., TARBELL, T. D., SCHMIDT, W., BALKE, C. & SCHARMER, G. 1992 On the differences between plage and quiet Sun in the solar photosphere. *Ap. J.* **393**, 782–794.

TOPKA, K. P., TARBELL, T. D. & TITLE, A. M. 1992 Properties of the smallest solar magnetic elements I. Facular contrast near Sun center. *Ap. J.* **396**, 351–363.

VRŠNAK. B., PLAČKO, D. & RUŽDJAK, V. 1991 Calcium plage intensity and solar irradiance variations. *Solar Phys.* **133**, 205–213.

WANG, H. & ZIRIN, H. 1987 The contrast of faculae near the solar limb. *Solar Phys.* **110**, 281–293.

WHITE, O. R. 1988 Ground-based surrogates for UV and EUV fluxes. In *Solar Radiative Output Variations* (ed. P. Foukal). pp. 88–112. NCAR, Boulder, CO, 1988.

WIENER, L. 1994 M. S. thesis, CSU, Northridge (in progress).

WILSON, P. R. 1981 Faculae, filigree, and calcium bright points. *Solar Phys.* **69**, 9–14.

WILLSON, R. C., GULKIS, S., JANSSEN, M., HUDSON, H. S. & CHAPMAN, G. A. 1981 Observations of solar irradiance variability. *Science* **211**, 700–702.

ZWAAN, C. 1987 Elements and patterns in the solar magnetic field. *Ann. Revs. Astron. Astrophys* **25**, 83–111.

Brightness Observations of the Sun

By JEFFREY R. KUHN

National Solar Observatory/Sunspot, NM 88349, USA

There are few solar surface brightness observations with sufficient accuracy to have *direct* bearing on the question of the origin of the Sun's bolometric luminosity and irradiance changes (the correlative observations have been reviewed elsewhere in these proceedings). This paper contrasts these observations with a model for the solar variability and illustrates our clear need for high precision surface photometry, in combination with helioseismic data and numerical experiments.

---

1. Introduction

It is important to distinguish between irradiance and luminosity. I use the term "irradiance" to refer to the brightness, in a given direction (in the plane of the ecliptic) from the Sun, or in a specific direction from a particular region of the photosphere. The "luminosity" is then the integrated flux over all outward directions and over some area. Thus, the net energy budget of a part of the photosphere or the entire Sun is then reflected in our understanding of the solar luminosity – not the irradiance. All measurements to date, whether spatially resolved or integrated full-disk observations, only directly yield the irradiance, and such changes, alone, do not imply luminosity changes. From the astrophysical perspective the reason to study these "total irradiance" variations is to learn about the mechanism that causes them. I believe we won't achieve this until we know more about how the solar cycle works, and it may be that we learn quite a bit about it from the irradiance measurements. In any case it is clear that a "good" physical model of the irradiance data must also account for the other solar cycle changes. We can expect the helioseismic cycle variations to be a very useful constraint in this regard. A previous review by Kuhn (1992) describes one perspective on this problem. This discussion elaborates on some of the ideas from that paper.

The short term (day-to-day) and long term (yearly) total irradiance variations have magnitudes of a few times 10^{-4} of the mean solar irradiance. Spatially resolved observations must have comparable accuracy to see these changes. Since the ground-based experiments must peer through an atmosphere with an overall extinction variation that is typically two orders of magnitude larger – we do well to give up on the task of seeing the total solar irradiance variability (at this level) from the ground. Yet, the ground-based data may very well tell us how the irradiance changes come about, since it is possible to obtain this level of photometric precision *differentially*. Then, from additional assumptions (e.g. that the surface brightness in quiet regions of the photosphere doesn't change), we can "bootstrap" an argument for spatially integrating the differential photometry data to obtain the total irradiance. Comparison with ERB or ACRIM satellite experiments, which do have good absolute precision, then allow a test of the assumptions that go into interpreting the differential photometry. Note that progress here depends on interpreting high precision differential measurements in combination with high precision absolute solar photometry.

We took this approach (Kuhn *et al.* 1988) to compare limb temperature observations to ACRIM data. With the assumption that the effective color temperature of the photosphere outside of the active latitude bands and the solar equator doesn't change during the cycle, it could be shown that the solar luminosity was changing over long (yearly) timescales. This limb photometry was limited, in that the spatial resolution was low and only the outer 20 arcseconds of the limb could be observed using the "oblateness" telescope. Also, the latitudinal photometric variations that were detected could not be uniquely identified, or easily associated with other solar phenomena (like the photospheric network, or "polar" faculae). On the other hand recent independent measurements (Kroll 1994) have described indirect evidence for effective limb-temperature variations comparable to these.

On shorter timescales the first non-correlative direct observations of irradiance changes were obtained by Nishikawa & Hirayama (1990) using a CCD camera system. Although they needed to fit and subtract largescale spatial trends from their full-disk photometry, they did achieve high photometric precision at the pixel scale. With the assumption that the quiet photosphere was not variable they were able to show that most of the short term solar irradiance changes are associated with sunspots and active regions.

Recently Lin (1993) and Lin & Kuhn (1992) used visible and IR sensitive array detectors to try to identify both the large and small scale contributions to the total irradiance signal. Lin also found a latitudinal variation similar to the Kuhn *et al.* (1988) results, but lacked sufficient spatial resolution to identify the latitudinal bolometric changes with, for example, a photospheric network irradiance signal. Topka (1993) has used very high spatial resolution observations, in relatively small fields-of-view, to argue that the network bright points do not account for the observed long term irradiance changes. His conclusions depend on the assumption that the local mean photosphere doesn't vary in brightness between regions where he measures the network contrast. It is notable that there is a large body of high precision, but small field-of-view, differential photometric data aimed at determining sunspot contrast (e.g. Fowler & Foukal 1983 for measuring CZ diffusivity) or facular contrast (e.g. Lawrence 1988 for pinning down facular models) but these observations do not directly yield constraints on the irradiance mechanism.

2. Local observations, global conclusion

The solar limb photometry implies that on a long timescale the solar luminosity changes. Although these photometric results need to be confirmed, the data imply that the Sun is about 0.1% brighter near solar maximum than near minimum. On shorter timescales we know that *irradiance* changes tend to be anticorrelated with surface magnetic fields. We cannot say on these timescales if there is a correlation between such magnetic fields and luminosity. The problem of course, is that the total luminosity associated with photospheric features is difficult both to measure and to model. The radiant energy budget of a flux tube obviously depends on the timescale over which you study it, while the direct measurement of the total emergent flux depends on simultaneous observations from many directions, or on waiting for it to rotate around the Sun (during which time it may evolve). Local observations of feature contrast (cf Topka 1993) versus angle can, in principle, be integrated to obtain the luminosity change due to the feature, but the calculation is flawed if the surrounding photosphere (which the contrast is measured with respect to) changes brightness.

Without high precision full-disk differential photometry, one method for measuring the magnetic feature luminosity is to use the ACRIM timeseries data, which is an integral of the full-disk brightness. This approach was used (Kuhn 1992) to show that there is a general *anticorrelation* between the average photospheric magnetic field and the irradiance on short timescales. Presumably this is just the effect seen in individual timeseries records where irradiance drops as sunspots rotate across the disk (cf. Willson & Hudson, 1988). There is a strong statistical signal in the ACRIM timeseries that shows an irradiance excess one quarter solar rotation before and after the irradiance

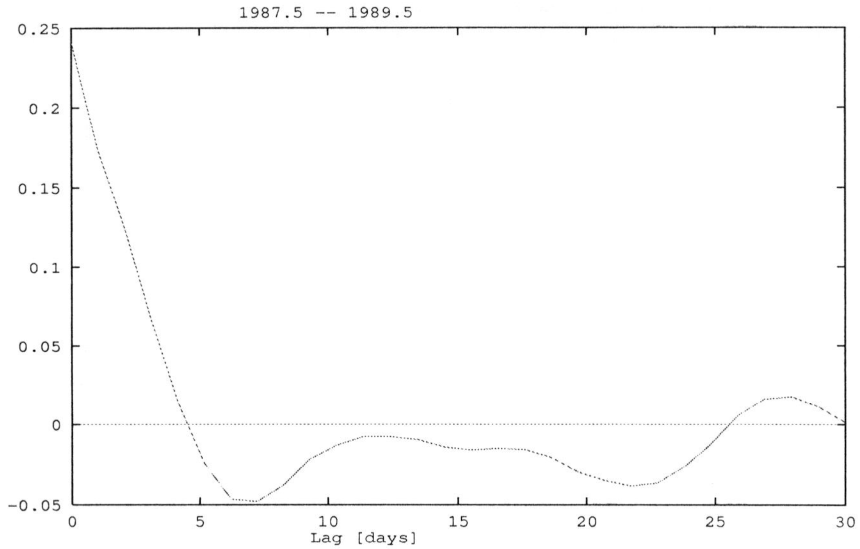

FIGURE 1. ACRIM autocorrelation.

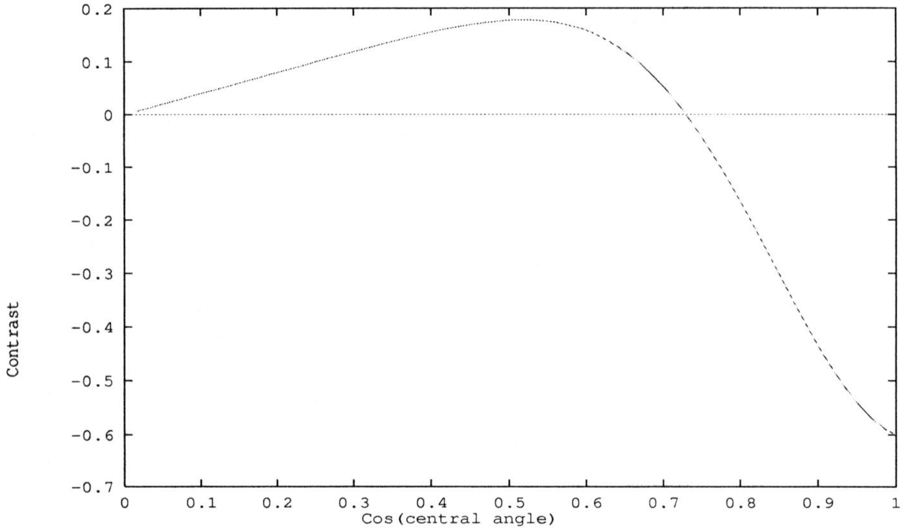

FIGURE 2. Active Region Contrast Function.

dips. This suggests that the short-term anticorrelated irradiance signal may not imply a corresponding decreased magnetic feature luminosity. The flux lost from the normal direction in active region magnetic features is redistributed at large angles from the local normal. This conclusion seems to describe the collective effect of all magnetic elements, without distingushing faculae and sunspots. Thus, individual spots and faculae may have very different radiation patterns, but the collective effect of active regions shows a clear flux redistribution signal in the ACRIM autocorrelation analysis.

Figure 1 shows the autocorrelation of ACRIM irradiance data after solar minimum until the end of the timeseries record in 1990. A 60 day moving mean was subtracted

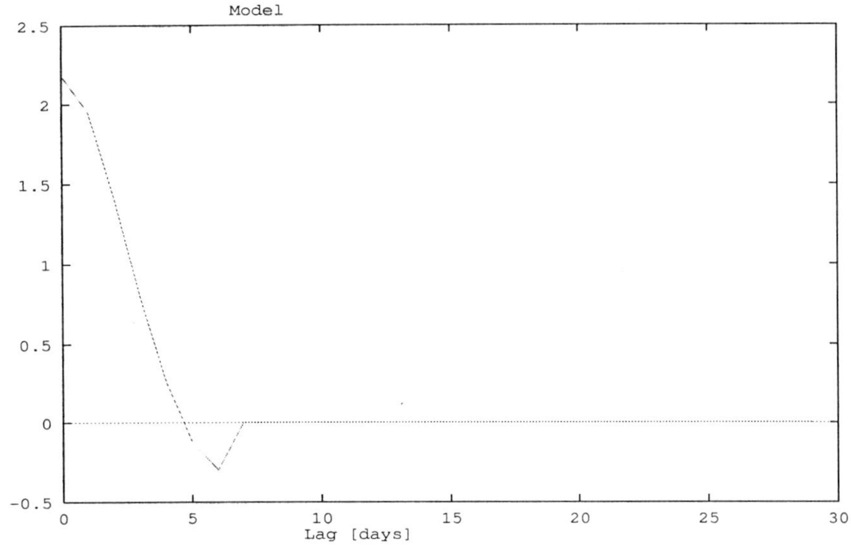

FIGURE 3. Model autocorrelation.

to remove the slow solar cycle trend. The graph shows the negative peak near a lag of 1/4 of a rotation (6-7 days). This data is probably too noisy to uniquely invert the autocorrelation function for the effective "active region contrast function" but Figure 2 illustrates a simple "forward" calculation where I hypothesised a 2 parameter contrast function and varied the parameters until I reproduced (by "eye") the autocorrelation displayed in Figure 1. The particular contrast function plotted in Figure 2 has the form $dI/I = \{c - \exp[-(\frac{(1-\mu)^2}{2\mu_0^2})]\} \times \mu$ and for parameters $c = 0.4$ and $\mu_0 = 0.3$ the angular integral of this function implies essentially no change in total luminosity, although this contrast function does exhibit an irradiance decrease near disk center ($\mu = 1$) and falls to zero at the extreme limb. It implies the autocorrelation shown in Figure 3, which is quite similar to the actual data plotted in Figure 1. The conclusion from this example is that the filtered ACRIM timeseries, even though it shows short-term irradiance drops, is consistent with an "active region" contrast function that implies no short-term solar luminosity changes. Testing this conclusion requires precise full-disk photometry over a long enough (several rotation periods) time series. A statistical measure of active region *luminosity* changes could be extracted from the rotationally modulated surface brightness records. To compare one timeseries record with the next, it is critical that the local active region brightness be known with respect to the solar surface brightness far from the region, i.e. global (full-disk) photometry is needed.

3. What causes long-term luminosity changes?

Considering the large fractional amplitude of the magnetic changes during the cycle it is likely that the luminosity variations are directly related to the magnetic field evolution. Does the luminosity vary as an incidental consequence of fields poking through the photosphere, which allow radiation to escape more easily through the lower gas density and opacity in flux tubes, or does the luminosity change reflect variations in entropy deeper in the convection zone (CZ), or below?

The rather small solar cycle change in the acoustic mode frequencies which are sensi-

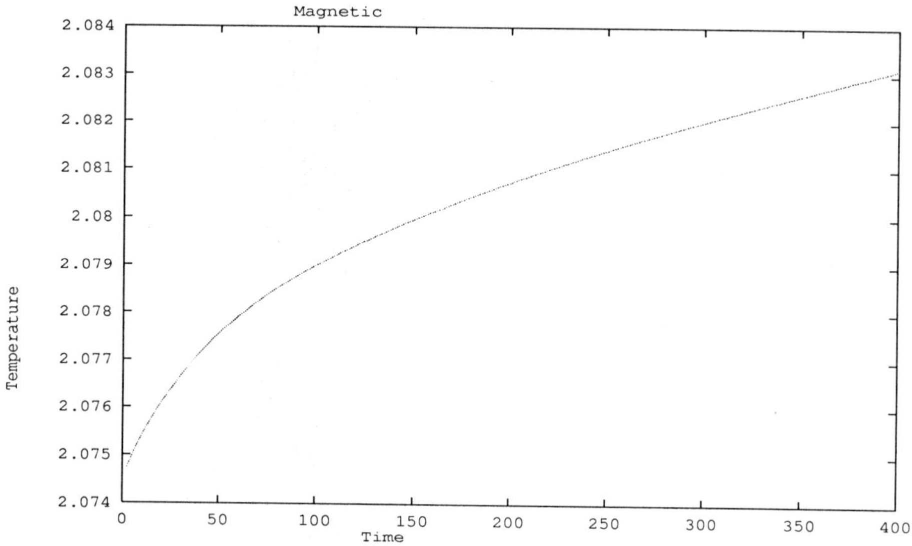

FIGURE 4. Temperature ($10^6 K$) vs. time (yrs).

tive to the solar core conditions rules out fundamental changes in the radiative interior temperature structure as the source of the yearly luminosity variations we see at the photosphere. The amplitude of the temperature variations in the energy generating core would need to be enormous (fractional changes of order unity) to appear as 11 year variations at the surface, since the radiative diffusion time exceeds 10^5 years.

In the simplest model photospheric magnetic fields may act as "shorts" to radiative diffusion. The magnetic field locally decreases gas density and increases the photon free path. In this simple picture most of the excess energy radiated through the "antiplug" is compensated by a deficit surrounding the short. Since the solar atmosphere is exponentially stratified most of the energy transported out the antiplug comes from horizontal diffusion. Thus in an exponentially stratified diffusive model the total outgoing energy flux is only marginally enhanced. In other words the radiation pattern of the region changes, but its total luminosity change is small (Kuhn 1992).

Of course the solar convection in and around faculae and sunspots makes a detailed solution a much more difficult problem. For example, if the effect of convective energy transport were to isotropically couple much of the convection zone (for example, a large "effective" conductivity) then the "antiplugs" described above could change the energy content of the convection zone while changing the solar luminosity (Spruit 1994). In conflict with this assumption, we find from numerical simulations of energy transport in a realistic outer convection zone that convective energy transport is very anisotropic, with entropy transport vertically being at least an order of magnitude larger than horizontally. This conclusion is quite consistent with the strong velocity anisotropy observed in the numerical simulations. Precise observations of the horizontal thermal structure around magnetic regions will tell us what the entropy transport properties of the upper convection zone are really like.

If the short term irradiance changes are due to flux redistribution from photospheric magnetic fields, and they do not produce significant luminosity changes, then what causes the solar cycle luminosity variation? A deeper mechanism is suggested by the longer timescale of these changes. Whether or not an $\alpha - \omega$ dynamo is at work to produce

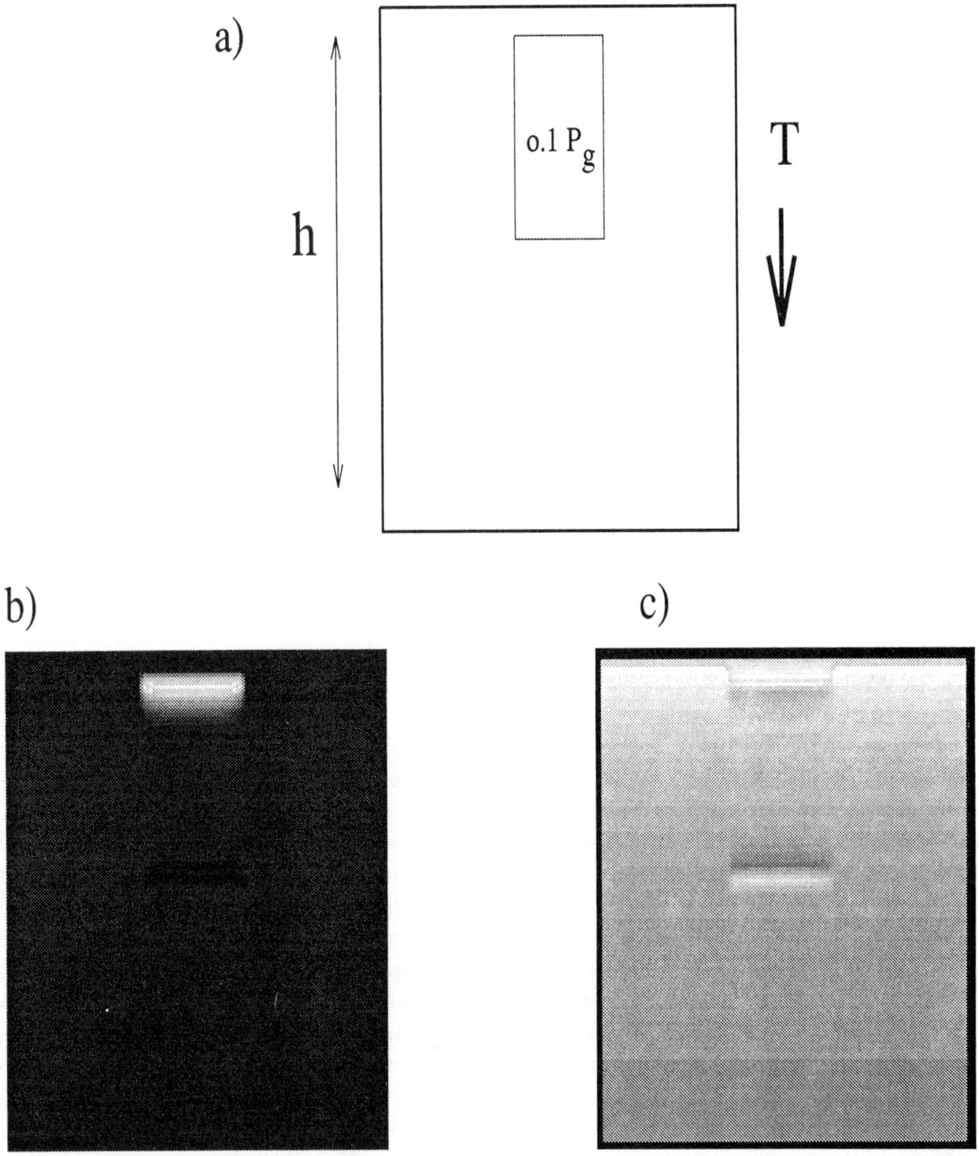

FIGURE 5. a) Simulation geometry, b) Fractional temperature change after 11 years, c) Temperature gradient after 11 years.

the magnetic cycle, it is likely that the surface magnetic flux originates from near the base of the convection zone. Since helioseismic observations of the interior differential rotation place the largest radial shear just below the CZ it seems that the toroidal field is amplified there. I argue below that magnetised fluid that originates in the radiative zone should have a higher entropy than surrounding layers. This excess entropy could emerge on a longer timescale than the magnetic flux and be responsible for the increase in the solar luminosity observed near the peak of the activity cycle.

A thin sheath of magnetic flux immediately below the base of the CZ must grow hotter than corresponding fluid with the same horizontal stratification. Radiation below the flux sheet diffuses through the magnetised region with a longer photon free-path because the local density is decreased by the local magnetic pressure. Thus the top of the magnetised fluid is hotter than the surroundings. Consequently the magnitude of the temperature gradient also increases above the top surface of the flux sheet. If the flux sheet is immediately below (and parallel to) the base of the convection zone the effect of the increased temperature gradient is to locally *lower* the bottom of the convection zone to the surface of the flux sheet. Convective overshoot or magnetic bouyancy can then operate to bring magnetic flux, which originated below the CZ boundary, into the CZ and finally into the photosphere. This radiative instability not only causes the magnetised fluid to have a higher entropy than unmagnetised fluid, but it also provides a mechanism for bringing flux from the RZ into the CZ. In effect the bottom surface of the CZ must be "corrugated" latitudinally by the magnetic perturbation. This effect can be looked for in helioseismic data, which already give very accurate measurements of the mean depth of the CZ (Christensen-Dalsgaard *et al.* 1991).

The timescale for heating the magnetised fluid depends on the vertical extent of the sheet and the field strength. The growth time τ is approximately

$$\tau = \frac{3\pi\kappa\lambda\rho}{16\sigma T\Delta|dT/dz|}$$

where κ is the mean opacity, λ is the vertical scale of the flux sheet, Δ is the fractional change in the density ρ due to the magnetic field, and dT/dz is the local temperature gradient. A numerical calculation illustrates how this effect may operate. We solve $dT/dt = kT/P\nabla(KT^4/P \ \ \nabla T)$ subject to the ideal gas law with $P = P_{gas} + P_B$ when P_B is a position dependent constant corresponding to the "local magnetic pressure". This nonlinear problem was first solved with P_B everywhere zero and constants k and K chosen to approximate a volume near the top of the RZ in the Sun. A 64×64 grid with constant temperature boundary conditions on the top and bottom surfaces and periodic horizontal surfaces was used with a standard relaxation technique. Into this equilibrium solution a magnetic field was added (corresponding to 10% of the gas pressure) near the surface of the volume. Figure 5a shows the relative size of the pressure scale height, h, and the magnetised volume. Figure 4 shows how the temperature at the top surface of the volume changes. Over 11 years the temperature increases by about 1000 degrees, or about 0.05%. Figure 5b shows a greyscale image of the fractional temperature change in the volume. The top surface gets warmer while the lower magnetic boundary is cooler. Figure 5c shows the vertical temperature gradient, also plotted as a greyscale image. The gradient increases toward the top boundary where the fluid becomes convectively unstable. Immediately above the magnetised volume the gradient also increases, tending to lower the base of the CZ above magnetic latitudes.

In this picture, photospheric magnetic flux causes the irradiance to drop as a magnetic region rotates toward disk center. This irradiance change is accompanied by an isotropic *luminosity* increase which is related to the total magnetic flux in the region. The excess entropy carried by magnetic flux tubes appears at the photosphere over a longer timescale than the active region growth time since the entropy excess is carried throughout the flux tube volume. How this energy diffuses to the photosphere is yet undetermined. This excess appears as a *luminosity* increase over longer timescales on top of the short term *irradiance* decrease. From the ecliptic plane we see the short term irradiance decrease with magnetic field while the overall magnetic activity defines an increasing irradiance "baseline" which is actually the solar-cycle luminosity change.

4. Future observations

Ground-based observations to come will measure small surface brightness variations over solar-cycle timescales. Interesting observations should achieve spatial resolution and photometric precision that is limited only by the atmosphere. We should expect such a dataset to help answer questions like "how do magnetic fields affect heat flow through the convection zone?" or "can we learn about how the solar cycle works from the luminosity of photospheric features?" Is the "valve" that modulates the total solar luminosity operating at the photosphere incidently as a consequence of the surface magnetic variation, or is the luminosity controlled from the base of the CZ? Answers will follow from sensitive measurements of, for example, the radiation temperature excess and its time dependence from active regions, or the latitudinal variation of the network luminosity. Such data will provide another "boundary condition" on the interpretation of helioseismic data that probe the solar interior, and another constraint on the interpretation of numerical simulations.

REFERENCES

Christensen-Dalsgaard, J., Gough, D.O. & Thompson, M.J. 1991 The depth of the convective zone. *Ap. J.* **378**, 413–437.

Fowler, L. & Foukal, P. 1983 Sunspot bright rings and the thermal diffusivity of solar convection. *Solar Phys.* **84**, 33–44.

Kroll, R. 1993 Latitudinal variation of the solar limb-darkening function. *Solar Phys.*, in press.

Kuhn, J.R. 1992 What causes cycle related global solar changes? *GONG 1992: Seismic Investigations of the Sun and Stars* (PASP: San Francisco), 27–40.

Kuhn, J.R., Libbrecht, K. & Dicke, R.H. 1988 The surface temperature of the Sun and changes in the solar constant. *Science* **242**, 908–911.

Lawrence, J.K. 1988 Multicolor photometric observations of facular contrast. *Solar Phys.* **116**, 17–32.

Lin, H.: 1993 Precision IR and visible photometry II. *Ap. J*, submitted.

Lin, H. & Kuhn, J.R. 1992 Precision IR and Visible solar photometry. *Solar Phys.* **141**, 1–26.

Nishikawa, N. & Hirayama, T. 1990 Facular structures derived from precise two-color contrast observations. *Solar Phys.* **74**, 315.

Spruit, H. 1994 Theoretical interpretation of solar and stellar irradiance variations. In *The Sun as a Variable Star: Solar and Stellar Irradiance Variations* (ed. J.M. Pap, C. Fröhlich, H.S. Hudson & S.K. Solanki). Cambridge University Press, in press.

Topka, K. 1993 High resolution solar photometry. *Solar Physics Division of AAS*, in press.

Willson, R.C. & Hudson, H.S.: 1988 Solar luminosity variation in cycle 21. *Nature* **332**, 810–812.

Observations of Global Solar Magnetic and Velocity Fields

By J. TODD HOEKSEMA

Center for Space Science & Astrophysics, Stanford, CA 94305, USA

The Sun's magnetic field varies on time scales of minutes, months, decades, and centuries. These changes drive variations in solar irradiance. Characteristics of the changing magnetic field reveal, at some level, how the solar cycle works. The basic 22-year magnetic cycle and corresponding 11-year activity cycle of the Sun are driven by the emergence of active regions. The frequencies of solar oscillations that probe the convection zone vary on the same time scale. While plausible explanations for the statistical properties of solar activity can be made, the details of flux emergence cannot be predicted. Global activity varies on other intermediate time scales as well.

Defining elements of a cycle as those sharing a common bipolar field orientation, the first signs of a cycle can appear at high latitudes up to 5 years before solar minimum and the final elements disappear at the equator several years after the following minimum – a span as long as 18 years. At a particular latitude waves of activity are spaced by 11 years. The torsional oscillation in solar rotation rate has similar characteristics.

The spatial distributions of total flux and large-scale polarity evolve quite differently: the total flux map resembles the butterfly diagram, moving from mid latitudes toward the equator during the cycle, while the net zonal flux has two maxima in each hemisphere that expand away from mid latitudes.

While each solar cycle shares these basic features, fixed by the action of the solar dynamo working within the convection zone, the details of each cycle are determined by the apparently random timing and placement of emerging flux and by the changing convective flow patterns.

1. Introduction

"Last night I had a curious dream ..." Most of us can remember a vivid dream – of flying, of embarrassment, or of a repeating noise. The elements of the dream seemed very real and made some kind of sense, but when considered in a broader context (often just after the alarm clock stopped) it became obvious that some important fact had been ignored. In some ways our understanding of the global solar magnetic and velocity fields is similar. We have increasingly accurate measurements of any number of parameters that vary more or less in phase or out of phase with the 22 year solar cycle. However, our understanding of the connections between the various features, of the detailed physics of their formation, lifetime, and decay, and of their common underlying causes is in many cases uncertain, inconsistent, or incomplete.

As we attempt to make sense of all these data, it is important to keep this fundamental question in mind: How much of what we observe is fundamentally related to the solar cycle mechanism? We see sunspots, measure radio flux, sense changes in total irradiance, puzzle over differential rotation in the photosphere and convection zone, and observe the reversal of the Sun's polar fields, but how are these related to the dynamo that drives the cycle in the interior? Do the surface field patterns migrate toward the pole and there provide the seed for the subsequent cycle? Or are the surface patterns, like the wake of a boat, merely an indicator of something larger, and perhaps more interesting, happening elsewhere? It is always tempting to conclude that the things we can see (and the questions we can begin to answer) are also the important ones.

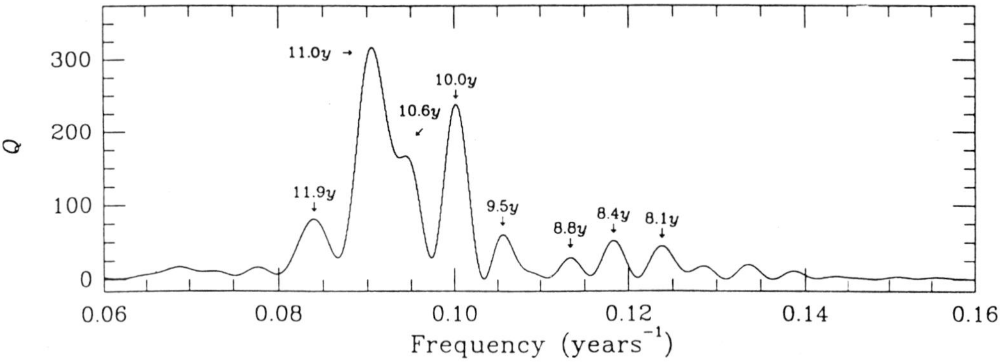

FIGURE 1. The periodogram of a 240-year time series of sunspot numbers shows the 11.04-year peak to be stronger than the other real periodicities in the record. "Q" is the intensity normalized to the total variance of the data (Donahue & Baliunas 1992).

In the remainder of this brief review, two main topics will be explored: the origin, distribution, and disappearance of flux in the photosphere, and the structure and variation of differential rotation and other large-scale solar motions. Continued progress on these questions coupled with improved understanding of the fundamental physical mechanisms operating on small spatial scales will enable the construction of a more complete picture of solar and stellar variations and of their influences on the Earth.

2. What is the life cycle of photospheric flux?

Magnetic flux patterns on the Sun vary with several characteristic time scales. The flux in all but the smallest elements appears to be concentrated entirely in intense kilogauss elements of small size, below the limits of spatial resolution. Associations of flux elements are seen as ephemeral regions, sunspots, active regions, and large-scale patterns including the polar field. The larger the scale of organization, the longer the lifetime. Individual sunspots and active regions live for days to months, nests of activity can persist as long as several years, the polar fields and gross level of activity vary with a 22-year period and it's 11-year harmonic. This paper presents observations of the largest scale features. Nevertheless, it is probably worth noting that the smallest intranetwork (IN) field elements are weak, but numerous and shortlived. Zirin's observations suggest that the total flux emerging in IN structures (10^{24} Mx per day) is about 100 times that in ephemeral regions and 10^4 that of active regions (Zirin 1987). About 30% of the IN elements in quiet regions disappear each day. Because the field orientation of INs is random, their contribution to the large-scale field patterns is probably negligible.

Sunspots, the most easily observed solar features, provide the longest direct observational record of solar activity. The record shows that each solar cycle is unique. Most recently Donahue & Baliunas (1992) demonstrated that solar cycles since 1750, as parameterized by the sunspot number, vary significantly in magnitude, profile, and period. Figure 1 shows the straightforward periodogram computed from the 240-year record. More recent quantitative observations of parameters such as radio flux, flare occurrence, and magnetic flux support this result. Though variations in period are smaller in recent cycles, even the earliest sunspot data are sufficiently accurate to show these variations are real. Observations of other stars show similar kinds of variations, including the simultaneous presence of multiple periodicities and Maunder-minimum like intervals.

The total solar flux varies with the solar cycle, but the strong field regions associated

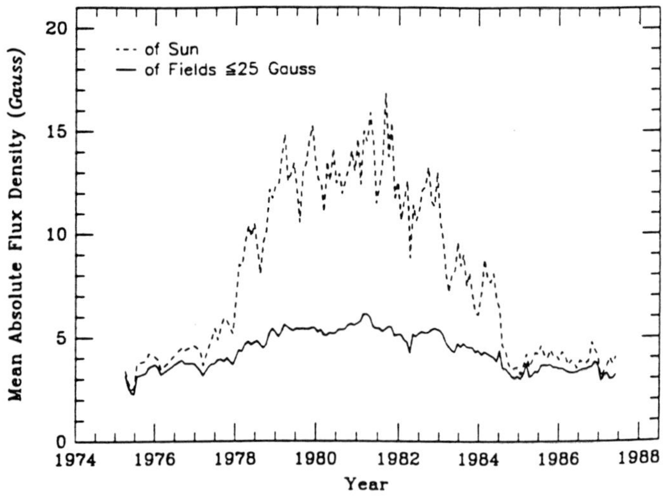

FIGURE 2. The variation of the strong-field and weak-field components of the total solar flux measured at KPNO. The total weak-field flux varies less than a factor of 2 from minimum in 1976 to maximum in 1979-80; the strong-field flux varies by more than a factor of 15. The total solar flux varies by at least a factor of 5. Some of the spiky peaks in the strong-field can be associated with individual activity complexes, but others are more globally distributed (Rabin *et al.* 1991).

with active regions account for almost the entire change (see Figure 2). This implies that at least 70% of the AR flux disappears in place (Rabin *et al.* 1991). Variations in strong-field flux can sometimes be associated with the evolution of particular active region complexes. The quasi-periodic pulses may be associated with the rates of flare occurrence. The north-south asymmetry of the flux is reflected most strongly in the strong-field regions, but is revealed in the weak-field regions as well.

The polar field provides an important component of the weak-field flux, particularly around solar minimum. The polar field is sometime regarded as an important indicator of the strength of the following sunspot cycle. One measure of the polar cap fields is shown in Figure 3. The polar fields have roughly constant magnitude near solar minimum (1976 and 1986) and reverse their polarities near solar maximum (1980 and 1990). Ignoring the annual variation due to the inclination of the Earth's orbit to the solar equator, the asymmetry in polar field strength in 1985–87 is quite apparent. In 1976–77 the asymmetry was not present and the average polar field strength was 25% less. The polar field appears to form as the result of the poleward migration of following-polarity flux from the active region zones (see Figure 5).

A one-dimensional parameter is not sufficient to describe the solar cycle. The latitudinal evolution of the flux patterns reveals something about the cycle mechanism. The progression of active region locations from mid-latitudes toward the equator through the cycle is well known. The "butterfly" diagram can be extended using the locations of plages. The plages appear earlier in the cycle and at higher latitudes than sunspots. Figure 4 shows how the field associated with two solar cycles overlaps for several years near solar minimum. Here elements from a cycle are identified by the orientation of the bipolar field structures. Observations of coronal brightness (Altrock 1988), polar facu-

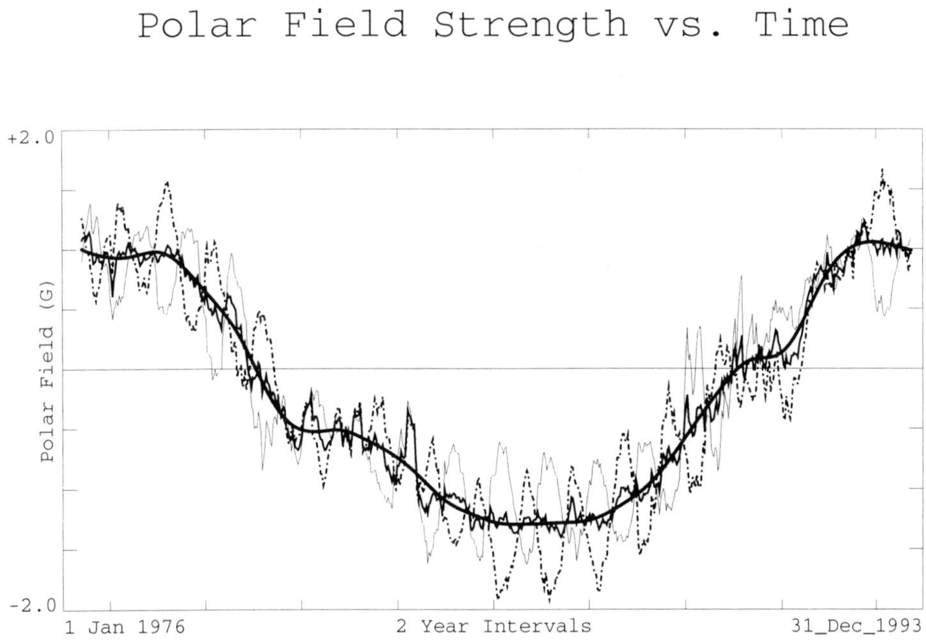

FIGURE 3. The Sun's polar field strength for a 17-year period showing two polarity reversals. This line-of-sight field measurement made in the polemost 3 arcmin aperture of daily magnetograms taken at the Wilcox Solar Observatory, covers latitudes from about 55 degrees to the pole. The light solid and dashed lines show the northern and southern pole strengths; the sign of the southern pole has been reversed to simplify intercomparison. The broader solid lines show the smoothed and unsmoothed averages of the two poles. The inclination of the ecliptic to the Sun's equator and the strongly peaked geometry of the polar field cause the annual variation of the individual polar caps.

lae (Callebaut & Makarov 1992), and the torsional oscillation in Doppler rotation rate (LaBonte & Howard 1982) appear to confirm that features associated with a particular solar cycle are present on the Sun for as long as 18 years, though the patterns at any given latitude repeat (with opposite sign) every 11 years.

Another feature of the large-scale field is the net polarity, shown in Figure 5. Panel 1 (bottom), showing total flux, has the same butterfly shape as Figure 4. The net flux (panel 2) emerges at mid latitude shortly after solar minimum in Cycles 21 and 22. Generally the leading polarity region expands toward the equator, while the following polarity region expands toward the poles. The boundary between the two remain roughly fixed. The polar field reverses when the following polarity region reaches high latitude. The reversal resulted from two components in Cycle 21: a gradually expanding antisymmetric flux pattern (panel 4, top) and a series of fast-moving surges that appear to have the same sign in the northern and southern hemisphere (panel 3). In Cycle 22 the gradually expanding term seems to dominate. The arrival times of the various features are associated with the changes in polar field strength shown in Figure 3. Similar polarity maps of previous cycles back to 1870 derived from Hα observations show very similar patterns (Makarov & Sivaraman 1989).

Stenflo and Gudel (1988) have decomposed the zonal solar field into harmonic coefficients. The amplitudes and phases of the first seven odd (antisymmetric) modes provide enough information to reproduce the shape of the low-latitude sunspot zone and the high-

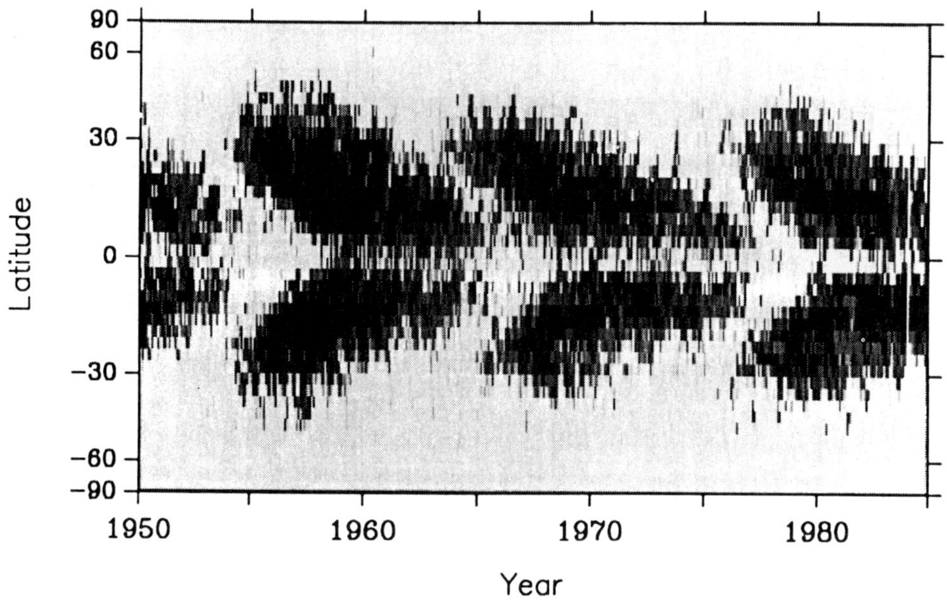

FIGURE 4. The butterfly diagram of calcium plage regions (courtesy J. Lean) indicates how solar cycles overlap for several years. Plage data are summed in bins of 0.05 of sine latitude for each month from 1950 to 1984. Darker shading indicates increased plage area (Harvey 1992).

latitude polar faculae from 1960–1985 (Callebaut & Makarov 1992). This may provide an important clue to the structure of the solar dynamo.

3. Large-scale velocity patterns

The most prominent feature of the large-scale solar velocity field is differential rotation (Figure 6). The variations of differential rotation with time are small and closely associated with magnetic field patterns. Various methods for determining the rotation rate give slightly different results. The main difficulties with tracers are temporal evolution and local flow fields. The problems with Doppler measurements are magnetic contamination of the line profile and systematic errors.

The presence of the torsional oscillation was first detected in Doppler measurements (LaBonte & Howard 1982) and later in magnetic tracers (Snodgrass 1991). The maximum shear of the plasma rotation occurs at the latitude of maximum magnetic flux emergence. The tracer pattern seems to be displaced equatorward by about 10 degrees. Recently the same signature has been found in tracer measurements (Komm *et al.* 1993) by correlating magnetic field maps obtained on consecutive days. Figure 7 shows that between 60 degrees and the equator a region of faster-than-average rotation moves equatorward from 1977 to 1986. Extended to higher latitudes this pattern supports the extended solar cycle concept.

Finally, if magnetic field patterns with only the largest size scales are considered, an interesting north-south asymmetry appears. Figure 8 (after Antonucci *et al.* 1990) shows the results of calculating the power spectrum of the time series of magnetic observations at each solar latitude. The frequency range plotted is only sensitive to hemispheric-scale patterns rotating with roughly the solar rotation period. During most of Cycle 21 the southern hemisphere rotated with a 28-day period, while the north had a 27-day rotation

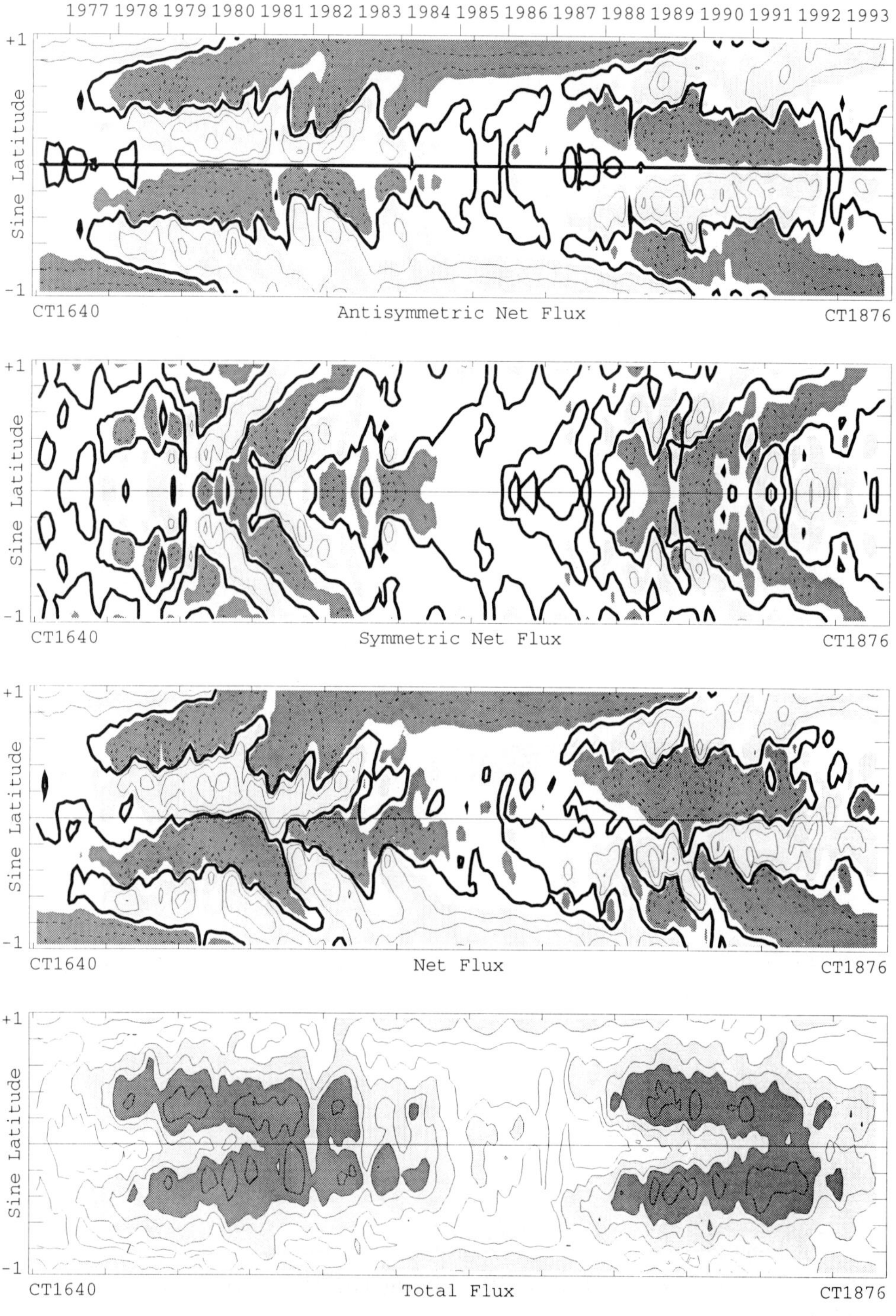

FIGURE 5. Zonal solar flux versus time since 1976. The bottom panel shows the total absolute flux computed for each Carrington rotation from WSO synoptic charts. The second panel shows the net flux computed for each rotation. The top two panels show the symmetric and antisymmetric components of the net flux. A three rotation triangular smoothing has been applied to the data. Note the precipitous decrease in flux in mid 1992.

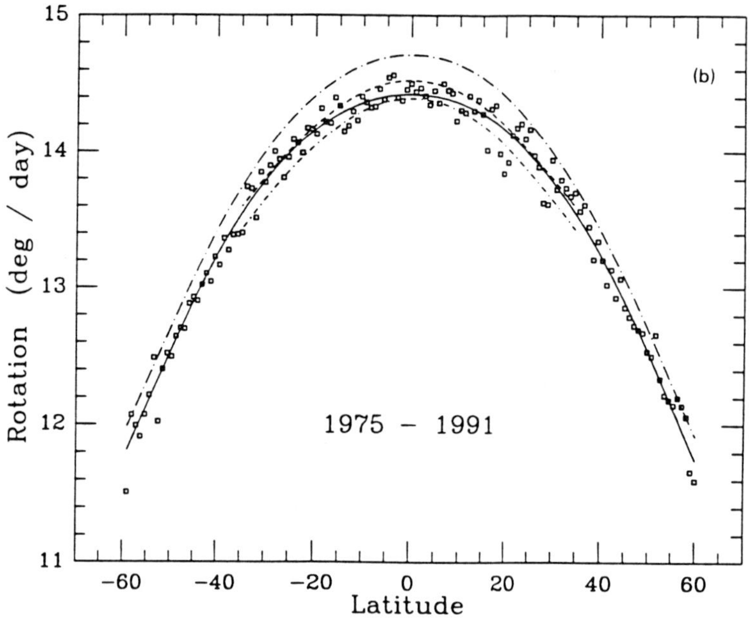

FIGURE 6. The average differential solar rotation rate (sidereal degrees/day) derived from correlations of the small-scale magnetic field pattern on consecutive days. Active regions have been excluded in this figure. The solid line shows a fit to the data. Rotation rates of other tracers are also shown: supergranules – dot-long dash; sunspot group – dot-short dash; and individual sunspots – short dash. The equatorial Doppler rotation is less than 3% faster than this magnetic rate (Komm *et al.* 1993).

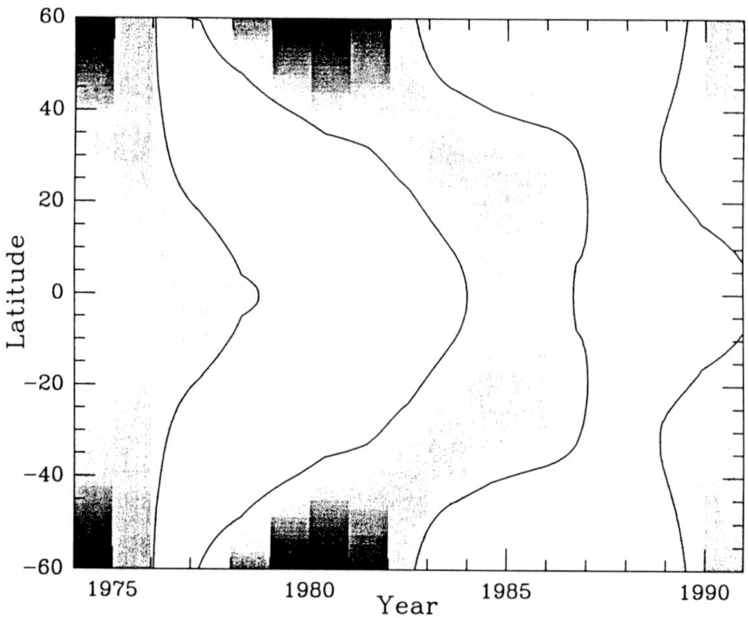

FIGURE 7. A map of the differences from the average differential rotation rate. Again active regions have been excluded (though the result is similar). The contour lines show 0 difference from the average curve shown in Figure 6. The darker regions rotate more rapidly than average (Komm *et al.* 1993).

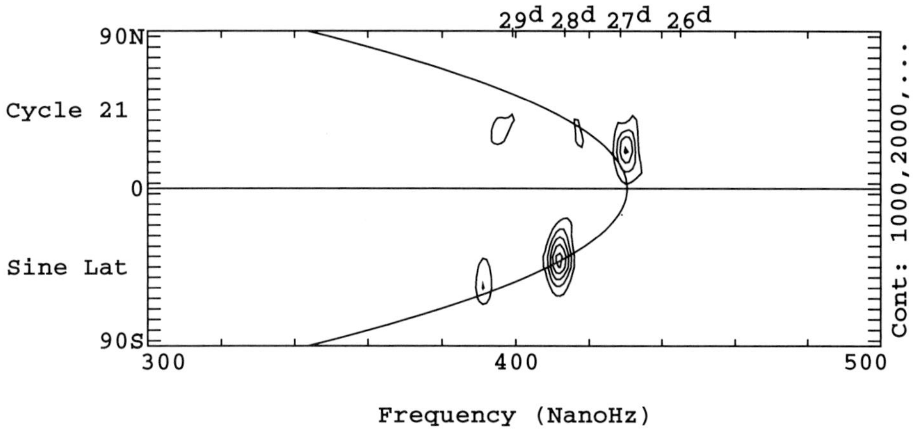

FIGURE 8. Solar rotation of the large-scale magnetic field for Solar Cycle 21. The power spectrum of the field at each latitude in the frequency range near the solar rotation period shows that the hemispheric scale patterns rotate at discrete rates that fall near the Newton & Nunn (1951) differential rotation curve. However the rates are different for Cycle 21 in the northern and southern hemispheres.

period. These are the most common periods observed in the interplanetary magnetic field since 1926. The same periods are seen in the photospheric field measured at Wilcox Solar Observatory (WSO), National Solar Observatory/Kitt Peak (KPNO), and Mt. Wilson Solar Observatory (MWSO) and in the computed coronal field. Cycle 22 has shown a similar strong asymmetry only during the declining phase of the cycle.

4. Conclusions

Unfortunately these data leave us with more questions than answers. Though progress is being made in characterizing the variable Sun, as shown throughout this volume, the key links between the cycle mechanism, the large-scale magnetic and velocity patterns, and the small-scale field elements are unclear. Evidence for the extended solar cycle makes the timeworn Babcock-Leighton model of the solar cycle appear more and more inadequate, even as a heuristic device. Dynamo models have problems reproducing the observations when accounting for helioseismologically determined differential rotation with depth. Until we understand physics of the generation and evolution of small-scale field, the connection to the mechanisms of solar variability will remain empirical at best. A real key for interpreting observations has to be determining which are truly causally related to the cycle mechanism. Somehow we must rouse ourselves enough to find the missing insight.

Acknowledgments. This work was supported by the NSF, NASA, and ONR.

REFERENCES

ALTROCK R.C. 1988 Variation of solar coronal Fe XIV 5303 emission during solar cycle 21. In *Solar and Stellar Coronal Structure and Dynamics* (ed. R.C. Altrock), 414–420. NSO/SP.

ANTONUCCI E., HOEKSEMA J.T. & SCHERRER P.H. 1990 Rotation of the photospheric B fields: a north-south asymmetry. *Astrophys. J.* **360**, 296–304.

CALLEBAUT D.K. & MAKAROV V.I. 1992 Latitude-time distribution of 3 types of B field activity in the global solar cycle. *Solar Phys.* **141**, 381–390.

DONAHUE R.A. & BALIUNAS, S.L. 1992 Periodogram analysis of 240 years of sunspot records. *Solar Phys.* **141**, 181–198.

HARVEY K.L. 1992 The cyclic behavior of solar activity. In *The Solar Cycle* (ed. K.L. Harvey). Astronomical Society of the Pacific Conference Series Vol **27**, pp. 335–267.

KOMM R.W., HOWARD, R.F. & HARVEY, J.W. 1993 Torsional oscillation patterns in photospheric magnetic features. *Solar Phys.* **143**, 19–40.

LABONTE B.J. & HOWARD R. 1982 Torsional waves on the Sun and the activity cycle. *Solar Phys.* **75**, 161–178.

MAKAROV V.I. & SIVARAMAN K.R. 1989 New results concerning the global solar cycle. *Solar Phys.* **123**, 367–380.

NEWTON, J.W. & NUNN, M.L. 1951 The Sun's rotation derived from sunspots 1934-1944 and additional results. *Monthly Notices of the Royal Astronomical Society* **111**, 413–421.

RABIN D., DEVORE C.R., SHEELEY, N.R. JR., HARVEY K.L. & HOEKSEMA J.T. 1991 The solar activity cycle. In *The Solar Interior and Atmosphere* (ed. A.N. Cox, W.C. Livingston & M.S. Matthews), 781–843. U. Arizona Press., Tucson, AZ, USA.

SNODGRASS, H.B. 1991 A torsional oscillation in the rotation of the solar magnetic field. *Astrophys.J.* **383**, L85–87.

STENFLO J.O. & GUDEL M. 1988 Evolution of solar magnetic fields: modal structure. *Astron. Astrophys.* **191**, 137–148.

ZIRIN H. 1987 Weak solar fields and their connection to the solar cycle. *Solar Phys.* **110**, 101–107.

The Variability of Stellar Magnetic Flux, Related Surface Structures and Activity

By STEVEN H. SAAR

Harvard-Smithsonian Center for Astrophysics, Cambridge, MA 02138, USA

I review information from a variety of sources, including magnetic field measurements, Doppler imaging, and various photometric and spectroscopic survey and monitoring programs for clues concerning the physical properties, spatial structure, variability and evolution of magnetic regions and associated activity on cool stars. Based on these data, I sketch a scenario of what the surfaces of cool stars look like as a function of spectral type and age/rotation/activity level.

1. Introduction

Recent analyses of long time series of photometry, Ca II fluxes, and absorption line profiles have yielded significant advances in our understanding of stellar magnetic fields, dynamos, surface structures, and their evolution in time. In this review, I briefly summarize these developments, constructing a scenario for how stars of different masses appear and behave as they evolve. To keep things tractable, I will focus on the evolution of cool solar-like (F-K) dwarf stars from about Pleiades age ($t \approx 7 \times 10^7$ years) to stars roughly twice the Sun's age ($t \approx 9 \times 10^9$ years). Stars will be divided into several broad age groups, and the typical magnetic activity characteristics of each group discussed in turn, in order to build up an overall picture of the evolution of magnetic-related phenomena during the life of a cool dwarf star. For a more detailed discussion of some of the issues raised, I refer the reader to Saar & Baliunas (1992a), Radick (1992) and Saar (1994).

2. Very young stars (10 Myr $\leq$ age $\lesssim$ 0.1 Gyr)

Although recent photometry of the Pleiades (Prosser *et al.* 1993) and other young stars has considerably increased our knowledge of these stars' rotational periods, there is unfortunately little information on either their magnetic properties or their long-term activity variations. Only a few stars this young are nearby enough to have been included in the Mt. Wilson H+K programs, and most have photometry spanning a few seasons at best. Some exceptional cases, however, offer vague glimpses of the dynamo properties of these youthful objects.

HD 17925, a K2 dwarf, is thought to be roughly Pleiades age (Cayrel de Strobel & Cayrel 1989). Detailed models of low and high Zeeman sensitive lines show that for this star, the product of the area filling factor of "bright" magnetic regions (plage/active network), f_p, and the mean field strength in these regions B, is $f_pB \approx 525$ G (Saar 1991a; consistent with Basri & Marcy 1994). Its Ca II HK record reveals $P_{\rm rot} \approx 6.8$ days and considerable long-term variability from 1967–present (with an overall downward trend), but no clear cycle (if present, $P_{\rm cyc} \gtrsim 30$ years). A magnetic flux measurement for the "naked" T Tauri star Tap 35 of $f_pB \approx 1000$ G (Basri *et al.* 1992) with a Li I based age of $t \approx 0.008$ Gyr, combined with HD 17925, suggest a "saturation" in magnetic flux and f_p for $t \lesssim 8.5$ Gyr (Saar 1991b; Figure 1). Analysis of f_p as a function of rotation (Figure 1) suggests a possible reason for this result: the surface area of plage/network has reached some maximum level. Indeed, if B on Tap 35 shows the pressure equipartition field strengths typical of the cool stars measured to date (e.g., Saar 1991b), it would have $B \sim 1400$ G, and thus $f_p \sim 70\%$, clearly a significant coverage.

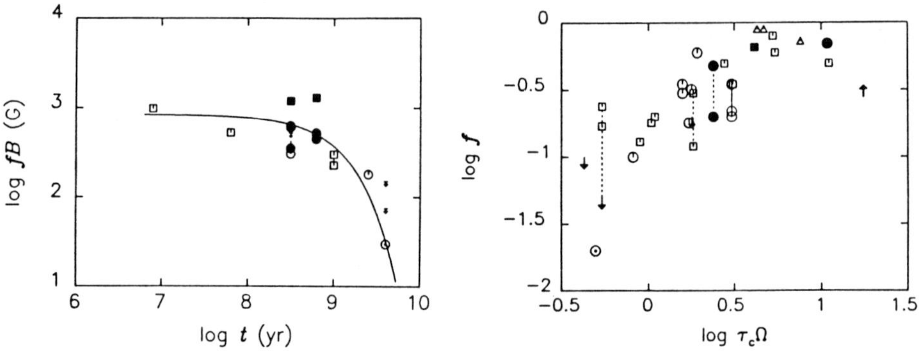

FIGURE 1. (*Left*): fB vs age t. Circles, squares and triangles represent G, K, and M stars; filled symbols represent cluster members, open symbols have ages based on Li I. An exponential fit is shown (solid; $fB \propto e^{-0.82t}$ with t in Gyr). HD 17925 and Tap 35 define the relation at young ages (from Saar 1991b). (*Right*): Magnetic filling factor f vs. inverse Rossby number $\tau_c/P_{\rm rot}$, where τ_c is the convective turnover time (e.g., Noyes *et al.* 1984). Here, open symbols are dwarfs and filled are lower gravity stars; the Sun is denoted by $\odot$; $\Omega \equiv P_{\rm rot}^{-1}$. Different measurements of the same object are connected by dotted lines. $f \propto (\tau_c\Omega)^\alpha$ $(1 \leq \alpha \leq 2)$ for low $\tau_c\Omega$; saturation sets in above $\tau_c/\Omega \gtrsim 4$ (from Saar 1991b).

HD 129333 (G0V; $P_{\rm rot} \approx 2.8$ d), one of the most active nearby G dwarfs, is also believed to be of Pleiades age. It displays a generally increasing trend in its Ca II flux, but no clear cycle (the earliest observations, however, date from 1978, so only short $P_{\rm cyc}$ are ruled out). Surface differential rotation (SDR) based on changes in photometric periods, is of order 0.1 of the solar value (E. Guinan 1993, private communication). The star does appear to show an anticorrelation between V band luminosity and Ca II flux, in the sense that the star is more active whan it is darker (Radick 1992), with $f_s \approx 5\%$ (Noyes *et al.* (1991). This trend is *opposite* of the solar case, but appears to be typical in young stars, and is discussed in more detail in the next section.

LQ Hya (=HD 82558; K2Ve; $P_{\rm rot}$ = 1.6 d) is also roughly Pleiades age, and one of the most active nearby K dwarfs. It has recently been Doppler imaged by two groups (Strassmeier *et al.* 1993; Saar *et al.* 1992, 1994), and shows an overall spot coverage of $f_s \sim 10\%$. Magnetic images, while only approximate, suggest the presence of significant flux almost everywhere and $f_p \approx 50\%$ if B = 2000 G. The images imply that dark spots on LQ Hya generally appear at higher latitudes than on the Sun, with significant spot coverage even near the poles. This picture of "polar" spots is supported by theory. Schüssler & Solanki (1992) suggest that such polar spots are expected on rapid rotators, where rotational effects dominate magnetic buoyancy and cause flux tubes to rise along equi-velocity surfaces rather than radially. From the study of the latitudinal extents of polar and equatorial spots, Saar *et al.* (1994) find an upper limit to surface differential rotation (SDR) rate of LQ Hya of $\lesssim 0.003$ of the solar value. Though no long-term Ca II data is available, there is some evidence for a 6.2 year photometric cycle (Jetsu 1993).

Finally, the well-studied BY Dra variable BD+26°730 (e.g., Bopp *et al.* 1983), though a member of the Hyades ($t \approx 0.7$ Gyr), rotates in just 1.8 days due to tidal synchronization with an unseen spectroscopic companion. It may thus be valid to compare BD+26°730 with other stars in this "very young" group, since it has the $P_{\rm rot}$ of a Pleiades star one tenth as old. BD+26°730 shows a clear long-term ($P_{\rm cyc} \sim 60$ year) photometric cycle. Magnetic field (Saar *et al.* 1990) and molecular band analysis (Saar & Neff 1990) imply that the star has $f_p \approx 50\%$ and $f_s \approx 20\%$. Since the stellar inclination is only

$i \approx 20°$, these results imply large magnetic flux and spot area near the pole. Little change in C IV 1550Å and Hα fluxes have been observed between 1981 and 1992, despite significant changes in photometric brightness (Saar *et al.* 1990). A possible explanation lies in the fact that BD+26°730, as well as HD 129333 and LQ Hya, are among the most active stars of their spectral types, with near-saturated levels of chromospheric and transition region emission (cf. Vilhu 1984). The dynamo production of magnetic flux in these stars can alter the relative amounts of f_s and f_p (and hence brightness) in time, it may be *unable* to significantly change the total fB (and hence the overall activity level) since $f_p + f_s \sim 1$.

Thus, although currently available data are sparse, preliminary indications are that very young stars have nearly saturated outer atmospheric emission, magnetic area filling factors, and perhaps surface magnetic flux. They have f_p on the order of 40% – 50% (or more) and $f_s \sim$ 5% to 20% (possibly larger for K stars than G stars), yielding spot-to-plage/network area ratios much larger than seen on the Sun. Spots are often found at or near the stellar poles, at least in K stars. The dynamo may show cycles as viewed in spot area (irradiance), but due to (near-)saturation in the chromosphere, other activity indicators can show a variety of behaviors, ranging from anticorrelations with brightness (HD 129333), to only irregular activity variability with no clear periodicity (BD+26°730). Thus, dynamo cycles may or may not be apparent, depending on the diagnostic used to search for them. Surface differential rotation appears to be greatly reduced (by factors of 10 to 100) relative to the Sun, while rotational periods (for single stars) span a broad range from $P_{\rm rot} \approx 1$ to 5 days, with $P_{\rm rot}$ typically increasing from F to K stars ($\tau_c/P_{\rm rot} \approx$ constant).

3. Young stars (0.3 Gyr $\lesssim$ age $\lesssim$ 0.8 Gyr)

With this group, we enter into a range of stellar ages which are reasonably well sampled by the long-term Ca II and photometric surveys. Considerably more data is therefore available. This particular group also encompasses the Hyades ($t \approx 0.7$ Gyr) and Ursa Major clusters ($t \approx 0.3$ Gyr), and thus many group members have well determined ages. Rotation periods at this epoch range from ~3 days (F stars) to ~12 days (K stars).

The bulk of the magnetic field data for G and K stars also come from this age range (younger stars often have too large $v \sin i$, older ones have too small f values for accurate measurement). Typically, one finds (Saar 1990) $f_p \approx$ 20% to 50%. These data are also primarily responsible in helping to establish the empirical relationships between magnetic flux and outer atmospheric emission (e.g., Schrijver *et al.* 1989). Chromospheric, transition region and coronal fluxes are correlated with fB as $F_{\rm chr} \propto (fB)^{0.5}$, $F_{\rm TR} \propto (fB)^{0.7}$, and $F_{\rm cor} \propto (fB)^{1.0}$ (Schrijver 1991), consistent with simple models (Montesinos & Jordan 1993). Few direct spot estimates exist; Campbell & Cayrel (1984) estimate $f_s \approx 3\%$ on the Hyades supercluster member HD 1835, and photometric variations suggest similar values for other Hyads and UMa cluster stars (Noyes *et al.* 1991). Thus, the f_s/f_p ratio (~ 0.1) is smaller here than for the very young stars (~ 0.3), but still larger than the solar value: $(f_s/f_p)_\odot \approx 0.1\%/2\% = 0.05$). Photometric variability drops below detection limits for stars hotter than about F7 in the Hyades (Radick *et al.* 1987), suggesting a more even distribution of spots (Giampapa & Rosner 1984) or an inhibition of large spot formation (Bünte & Saar 1993) may occur as $T_{\rm eff}$ increases.

Many of the basic characteristics and trends tentatively identified above in the very young stars are more clearly evident in this group. Analysis of yearly variability in the combined Mount Wilson Ca II (e.g., Baliunas *et al.* 1985) and Lowell/Cloudcroft photometry (e.g., Radick 1992) clearly shows that stars of Hyades age and younger show

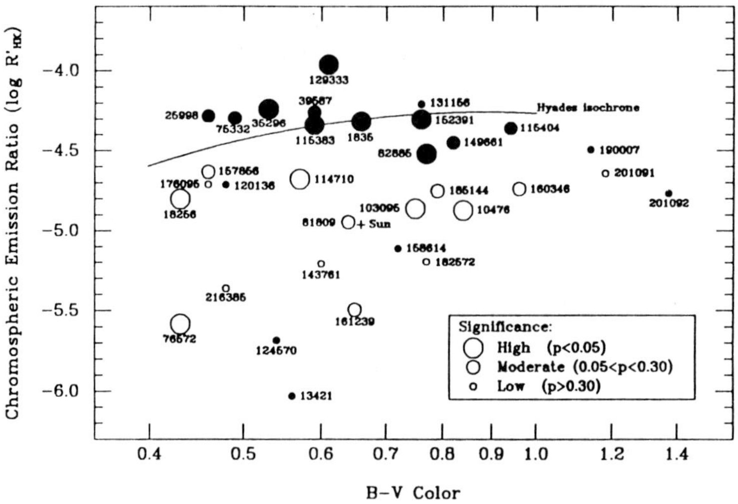

FIGURE 2. Correlation (open symbol = positive, filled = negative) between yearly brightness and Ca II variations, plotted in a color vs. normalized Ca II flux $R'_{\rm HK}$ ($= F'_{\rm HK}/\sigma T^4_{\rm eff}$) diagram (from Radick 1992).

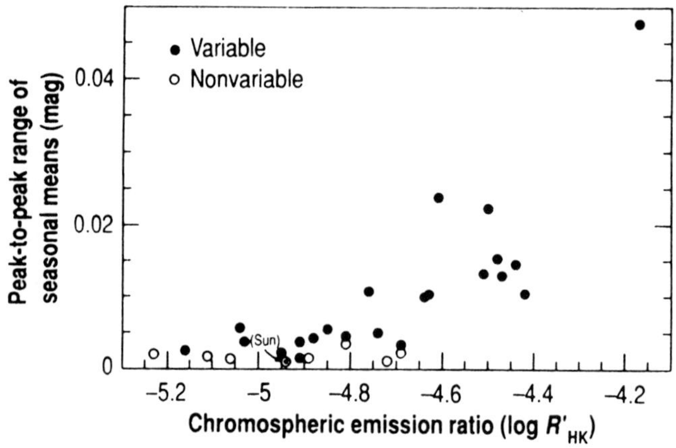

FIGURE 3. Seasonal photometric amplitude vs. $R'_{\rm HK}$ (from Radick *et al.* 1989).

anticorrelations between brightness and chromospheric activity – stars darken with increased activity (Radick *et al.* 1987; Figure 2). This fact, combined with evidence that the amplitude of the brightness variations shows a non-linear increase as a function of the Ca II variations (Figure 3), suggests that dark spots dominate brightness variations in active stars and increase more rapidly than plage with activity (e.g., Radick *et al.* 1989). These anticorrelations are even seen during the most active solar cycles (Foukal 1993); although more typically, solar plage, network and faculae conspire to overcome the irradiance decrease due to spots and produce a *positive* correlation between brightness changes and activity (e.g., Foukal & Lean 1988 and other papers in this volume: e.g., Chapman 1994; Dorren & Guinan 1994; Fröhlich 1994; Radick 1994).

A casual study of the Mount Wilson Ca II timeseries suggests they may be grouped into several broad morphological classes: irregular variables, cyclic variables (including a

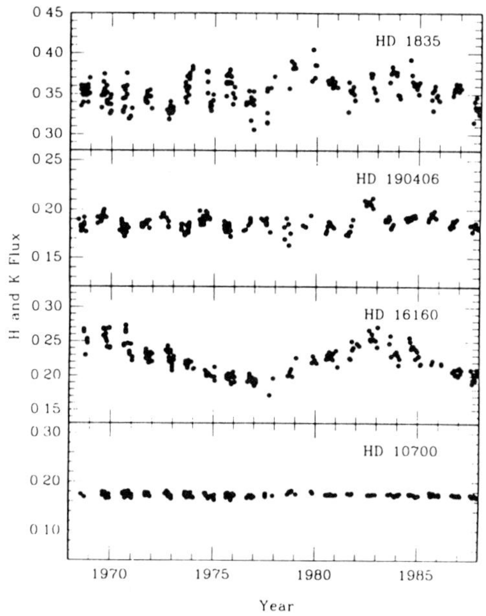

FIGURE 4. Timeseries of the Mount Wilson Ca II S index for four dwarfs (Baliunas 1991), showing the main types: irregular (HD 1835; $t \approx 0.8$ Gyr), multicyclic (HD 190406; $t \sim 2$ Gyr), cyclic (HD 16160; $t > 5$ Gyr), and non-variable (HD 10700; $t > 5$ Gyr).

few showing two periods simultaneously), and non-variables (e.g., Wilson 1978; Baliunas & Vaughan 1985; Figure 4). If a combination of cluster membership and an approximate Ca II flux vs. age relationship (Soderblom *et al.* 1991; Donahue 1993) is used to approximately date the stars in the sample, we find several general trends. First, there is a general progression from irregular variability, to cyclic and multicyclic behavior, to non-variability, with advancing age (Figure 5). In the age range considered here, most of the stars are irregular variables, although a few cyclic stars exist for $\log t \lesssim 8.6$. The interpretation of this irregular variability is somewhat uncertain. It could be that stars of Hyades/UMa age are still so active that the large changes in f_p needed to produce large amplitude Ca II variations are not possible. Flares and active region growth and decay may be so frequent that these mask the Ca II variations. Perhaps several dynamo "modes" of different amplitude and period $P_{\rm cyc}$ are operating simultaneously, which when noise is added yields seemingly chaotic timeseries. Throughout, K stars have larger fractional Ca II variations $A_{\rm HK} \equiv \Delta F_{\rm HK}/F'_{\rm HK}$, where $F'_{\rm HK}$ is the Ca II HK flux with the photospheric/basal component removed; see e.g., Noyes *et al.* 1984) than G stars (Saar & Baliunas 1992a). Support for the hypothesis that the Ca II cycles *are* indeed magnetic in origin comes from a four year study of fB and $F'_{\rm HK}$ in κ Ceti (Saar & Baliunas 1992b), in which the two quantities vary in phase during the declining part of a stellar cycle.

More information on the SDR rates is also available from stars in this group. Donahue *et al.* (1994) compute seasonal values of $P_{\rm rot}$ from the S dataset, and take the range of these, ΔP, to be a measure (actually a lower limit) to the surface differential rotation rate. Figure 6 shows the resulting ΔP values plotted against $P_{\rm rot}$. A clear correlation is seen, with $\Delta P \propto P_{\rm rot}^{1.35}$, independent of spectral type (i.e., mass or convective zone depth, d) on the main sequence. SDR rates for the "young" group are therefore on the order of 0.1 of the solar value (larger in K stars, smaller in F stars). Although several selection effects complicate matters, nevertheless, the non-linear relationship and lack of

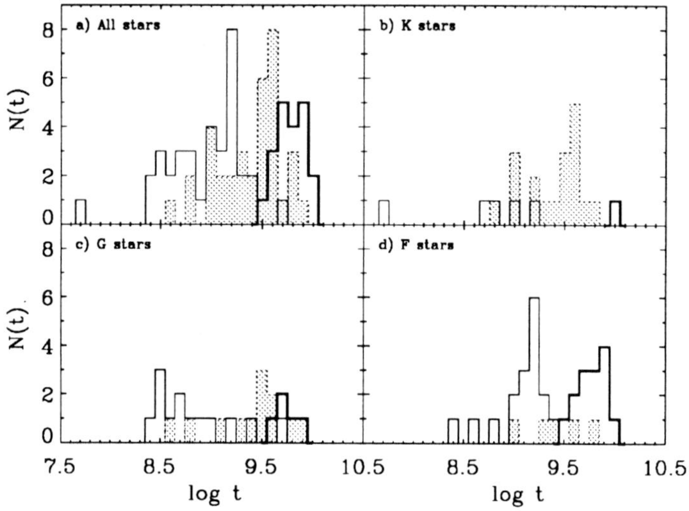

FIGURE 5. Histogram of stellar ages by spectral class and morphology type (thin = irregular, shaded = cyclic/multicyclic, thick=non-variable). Note the trends with spectral type and age (from Saar *et al.* 1994).

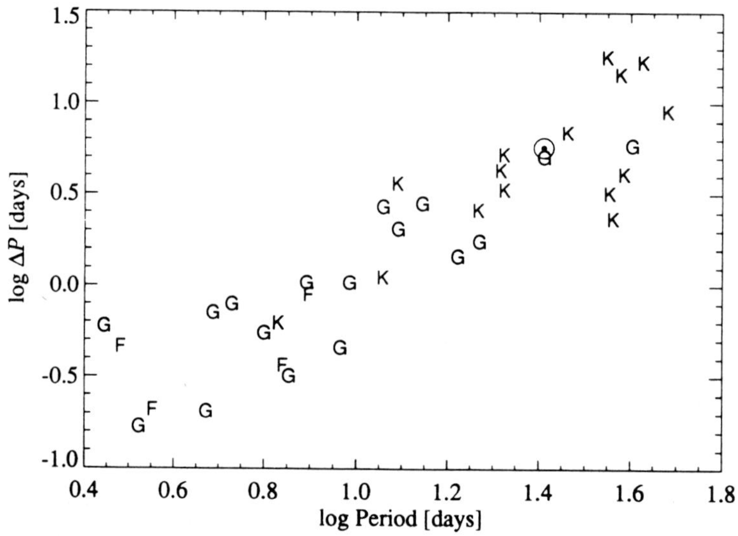

FIGURE 6. The range of seasonally derived $P_{\rm rot}$ values, ΔP, vs. $P_{\rm rot}$ (spectral types as marked; Sun=$\odot$). The best fit yields $\Delta P \propto P_{\rm rot}^{1.35}$ (Donahue *et al.* 1994).

spectral type dependence are both at odds with simple theoretical models (e.g., Belvedere *et al.* 1980; Küker *et al.* 1993).

Finally, there is evidence that convection is altered on young, active stars. Convective velocities can be estimated from the so-called macroturbulent velocity ($v_{\rm mac}$) derived from detailed models of high resolution absorption line spectra (e.g., Gray 1988). Gray (1984) and Toner & Gray (1988) already noted enhanced $v_{\rm mac}$ on active stars, especially ξ Boo A. Saar & Osten (1994, in preparation) have modeled high resolution ESO coude

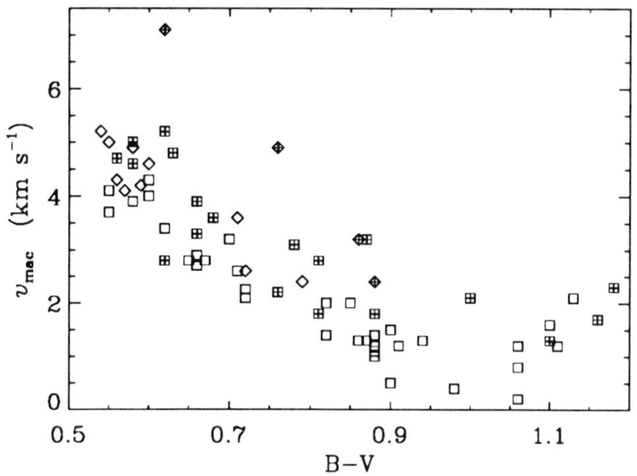

FIGURE 7. Radial-tangential macroturbulence $v_{\rm mac}$ vs. B–V color; squares are derived from ESO data, diamonds are from Gray (1984), crossed symbols denote stars with significant chromospheric and/or coronal emission (typically with $t \lesssim 2$ Gyr). Note that the mean $v_{\rm mac}$ for active stars is roughly 1 km s^{-1} more than for inactive stars at fixed color (i.e., fixed $T_{\rm eff}$). From Saar & Osten (1994, in preparation).

spectra of a large sample of southern dwarfs, and confirm Gray's hypothesis (Figure 7): active stars show $v_{\rm mac} \approx 1-2$ km s^{-1} higher than inactive stars of the same color ($T_{\rm eff}$). Line bisector spans also show variation with stellar cycle (Gray 1992; Livingston 1991). Since convective energy transport is inhibited in magnetic regions (e.g., Livingston 1991), stars where f is large may force convection outside magnetic regions to be that much more violent (e.g., Gray 1992). The enhancement in $v_{\rm mac}$ is likely present in the very young stars as well, but is difficult to observe due to increased rotational broadening.

4. Intermediate age stars (age $\approx$ 1 to 2 Gyr)

Intermediate age stars have rotation periods (based on HK data) of 6 days (F stars) $\lesssim P_{\rm cyc} \lesssim 20$ days (K stars). There are no nearby clusters in this age range, so t must be estimated indirectly (e.g., Soderblom *et al.* 1991). As f_p is declining with t (Figure 1), only a few stars in this group have magnetic measurements. One of the most recent (and accurate) measurements is for the K2 dwarf ϵ Eri (Valenti & Basri 1994), which shows $f_p \sim 12\%$ and $B \sim 1900$ G. Dorren & Guinan (1982) estimate f_s of a few percent for several of these stars. For the class as a whole, I estimate $f_p \approx 5\%$ to 20%, and $f_s \lesssim 2\%$, suggesting spot-to-plage ratios only somewhat larger than the Sun. Convective velocities here are subsiding to their "typical" inactive values (Figure 7).

Intermediate age stars are transitional in several respects. Cyclic variability in Ca II becomes more common, especially in G and K stars (Figure 5). The paucity of cyclic F stars may partly be a threshold problem, however (i.e., $A_{\rm HK}$ too small to detect). Several of these young cyclic stars are of the multicyclic variety, displaying two distinct $P_{\rm cyc}$ values. As was the case with the irregular variables, the fractional amplitude $A_{\rm HK}$ of the cycles increases with decreasing $T_{\rm eff}$ and increasing d (Saar & Baliunas 1992a). SDR rates are perhaps a factor of 2 to 3 faster than the Sun (Figure 6) and they can show distinctly non-solar SDR behavior. HD 114710, for example, displays evidence for *two* activity belts and possibly a reversed pole-to-equator angular velocity gradient

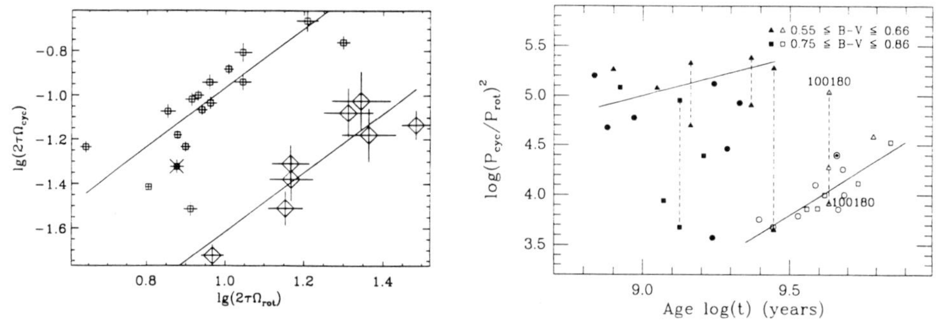

FIGURE 8. *Left*: Cycle frequency vs. rotational frequency, both normalized to τ_c (large symbols are young/active, small symbols are old/inactive, the Sun is marked with a $\times$). All cycles plotted are primary $P_{\rm cyc}$ values with FAP $\leq 10^{-6}$. Note the segregation of stars into active and inactive branches, each with $\tau_c\Omega_{\rm cyc} \propto (\tau_c\Omega_{\rm rot})^{1.3}$ (solid). *Right*: $(P_{\rm cyc}/P_{\rm rot})^2$ vs. t (from Soderblom *et al.* 1991) for stars with a less restrictive FAP $\leq 10^{-2}$ (solid line denotes fits to subsets with FAP $\leq 10^{-5}$). Dashed lines connect two $P_{\rm cyc}$ for multicyclic stars, and filled symbols are active stars. The young/old dichotomy is again apparent, with the transition between 1 and 3 Gyr (from Soon *et al.* 1993).

(Donahue & Baliunas 1992). The sense of the brightness-activity relation changes as well, with some G stars showing solar-like positive correlations (e.g., HD 114710; Figure 2) while some K dwarfs continuing to display young-star-type anticorrelations (e.g., HD 149661; Figure 2). One of the most striking transitional aspects of the intermediate stars however, involves their cycle periods.

Many dynamo theories predict relationships between $P_{\rm cyc}$ and $P_{\rm rot}$ (e.g, Robinson & Durney 1982). Noyes *et al.* (1984) first noted a relationship between $P_{\rm cyc}$ and $P_{\rm rot}$ for inactive, older stars (see next section). Saar & Baliunas (1992a) studied this issue again, with the benefit of almost 10 more years of data (Baliunas *et al.* 1994, in preparation). They demonstrated that both inactive and active stars show clear and similar correlations when the best determined $P_{\rm cyc}$ are used (those with "false alarm probabilities" or FAP $\leq 10^{-6}$), and when $P_{\rm cyc}$ and $P_{\rm rot}$ are properly normalized (they used the magnetic diffusion timescale and τ_c in two examples). If these steps are taken, the inactive and active stars are displaced into nearly parallel branchs, with inactive stars showing lower $P_{\rm cyc}$ values (Figure 8). If τ_c is used in the normalization, $\tau_c\Omega_{\rm cyc} \propto (\tau_c\Omega_{\rm rot})^{1.3}$. Note that then (approximately), $\Omega_{\rm cyc} \propto (\Delta P)^{-1}$, suggesting a possible connection between SDR and the cycle period. Since $\tau_c/P_{\rm rot}$ is, to first order, correlated with stellar age, a transition $P_{\rm cyc}/P_{\rm rot}$ is implied at some t. Several of the multicyclic stars have a $P_{\rm cyc}$ value in each regime, suggesting they may represent a transitional phase in evolution between branches. Soon *et al.* (1993) confirmed this idea using a somewhat different parameterization (Figure 8), and showed that the transition takes place between 1 and 3 Gyr. Perhaps some critical change in the dynamo mechanism (e.g., Knobloch *et al.* 1981) occurs at this time. Soon *et al.* (1993) also show that $P_{\rm cyc}/P_{\rm rot}$ decreases with $T_{\rm eff}$ (and d) in both branches.

5. Old stars (age $\gtrsim$ 3 Gyr)

This group encompasses roughly solar age and older stars, and consistent with this fact, most behavior observed is truly "solar-like". I discuss stars with $t \lesssim t_\odot$ first. Typical $P_{\rm rot}$ values range from 15 days (late F stars) to 40 days (K stars). Of this class, only the

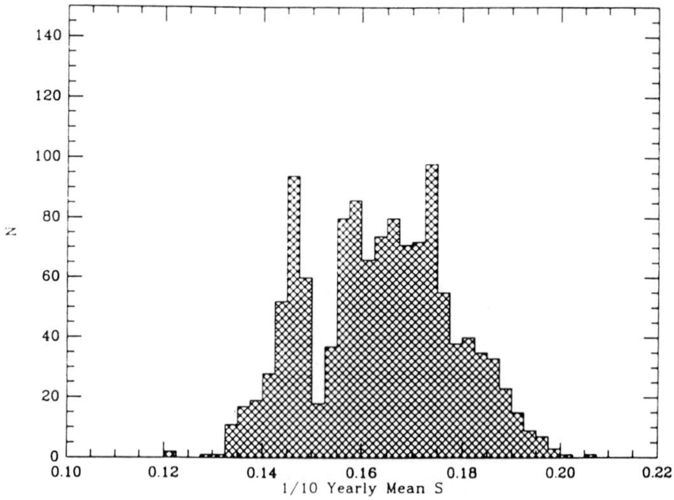

FIGURE 9. Frequency distribution of the S index (0.1 year averages) for inactive ($0.13 \leq \langle S \rangle \leq 0.20$) solar–like ($0.60 \leq$ (B-V) ≤ 0.76) dwarfs (from Baliunas & Jastrow 1990).

Sun has reasonably accurate estimates of spot and plage area: $f_p \sim 2\%$ (Montesinos & Jordan 1993) and $f_s \sim 0.1\%$ SDR rates are similar to the Sun, though proportionately larger in K stars and smaller in F stars. Very few irregular Ca II variables remain by this epoch; cycles dominate in the K stars, while a "new" class, the non-variables, becomes important in the G and especially the F stars (Figure 5). Well determined correlations between brightness and Ca II variations are almost invariably positive in sign (solar-like), indicating the dominance of plage and network in the irradiance variations. The cyclic variables are found only on low $P_{\rm cyc}$ branch in the $P_{\rm cyc}$-$P_{\rm rot}$ diagrams (Figure 8).

The non-variable stars deserve special attention, since they have characteristics similar to what we infer the Sun was like during the Maunder and similar "magnetic minima". These are epochs when it is thought that cycles, but not all magnetic activity, temporarily cease. Baliunas & Jastrow (1990) compiled S values for 74 dwarfs, similar to the Sun in color and age, from the combined Wilson and solar neighborhood surveys. After averaging the data into 0.1 year bins, they found that a histogram of the results showed a sharply bimodal distribution (Figure 9): a narrowly peaked distribution for $S < 0.15$, separated by a sharp gap from a broad distribution produced by cyclic variables. Stars below the gap have no cycles, and lie at the minimum S levels for their spectral types. Baliunas & Jastrow (1990) suggest that the low activity, non-varying stars in the low S group are, in fact, experiencing the stellar equivalent of the Maunder Minimum, noting that the fraction of stars in this group is similar to the fraction of time the Sun spends in a quiescent magnetic state. If this hypothesis is correct, and the non-variable stars are in Maunder-like states, several statements can be made: 1) The frequency and/or duration of Maunder-type minima drops sharply with increasing d (Figure 5). 2) The Maunder phenomenon begins rather abruptly at $\log t \approx 9.5$. 3) For stars younger than this, it is possible to have cyclic dynamos without Maunder-type minima. This suggests the possibility that Maunder-like states increase in frequency and/or duration with t, though there is little *direct* evidence for this.

In stars older than the Sun, the F dwarfs will have mostly evolved off the main sequence. For the G and K stars, an extrapolation of the above trends suggests $P_{\rm rot}$ will range from

30 to 50 days, SDR will increase above solar values, f_p and f_s will decrease further, and cyclic variability and Maunder minima will dominate the Ca II timeseries morphology (with K stars more often showing cycles and G stars Maunder minima). $P_{\rm cyc}/P_{\rm rot}$ should increase with time along the inactive tracks in Figure 8. There is no concrete evidence that the dynamo ever completely "shuts off".

Acknowledgments. I gratefully acknowledge preprints from and useful and informative discussions with many of those referenced below! Special thanks are due S. Baliunas, A. Brandenburg, R. Donahue, and E. Guinan for permission to discuss unpublished results and to R. Osten for help with the references. I thank ESO and NOAO for generous allocations of telescope time.

REFERENCES

BALIUNAS S.L. 1991 The past, present, and future of solar magnetism: stellar magnetic activity. In *The Sun in Time* (ed. M. Matthews, C.P. Sonnett & M.S. Giampapa). pp. 809–831. Univ. Arizona Press, Tucson, AZ, USA.

BALIUNAS, S.L. & VAUGHAN, A.H. 1985 Stellar activity cycles. *Ann. Rev. Astron. Astrophys.* **23**, 379–412.

BALIUNAS, S.L. & JASTROW, R. 1990 Evidence for long-term brightness changes of solar-type stars. *Nature* **348**, 520–523.

BALIUNAS, S.L., HORNE, J.H., PORTER, A., DUNCAN, D.K., FRAZER, J., LANNING, H., MISCH, A., MUELLER, J., NOYES, R.W., SOYUMER, D., VAUGHAN, A.H. & WOODARD, L. 1985 Time-series measurements of chromospheric Ca II H and K emission in cool stars and the search for differential rotation. *Astrophys. J.* **294**, 310–325.

BASRI, G.S. & MARCY, G.W. 1994 Zeeman enhancement of lines in extremely active K dwarfs. *Astrophys. J.*, in press.

BASRI, G.S., MARCY, G.W. & VALENTI, J. 1992 Limits on the magnetic flux of pre-main-sequence stars. *Astrophys. J.* **390**, 622–633.

BELVEDERE, G., PATERNO, L. & STIX, M. 1980 Differential rotation along the lower main sequence: A theoretical investigation. *Astron. Astrophys.* **88**, 240–247.

BOPP, B.W., AFRICANO, J.L., STENCEL, R.E., NOAH, P.V. & KLIMKE, A. 1983 Observations of active chromosphere stars. *Astrophys. J.* **275**, 691–703.

BÜNTE, M. & SAAR, S. H. 1994 The interchange instability in stellar magnetic flux tubes. *Astron. Astrophys.* **271**, 167–174.

CAMPBELL, B. & CAYREL, R. 1984 Spectroscopic evidence for starspots in the G dwarf HD 1835. *Astrophys. J.* **283**, L17–L20.

CAYREL DE STROBEL, G. & CAYREL, R. 1989 Strong lithium in the very nearby K dwarf HD 17925. *Astron. Astrophys.* **218**, L9–L11.

CHAPMAN, G.A. Photometric observations of the Sun. In *The Sun as a Variable Star: Solar and Stellar Irradiance Variations* (ed. J.M. Pap, C. Fröhlich, H.S. Hudson & S.K. Solanki). Cambridge Univ. Press, in press.

DORREN J.D. & GUINAN, E.F. 1982 Evidence for starspots on single solar-like stars. *Astron. J.* **87**, 1546–1557.

DONAHUE, R.A. 1993 PhD thesis, New Mex. State Univ.

DONAHUE, R.A.& BALIUNAS, S.L. 1992 Evidence of differential surface rotation in the solar-type star HD 114710. *Astrophys. J.* **393**, L63–66.

DONAHUE, R.A., SAAR, S.H. & BALIUNAS, S.L. 1994 Differential rotation in cool dwarfs: A power law relationship between mean rotation period and its observed range. *Astrophys. J.*, submitted.

DORREN, J.D. & GUINAN, E.F. 1994 The Sun in time: Detecting and modelling magnetic inhomogenities on solar-type stars. In *The Sun as a Variable Star: Solar and Stellar Irradiance Variations* (ed. J.M. Pap, H.S. Hudson, C. Fröhlich & S.K. Solanki). Cambridge Univ. Press, in press.

FOUKAL, P. 1993 The curious case of the Greenwich faculae. *Solar Phys.* **148**, 219–232.

FOUKAL, P. & LEAN, J. 1988 Magnetic modulation of solar luminosity by photospheric activity. *Astrophys. J.* **328**, 347–357.

FRÖHLICH, C. 1994 Irradiance observations of the Sun. In *The Sun as a Variable Star: Solar and Stellar Irradiance Variations* (ed. J.M. Pap, C. Fröhlich, H.S. Hudson & S.K. Solanki). Cambridge Univ. Press, in press.

GIAMPAPA, M.S. & ROSNER, R. 1984 The appearance of magnetic flux on the surfaces of the early main-sequence F stars. *Astrophys. J.* **286**, L19–L22.

GRAY, D.F. 1984 Measurements of rotation and turbulence in F, G, and K stars. *Astrophys. J.* **281**, 719–722.

GRAY, D.F. 1988 *Lectures on Spectral Line Analysis: F, G, and K Stars* (Arva).

GRAY, D.F. 1992 Stellar convection: the observations. In *Cool Stars, Stellar Systems, and the Sun* (ed. M. Giampapa & J. Bookbinder). ASP **Vol. 26**, 127–138.

JETSU, L. 1993 A decade of photometry of LQ Hydrae. *Astron. Astrophys.* **276**, 345–352.

KNOBLOCH, E., ROSNER, R. & WEISS, N.O. 1981 Magnetic fields in late-type stars. *Mon. Not. R. A. S.* **197**, 45P–49P.

KÜKER, M., RÜDIGER, G. & KITCHATINOV, L.L. 1993 An $\alpha\Omega$ model of the solar differential rotation. *Astron. Astrophys.* **279**, L1–L4.

LIVINGSTON, W. 1991 Sun as a star: its convective signature and the activity cycle. In *IAU Colloq. 130, The Sun and Cool Stars: Activity, Magnetism, Dynamos* (ed. I. Tuominen, D. Moss & G. Rüdiger). pp. 246–251. Springer.

MONTESINOS, B.M. & JORDAN, C. 1993 On magnetic fields, stellar coronae, and dynamo action in late-type dwarfs. *Mon. Not. R. A. S.* **264**, 900–918.

NOYES, R.W., HARTMANN, L.W., BALIUNAS, S.L., DUNCAN, D.K. & VAUGHAN, A.H. 1984 Rotation, convection, and magnetic activity in lower main-sequence stars. *Astrophys. J.* **279**, 763–777.

NOYES, R.W., WEISS, N.O. & VAUGHAN, A.H. 1984 The relation between stellar rotation rate and activity cycle periods. *Astrophys. J.* **287**, 769–773.

NOYES, R.W., BALIUNAS, S.L. & GUINAN, E.F. 1991 What can other stars tell us about the Sun? In *The Solar Interior and Atmosphere* (ed. A.N. Cox, W.C. Livingston & M.S. Matthews). pp. 1161–1185. Arizona Univ. Press, Tucson, AZ, USA.

PROSSER, C.F., SHETRONE, M.D., MARILLI, E., CATALANO, S., WILLIAMS, S.D., BACKMAN, D.E., LAAKSONEN, B.D., ADIGE, V., MARSCHALL, L.A. & STAUFFER, J.R. 1993 Rotation periods of open cluster stars II. *Publ. Astron. Soc. Pac.* **105**, 1407–1414.

RADICK, R.R. 1992 Luminosity variability of lower main-sequence stars. In *The Solar Cycle* (ed. K. Harvey). ASP **Vol. 27**, 450–464.

RADICK, R.R. 1994 Photometric variations of solar type stars. In *The Sun as a Variable Star: Solar and Stellar Irradiance Variations* (ed. J.M. Pap, C. Fröhlich, H.S. Hudson & S.K. Solanki). Cambridge Univ. Press, in press.

RADICK, R.R., THOMPSON, D.T., LOCKWOOD, G.W., DUNCAN, D.K. & BAGGETT, W.E. 1987 The activity, variability, and rotation of lower main-sequence Hyades stars. *Astrophys. J.* **321**, 459–472.

RADICK, R.R., LOCKWOOD, G.W. & BALIUNAS, S.L. 1989 Stellar activity and brightness variations: A glimpse at the Sun's history. *Science* **247**, 39–44.

ROBINSON, R.D. & DURNEY, B.R. 1982 On the generation of magnetic fields in late-type stars: a local time–dependent dynamo model. *Astron. Astrophys.* **108**, 322–325.

SAAR, S.H. 1990 Magnetic fields on solar-like stars: The first decade. In *IAU Symposium 138, The Solar Photosphere: Structure, Convection, and Magnetic Fields* (ed. J. O. Stenflo). pp. 427–441. Kluwer.

SAAR, S.H. 1991a Recent advances in the observation and analysis of stellar magnetic fields. In *IAU Colloquium 130, The Sun and Cool Stars: Activity, Magnetism, Dynamos* (ed. I. Tuominen, D. Moss & G. Rüdiger). pp. 389–401. Springer.

SAAR, S.H. 1991b Recent measurements of stellar magnetic fields. In *Mechanisms of Chromospheric and Coronal Heating* (ed. P. Ulmschneider, E. Priest & R. Rosner). pp. 273–278. Springer.

SAAR, S.H. 1994 The time evolution of magnetic flux, dynamos, and surface structures on cool stars. In *The Eighth Cambridge Workshop on Cool Stars, Stellar Systems, and the Sun* (ed. J.-P. Caillault). ASP, in press.

SAAR, S.H. & NEFF, J.E. 1990 Spot temperatures and area coverages on active dwarf stars. In *Cool Stars, Stellar Systems, and the Sun* (ed. G. Wallerstein). ASP *Vol. 9*, 171–173.

SAAR, S.H. & BALIUNAS, S.L. 1992a, Recent advances in stellar cycle research. In *The Solar Cycle* (ed. K. Harvey). ASP **Vol. 27**, 150–167.

SAAR, S.H. & BALIUNAS, S.L. 1992b The magnetic cycle of Kappa Ceti (G5V). *Ibid*, 197–202

SAAR, S.H., GOLUB, L., BOPP, B.W., HERBST, W. & HUOVELIN, J. 1990 Long and short timescale variability of magnetic activity on the BY Dra star BD +26° 730. In *Evolution in Astrophysics: IUE Astronomy in the Era of New Space Missions* (ed. E. Rolfe). ESA SP–310, 431–434.

SAAR, S.H., PISKUNOV, N.E. & TUOMINEN, I. 1992 Magnetic surface images of the BY Dra Star HD 82558. In *The Seventh Cambridge Workshop on Cool Stars, Stellar Systems, and the Sun* (ed. M.S. Giampapa & J.A. Bookbinder). ASP **Vol. 26**, 252–255.

SAAR, S.H., PISKUNOV, N.E. & TUOMINEN, I. 1994a Multiepoch magnetic surface images of LQ Hya In *The Eighth Cambridge Workshop on Cool Stars, Stellar Systems, and the Sun* (ed. J.-P. Caillault). ASP, in press.

SAAR, S.H., BRANDENBURG, A., DONAHUE, R.A. & BALIUNAS, S.L. 1994b The evolution of stellar dynamo variations. In *The Eighth Cambridge Workshop on Cool Stars, Stellar Systems, and the Sun* (ed. J.-P. Caillault). ASP, in press.

SCHRIJVER, C.J. 1991 Relations between magnetic fields and activity. In *Mechanisms of Chromospheric and Coronal Heating* (ed. P. Ulmschneider, E. Priest & R. Rosner). pp. 257–272. Springer.

SCHRIJVER, C.J., COTÉ, J., ZWAAN, C. & SAAR, S.H. 1989 Relations between the photospheric magnetic field and the emission from the outer atmospheres of cool stars. I. The solar Ca II K line core emission. *Astrophys. J.* **337**, 964–976.

SCHÜSSLER, M. & SOLANKI, S.K. 1992 Why rapid rotators have polar spots. *Astron. Astrophys.* **264**, L13–L16.

SODERBLOM, D.R., DUNCAN, D.K. & JOHNSON, D.H.R. 1991 The chromospheric emission-age relation for stars of the lower main sequence and its implications for the star formation rate. *Astrophys. J.* **375**, 722–739.

SOON, W.H., BALIUNAS, S.L. & ZHANG, Q. 1993 An interpretation of cycle periods of stellar chromospheric activity. *Astrophys. J.* **414**, L33–L36.

STRASSMEIER, K., RICE, J.B., WEHLAU, W.H., HILL, G.M. & MATTHEWS, J.M. 1993 Surface features in the lower atmosphere of HD 82558 = LQ Hya. *Astron. Astrophys.* **268**, 671–683.

TONER, C.G. & GRAY D.F. 1988 The starpatch on the G8 dwarf ξ Bootis A. *Astrophys. J.* **334**, 1008–1020.

VALENTI, J. & BASRI, G. 1994 Multiline Zeeman analysis of ϵ Eri in the near infrared. In *The Eighth Cambridge Workshop on Cool Stars, Stellar Systems, and the Sun* (ed. J.-P. Caillault). ASP, in press.

VILHU, O. 1984 The nature of magnetic activity in lower main sequence stars. *Astron. Astrophys.* **133**, 117–126.

WILSON, O.C. 1978 Chromospheric variations in main sequence stars. *Astrophys. J.* **266**, 379–396.

Variations of the Coronal Radiation in X-ray Related to Coronal Holes, Active Region Loop Systems, Bright Points

By ESTER ANTONUCCI

Istituto di Fisica Generale, University of Torino, 10125 Torino, Italy

The coronal features observed in X-ray emission, varying from the small-scale, short-lived bright points to the large-scale, long-lived coronal holes, are closely associated with the coronal magnetic field and its topology, and their variability depends strongly on the solar cycle. Here we discuss the spatial distribution of the coronal structures, the frequency distribution of the brightness variations in active regions, and the role of magnetic reconnection in determining the variability of the coronal features, on the basis of the new observations of the soft X-ray emission recently obtained with the *Yohkoh* satellite and the NIXT experiment.

1. Introduction

The highly variable character of the solar X-ray emission through the solar cycle and the tendency to observe an increasing variability with photon energy became immediately evident since the first X-ray observations performed in the fifties. The X-ray emission appeared to be not only variable in time but also highly inhomogeneous: the first X-ray image of the Sun obtained on April 19, 1960, by the Naval Research Laboratory researchers, showed that approximately 80% of the emission was coming from 5% or less of the area of the disk.

Only with the *Skylab* experiments, in the early 70's, however, were the characteristics of the highly variable nature and the detailed structure of the corona definitely established. Coronal loops were recognized as the basic elements of the solar corona (Rosner *et al.* 1978a), since the X-ray emission comes principally from coronal regions of closed magnetic fields, while it becomes negligible in coronal holes where the field is believed to be open. A wealth of interconnections and interactions between the different coronal regions was discovered. It was also possible to establish the existence in the transition region and in the corona of structures down to a few arcseconds, that is down to the instrumental resolution.

An extremely important discovery, during *Skylab*, was that magnetic fields are not only acting to confine the solar plasma in the corona, but they are also playing a fundamental role in coronal heating. Rosner *et al.* (1978b) pointed out that two elements are very likely necessary to produce a hot corona: a relatively strong magnetic field coupled with an input of mechanical energy into the field. At approximately the same time, Athay & White (1978; 1979) were establishing on the basis of the OSO VIII observations, that the acoustic waves originating in the photosphere do not represent an effective agent for coronal heating, being stopped before reaching the corona, thus supporting the view of a coronal heating process strictly related to the presence of the magnetic field.

The picture of the solar corona emerging from *Skylab* and the OSO satellites was that of an inhomogeneous atmosphere, consisting of features of different scales, each with a different role in different phases of the solar cycle. Active regions are intermediate structures, dominant in the rising phase of the cycle and at solar maximum, when with the beginning of the new cycle the relatively simple configuration of the magnetic field typical of solar minimum is disrupted by the emergence of the new toroidal field. Active regions originate non-uniformly over the solar disk, and they are typically organized in latitude belts migrating toward the equator. Coronal holes, reflecting the long-lived large-scale pattern of the solar magnetic fields, are dominant in the declining phase of solar activity, when the solar magnetic field is dominated by the lower order multipoles. They appear as large structures extending over 1-5% of the solar disk, with a fast growth, followed by a relatively slow evolution, and an overall lifetime of several solar rotations. In the declining phase, coronal holes are typically longitudinal structures extending from low latitudes to the poles. At solar minimum the X-ray bright points (XBP's), which appear as small ($\leq$ 1 arcmin), short-lived ($\leq$ 2 days) active regions, are the dominant structures, with a rather uniform distribution on the solar disk (Golub et al. 1979).

The overall picture of the solar corona emerging with *Skylab* remained substantially unchanged till the late 80's, when a new potentiality for coronal studies was provided by achieving the capability of subarcsecond imaging in the extreme ultraviolet and soft X-rays, and with the launch in 1991 of the Japanese satellite, *Yohkoh.* The Soft X-ray Telescope (SXT) on *Yohkoh* is almost continuously imaging the solar corona in the region from 2 Å to 40 Å with a wide dynamic range and an excellent spatial resolution (2.5 arcsec pixels). The almost continuous monitoring of the Sun provided by *Yohkoh* is revealing in detail the extremely dynamic behavior of the corona, which is in continuous evolution from the small to the large-scale. The sub-arcsecond imaging of the solar corona obtained with NIXT, made possible by the advent of telescopes carrying multilayer-coated optics responding in the XUV range, has revealed a complex coronal fine structure down to 600 km, the instrument resolution (Golub *et al.* 1990).

2. Spatial distribution of coronal structures in active regions

According to the picture emerging from *Skylab* observations, the large majority of the material of an active region above the chromosphere is emitting at a temperature above 10^6 °K. The emission comes mainly from compact hot loops, $10^4 - 10^5$ km in length scale, with their footpoints in bright plages of opposite magnetic polarity, sites for strong fields. Hot bright loops, visible in hotter ions such as Mg IX, are not the only constituents of an active region. They coexist with cool loops which are visible in cooler ions formed below 10^6 °K, such as Ne VII, and they are typically thinner, longer, less stable and less numerous than hot loops. Hot and cool loops are structures forming in different parts of an active region; hot loops are clustered across the neutral line, while the cooler longer loops are rooted in the periphery of active regions (Webb 1981).

The coronal portion of the loop is inferred to be quasi-isothermal while the cool emission is confined near the footpoints due to the high conductivity of the solar plasma which tends to smooth the temperature variations along the loop. Typical quiescent regions appear to be formed by individual plasma loops whose temperatures cluster in a narrow range within 2 and 3×10^6 °K and with a temperature gradient at the loop base inferred to be above 2×10^{-2} °K cm^{-1}.

The largest brightness variations between the loops, or within the same loop, are principally due to density rather than temperature variations. Neupert (1968) suggested chromospheric evaporation as the most likely explanation for density enhancements in the solar corona. That is, flows of plasma from the dense chromosphere to the rarefied corona can be induced whenever during transient energy release phenomena only a negligible amount of the energy deposited in the upper chromosphere can be dissipated. In this case, the energy in excess is transferred to the corona in form of an upward enthalpy flux.

The NIXT experiment, initiated successfully with the rocket launched on September 11, 1989, provided sub-arcsecond (0.75 arcsec) narrow band images at 63.5 Å, which include the emission from the Fe XVI and Mg X ions (Golub *et al.* 1990). NIXT observations clearly evidenced the existence of fine structure, down to the instrumental resolution, at all locations in the corona, showing that the magnetic complexity observed in the photosphere is probably reflected at coronal levels (Golub *et al.* 1990). These observations are strongly suggesting that coronal loops are further structured into even more elementary features.

A possible filamentary nature of the corona was previously suggested through indirect evidence derived from the *Solar Maximum Mission* (SMM) observations, which led to infer a width of the filaments of the order of 100-300 km (Linford & Wolfson 1988; Martens *et al.* 1985). In fact the density of the coronal plasma derived by direct methods, such as by using density-sensitive line ratios, is usually found to exceed the density inferred by combining the plasma emission measure and the scale of the soft X-ray emission region. This can be interpreted as an indication of the non-uniform distribution of the plasma in the coronal region which is concentrated in filamentary structures, not yet resolved in the observations.

When studying the scale distribution of the coronal structures with NIXT images, which resolve magnetic features down to about 600 km, Gomez *et al.* (1993a) conclude that no preferred length-scale can be found for X-ray coronal structures in active regions. The two-dimensional power spectra on the wavenumber plane are substantially isotropic with no significant individual peaks as one would expect if active regions were consisting of a number of magnetic fluxtubes of given size with no internal structures. In this case, in fact, power spectra would exhibit peaks corresponding to the characteristic size of the fluxtubes.

The omnidirectional spectra, obtained by calculating the average spectrum in rings of increasing radius, are power law with the same power law index α constant for the different active regions. This conclusion suggests that the process generating the X-ray emission is substantially the same for active regions. Dark regions behave in a qualitatively different manner and do not share the same power law distribution found for active regions. NIXT results are confirmed by a similar analysis performed on the *Yohkoh* images (Martens & Gomez 1992), which however yields a power law index, $\alpha = -2.4$, which is somewhat different from the NIXT value, $\alpha = -3$ (Gomez *et al.* 1993a). The difference, if interpreted in the light of the different responses of the instruments, would imply that hotter structures tend to be smaller in size, the *Yohkoh* soft X-ray telescope SXT being more sensitive to higher temperatures.

Gomez *et al.* (1993b) have been able to model the observed spatial distribution of the X-ray emission, by assuming that the Poynting flux entering the corona is proportional to the radiative losses in the coronal plasma and by relating it to the fluctuating velocity and magnetic fields in the photosphere which can be assumed to be also distributed according to a power law.

3. Temporal variability in active regions

3.1. *Transient brightenings*

Emission in active regions is highly variable not only during the most energetic phenomena such as flares and subflares, but also when active regions are in their 'quiescent' conditions. The amplitude of the brightness variations in active regions is much larger in the less frequent and more energetic events and decreases with increasing frequency of the variation. According to the *Skylab* observations (e.g. Sheeley 1980; Webb 1981), obtained in years of declining activity, with typical time resolution within 1.5 minutes to 3 hours, in an active region in 'quiescent conditions' the cooler emission from the legs of the bright hot loops was typically changing on time scales of minutes, while the brightness of the hotter coronal portion of the loop was found to be constant over periods comparable to the radiative lifetime (Gerassimenko *et al.* 1978). Individual loops were found to evolve on a time scale of several hours, while the overall pattern might persist for several days. This fact was interpreted as evidence for steady-state heating.

The recent results obtained on an almost continuous basis with the SXT on *Yohkoh* reveal in active regions a high variability also on a short term, with frequent brightenings in compact loops. 'Active' active regions, as observed in the years of maximum activity, show a coronal variability of the order of 10^2 s within the overall active region configuration which evolves on a much longer time scale of 10^5 s.

Transient brightenings indicating episodes of energy release in compact loops in the low corona, with lengthscale from a few 10^8 cm to 5×10^9 cm, localized in regions of 10 arcseconds in size in the hottest portions of the loops, occur with a timescale of about 3 minutes and tend to be recurrent (Shimizu *et al.* 1992). A longer time scale, about 1 hour, is observed for the occurrence of transient brightenings in 'quieter' active regions. A typical transient brightening is characterized by a temperature of 7×10^6 °K and an emission measure of about 7×10^{28} cm^{-5}, by a factor of 1.3 and 6 larger than in quiet conditions, respectively, an electron density approximately doubled, of 10^{10} cm^{-3}, and an energy release of 5×10^{28} ergs (Shimizu *et al.* 1992), of the same order of magnitude of the energy associated with microflares. Interestingly enough, the total soft X-ray flux of an active region is found to be correlated with the frequency of occurrence of the transient brightenings in the active region.

These new findings confirm the existence of small-scale high-temperature ($\approx 10^7$ °K) regions in non-flaring active regions discovered during SMM (Martens *et al.* 1985), and point to the episodic nature of the heating process in active regions. Heating by discrete events was proposed on a theoretical ground in a series of papers by Parker, who has introduced the concept of nanoflares, consisting in localized releases of magnetic energy due to small-scale magnetic reconnection events in the process of topological dissipation (e.g. Parker 1988). The energy released in a transient brightening is however much larger than that predicted for nanoflares ($\ll 10^{26} - 10^{27}$ ergs).

3.2. *Variability related to flares*

The occurrence rate of transient brightenings in active regions as a function of the total energy released in the brightening has not yet been established. This relation is instead well established for the more energetic transient brightness variations in active regions which form the large class of flare events: the rate of occurrence of the brightness variations in active regions is decreasing with increasing amplitude of the variation, that is with the rate of energy release.

The observations of microflares in hard X-rays (13-600 keV) by Lin *et al.* (1984) confirmed the tendency for less energetic events to occur more frequently. Microflares are also observed in the UV emission (e.g., C IV, λ=1548 Å, T=10^5 °K) which is increasing of 30%–100% on a time scale of 10-60 sec, with an energy release rate of 10^7 erg s^{-1} cm^{-2}, two orders of magnitude larger than in quiescent active regions (Porter *et al.* 1984, 1987; Porter 1992). The fluctuating component represents about 60% of the total UV emission in years of maximum activity and is also correlated with the soft X-ray (3.5-5.5 keV) variability.

The *Solar Maximum Mission* detection in hard X-rays of thousands of flares collected over a decade of observations (1980–1989) has allowed a very accurate determination of the distribution of the flare occurrence rate. Hard X-ray flares are distributed over four orders of magnitude in peak flux, ranging from microflares at the limit of instrumental sensitivity to events which are so intense to saturate the detectors flown up to now.

All of the relevant parameters characterizing the energy release as observed in hard X-rays (25-500 keV) can be described by a power law distribution $dN\,/\,dx \propto x^{-\alpha}$ (Crosby *et al.* 1993). These SMM results confirm previous findings obtained by investigating the hard X-ray (Datlowe *et al.* 1974; Lin *et al.* 1984) and soft X-ray flare emission (1-2 keV) (e.g., Drake 1971) as well as the interplanetary particle events. The power law index remains fairly constant for the different flare parameters, $1.4 \leq \alpha \leq 2$; a flatter distribution is however found for interplanetary particle events, $1.1 \leq \alpha \leq 1.3$ (Van Hollebeke *et al.* 1975; Cliver *et al.* 1991). The slope of the frequency distribution of UV microflares is also consistent with the values found for hard X-ray microflares.

By combining the integral rate of flare occurrence for the data sets obtained by Datlowe *et al.* (1974) with OSO-VII and by Crosby *et al.* (1993) with the *Solar Maximum Mission*, and of microflares obtained during the rocket flights by Lin *et al.* (1984, 1991), it can be shown that while the flare frequency can vary considerably between the solar cycle maximum and minimum, the slope of the frequency distribution remains substantially invariant, although fluctuating on a yearly basis.

The frequency distribution of flare energy is fully consistent with the model developed by Rosner & Vaiana (1978), which assumes an exponential growth for the flare energy build up, followed by a stochastic flaring, when all the free energy built up between two consecutive flares is released. The slope of the frequency distribution in this model can be related to the characteristic energy build up time, to the average time between flares as well as to the average dissipated energy per flare. The quasi-invariance through the solar cycle of the slope of the flare frequency distribution implies that the average energy dissipation rate per flare and the magnetic build up time remain approximately constant through the solar cycle. Hence, the solar cycle variability of the flaring rate (which can vary by a factor of 20) is suggested to be principally caused by a variation in the number of flaring active regions more than in the average flaring rate per active region (Crosby *et al.* 1993).

The power law distribution for flare occurrence is also found to be consistent with avalanche models. The hypothesis at the basis of these models is that the magnetic energy released by reconnection at a given location is transported to the nearby sites inducing further reconnection. A simulation where magnetic stress is injected periodically at random sites yields a power law with slope consistent with the observations (Lu & Hamilton 1991). Zirker & Cleveland (1993) introducing in the avalanche model specific physical mechanisms causing the energy release have been capable to reproduce a frequency distribution which might correspond to the nanoflare distribution with energy per event up to 10^{26} ergs, by twisting the coronal loop footpoints by granular velocity, and the flare distribution up to 10^{28}-10^{29} ergs, by braiding of neighboring loops.

When integrating the flare distribution over flare energy within 10^{26}-10^{32} ergs, the total dissipated power in the solar corona is of the order of 10^{25} erg s^{-1}. This quantity is insufficient to explain the solar soft X-ray luminosity of $7{\times}10^{27}$ erg s^{-1} (Mewe 1972). These considerations on the energetics of the corona has led Hudson (1991) to invoke a separate soft X-ray distribution for episodic events, such as nanoflares, in order to support the idea of general coronal heating sustained by episodic events proposed by Parker (1988).

3.3. *Coronal variability during flares*

During the individual events of flare energy release no significant periodic variability has been found, except in very few cases: for example, a 1.6 s periodicity was observed for as long as 700 cycles in soft X-rays during the August 7, 1972 flare (Thomas *et al.* 1987). Flares are, however, often characterized by a quasi-periodic variability which can be observed at different wavelengths. Characteristic time delays between quasi-periodic events are typically within 1.6 s and 36 s in microwave, hard X-ray and γ emission (e.g., Kiplinger *et al.* 1983; Kane *et al.* 1983; Nakajima *et al.* 1983; Forrest & Chupp 1983; Desai *et al.* 1987; Kosugi & Kiplinger 1987). Characteristic rise and decay times of 10 s are found for the soft X-ray emission fluctuations (Zarro *et al.* 1987), and fluctuations above 20 s are observed in the UV, namely, in the Fe XXI emission line (Cheng 1987). The quasi-periodic variability during flares on time scales of 1-10 s might be interpreted as consistent with the concept introduced by de Jager and de Jonge (1978) that the flare energy is released in a large number of elementary flare processes, releasing 10^{27}-10^{28} ergs. Quantities of energy of the same order can be easily stored in an elementary flux tube due to the photospheric twisting motions of the footpoints (e.g. Sturrock *et al.* 1984).

Episodic energy release below or of the order of 10^{28} ergs in sporadic events is continually present on the Sun in form of hard X-ray microflares (Lin *et al.* 1984), soft X-ray brightenings (Shimizu *et al.* 1992), and UV microflares (Porter 1992). It remains to be established whether a flare can be produced by an integration of several small-scale localized heating processes (as in the case proposed by Parker 1987, in which heating occurs for dissipation of magnetic energy at several small-scale tangential discontinuities scattered throughout the region interested in the energy release), as suggested to explain coronal heating, or whether, to account for the high flare energy release rates $\geq 10^{29}$ erg s^{-1}, the magnetic dissipation has to occur predominantly on larger scales (e.g. determined by the dimension of the surface separating magnetic fluxes of opposite polarities where magnetic reconnection can occur). If the agent of energy release is indeed magnetic reconnection, the question is then whether in a flare the energy is released preferentially by reconnecting current sheets on a scale size of the order of the flare region or in numerous elementary current sheets scattered throughout the flare region.

The high temporal resolution observations of the hard X-ray emission during the SMM period have also definitely established the existence of a subsecond variability, first observed by Hoyng *et al.* (1976). Sub-second fluctuations are observed in hard X-rays by Kiplinger *et al.* (1983), Hurley *et al.* (1983), Bai (1987), and in microwaves, at frequencies of 22-44 GHz, by Kaufmann *et al.* (1984). The most plausible origin of the hard X-ray variability on a sub-second timescale can be found in rapid changes in the production of non-thermal electrons. Microbursts on time-scales of 10-10^2 ms have been suggested to be related to the magnetic islands developing in a current sheet during magnetic reconnection (Sturrock 1987).

4. Coronal hole evolution

Coronal holes, defined as the sites where the soft X-ray emission is drastically reduced, have been studied in detail during *Skylab*. They are dominant on the solar disk in the years preceding solar minimum, when the magnetic dipole is well-established. At this time their size is more extended and their lifetime is more prolonged. They develop according to the pattern of the large-scale coronal magnetic fields and they are associated with the large unipolar magnetic regions of open field lines as inferred by the extrapolation of the surface fields (Levine *et al.* 1977). Coronal holes first appear at low latitudes, and then tend to connect with the poles of the same polarity. Following a rapid growth, they evolve slowly and eventually shrink and disappear. Active regions can drift across the coronal hole without substantial alteration of that pattern.

4.1. *Coronal hole rotation*

One aspect which confirms the close connection of coronal holes to the large-scale magnetic fields in the corona is the rotational behavior of the coronal holes. Both coronal holes (Timothy *et al.* 1975; Wagner 1975) and the large-scale coronal magnetic pattern (Antonucci & Svalgaard 1974; Hoeksema & Scherrer 1987) are not significantly sheared by differential rotation, and this phenomenon is more evident when activity is declining. Furthermore, the north-south asymmetry found by Antonucci *et al.* (1990) for the large-scale photospheric field rotation during cycle 21 (that is the northern hemisphere is rotating faster, 26.9 days, than the southern hemisphere, 28.1 days) is consistent with the eastward migration of coronal holes, following a 28.5 days pattern in the south and a 27 days one in the north observed by Sheeley & Harvey (1981). The coronal hole study was performed by Sheeley and Harvey by using the He I (10830 Å) emission as a proxy of the soft X-ray emission, not available during the solar maximum of cycle 21.

Direct systematic investigations of coronal holes in soft X-rays have been resumed only recently. The SXT/*Yohkoh* observations have established that the rigid rotation of individual soft X-ray coronal holes is not only a characteristic of the declining phase of activity, but it is preserved throughout the solar cycle. Thin long coronal holes are observed with the SXT to rotate at a rate close to the equatorial one during solar maximum (Tsuneta & Lemen 1993).

4.2. *Evolution of coronal hole boundaries*

The investigation of the evolution of coronal hole boundaries is essential in order to understand the coronal hole rotation properties. *Skylab* observations led to the determination that the average areal changes, $dA/dt = 1.5\ 10^4\ \mathrm{km}^2\ \mathrm{s}^{-1}$, are consistent with the diffusion by supergranular cells (Bohlin 1977; Nolte *et al.* 1978a). Transient changes of coronal hole boundaries were also put in relation to the occurrence of X-ray coronal transients and filament eruptions (Webb *et al.* 1978; Rust 1983). Evidence for a statistical relationship between the X-ray bright point density and the rate of contraction of coronal holes was put forward by Nolte *et al.* (1978b), but also evidence against a simple relationship between XBP's and coronal holes was found (Davis 1985), therefore this relationship remained in doubt.

A recent re-examination of the *Skylab* observations of the extended, long-lived coronal hole CH 1 by Kahler & Moses (1990), however, has established a scale of boundary changes, 1.7×10^4 km, below the supergranular cell size, leading to exclude a relationship between the evolution of coronal hole boundaries and supergranular diffusion. This same analysis suggests that X-ray bright points are responsible both for contractions and expansions of CH 1, and the boundary variations associated with bright points are contributing in such a way to preserve rigid rotation of the coronal hole pattern, being predominantly eastward below 20° and predominantly westward above 20° in latitude. Kahler & Moses (1990) are therefore suggesting that the closed magnetic fields of bright points and the open field lines in coronal holes reconnect at coronal hole boundaries in such a way to maintain the rigid rotation behavior. According to this interpretation, therefore, no systematic flux emergence is required to maintain rigid rotation.

The well-established rotational properties of coronal holes and this new evidence for the role played by magnetic reconnection in contributing to maintain the coronal hole boundaries unsheared by differential rotation are substantially consistent with a model of coronal magnetic fields developed by Nash *et al.* (1988). The corona is assumed as a current-free region within the source surface for the interplanetary magnetic field (as defined by Schatten *et al.* 1969). The coronal field is determined as follows. Its radial component is matching the photospheric flux distribution, which is differentially rotating, while its non-radial component has to be equal to zero at the source surface. Coronal holes being the footpoint locations of open field lines, are connected to the source surface where the lower order multipoles tend to dominate. Therefore they share the rotational properties of the lower order multipoles and then rotate quasi rigidly. The hole is allowed to follow the source surface field through reconnection of the magnetic field lines. This process makes it possible to uncouple the motion of the hole boundaries from that of the photospheric fields sheared by differential rotation.

According to the model by Nash *et al.* (1988), then, large-scale magnetic fields are not deriving their fluxes from the dispersal of active regions, but their pattern is an intrinsic property of the organization in multipoles of the solar magnetic field. Polar holes with meridional extensions and prolonged rigid rotation, as observed during the *Skylab* period in the declining phase of cycle 20, are expected on the basis of this model when a strong axisymmetric field component is assumed to be dominant. When, on the other hand, non-axisymmetric field components dominate, a less-rigid rotation is expected for coronal holes.

4.3. *Mass flux in coronal holes*

An analysis of C IV profiles, obtained with the High Resolution Telescope and Spectrograph (HRTS), confirms that downflows are observed at all locations of the solar atmosphere in the transition region; however there is a greater proportion of outflows in coronal holes than in the quiet Sun (Dere *et al.* 1989). That is, in coronal holes there are more regions than in the quiet Sun characterized by outflows. The average outflow velocity in coronal hole regions is 5 km s^{-1}, while the average downflow velocity is within 5 and 12 km s^{-1} in the quiet Sun and decreases to 2 km s^{-1} in coronal holes. On the other hand, the average non-thermal velocities are in both cases comparable, $\sim$ 28 km s^{-1} (Doschek *et al.* 1976; Roussel-Dupré & Shine 1982; Dere *et al.* 1984, 1989).

Assuming only 1% of the solar surface covered by coronal holes, with a transition region density of 10^{10} cm^{-3}, the observed outflow velocities yield a solar wind mass flux of 1.3 10^{15} protons cm^{-2} s^{-1} (Dere *et al.* 1989), to be compared with a proton flux of about 10^{14} protons cm^{-2} s^{-1} inferred at the base of the solar wind streams (e.g. Zirker 1976). The HRTS observations are then strongly supporting the current idea of coronal holes as source of the fast streams in the solar wind.

A surprising result comes from the recent *Yohkoh* observations which have been used to establish a mass loss in the solar wind coming from the outward expansion of the coronal active regions (Uchida *et al.* 1992). This might be a relevant observational evidence, in view of a possible identification of the source of the low-speed solar wind. The active region outward expansion is conjectured to be related to the occurrence of brightenings in active regions. For a single active region expansion, Uchida *et al.* (1992) have determined a rate of mass loss of 10^{10}-10^{11} g s^{-1}, assuming coronal densities of 10^9 cm^{-3}, and a magnetic flux loss of $10^{16.5}$ Mx s^{-1}.

5. X-ray bright points

X-ray bright points were first studied in detail during *Skylab* (Golub *et al.* 1974, 1977). They are uniformly distributed on the solar disk and they have been associated with elementary bipolar magnetic regions. The He I 10830 Å line has usually been assumed as a proxy in the bright point studies performed after *Skylab*, due to the lack of direct soft X-ray emission observations.

With the discovery of bright points it became clear that the size of active regions on the Sun could be considered as forming a continuous distribution, where the number of active regions is a monotonically decreasing function of either size or lifetime (Golub 1980). Bright points were considered at the extreme of the size distribution of the regions observed in soft X-rays, being defined as elementary active regions with dimensions below 60 arcsec (Webb 1981).

Owing to the significant improvement in instrumental resolution, the loop structure of X-ray bright points, also inferred from the association of XBP's with magnetic bipolar regions, has been better resolved in the NIXT observations (Golub *et al.* 1990). The investigation of the NIXT images has led to identify a number of X-ray bright points which is below the expectation on the basis of the Golub's empirical law, which predicts a cyclic variation in the XBP number in anticorrelation with the active region cycle.

The systematic study of soft X-ray bright points, as for coronal holes, has been resumed with the SXT experiment on *Yohkoh* (Strong *et al.* 1992). According to a recent investigation of XBP's (Harvey *et al.* 1993), only 30% of magnetic bipoles coincide with XBP's, but the majority of XBP's are associated with bipolar magnetic areas of decreasing magnetic flux, with a decrease exceeding on the average 3 10^{14} Mx s^{-1}, in absolute value. However, the association with regions where the magnetic flux is decreasing is neither a necessary nor a sufficient condition for the occurrence of XBP's (Harvey *et al.* 1993). XBP's are rather associated with chance encounters of features of opposite magnetic polarity, conclusion that was also reached by considering the He I dark points.

Large amplitude fluctuations in the soft X-ray emission of bright points, within 30% and 200%, have characteristic time scales of a few minutes and are superimposed to a slower evolution in the emission (Strong *et al.* 1992). The largest fluctuations are defined as XBP flares. A similar variability was found for EUV bright points by Habbal & Withbroe (1981). The observations strongly suggest that the XBP variability can be related to magnetic reconnection of features of opposite polarity. The occurrence of magnetic reconnection is further supported by an excellent observation of a plasma jet at a velocity of 1000 km s^{-1}, associated with a bright point flare which was observed to be related to magnetic flux emergence (Harvey *et al.* 1993).

The rate of XBP's emergence observed with the SXT is exceeding that expected from Golub's empirical law, result which is inconsistent with the NIXT results. The responses of NIXT and SXT to temperature are however different, and this might be at the basis of this apparent contradiction.

6. Summary

New results on the spatial and temporal distributions of the soft X-ray emission have been obtained recently with the SXT telescope on *Yohkoh*, which is monitoring on an almost continuous basis the solar corona at high resolution, and with the introduction of sub-arcsecond imaging of the soft X-ray corona with the NIXT experiment.

The corona is extremely structured also on the small-scale down to 600 km, the best present instrumental resolution. Possible structuring on still finer scales is suggested on the basis of the NIXT observations and the density measurements obtained with SMM. No preferred scale sizes are observed in active regions and the slope of the size distribution is approximately invariant.

Frequent transient brightenings are observed in active region loops which may support the concept of a corona heated by a large number of elementary energy release processes. The distribution of the flare occurrence rate is characterized by a slope which remains approximately constant during a solar cycle, implying that the average energy dissipation rate per flare remains constant through the cycle, while the number of flaring active regions is considerably varying.

Very likely the dissipation of magnetic energy responsible for the release of energy observed in the coronal fluxtubes, in moderately active conditions as well as in flares, occurs through magnetic reconnection, a process which might also be in general responsible for the evolution of coronal features. In this connection, reconnection of magnetic field lines is suggested to be the most probable candidate to explain the evolution of the coronal hole boundaries, and also to play an essential role in the variability of the X-ray emission of bright points.

REFERENCES

ANTONUCCI, E. & SVALGAARD, L. 1974 Rigid and differential rotation of the solar corona. *Solar Physics* **34**, 3–10.

ANTONUCCI, E., HOEKSEMA, J. T. & SCHERRER, P.H. 1990 Rotation of the photospheric magnetic fields: a north-south asymmetry. *The Astrophysical Journal* **360**, 296–304.

ATHAY, R. G. & WHITE, O. R. 1978 Chromospheric and coronal heating by sound waves. *The Astrophysical Journal* **226**, 1135–1139.

ATHAY, R. G. & WHITE, O. R. 1979 Chromospheric oscillations observed with OSO 8. II. Average power spectra for Si II. *The Astrophysical Journal Suppl.* **39**, 333–346.

BAI, T. 1987 Sub-second variations of high-energy ($\geq$ 300 keV) hard X-ray emission from solar flares. In *Rapid Fluctuations in Solar Flares* NASA CP-2449, 29–33.

BOHLIN, J.D. 1977 Extreme-ultraviolet observations of coronal holes. I: Locations, sizes and evolution of coronal holes, June 1973 - January 1974. *Solar Physics* **51**, 377–398.

CHENG, C.-C. 1987 A search for fast variations in the Fe XXI emission during solar flares. In *Rapid Fluctuations in Solar Flares* NASA CP-2449, 263-269.

CLIVER, E., REAMES, D., KAHLER, S. & CANE H. 1991 Size distribution of solar energetic particle events. *Proc. 22nd Int. Cosmic Ray Conf.* **3**, 25–28.

CROSBY, N. B., ASCHWANDEN, M. J. & DENNIS B. R. 1993 Frequency distributions and correlations of soft X-ray flare parameters. *Solar Physics* **143**, 275–299.

DATLOWE, D. W., ELCAN, M. J., & HUDSON, H. S. 1974 OSO-7 observations of solar X-rays in the energy range 10 - 100 keV. *Solar Physics* **39**, 155–174.

DAVIS, J. M. 1985 Small-scale flux emergence and the evolution of equatorial coronal holes. *Solar Physics* **95**, 73–82.

DE JAGER, C. & DE JONGE G. 1978 Properties of elementary flare bursts. *Solar Physics* **58**, 127–137.

DERE, K. P., BARTOE, J.-D. F. & BRUECKNER, G. E. 1984 High-resolution telescope and spectrograph observations of the quiet solar chromosphere and transition zone. *The Astrophysical Journal* **281**, 870–883.

DERE, K. P., BARTOE, J.-D. F., BRUECKNER, G. E. & RECELY, F. 1989 Transition zone flows observed in a coronal hole on the solar disk. *The Astrophysical Journal* **345**, L95–L97.

DESAI, U. D., KOUVELIOTOU, C., BARAT, C., HURLEY, K., NIEL, M., TALON, R. & VEDRENNE G. 1987 Persistent time intervals between features in solar flare hard X-ray emission. In *Rapid Fluctuations in Solar Flares* NASA CP-2449, 63–74.

DOSCHEK, G. A., FELDMAN, U. & BOHLIN, J. D. 1976 Doppler wavelength shifts of transition zone lines measured in Skylab solar spectra. *The Astrophysical Journal* **205**, L177–L180.

DRAKE, J.F. 1971 Characteristics of soft solar X-ray bursts. *Solar Physics* **16**, 152–185.

FORREST, D. J. & CHUPP, E. L. 1983 Simultaneous acceleration of electrons and ions in solar flares. *Nature* **305**, 291–292.

GERASSIMENKO, M., SOLODYNA, C. V. & NOLTE J. T. 1978 Observational evidence of continual heating in X-ray emitting coronal loops. *Solar Physics* **57**, 103–110.

GOLUB, L. 1980 X-ray bright points and the solar cycle. *Proc. Roy. Soc. A.* **297**, 595–604.

GOLUB, L., KRIEGER, A. S., SILK, J. K., TIMOTHY, A. F. & VAIANA, G. S. 1974 Solar X-ray bright points. *The Astrophysical Journal* **189**, L93–L97.

GOLUB, L., KRIEGER, A. S., VAIANA, G. S. & HARVEY, J. W. 1977 Magnetic properties of X-ray bright points. *Solar Physics* **53**, 111–121.

GOLUB, L., DAVIS, J. M. & KRIEGER, A. S. 1979 Anticorrelation of X-ray bright points with sunspot number, 1970 - 1978. *The Astrophysical Journal* **229**, L145–L150.

GOLUB, L., ET AL. 1990 Subarcsecond observations of the solar X-ray corona. *Nature* **344**, 842–844.

GOMEZ, D. O., MARTENS, P. C. H. & GOLUB, L. 1993a Normal incidence X-ray telescope power spectra of X-ray emission from solar active regions. I. Observations. *The Astrophysical Journal* **405**, 767–772.

GOMEZ, D. O., MARTENS, P. C. H. & GOLUB, L. 1993b Normal incidence X-ray telescope power spectra of X-ray emission. II. Theory. *The Astrophysical Journal* **405**, 773–781.

HABBAL, S. R. & WITHBROE, G. L. 1981 Spatial and temporal variations of EUV coronal bright points. *Solar Physics* **69**, 77–97.

HARVEY, K. L., NITTA, N., STRONG, K. T. & TSUNETA, S. 1993 The relationship of X-ray bright points to the photospheric magnetic fields. In preparation.

HOYNG, P., BROWN J. C. & VAN BEEK, H. F. 1976 High time resolution analysis of solar hard X-ray flares observed on board the ESRO TD-1A satellite. *Solar Physics* **48**, 197–254.

HOEKSEMA, J. T. & SCHERRER, P. H. 1987 Rotation of the coronal magnetic field. *The Astrophysical Journal* **318**, 428–436.

HUDSON, H. S. 1991 Solar flares, microflares, nanoflares, and coronal heating. *Solar Physics* **133**, 357–369.

HURLEY, K., NIEL, M., TALON, R., ESTULIN, I. V. & DOLIDZE V.CH. 1983 Multispacecraft hard X-ray observations of fine time structure in two solar flares. *The Astrophysical Journal* **265**, 1076–1983.

KAHLER, S. W. & MOSES, D. 1990 Discrete changes in solar coronal hole boundaries. *The Astrophysical Journal* **362**, 728–733.

KANE, S. R., KAI, K., KOSUGI, T., ENOME, S., LANDECKER, P. B. & MCKENZIE, D. L. 1983 Acceleration and confinement of energetic particles in the 1980 June 7 solar flare. *The Astrophysical Journal* **271**, 376–387.

KAUFMANN, P., CORREIA, E., COSTA, J. E. R., DENNIS, B. R., HURFORD, G. J. & BROWN, J. C. 1984 Multiple energetic injections in a strong spike-like solar burst. *Solar Physics* **91**, 359–376.

KIPLINGER, A. L., DENNIS, B. R., FROST, K. J. & ORWIG, L. E. 1983 Recurrent pulse trains in the solar hard X-ray flare of 1980 June 7. *The Astrophysical Journal* **273**, 783–794.

KOSUGI, T. & KIPLINGER, A. L. 1987 Quasi-periodic pulsations in solar hard X-ray and microwave flares. In *Rapid Fluctuations in Solar Flares* NASA CP-2449, 185–191.

LEVINE, R. H., ALTSCHULER, M. D., HARVEY, J. W. & JACKSON, B. V. 1977 Open magnetic structures on the Sun. *The Astrophysical Journal* **215**, 636–651.

LIN, R. P., SCHWARTZ, R. A., KANE, S. R., PELLING, R. M. & HURLEY, K. C. 1984 Solar hard X-ray microflares. *The Astrophysical Journal* **283**, 421–425.

LIN, R. P., HURLEY, K. C., SMITH, D. M. & PELLING, R. M. 1991 A search for hard X-ray microflares near solar minimum. *Solar Physics* **135**, 57–64.

LINFORD, G. A. & WOLFSON, C. J. 1988 Properties of an impulsive compact solar flare determined from Solar Maximum Mission X-ray measurements. *The Astrophysical Journal* **331**, 1036–1046.

LU, E. T. & HAMILTON, R. J. 1991 Avalanches and the distribution of solar flares. *The Astrophysical Journal* **380**, L89–L92.

MARTENS, P. C. H., VAN DEN OORD, G. H. J. & HOYNG, P. 1985 Observations of steady anomalous magnetic heating in thin current sheets. *Solar Physics* **96**, 253–275.

MARTENS, P. C. H. & GOMEZ, D. O. 1992 Spatial power-spectra from *Yohkoh* soft X-ray images. *PASJ* **44**, L187–L191.

MEWE, R. 1972 Calculated solar X-ray radiation from 1 to 60 Å. *Solar Physics* **22**, 459–491.

NAKAJIMA, H., KOSUGI, T., KAI, K. & ENOME, S. 1983 Successive electron and ion accelerations in impulsive solar flares on 7 and 21 June 1980. *Nature* **305**, 292–294.

NASH, A. G., SHEELEY, N. R., JR. & WANG, Y.-M. 1988 Mechanisms for rigid rotation of coronal holes. *Solar Physics* **117**, 359–389.

NEUPERT, W. M. 1968 Comparison of solar X-ray line emission with microwave emission during flares. *The Astrophysical Journal* **153**, L59–L64.

NOLTE, J. T., KRIEGER, A. S. & SOLODYNA, C. V. 1978a Short term evolution of coronal hole boundaries. *Solar Physics* **57**, 129–139.

NOLTE, J. T., DAVIS, J. M., GERASSIMENKO, M., KRIEGER, A. S., SOLODYNA, C. V. & GOLUB, L. 1978b The relationship between solar activity and coronal hole evolution. *Solar Physics* **60**, 143–153.

PARKER, E. N. 1987 Stimulated dissipation of magnetic discontinuities and the origin of solar flares. *Solar Physics* **111**, 297–308.

PARKER, E. N. 1988 Nanoflares and the solar X-ray corona. *The Astrophysical Journal* **330**, 474–479.

PORTER, J. G. 1992 Microflares observed with UVSP and HXIS. In *Coronal Streamers, Coronal Loops, and Coronal and Solar Wind Composition* ESA, SP, **348**, 289–292.

PORTER, J. G., TOOMRE, J. & GEBBIE, K. B. 1984 Frequent ultraviolet brightenings observed in a solar active region with Solar Maximum Mission. *The Astrophysical Journal* **283**, 879–886.

PORTER, J. G., MOORE, R. L., REICHMANN, E. J., ENGVOLD, O. & HARVEY, K. L. 1987 Microflares in the solar magnetic network. *The Astrophysical Journal* **323**, 380–390.

ROSNER, R. & VAIANA, G. S. 1978 Cosmic flare transients: constraints upon models for energy storage and release derived from the event frequency distribution. *The Astrophysical Journal* **222**, 1104–1108.

ROSNER, R., TUCKER, W. H. & VAIANA, G. S. 1978a Dynamics of the quiescent solar corona. *The Astrophysical Journal* **220**, 643–665.

ROSNER, R., GOLUB, L., COPPI, B. & VAIANA, G. S. 1978b Heating of coronal plasma by anomalous current dissipation. *The Astrophysical Journal* **222**, 317–332.

ROUSSEL-DUPRÉ, D. & SHINE, R. A. 1982 Evidence of redshifts in the average solar line profiles of C IV ans Si IV from OSO-8 observations. *Solar Physics* **77**, 329–340.

RUST, D. M. 1983 Coronal disturbances and their terrestrial effects. *Space Sci. Rev.* **34**, 21–36.

SCHATTEN, K. H., WILCOX, J. M. & NESS N. F. 1969 A model of interplanetary and coronal magnetic fields. *Solar Physics* **6**, 442–455.

SHEELEY, N. R., JR. 1980 Temporal variations of loop structures in the solar atmosphere. *Solar Physics* **66**, 79–87.

SHEELEY, N. R., JR. & HARVEY, J. W. 1981 Coronal holes, solar wind streams, and geomagnetic disturbances during 1978 and 1979. *Solar Physics* **70**, 237–249.

SHIMIZU, T., TSUNETA, S., ACTON, L. W., LEMEN, J. R. & UCHIDA, Y. 1992 Transient brightenings in active regions observed by the Soft X-ray Telescope on *Yohkoh. PASJ* **44**, L147-L153.

STRONG, K. T., HARVEY, K. L., HIRAYAMA, T., NITTA, N., SHIMIZU, T. & TSUNETA, S. 1992 Observations of the variability of coronal bright points by the Soft X-ray Telescope on *Yohkoh. PASJ* **44**, L161–L166.

STURROCK, P. A. 1987 Rapid fluctuations in solar flares. In *Rapid Fluctuations in Solar Flares* NASA CP-2449, 1–28.

STURROCK, P. A., KAUFMANN, P., MOORE, R. L. & SMITH, D. F. 1984 Energy release in solar flares. *Solar Physics* **94**, 341–357.

TIMOTHY, A. F., KRIEGER, A. S. & VAIANA, G. S. 1975 The structure and evolution of coronal holes. *Solar Physics* **42**, 135–156.

THOMAS, R. J., NEUPERT, W. M. & THOMPSON, W. T. 1987 Soft X-ray oscillations during the flare of 7 August 1972. In *Rapid Fluctuations in Solar Flares* NASA CP-2449, 299–300.

TSUNETA, S. & LEMEN J. R. 1993 Dynamics of the solar corona observed with the *Yohkoh* Soft X-ray Telescope. In *Physics of Solar and Stellar Coronae: G.S. Vaiana Memorial Symposium* (eds. J. F. Linsky and S. Serio). pp. 113–130. Kluwer Academic Publishers.

UCHIDA, Y., MCALLISTER, A., STRONG, K. T., OGAWARA, Y., SHIMIZU, T., MATSUMOTO, R. & HUDSON, H. S. 1992 Continual expansion of the active-region corona observed by the *Yohkoh* Soft X-ray Telescope. *PASJ* **44**, L155-L160.

VAN HOLLEBEKE, M. A. I., MA SUNG, L. S. & MCDONALD, F. B. 1975 The variation of solar proton energy spectra and size distribution with heliolongitude. *Solar Physics* **41**, 189–223.

WAGNER, W. J. 1975 Solar rotation as marked by extreme-ultraviolet coronal holes. *The Astrophysical Journal* **198**, L141–L144.

WEBB, D. F. 1981 Active region structures in the transition region and corona. In *Solar Active Regions. A Monograph from Skylab Solar Workshop III* (ed. F.Q. Orrall). pp. 165–198. Colorado Associated University Press, Boulder, CO., USA.

WEBB, D. F., MCINTOSH, P. S., NOLTE, J. T. & SOLODYNA, C. V. 1978 Evidence linking coronal transients to the evolution of coronal holes. *Solar Physics* **58**, 389–396.

ZARRO, D. M., SABA, J. L. R. & STRONG, K. T. 1987 Rapid soft X-ray fluctuations in solar flares observed with the X-ray Polychromator. In *Rapid Fluctuations in Solar Flares* NASA CP-2449, 289–298.

ZIRKER, J. B. 1976 Energy balance in coronal holes and solar wind streams. In *The Energy Balance and Hydrodynamics of the Solar Chromosphere and Corona* (ed. R.-M. Bonnet & P. Delache). 421–445.

ZIRKER, J. B. & CLEVELAND, F. M. 1993 Avalanche models of active region heating and flaring. *Solar Physics* **145**, 119-128.

Variations of Coronal Radiations at Optical Wavelengths

By RICHARD C. ALTROCK

Phillips Laboratory (AFMC), Geophysics Directorate
National Solar Observatory/Sacramento Peak
National Optical Astronomy Observatories Sunspot, NM 88349, USA

This paper reviews observations of the variation over long and short time scales of solar coronal flux as observed from ground-based optical observatories. This includes data from Fe X, Fe XIV, Ca XV and the white-light electron corona. The various parameters are compared and contrasted as indicators of global coronal structure and output. The overall envelope of Ca XV minimizes before and during sunspot minimum, begins its slow rise in phase with the solar cycle, but maximizes late with strong 150-day oscillations resulting in values of zero at the local minima. Within the sensitivity of the observations, the increase of the envelope from Cycle minimum to maximum is over a factor of 30. Excluding possible secondary maxima and ignoring strong 150-day oscillations, the general envelope of Fe X appears to indicate a slow decline in intensity following sunspot maximum. Earlier observations indicate a Cycle minimum-to-maximum variation of a factor of 2 to 3. Later observations indicate a factor of 4 to 5. The overall picture for Fe XIV indicates that it varies in phase with the sunspot cycle. However, multiple maxima occur. The amplitudes of all observed Cycles are probably approximately equal, and the increase from Cycle minimum to maximum is approximately a factor of 10. The variation of the polarized and unpolarized white light corona is in phase with the sunspot cycle, although it may have a slower decline to minimum than the sunspot number. The increase from Cycle minimum to maximum is a factor of 2 to 3. Notable periodicities seen in the emission-line corona are (approximately) 27, 150, 180, 220 and 340 days and 3.4 years.

---

1. Introduction

1.1. *Restrictions*

In the context of this colloquium (*The Sun as a Variable Star: Solar and Stellar Irradiance Variations*) I will only consider variations in global coronal properties, mainly irradiance. As we will see, coronal "irradiance" in the context of this article may have several different definitions. Another restriction that I will place on this article is to consider only ground-based observations. Although there are space-based observations that could bear on this topic, space limitations will not permit consideration of them. A further restriction will be no consideration of correlations with other global solar parameters. Finally, no discussion of latitudinal variations of the corona will occur.

1.2. *Types of observations*

I will consider variations of the **K** (white light) and **L** (emission line) coronas (but not the **F**, or dust, corona). Coronal "proxy" lines, such as He I 1083.0 nm, will not be discussed. Methods of observation considered will include K observations at eclipse or with a coronagraph or "K-Coronameter" (which observes only the component of the K corona that is polarized parallel to the limb) and L observations with a coronagraph (I am unaware of any studies of the temporal variation of the corona utilizing observations of the L corona at eclipse). For clarification, all observations of the optical corona must be made above the limb due to the fact that the signal on the disk would be masked by the "noise" of the disk continuum, which is a factor of 10^5 to 10^6 more intense than the signal. Even with this restriction, many types of observations are severely limited by the brightness of the sky near the sun, which often exceeds the brightness of the corona.

Briefly, the K corona is due to Thomson scattering of the disk continuum by electrons in the corona; the radiation is optically thin, and the intensity is proportional to the integral of the electron density along the line of sight. The optical L corona is optically-thin radiation produced by forbidden transitions in highly-ionized species, which are excited by collisions and/or high-energy radiation, and the intensity is proportional to the integral of the electron density along the line of sight raised to a power between 1 and 2 (e.g., Guhathakurta *et al.* [1992] obtain a power of 1.68 for Fe XIV 530.3 nm). Only three lines will be considered in this article, as I am aware of no studies of the temporal variation of other lines. These lines are Fe X 637.4 nm, formed near 1 MK; Fe XIV 530.3 nm, formed near 2 MK; and Ca XV 569.4 nm, formed near 3 MK. These lines are often referred to as the "red", "green" and "yellow" coronal lines, respectively.

2. Problem areas

As might be expected from a discipline whose observations have been made over hundreds of years (the K corona) or for more than 50 years (the L corona), observational techniques and observational programs have varied widely. The types of observations that I will consider have been obtained by spectrographs using film or filters using film, photoelectric sensors or the eye. The observational techniques utilized have been at a non-uniform multitude of heights above the limb, spatial resolutions, latitudes and sensitivities to sky brightness. The accuracy of the definitions of the above parameters has also varied with time and location.

Further problems that bear on attempting to use observations from different locations come from the definitions of the global parameters. A widely-used parameter is one I will generally define as "Limb Flux". This is an average intensity at a given height taken over different coronal features. Definition variations include averages over the entire limb or over restricted latitudes, such as $\pm 60°$ or $\pm 30°$. Another parameter is "Pseudo-Full-Disk Flux", in which a synoptic map extending from the North pole to the South pole and covering 15 days is projected onto a sphere and then integrated over the visible disk. A variation of that technique is the Slovakian Coronal Index, which includes not only the intensities projected onto the solar disk but also the contribution from the corona above the limb. Obviously the last two techniques only obtain information about what the radio investigators term the "slowly-varying" component of the corona, due to the reliance on observations from 15 consecutive days.

Finally, the observations reported here will use a variety of temporal-averaging window-lengths and methods (e.g., block averages vs. running mean) and may incorporate observations from more than one observatory to form a single parameter.

The single greatest problem area is that of calibration. In order to deduce the temporal variation of coronal radiation, it is necessary to reduce all observations to absolute units. This has proved to be extremely difficult. Although the most-widely used calibration source, the solar disk, may be assumed to be of constant intensity for this purpose, other sources, techniques, step-wedges, neutral-density filters, etc., have evidently been variable and/or themselves poorly calibrated. This will be demonstrated later in a discussion of the intensities obtained by different observatories of the same coronal features at the same time. But suffice it to say that this is a **major** limitation in the attempt to infer long-term, and even short-term variations of the corona.

3. Solar-cycle variations of Ca XV 569.4 nm

There is only one set of observations that bears on this subject: those obtained at the National Solar Observatory at Sacramento Peak, New Mexico, with the Emission-Line Coronal Photometer (**ELCP**). This instrument scans a 1.1′ aperture around a circle of radius 1.13 solar radii ($R_\odot$) (for this line), removes the sky background through use of a lockin amplifier and records the intensity every 3° of latitude. The general character of the observations is that Ca XV emission is normally absent except in isolated regions. Sources generally overlie active regions that are or have been flaring. It is not appropriate to study the long-term variation of this line by producing a Pseudo-Full-Disk Flux, since most of the disk will have zero intensity. A better parameter is the maximum value of the emission on each day, averaged over some time interval. This should give a quantitative measure of the average Ca XV activity. I have produced such a parameter and plotted it from the time observations began in 1984 to the present. The 27-day-block-average (i.e., the average over the first 27 days, then the average over days 28-54, etc.) maximum intensity decreased rapidly to zero through 1984, fluctuated from zero to low values through 1987, began a slow increase with noise until 1991, when it shot up to a value of 3 millionths of the intensity of disk center at 569.4 nm, oscillated over the next 1.5 years, dropped to zero in mid-1992 and recovered to 1 millionth in late 1992. Considering that sunspot maximum occurred in 1989, it is remarkable that only low levels of activity were observed until 1991. Since major flares maximize a couple of years after sunspot maximum, it seems likely that this parameter is connected with the occurrence of major flare activity. Although only 4 maxima occurred in 1991 and 1992, they were separated by roughly 150 days, which may indicate a connection with the previously-found 153 day period in gamma-ray flares.

We thus see an overall envelope that minimizes before and during sunspot minimum, begins its slow rise in phase with the solar cycle, but maximizes late with strong 150-day oscillations resulting in values of zero at the local minima. Within the sensitivity of these observations, the increase of the envelope from Cycle minimum to maximum is over a factor of 30.

4. Solar-cycle variations of Fe X 637.4 nm

The earliest observations of this line were reported by Trellis (1957) and Waldmeier (1957). Waldmeier reports on Arosa observations of Limb Flux going back to 1940. During the period 1940 to 1952 the data are very noisy. There is a slight indication of a maximum near the time of sunspot maximum in 1947 followed by a slow decline into 1952. There is no indication of a minimum during sunspot minimum in 1944, but there is a *major maximum in 1943* preceded by a rise from 1941 through 1942. The overall variation from a minimum in 1941 through 1943 is a factor of 2 to 3, and from 1944 to 1947 a factor of 1.5 to 2.

Trellis (1957) presents Pic du Midi observations from 1943 through 1955. He does not show any indication of the 1943 maximum. He shows smoothed curves representing annual averages and annual maxima of Full-Limb Flux. His curves minimize at or near sunspot minima in 1944 and 1954 and maximize during sunspot maximum in 1947. Each of his curves also shows a secondary maximum in 1950, with a slow decline from then to sunspot minimum. The variation from Cycle minimum to maximum is a factor of 2 in the average values and a factor of 3 in the maximum values.

Rušin *al.* (1988) report on Lomnický Štít Limb Flux values from 1976 to 1986. They find a maximum near the time of sunspot maximum in early 1980 but also find secondary maxima just after sunspot minimum in 1977 and just before sunspot minimum in 1986, the latter thus tending to confirm the results of Waldmeier (1957). There is no indication of a secondary maximum soon after sunspot maximum as found by Trellis (1957). Other than that, the general form of a slow decline towards sunspot minimum confirms Trellis. The range of intensities covered during this epoch is a factor of 4.

Sacramento Peak *ELCP* Full-Limb Flux at 1.15 $R_\odot$ averaged over 27 days shows yet another story. These values are completely in phase with the solar cycle from the time observations began in 1984 up to the present. The minimum occurred in 1987 and the maximum in 1989. Since then there has been a slow oscillating decline. The nature of these oscillations will be discussed later. The variation from Cycle maximum to minimum is a factor of 5.

The general nature of the solar cycle variation of Fe X remains uncertain if all of these observations are given equal weight. The earlier photographic and visual observations taken 1.04 to 1.06 $R_\odot$ above the limb are inconsistent among themselves and with the later photoelectric observations taken higher above the limb. There is an indication of a possibility of secondary maxima at low altitudes at various times, particularly near sunspot minimum. There is almost certainly a maximum near the time of sunspot maximum. Excluding possible secondary maxima and ignoring oscillations, the general envelope appears to indicate a slow decline in intensity following sunspot maximum. Earlier observations indicate a Cycle minimum-to-maximum variation of a factor of 2 to 3. Later observations indicate a factor of 4 to 5.

5. Solar-cycle variations of Fe XIV 530.3 nm

The earliest observations reported by Waldmeier (1957) go back to 1939. In observations similar to those reported for Fe X he reports very noisy Full-Limb Flux values up through 1943. Interestingly, there is a local maximum in 1943 similar to that reported by him for Fe X. After 1943, the noise level appears to go down, and the intensity rises relatively smoothly from a minimum near sunspot minimum in 1944 to a maximum near sunspot maximum in 1947. The increase from Cycle minimum to maximum is approximately a factor of 4.

Trellis (1957) reports observations similar to those he reports for Fe X. The Full-Limb Flux values minimize at sunspot minima and maximize at sunspot maximum. Whereas Fe X showed a secondary maximum in 1950, Fe XIV shows a secondary maximum in 1949, closer to the major maximum. The larger relative intensity of the secondary maximum in this case combined with the smaller temporal separation from the major maximum gives the first indication of a double maximum in Fe XIV. The increase from Cycle minimum to maximum is approximately a factor of 10.

Leroy & Trellis (1974) display annual averages of $\pm 30°$–Limb Flux from Pic du Midi, Kislovodsk, Arosa and Norikura during the interval 1957 to 1973. The last three stations were renormalized to match the Pic du Midi scale. The relative variations of the four stations match fairly well from 1959 through approximately 1966. From 1967 through 1969, Norikura falls significantly below the average of the other three (Arosa and Norikura are reported only through 1969). The overall form is that of maxima one or two years after sunspot maxima in 1958 (there may have been earlier maxima), followed by minima in 1964 or 1965 near sunspot minimum and then broad and/or multiple maxima extending from 1968 through 1972 during Solar Cycle 20. Pic du Midi in particular gives the impression of having two or three maxima, although the intervening minima could be statistical fluctuations (no uncertainty bars are given).

Sýkora (1992) presents a similar analysis covering 1943 through 1990. He reports semi-annual averages of Full-Limb Flux for Pic du Midi, Norikura, Kislovodsk, Lomnický Štít and Sacramento Peak *ELCP*. Major features of the Pic du Midi curve are double peaks in Cycles 18 and 19 and a triple peak in Cycle 20. Norikura begins in 1951 and shows a double peak in Cycle 19 similar to that of Pic du Midi. After that, the intensity of Norikura drops substantially below the average of the other observatories. It shows single peaks in Cycles 20 and 21. Kislovodsk begins in 1957 and shows the possibility of a double peak in Cycle 19, but only single peaks in Cycles 20 and 21. Lomnický Štít begins in Cycle 20. From 1970 through 1982 it shows a substantially lower intensity than Pic du Midi and Kislovodsk. From then on it tracks Kislovodsk well. Pic du Midi ended observations in 1974, but Sýkora produces a unification of Kislovodsk, Norikura and Pic du Midi before then, based on the Pic du Midi scale, and uses the unified Kislovodsk and Norikura observations after 1973 to extend the Pic du Midi curve. Obviously, this curve is single-peaked in Cycle 21.

Noëns (1990) unifies Pic du Midi and Lomnický Štít observations to obtain a Limb Flux from 1945 through 1985. He also obtains double peaks in Cycle 18 and 19 but further double peaks in Cycles 20 and 21. In 18 and 19 the first peak is higher; they are equal in 20 and reversed in 21.

Rybanský *et al.* (1993) present values of the Coronal Index (**CI**), which is similar to a Pseudo-Full-Disk Flux with extensions above the limb. It is based on the intensity scale of Lomnický Štít. They find a single peak in Cycle 22. The question of single or multiple peaks in cycles 20 and 21 must await a resolution of the reality of a deficit in *CI* between 1970 and 1982, as first shown by Sýkora (1992).

Ambrož (1993) shows statistically that the Lomnický Štít *CI* is lower than normal relative to the sunspot number for the first half of Cycle 21, thus confirming the above speculation.

Rybanský *et al.* (1994) produce a unification of several stations to obtain the run of *CI* from 1939 to 1963. They find a solar-cycle envelope that is dramatically increasing in amplitude from Cycle 18 onwards (observations of Cycle 17 were infrequent), such that Cycle 22 is twice as strong as Cycle 18. This increase is not found in any other data set. For example, Sýkora *et al.* (1993), using similar data, find no such increase.

Altrock (1990) demonstrated that Sacramento Peak *ELCP* 27-day averages of Full-Limb Flux show four peaks in Cycle 21 of roughly equal amplitude, separated by 1 to 1.5 years. However, if a 13-month running mean is produced, the four peaks turn into a double peak! In this case, the first peak is the higher. This result opens the possibility that previously-seen double peaks could be an artifact of the averaging technique. The double peak presented by Sýkora (1992) for Lomnický Štít has the second peak higher, strengthening the case that the Lomnický Štít *CI* is depressed prior to 1982.

Sacramento Peak *ELCP* 27-day averages of Full-Limb Flux rise by a factor of 10 from minimum in 1976 to the maximum in 1979 and by a factor of 8 from minimum in 1987 to maximum in 1989, thus coinciding with the sunspot minima and maxima. However, as noted above, there is a distinct departure from the sunspot curve in Cycle 21 and also in Cycle 22. Both Cycles show pulses of activity spaced by 1 to 1.5 years. In these data, this appears to be a constant feature of the Fe XIV corona and must be somehow connected with some basic characteristic of the solar dynamo. The intensity at solar minimum has the same value between Cycles 20 and 21, and 21 and 22. This implies the existence of a constant, basal corona that never disappears.

The overall picture presented by the above investigations indicates that Fe XIV varies in phase with the sunspot cycle. There may have been double peaks in Cycles 18 and 19. However, Sacramento Peak *ELCP* data indicate that use of a 12 or 13 month running mean, as used in some of the above studies, can mask a greater number of maxima and artificially produce double peaks. Cycle 20 was characterized by either a broad maximum or up to 3 peaks. Cycle 21 had either a single maximum (Kislovodsk and Norikura) or 4 maxima (Sacramento Peak). Cycle 22 shows peaks in the Sacramento Peak data similar to Cycle 21, but with a gradually decreasing envelope. The amplitudes of the Cycles are probably approximately equal, in disagreement with the results of Rybanský *et al.* (1994). In many epochs there were great dissimilarities between observations from different observatories. The increase from Cycle minimum to maximum is approximately a factor of 10.

6. Solar-cycle variations of the white-light corona

Trellis (1957) displays Limb Flux from 1943 through 1955. The average and maximum curves (see Section 4) are in phase, or nearly so, with the sunspot cycle. After maximum, there is a sharp decline of about 10%, followed by a long, monotonic, slow decline until just before minimum, at which time there is a more rapid decline. The increase from minimum to maximum is about a factor of two.

Rušin & Rybanský (1985) use eclipse observations to obtain total emission of the corona as a function of the phase of the sunspot cycle. They find a curve similar to the sunspot number but with a slight rise or plateau during the last couple of years prior to minimum. There is substantial scatter in their data but also evidence that the total emission is independent of the Cycle number. The increase from minimum to maximum is about a factor of 1.5.

Hansen *et al.* (1969) show the rise to the maximum of Cycle 20 from 1964 through 1967 of the polarized Limb Flux (**pB**) from the High Altitude Observatory (HAO) Mauna Loa Solar Observatory K-Coronameter. Normalized to unity in 1964, the functional dependences of the Limb Flux at four heights from 1.125 $R_\odot$ to 1.75 $R_\odot$ are nearly identical and equal to the 10.7 cm radio-flux dependence. The increase from minimum to maximum is a factor of ≥ 2, and probably 3. They also list other historical determinations of factors ranging from 1.8 to 2.5.

Fisher & Sime (1984) find that monthly means of pB at 1.5 $R_\odot$ are well-correlated with the sunspot number from 1965 through 1983. Although they do not display pB vs. time, one may infer from their Figure 1 that pB increases by a factor of 2 during Cycles 20 and 21. They also conclude the existence of a basal corona, as was found from the Sacramento Peak *ELCP* Fe XIV observations above.

Altrock & Sime (1991) compare Pseudo-Full-Disk Flux values of pB at 1.35 $R_\odot$ and Sacramento Peak *ELCP* Fe XIV at 1.15 $R_\odot$. The solar-cycle variation of pB is about a factor of 3. The maximum phase is long and wandering with isolated outbursts of density fluctuations lasting sometimes only on the order of one rotation. Maxima do not appear to occur at the time of sunspot maximum. Differences between the two data sets are, in part, due to the different density and temperature sensitivities. pB is approximately proportional to the square root of the Fe XIV flux. The short-term activity differences are most probably due to temperature fluctuations.

So we see that the variation of the polarized and unpolarized white light (K) corona is in phase with the sunspot cycle. Although pB appears to be well-correlated with the sunspot number, other observations have indicated some unusual features, including a slower decline to minimum than the sunspot number. pB is approximately proportional to the square root of the Sacramento Peak *ELCP* Fe XIV Pseudo-Full-Disk Flux. The increase from sunspot minimum to sunspot maximum is a factor of 2 to 3.

7. Higher-frequency periodicities in the *L* corona

Altrock *et al.* (1990) analyzed periodicities in the Sacramento Peak *ELCP* Full-Limb Flux data from 1973 through 1989. Persistent signals were seen at periods near one year in all three ions. These signals were identified in Fe XIV as being due to surges of activity every 1.0 to 1.5 years during the maximum phase of Cycle 21. Persistent signals were seen at periods near 7-10 months in Fe XIV and Fe X. Periods in this range have been identified in a number of other data sets. Periods near 155 days were observed in Fe XIV during the maximum phase of Cycle 21 and in Fe X from 1984 onwards. Such oscillations in the Fe X flux are clearly seen over the last two years, a result that was not expected. Possible other periodicities occur near 2.4, 2.8, 3.4, 4.0, 4.9, 5.8 and 6.6 years. Periodogram power also appears briefly near 57, 59, 76, 89 and 101 days. No persistent signals were seen at a period of 51 days. Rotational signals are clearly seen; these signals were not investigated.

Rušin & Zverko (1990) obtained periodicities from *CI* during Cycles 20 and 21. Persistent periods were 182 and 346 days and 2.2, 3.0 and 5.0 years.

Özgüç (1992) analyzed the multi-station unified data set of Sýkora (1983). He found persistent widespread periods during Cycles 18-20 at 2.1 – 2.3, 2.5 and 3.4 – 3.5 years. He also found some evidence for periodicity at 15 - 18 months.

Altrock & Henry (1992) further analyzed periodicities in Sacramento Peak *ELCP* 27-day block averages of Fe X Full-Limb Flux during 1988 to 1992. They found an extremely regular, almost sinusoidal, oscillation of this parameter with a 148 $\pm$ 12 day period. In the power spectrum this feature was 2.5 times the magnitude of any other period. Figure 1 demonstrates this property.

The analysis of Sacramento Peak *ELCP* Ca XV data discussed in Section 3 showed a 150-day periodicity during Cycle 22.

Rybanský & Rušin (1988) and Altrock (1990) have shown extremely regular 27-day rotational fluctuations in Fe XIV during certain epochs; e.g., 1986.

In conclusion, notable periodicities seen in the *L* corona are (approximately) 27, 150, 180, 220 and 340 days and 3.4 years.

8. Conclusions

I have reviewed the solar cycle and higher-frequency variations of the emission-line and electron corona. The major conclusions of the study are as follows.

The overall envelope of Ca XV minimizes before and during sunspot minimum, begins its slow rise in phase with the solar cycle, but maximizes late with *strong 150-day oscillations* resulting in values of zero at the local minima. Within the sensitivity of these observations, the increase of the envelope from Cycle minimum to maximum is over a factor of 30.

The general nature of the solar cycle variation of Fe X remains uncertain if all of the observations are given equal weight. The earlier photographic and visual observations taken 1.04 to 1.06 $R_\odot$ above the limb are inconsistent among themselves and with the later photoelectric observations taken higher above the limb. There is an indication of a possibility of secondary maxima at low altitudes at various times, particularly near sunspot minimum. There is almost certainly a maximum near the time of sunspot maximum. Excluding possible secondary maxima and ignoring *strong 150-day oscillations*, the general envelope appears to indicate a slow decline in intensity following sunspot maximum. Earlier observations indicate a minimum-to-maximum variation of a factor of 2 to 3. Later observations indicate a factor of 4 to 5.

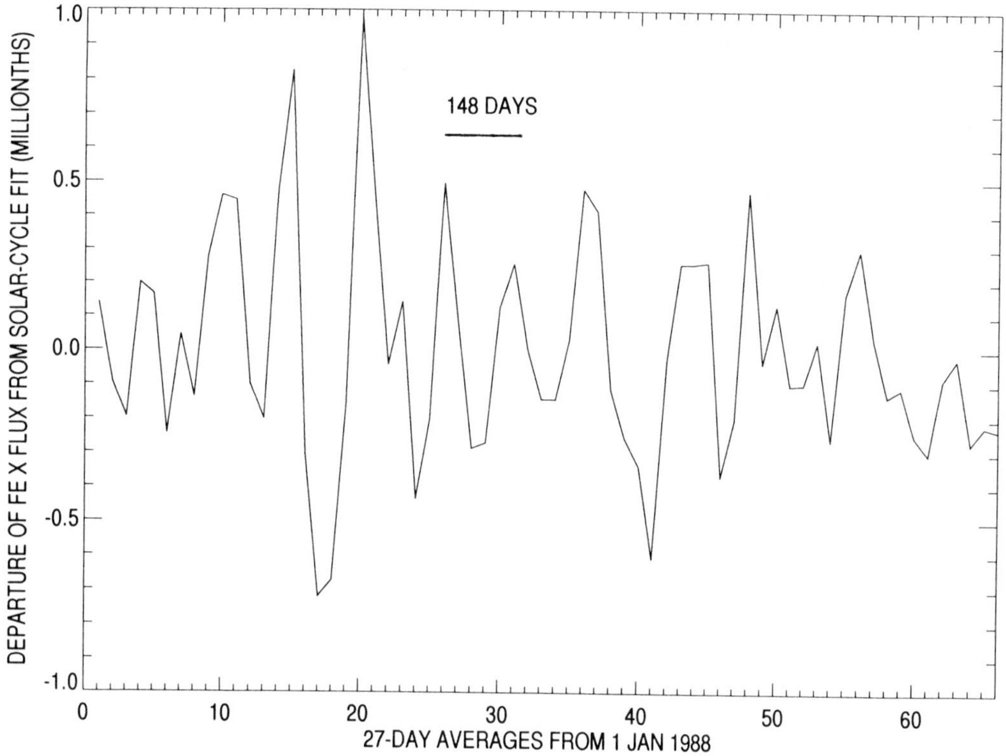

FIGURE 1. Sacramento Peak *ELCP* 27-day block averages of detrended (solar-cycle variation removed) Fe X Full-Limb Flux from 1988 to 1992 demonstrating a 148 ± 12 day period. The values are in millionths of the intensity of disk center at 637.4 nm.

The overall picture for Fe XIV indicates that it varies in phase with the sunspot cycle. There may have been double peaks in Cycles 18 and 19. However, Sacramento Peak *ELCP* data indicate that use of a 12 or 13 month running mean, as used in some of the studies, can mask a greater number of maxima and *artificially produce double peaks.* Cycle 20 was characterized by either a broad maximum or up to 3 peaks. Cycle 21 had either a single maximum (Kislovodsk and Norikura) or *4 maxima* (Sacramento Peak). Cycle 22 shows similar peaks to Cycle 21 in the Sacramento Peak data, but with a gradually decreasing envelope. The *amplitudes of the Cycles are probably approximately equal,* in disagreement with the results of Rybanský *et al.* (1994). In many epochs there were great dissimilarities between observations from different observatories. The increase from Cycle minimum to maximum is approximately a factor of 10.

H. Hudson (private communication) points out that there is an interesting result in the pattern of Cycle amplitudes. That is, the Fe X, Fe XIV and Ca XV amplitudes are approximately 4-5, 10 and > 30, respectively. If these variations are scaled to unity at Cycle minimum, the increase in the Cycle-maximum value with ionization temperature consistently implies a hotter corona at solar maximum. Guhathakurta & Altrock (1992) came to a similar conclusion from the Sacramento Peak *ELCP* Fe X and Fe XIV data alone, although their study implies that the maximum temperature may occur on the declining phase of the sunspot cycle. They found a minimum temperature of 1.52 ± 0.03 MK during 1988 and maximum temperatures of 1.64 ± 0.06 MK during 1984 (the first year of joint Fe X and Fe XIV observations) and 1.57 ± 0.03 MK during 1989 to

1991. These temperatures are not consistent with the ionization temperature of Ca XV, however, and they either represent a non-flaring component of the corona or the cooler portion of a dual-temperature corona. Altrock & Sime (1991) found evidence of solar-cycle temperature fluctuations through comparison of K and Fe XIV observations (see Section 6).

The variation of the polarized and unpolarized white light (K) corona is in phase with the sunspot cycle. Although the polarized brightness (pB) appears to be well-correlated with the sunspot number, other observations have indicated some unusual features, including a slower decline to minimum than the sunspot number. pB is approximately proportional to the square root of the Sacramento Peak *ELCP* Fe XIV Pseudo-Full-Disk Flux. The increase from Cycle minimum to maximum is a factor of 2 to 3.

Notable periodicities seen in the optical emission-line corona are (approximately) 27, 150, 180, 220 and 340 days and 3.4 years.

Acknowledgments. Thanks to Hugh S. Hudson for several helpful comments. I am grateful for support from the U. S. Air Force Office of Scientific Research under Grant PL011. I wish to thank Timothy W. Henry of Sacramento Peak for programming and data reduction and analysis assistance, and Lou B. Gilliam, Chief Observer of the Sacramento Peak John W. Evans Solar Facility for obtaining the observations. The National Optical Astronomy Observatories are operated by the Association of Universities for Research in Astronomy, Inc., (AURA) under cooperative agreement with the National Science Foundation (NSF). The National Solar Observatory is partially supported by the U.S. Air Force under a Memorandum of Understanding with the NSF.

REFERENCES

ALTROCK, R.C. 1990 The variation of solar Fe XIV and Fe X flux over 1.5 solar activity cycles. In *Climate Impact of Solar Variability* (ed. K. H. Schatten & A. Arking). NASA CP **3086**, pp. 287–292.

ALTROCK, R.C., RADICK, R.R. & HENRY, T.W. 1990 A search for non-rotational periodicities in solar Fe XIV and Fe X flux during Solar Cycles 21 and 22 (Abstract). *Bull. Am. Astron. Soc.* **22**, 873.

ALTROCK, R.C. & HENRY, T.W. 1992 Solar Fe XIV and Fe X fluxes as proxies for global solar parameters (Abstract). *Trans. Am. Geophys. Union* **73**(43), Fall Meeting Supplement, 433.

ALTROCK, R.C. & SIME, D.G. 1991 Long-term variations in the solar global coronal flux (Abstract). *Trans. Am. Geophys. Union* **72**, Spring Meeting Supplement, 224.

AMBROŽ, P. 1993 Variability of solar irradiance related to the large-scale convective patterns. Contributed paper presented at *IAU Colloquium No. 143 "The Sun as a Variable Star: Solar and Stellar Irradiance Variations*, June 20–25, 1993, Boulder, CO.

FISHER, R. & SIME, D.G. 1984 Solar activity cycle variation of the K-corona. *Astrophys. J.* **285**, 354–358.

GUHATHAKURTA, M. & ALTROCK, R.C. 1992 The solar cycle variation of coronal temperature during Cycle 22. In *The Solar Cycle:* Workshop Proceedings, National Solar Observatory/ Sacramento Peak (12th), 15-18 October, 1991 (ed. K.L. Harvey). pp. 395–403. Astronomical Society of the Pacific.

GUHATHAKURTA, M., ROTTMAN, G.J., FISHER, R.R., ORRALL, F.Q. & ALTROCK, R.C. 1992 Coronal density and temperature structure from coordinated observations associated with the total solar eclipse of 1988 March 18. *Astrophys. J.* **388**, 633–643.

HANSEN, R.T., GARCIA, C.J., HANSEN, S.F. & LOOMIS, H.G. 1969 Brightness variations of the white light corona during the years 1964–67. *Solar Phys.* **7**, 417–433.

LEROY, J.L. & TRELLIS, M. 1974 Trois cycles d'activite dans la basse couronne. *Astron. Astrophys.* **35**, 283–288.

NOËNS, J.-C. 1990 Le cycle d'activite dans la couronne solaire. *Astronomie* November-December 1990, 351–355.

ÖZGÜÇ, A. 1992 Intermediate-term periodicities in the green corona brightness of the Sun. *Astrophys. and Space Sci.* **187**, 197–207.

RUŠIN, V., RYBANSKÝ, M. & STREČKO, J. 1988 Variations of the red corona over Cycle 21. In *Solar and Stellar Coronal Structure and Dynamics: a Festschrift in Honor of Dr. John W. Evans*. Proceedings of the Ninth Sacramento Peak Summer Symposium, Sunspot, NM, 17–21 August, 1987, (ed. R.C. Altrock). pp. 288–291. National Solar Observatory.

RUŠIN, V. & RYBANSKÝ, M. 1985 Variations of the total brightness of the white-light corona with the phase of the solar cycle. *Bull. Astron. Inst. Czechosl.* **36**, 77–81.

RUŠIN, V. & ZVERKO, J. 1990 Periodicities in the green corona for the Sun as a star. *Solar Phys.* **128**, 261–268.

RYBANSKÝ, M., RUŠIN, V., GAŠPAR, P. & ALTROCK, R.C. 1993 Coronal Index of solar activity VII, years 1988–1991. *Solar Phys.*, submitted.

RYBANSKÝ, M., RUŠIN, V., MINAROVJECH, M. & GAŠPAR, P. 1994 Coronal Index of solar activity: 1939–1963. *Solar Physics*, in press.

RYBANSKÝ, M. & RUŠIN, V. 1988 Solar 530.3 nm irradiance variations (1964 - 1986). In *Solar and Stellar Coronal Structure and Dynamics: a Festschrift in Honor of Dr. John W. Evans*. Proceedings of the Ninth Sacramento Peak Summer Symposium, Sunspot, NM, 17–21 August, 1987. (ed. R.C. Altrock). pp. 342–351. National Solar Observatory.

SÝKORA, J. 1983 Time-latitude distribution of green corona brightness during solar Cycles 18, 19 and 20. *Contrib. Astron. Obs. Skalnate Pleso* **11**, 25–62.

SÝKORA, J. 1992 Intensity variations of the solar corona over 4.5 solar activity cycles. *Contrib. Astron. Obs. Skalnate Pleso* **22**, 55–67.

SÝKORA, J., BAVASSANO, B., STORINI, M. & PARISI, M. 1993 Long-term variations in the solar corona: a study of latitudinal asymmetries. In Proceedings *1992 STEP Symposium*, (ed. M. Teague). in press.

TRELLIS, M. 1957 Contribution a l'etude de la couronne solaire. *Ann. D'Ap.*, Suppl. No. 5.

WALDMEIER, M. 1957 *Die Sonnenkorona* Vol. II, Birkhauser, Basel.

Slowly-Varying Microwave Emissions from the Solar Corona

By KEN F. TAPPING[1] AND KAREN L. HARVEY[2]

[1]Dominion Radio Astrophysical Observatory, P.O. Box 248, Penticton, British Columbia,V2A 6K3, Canada

[2]Solar Physics Research Corporation, 4270 Calle Desecada, Tucson, AZ 85718, USA

Measurements of the total microwave† emission from the solar disc at single and multiple wavelengths provide a useful indicator of solar activity, and correlate well with other indices of solar activity and with solar luminosity.

Since activity in the solar atmosphere is driven by the emergence, redistribution and disappearance of magnetic flux, the close and linear correspondence between the microwave flux densities, as observed at 10.7 and 21 cm wavelengths, provide a convenient and objective indicator of solar magnetic activity.

Microwave emissions from the "non-flaring" solar corona originate in a wide range of sources. These generally lie in active regions, and comprise two main types: bright, *compact* sources a few arc-seconds across and brightness temperatures of the order of 10^6K, and weaker, *diffuse* sources arc-minutes in size and having brightness temperatures of about 10^4K. There is also evidence of a weaker, widespread emission covering large areas of the solar disc. The *compact* sources are probably due to gyroresonance, although some could be non-thermal. The *diffuse* widespread emission are almost certainly free-free thermal emission from plasma concentrations in the corona.

The strong correlation between the 10.7 cm flux and other activity indices, even when they are produced by disimilar processes and at different heights in the solar atmosphere, may at least partially be due to the overiding role played by magnetic activity in dictating the structure of the chromosphere and corona. However, the consistent correspondences over wide ranges of activity suggest dependences upon relatively few quantities and that the microwave emission from the non-flaring Sun is dominated by a single process, which is closely and simply-related to density and temperature, such as free-free (bremsstrahlung) emission from plasma concentrations trapped in coronal and chromospheric magnetic structures. Attempts to "budget" the total microwave emission at 2.8, 10.7 and 21 cm wavelengths suggest that this is the case.

Where ambient magnetic fields exceed $3000/\lambda$ Gauss, gyroresonance can produce much higher optical depths than free-free emission, and accounts for the bright emission observed in the vicinity of sunspots. Although not the dominant contributor to the total emission from the solar disc, these sources stand out brightly in high-resolution maps of active regions.

To examine the assumptions of dominance of free-free thermal emission, and simple dependences upon few parameters, we compute the spectrum for a range of coronal density models, and compare the results with observations. The agreement is good, suggesting that the total disc flux is dominated by thermal, free-free emission, and that the intensity is controlled mainly by the density at the base of the corona, but is otherwise fairly tolerant of the density model used.

---

1. Introduction

The solar corona is an extended, inhomogeneous and dynamically-disturbed plasma, pervaded by magnetic fields which are strong enough to control its distribution and activity. The spatial scales of the inhomogeneities range from structures larger than the

† We use the term 'microwave' here to refer to electromagnetic waves in the wavelength range 1–100 cm

photospheric diameter, probably down to as small as an electron gyroradius. A review of the properties of the corona and its activity is given by Koutchmy *et al.* (1991), and Rabin *et al.* (1991).

The inhomogeneity and dynamism make the corona a treasure house of radio emissions. These range from compact, bright sources having durations of milliseconds (Slottje 1978; Slottje *et al.* 1980; Zhao & Jin 1981; Tapping *et al.* 1983) through the menagerie of transient emissions associated with flares, to the slow variations following the ebb and flow of magnetic activity inside and outside active regions, and underlying these, a base level known as the *quiet Sun*. The complex field of solar radio astronomy is surveyed in detail in books by Kundu (1965) and Krüger (1979).

Early solar radio telescopes had small antennas, and could only measure the integrated flux from all sources present on the solar disc. Using "characteristic timescale" as a criterion, the emission was divided, rather arbitrarily, into three components: a *rapidly-varying* component associated with flares, typically having durations shorter than about an hour, a *slowly-varying* or *S-component* embracing all slower variability, and underlying these emissions, corresponding to an extrapolated zero level of activity, a constant, *quiet Sun* level. The S-component was first defined as a particular subset of solar radio emissions independently by Covington at 10.7 cm wavelength (1947, 1948) and Lehaney & Yabsley at 21 cm (1949).

The emissions from the non-flaring Sun vary with the ebb and flow of activity in the solar atmosphere, and are connected with both localized and large-scale changes in the structure of the chromosphere and corona as the level of activity varies. The *quiet Sun* is not different from the *S-component*, it merely reflects the baseline of a continuum of states which reflect increases in coronal density and magnetic activity. The full-disc microwave fluxes are indices of the Sun's "stellar" properties, rather than the localized dynamics of individual active regions and magnetic structures, even though the sources in these contribute much of the total microwave emission from the non-flaring Sun.

Radio indices of solar activity, such as the 10.7 cm flux, consist of measurements of the total emission with the flare contributions filtered out. They comprise the sum of the quiet Sun and S-component emissions, with the quiet Sun contributing a constant, base level, and the variability attributed to the S-component.

Compared with solar radio emissions other than the quiet Sun component, the S-component has remarkably-stable properties (Krüger 1979). It is most easily observed at wavelengths between 1 and 100 cm, and has a spectrum peaking at a wavelength of about 10 cm. The fortuitous choice by Covington and his co-workers of 10.7 cm for patrol observations of the S-component has provided an effective estimate of the S-component intensity and a good index of solar activity (the 10.7 cm flux† $F_{10.7}$) used both in its own right, and since it correlates well with other indices (Tapping 1987), as a proxy for other quantities which are more difficult to measure with the desired continuity.

Research on solar microwave emissions covers a wide range of topics, including: the relationship between the integrated emission and other activity indices, and with large-scale structural changes in the solar atmosphere; what magnetic structures house the sources; what emission mechanisms are involved, and what are their relative importances. In this paper we will discuss these points and investigate how the integrated emission relates to large-scale structure of the solar atmosphere.

† what has become widely referred to as a *flux* is reality a *flux density*, having the dimensions of $W.m^{-2}.Hz^{-1}$

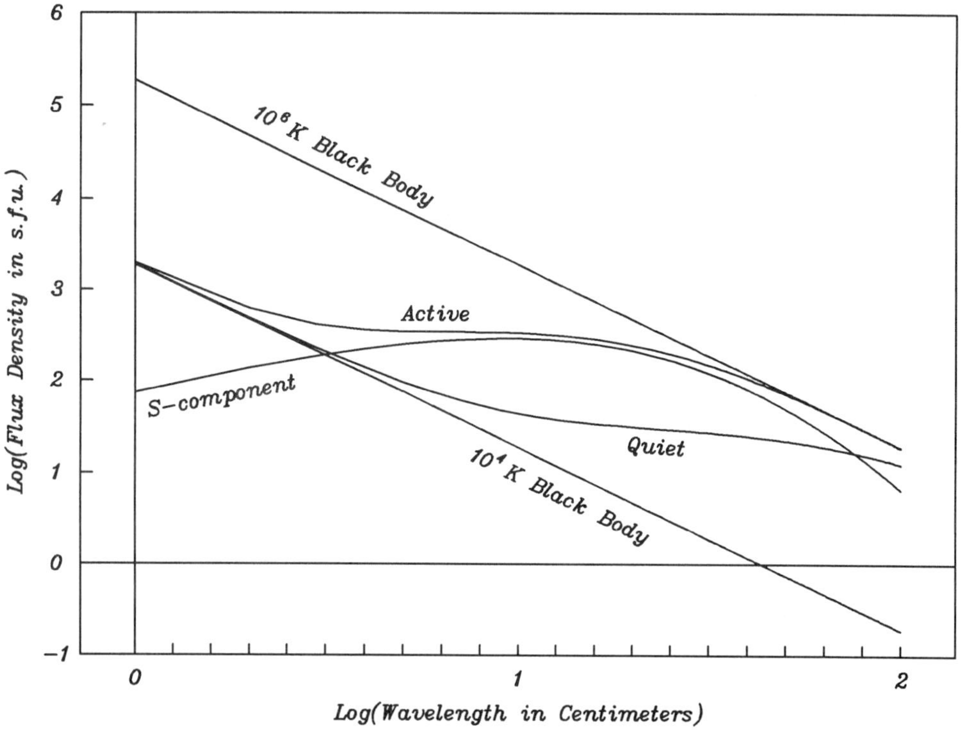

FIGURE 1. The spectrum of the integrated microwave emission from the solar disc. The S-component is the difference between the spectra of the quiet and active Sun.

2. The total disc emission

Figure 1 shows the spectrum of the integrated emission from the non-flaring Sun over the wavelength range 1-200 cm. It is a composite, constructed from plots in Krüger (1979), Kundu (1965), and from a large collection of multi-wavelength flux measurements published in *Aeronautica Acta* (Nicolet & Bossy 1984). In general, the spectral curve of solar radio emissions lie between the two curves marked "active" and "quiet". At activity minimum, the spectrum lies along the lower curve, marked "quiet", and at activity maximum, it approaches the upper curve.

The area between the two curves represents the S-component, the spectrum of which is estimated by subtracting the "quiet" spectrum from the spectrum of the integrated emission. The S-component is strongest at about 10 cm wavelength.

In the solar corona, the contribution to the total plasma pressure by magnetic fields far exceeds that due to particle motions (i.e. $B^2/2\mu \gg NkT$), and the electrical conductivity is generally high. Therefore, the distribution and movements of magnetic flux usually control and support the concentrations of plasma which give rise to the S-component. This interpretation was proposed soon after the S-component was discovered (Denisse 1948; Lehaney & Yabsley 1949).

The two straight lines represent the emission curves for black bodies at temperatures of 10^4 and 10^6K. The shapes of the curves suggests a thermal emission mechanism, where the 10^4K chromosphere is viewed through an attenuating, 10^6K corona:

$$T_b(\lambda) = \alpha(\lambda)T_{chromosphere} + (1 - \alpha(\lambda))T_{corona} \tag{2.1}$$

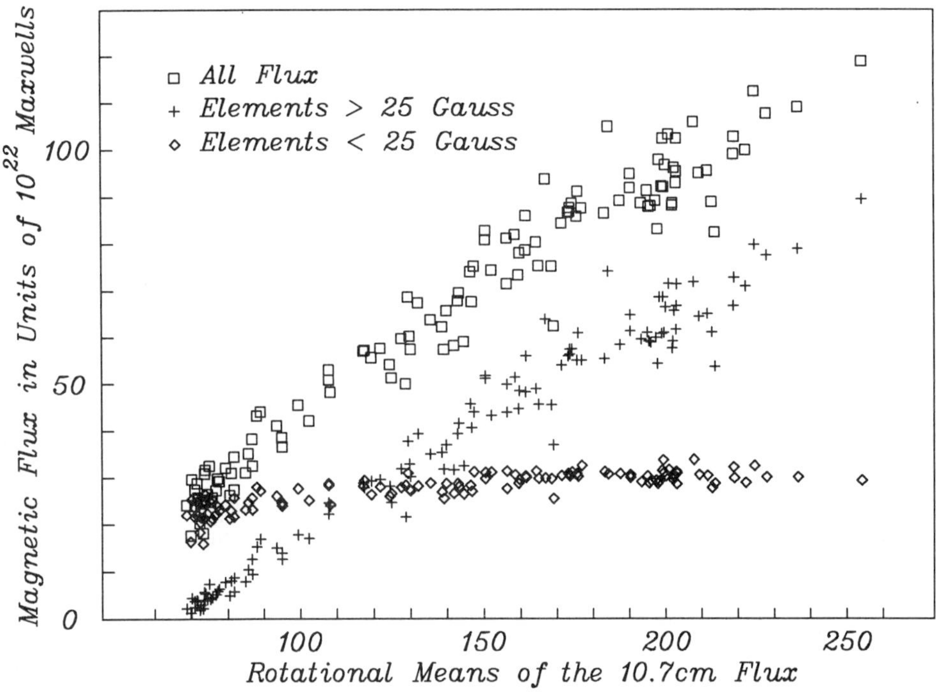

FIGURE 2. Total photospheric magnetic flux plotted against the cotemporal rotational means of the 10.7 cm flux for Solar Cycle 21.

where T_b is the observed brightness temperature, and α is the transmissivity of the corona. Both these quantities are functions of wavelength.

Felli *et al.* (1977) observed a correlation between the brightness of the emission at 2.8 cm wavelength and the total amount of magnetic flux, and Zwaan (1987) found a linear relationship between coronal emissions and magnetic flux in individual active regions. In general the solar photosphere can be separated into weak and strong-field areas (Rabin *et al.* 1991; Harvey 1992), where the average average magnetic field in the map elements is significantly weaker or stronger than about 25 Gauss. Over the whole of Cycle 21, the magnetograph at the National Solar Observatory at Kitt Peak was used to measure the total magnetic flux over the solar sphere. Figure 2 shows a plot of total magnetic flux over the solar sphere against cotemporal rotational averages of the 10.7 cm flux.

The upper plot (squares) is the total magnetic flux. The lower plot (crosses) is the total magnetic flux contained in elements stronger than 25 Gauss. Subtracting the values gives the total magnetic flux in elements weaker than 25 Gauss (diamonds). The 10.7 cm flux correlates well with the total magnetic flux in elements stronger than 25 Gauss, but not with the magnetic flux in elements weaker than that value. Moreover, the total flux in the "weak" elements does not vary over the solar cycle, and is present even at solar minimum. According to Figure 2, the total photospheric magnetic flux contained in "weak" elements is about 2.5×10^{23} Maxwells, corresponding to an average photospheric magnetic field strength of about 4 Gauss. Although playing no visible role in the S-component, the weak magnetic fields may play a role in maintaining the plasma distributions which produce the quiet Sun spectrum.

The rotational means of the 10.7 cm flux are an excellent proxy for the total photospheric magnetic flux over the solar sphere:

$$\Phi(\text{in units of } 10^{22} \text{ Maxwells}) = aS + b \tag{2.2}$$

where $a = 0.5$, $b = -10$, and S is the appropriate mean value of $F_{10.7}$, measured in solar flux units†.

The strong correlation between the S-component intensity and total magnetic flux has been observed at 21 cm wavelength, in 11 measurements made during Cycle 22 by Burke (1993).

3. Contributors to the total microwave flux

The advent of high resolution radio telescopes has made it possible to map the S-component, firstly over the solar disc and then within individual active regions. Concentrations of S-component emission have been found in association with sunspots (Alissandrakis *et al.* 1980; Chiuderi Drago *et al.* 1982; Shibasaki *et al.* 1983; Alissandrakis & Kundu 1984), bright plage (Gaizauskas & Tapping 1980; Felli *et al.* 1981; Shevgaonkar & Kundu 1985; Gaizauskas & Tapping 1988), filaments overlying magnetic polarity reversal lines (Kundu *et al.* 1981; Erskine *et al.* 1983; Gaizauskas & Tapping 1988), and strong magnetic field strength gradients (Alissandrakis & Kundu 1984). However, none of these associations are consistent. Examples of all these magnetic features have been observed which are devoid of detectable microwave emission (Felli *et al.* 1981; Webb *et al.* 1983; Gaizauskas & Tapping 1988). In addition, sources are sometimes observed where no concentrations of magnetic flux are observed (Pallavicini *al.* 1979; Felli *et al.* 1981; Kundu *et al.* 1981; Webb *et al.* 1983; Gaizauskas & Tapping 1980).

However, despite this complexity, Pallavicini *et al.* (1979) show that on the basis of observations at 2.8 cm and in soft X-rays, the S-component sources can be divided into two broad classes: (a) *diffuse* sources, having brightness temperatures of the order of 10^4 K and angular sizes of arc-minutes, and (b) *compact* sources, with brightness temperatures of 10^6 K or more and angular sizes of only a few arc-seconds. The *diffuse* sources are also associated with bright soft X-ray emission, whereas the *compact* ones are not.

Through examining full-disc maps and scans made at 2.8 cm, 10.7 cm and 21 cm, Tapping (1987), and Burke (1993), identify a further contribution, a weak, widely-distributed emission which is distributed over large areas of the solar disc, and which can produce a significant contribution to the total microwave flux. From maps made at 2.8 cm wavelength in 1981, and at 21 cm in 1992 and 1993, it appears that this contribution originates in areas of "magnetic network".

4. Emission mechanisms

This is a short discussion only; more comprehensive treatments can be found in Kundu (1965) and Krüger (1979). Microwave emissions by the non-flaring Sun are mostly thermal, with possibly a contribution by non-thermal processes.

The thermal emission is due to the deflection of thermal electrons by interactions with ions (free-free, bremsstrahlung emission) or magnetic fields (gyroresonance). For convenience's sake, they are usually discussed as two separate processes, in the limits of "weak" and "strong" magnetic fields.

† 1 solar flux unit (s.f.u.) $= 10^{-22} W.m^{-2}.Hz^{-1}$

The two distinct classes of source evident in the study by Pallavicini *et al.* (1979) indicate at least two emission processes. The *diffuse* sources, having brightness temperatures of the order of 10^4K and producing soft X-ray emission, can be explained in terms of free-free emission from density enhancements supported in the corona by active region magnetic fields.

The *compact* sources are brighter (10^6K), but not producers of X-rays. They cannot be accounted for using free-free emission models. Some of them might be non-thermal, but the majority can be accounted for in terms of gyroresonance. This process can produce bright, optically-thick thermal sources in the corona if the ambient magnetic fields are stronger than about $3000/\lambda$ Gauss (Kakinuma & Swarup 1962; Zheleznyakov 1962), where λ is the observing wavelength in centimeters.

4.1. *Thermal free-free emission*

The absorption coefficient for free-free interactions between thermal electrons and "heavy" ions is given by (Kundu 1965):

$$\kappa_{ff} = \frac{a\lambda^2 N^2}{nc^2 T^{3/2}} \tag{4.3}$$

where λ is the observing wavelength, N the electron density, n the refractive index, c the speed of light and T the ambient electron temperature. The constant a is 1.1×10^{-11} in the chromosphere and 1.6×10^{-11} in the corona.

The refractive index is:

$$n = \left(1 - \left(\frac{f_0}{f}\right)^2\right)^{\frac{1}{2}} \tag{4.4}$$

where $f = c/\lambda$ is the observing frequency, and f_0 is the plasma frequency:

$$f_0 = 9\sqrt{N} \text{ (Hz)} \tag{4.5}$$

We assume here that ambient magnetic fields are sufficiently weak for the electron gyrofrequency (2.8B MHz, where B is in Gauss) to be roughly at least an order of magnitude smaller than the observing frequency.

4.2. *Thermal gyroresonance*

Where the electron gyrofrequency is larger than about a quarter of the observing frequency, the absoprtion coefficient by gyroresonance becomes significant. The exact formula for the absorption coefficient is very complicated. However, a useful approximate form is given by Kundu (1965):

$$\kappa_{gr} = 1.5 \times 10^{-4} \Phi_s \frac{N}{f_H} \left(\frac{2kT}{mc^2}\right)^{s-3/2} \zeta_s^{\pm}(\theta) \exp\left(\frac{(1 - s\frac{f_H}{f})^2}{\frac{2kT}{mc^2}\cos^2\theta}\right) \tag{4.6}$$

where

$$\Phi_s = \frac{2}{s!}\left(\frac{s}{2}\right)^{2s-1} \tag{4.7}$$

and

$$\zeta_s^{\pm}(\theta) = (\sin\theta)^{2s-2}\frac{(C_1 \pm C_2\cos\theta)^2}{\cos\theta} \tag{4.8}$$

where N = electron density, f_H = electron gyrofrequency, f = observing frequency = c/λ, λ = observing wavelength, c = speed of light, k = Boltzmann's constant, T =

ambient electron temperature, m = electron rest mass, s = harmonic number (2,3,4,5 etc.), $C1$ and C_2 = axes of polarization ellipse (where $1 = C_1^2 + C_2^2$), and θ = angle between magnetic field and line of sight.

The plus and minus signs indicate the *ordinary* and *extraordinary* wave modes. Gyroresonant emission is generally highly polarized.

4.3. *Non-thermal emission*

Bright microwave sources which are devoid of X-ray emission cannot be due to free-free thermal emission, but, if the ambient magnetic fields are strong enough, can be caused by gyroresonance. Sources having brightness temperatures of the order of 10^6K which are located in areas where the magnetic fields are estimated to be too weak for gyroresonance at that wavelength have been suggested to be non-thermal.

To drive non-thermal emission processes, such as the gyrosynchrotron mechanism, accelerated (non-thermal) electrons are needed. Where these come from is not yet clear, but Chiuderi Drago & Melozzi (1984) have demonstrated that sources having the observed properties can be driven by gyrosynchrotron emission produced by similar populations of accelerated electrons to those believed to give rise to Type III bursts.

Non-thermal emissions are believed to be a minor contribution to the total microwave flux and will not be further discussed here.

5. Propagation of microwaves in the solar atmosphere

The solar atmosphere itself can affect the polarization, intensity and direction of radio waves passing through it, and even if the emissions can be detected at the Earth at all.

The absorption by free-free processes and gyroresonance attenuate the incident radio emissions and then add a thermal noise contribution from the ambient plasma, so that in passing through a thin slab of plasma, the brightness temperature of the incident waves is modified:

$$T_b(new) = \alpha T_b(old) + (1 - \alpha)T_0 \tag{5.9}$$

where α is the *transmissivity* of the layer, and T_0 is the electron temperature in it.

The propagation of radio waves in the solar atmosphere is discussed in some detail by Kundu (1965), here we will just mention some of the main points. In the absence of magnetic fields, the refractive index is given by equation 4.4. Waves can only propagate when $f > f_0$.

In the presence of magnetic fields, the situation becomes more complicated. A wave propagating in any direction with respect to the magnetic field is split into two polarized waves, an *ordinary* wave and an *extraordinary* wave, each of which has its own refractive index and absorption coefficient. The presence of a magnetic field in the medium between an unpolarized source and the observer can lead to a polarized wave being observed. Changes in the distribution of magnetic flux in a magnetic structure which houses sources of radio emission can produce changes in both the spectrum and polarization of the emission being observed.

In general, if the frequency of the radio emissions is less than about 10 times the electron gyrofrequency in the source or in the intervening medium through which the waves are travelling, the magnetic fields are playing a significant role in their generation and propagation, and must be taken into account in any investigation of the variability of the emission.

6. Models for solar microwave emissions

The gyroresonance process has been used extensively in studying emission from individual magnetic structures. However attempts to model the Sun's total microwave emission are largely based upon free-free, thermal emission (Denisse 1948; Smerd & Westfold 1949; Christiansen & Mathewson 1959; Chiuderi Drago *et al.* 1982; Tapping & DeTracey 1990).

The two processes are usually invoked separately; the bright sources close to concentrations of magnetic flux are usually interpreted in terms of gyroresonance, and emission from elsewhere is assumed to be free-free. However, with the magnetic fields being the dominant contributor to the total pressure in the corona, it is probably more realistic (but much more complicated) to combine the two processes (Zlotnik 1968; Lantos 1968; Krüger *et al.* 1985). However, simple free-free models can yield useful results and within their limitations, are much more tractable.

An indication of the average relative importance of the two processes can be obtained from making a solar emission budget. At 2.8 cm, the quiet Sun has an average disc brightness temperature of about 10,000K. A single *compact* source having an angular diameter of 10 arc-sec and a brightness temperature of 10^6K increases the brightness temperature of the disc by about 30K, which, if there are no other sources on the (quiet) disc, corresponds to an increase in the average brightness temperature of the disc of 0.3%. In reality, such sources are unlikely to occur without sources of additional thermal, free-free emission, so the contribution by *compact*, gyroresonant or perhaps non-thermal sources to the total disc emission at 2.8 cm is likely to be small. At longer wavelengths, the quiet Sun brightness temperature increases, so the *compact* sources probably contribute relatively even less to the total emission.

In their estimates of the S-component budget, Tapping (1987), and Burke (1993), separate three contributions to the Sun's emission at 2.8, 10.7 and 21 cm wavelength: active region emission, a widely-distributed background, and the quiet Sun. After the quiet Sun component is subtracted, the contributions from active regions and the widely-distributed emission are comparable at all three wavelengths.

Apart from the contributions by *compact* sources, the rest of the centimetric emission from active regions is thermal bremsstrahlung, and the widely-distributed contribution is almost certainly produced by the same mechanism. The dominance of free-free emission, together with the multiplicity of sources, both inside and outside active regions, is probably the reason for the stable properties of total disc emission, and its consistent correlations with other indices of solar activity, which are produced by different processes.

7. A thermal, free-free emission model

Most models for the microwave emission from the non-flaring Sun are based upon the assumption that thermal free-free emission dominates other contributions. Based upon measurements or assumptions, a model for the spatial distribution of coronal density is constructed, and the resulting free-free emission intensity and spectrum calculated. One of the earliest (Waldmeier & Müller 1950) invoked "coronal condensations", which are concentrations of plasma over active regions. The agreement between 5303Å coronagrams and maps made at 20 cm wavelength support this (Waldmeier 1959). Other atmosphere models, such as the Newkirk streamer model, also could produce the observed spectrum and intensity. One of the more recent attempts is by Tapping & DeTracey (1990). This study suggested the total emission to be primarily sensitive to the density at the base of the corona, and small changes in the S-component spectrum during the solar cycle might be explained in terms of changes in the structure of the interface between the chromosphere and corona. However, this model, like many others, does not take into account the very significant role of coronal magnetic fields in controlling the plasma density, or else handles them empirically.

It is very interesting how a variety of density models, based upon measurements, empirical concepts or crude approximations, all account for the centimeter wavelength spectrum of the quiet and active Sun with a good degree of success. This suggests a degree of independence of the actual density model used.

7.1. *The density models*

The spatial distribution of plasma and magnetic flux in the solar atmosphere is inhomogeneous over a wide range of spatial scales and varies with time. It would be very difficult to construct an accurate distribution for any particular moment, and the validity of that distribution at a different time would not be clear. We therefore use a different approach.

We define two density models, $f_{min}(z)$ and $f_{max}(z)$, where f_{min} lies below all "realistic" density distributions and f_{max} lies above them. Since magnetic fields generally increase the amount of plasma supported in the solar atmosphere, $f_{min}(z)$ is assumed to be defined by vertical balance of plasma pressures, with no magnetic fields, as in the "hydrostatic model" used by Tapping & DeTracey (1990):

$$N = N_0 \exp\left(-\frac{m_0 g_0}{kT}\frac{R+Z_0}{R+Z_0+z}z\right) \tag{7.10}$$

where k = Boltzmann's constant, N = the ambient electron density, N_0 = the electron density at the base of the corona, m_0 = the mass of the hydrogen atom, g_0 = the gravitational acceleration at the photosphere, R = the photospheric radius, Z_0 = the height of base of corona above photosphere, and z = the height measured from the base of the corona.

The upper envelope model, $f_{max}(z)$ is more difficult to estimate. However, it is extremely unlikely that magnetic fields will enhance coronal densities to the point where the density will not fall with increasing height above the photosphere. For the upper-envelope model "$f_{max}(z)$", we use "uniform-density slab", $N = N_0$, with the density in the corona being constant with height and equal to the value at its base.

For comparison purposes, we include two density models which are based upon measurements: the Baumbach-Allen model and the Newkirk coronal streamer model, both of which have been modified to accommodate a density N_0 at the base of the corona.

The Baumbach Allen model is

$$N = N_0\frac{1}{a_1+a_2}(a_1x^6 + a_2x^{16}) \tag{7.11}$$

where $x = R/(R+Z_0+z)$, $a_1 = 1.55$ and $a_2 = 2.99$, and the scaled version of the Newkirk streamer model:

$$N = a_1 N_0 \exp(a_2 x) \tag{7.12}$$

where $a_1 = 0.014$ and $a_2 = 4.32$.

The only degree of freedom in these models in the density at the base of the corona. They can be fixed using a single flux density or brightness temperature at a single wavelength.

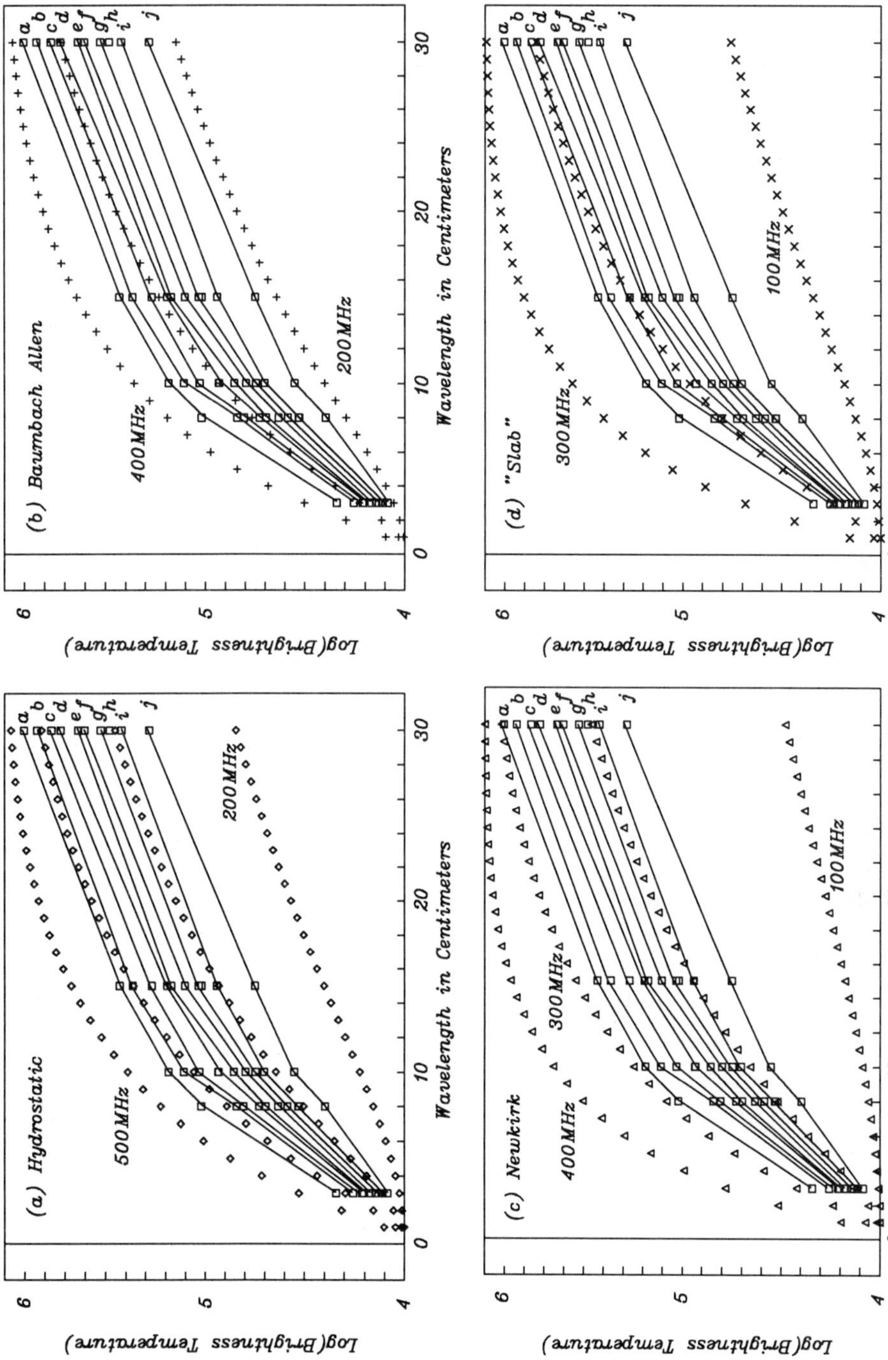

FIGURE 3. Plots of observed solar microwave spectra against those predicted assuming free-free thermal emission and four density models (a: hydrostatic model, b: Baumbach Allen, c: Newkirk, and d: slab). The observed spectra were recorded on (a) 1957/10/19, (b) 1957/06/09, (c) 1960/06/30, (d) 1959/12/14, (e) 1960/12/04, (f) 1961/08/11, (g) 1960/11/27, (h) 1961/12/01, (i) 1964/07/27.

7.2. *Calculating the brightness temperature*

The brightness temperature is calculated by assigning an initial brightness temperature of 10,000K (the chromospheric temperature at the base of the transition region), and integrating the brightness temperature outwards along a vertical ray path through the corona, which is assumed to be isothermal, with an electron temperature of 10^6K. If the the refractive index is imaginary at the start of the integration, the brightness temperature is set equal to the coronal temperature at the height where the refractive index first becomes real.

For the density models, we assign a plasma frequency (f_{00}) at the base of the corona. It is related to the electron density N_0 by the relation $N_0 = f_{00}^2/81$. We assume that the corona starts at 10,000km above the photosphere.

We compare the results with measurements of the solar flux density at 3, 8, 10, 15 and 30 cm wavelength during Cycle 19, published in Nicolet & Bossy (1984). The values were chosen to cover different parts of the falling phase of the cycle. Other than that, the dates were chosen randomly. Figure 3 shows the observed spectra of solar disc emission, together with the spectra predicted by the free-free emission model for the various atmospheric density models. The frequencies on the plots identify the plasma frequencies (f_{00}) at the base of the corona used in the density models. We assume that f_{00} is a function of solar activity.

The observed data well fit the free-free emission model of all the density distributions used.

The "quiet Sun" flux densities are extrapolated values corresponding to the minimum of solar activity. Estimates of these are 255, 74, 66, 64, 49, 46 and 41 solar flux units at 3, 8, 10, 10.7, 15, 21 and 30 cm wavelength respectively. If for each density model we pick the value of the f_{00} producing a flux density of 64 solar flux units at 10.7 cm, the errors in the predicted flux densities is better than 10% for all density models.

The intensity and spectrum of the total microwave emission from the non-flaring Sun is relatively independent of the density model used, but sensitive to the density at the base of the corona, where most of the optical thickness is located.

8. Discussion and conclusions

The total centimeter-wavelength emission from the solar disc contains contributions from an often large number of sources. In general the sources are located in active regions, although there is evidence of a weak, widely-distributed emission from outside them. A number of radiative processes are involved, but the dominant one is probably thermal, free-free emission from density concentrations supported in the corona by magnetic fields. Thermal gyroresonance is the main explanation for the bright, compact sources which are often observed close to sunspots. There is evidence for a non-thermal contribution, but if present, it is a probably a minor component which need not be included in models for the total flux.

The correlation between the total disc emission at 10.7 cm wavelength (the 10.7 cm flux) and the total magnetic flux over the solar disc, which has been observed to persist for more than a solar cycle, is particularly striking. Only the flux in elements stronger than about 25 Gauss seems to be associated with the emission at 10.7 cm. The total magnetic flux in weaker elements remains reasonably constant over the solar cycle, and shows little correlation with the 10.7 cm flux.

The correspondence between the total disc emission, which we assume to be predominantly free-free thermal emission, and the total magnetic flux is probably simple. The more coronal magnetic flux, the more plasma is supported by it, producing in turn more free-free emission. The contribution by gyroresonance is more direct. The more magnetic flux in the corona, the higher the probability that areas will contain strong enough magnetic fields for gyroresonance to occur at centimeter wavelengths.

The free-free emission from the solar corona is remarkably insensitive to the density models used. It is largely controlled by the density at the base of the corona. This simple connection might account for the good correspondence between the 10.7 cm flux (for example) and other activity indices. However, the strong dependence of all manifestations of solar activity upon the distribution and movement of magnetic flux, which are in turn driven by the "magnetic activity engine" located below the photosphere might mean that no matter what the mix of emission processes contributing to the total microwave emission from the solar disc might be, there will be a correlation with other emissions driven by the same "magnetic engine".

The observational and theoretical evidence supports a tight, unambiguous relationship between the variations in coronal magnetic fields and the total microwave emission from the solar disc, at least at wavelengths between 2.8 and 21 cm. There has been a considerable amount of research into the localized sources of centimetric emissions in active regions. There is a need for more study of the integrated centimetric emission from the solar disc, and the various sources as contributors to the whole. The total microwave emission from all sources on the solar disc provides a powerful tool for studying the Sun as a magnetic star.

Acknowledgments. The National Solar Observatory/Kitt Peak data used here are produced cooperatively by the National Solar Observatories/National Optical Astronomical Observatory, the National Aeronautical and Space Administration/Goddard Space Flight Center, and the National Atmospheric and Oceanic Administration/Space Environment Laboratory. One of us (KLH) was supported in this work by NASA Contract NASW-4721, while a visitor at the National Solar Observatory (The National Solar Observatories are operated by the Association of Universities in Research in Astronomy Inc. (AURA), in cooperative agreement with the National Science Foundation). We would like to thank O.R. White and R.F. Donnelly for suggestions and comments which helped form the content of this paper, and I. Burke for proof reading this paper a number of times and making some very useful suggestions. Thanks are also due to the Referee, who contributed some valuable points.

REFERENCES

ALISSANDRAKIS, C.E., KUNDU, M.R. & LANTOS, P. 1980 A model for sunspot-associated emission at 6-cm wavelength. *Astronomy and Astrophysics* **82**, 30–40.

ALISSANDRAKIS, C.E. & KUNDU, M.R. 1984 Center-to-limb variation of a sunspot-associated microwave source. *Astron. Ap.* **139**, 271–284.

BURKE, I. 1993 Imaging the Sun with the DRAO Synthesis Telescope. MSc thesis, University of British Columbia, Vancouver, Canada.

CHIUDERI DRAGO, F. & MELOZZI, M. 1984 Non-thermal radio sources in solar active regions. *Astronomy and Astrophysics* **131**, 103–119.

CHIUDERI DRAGO, F., BANDIERA, R., FALCIANI, R., ANTONUCCI, E., LANG, K.R., WILLSON, R.F., SHIBASAKI, K. & SLOTTJE, C. 1982 Multiple wavelength observations of a solar active region. *Solar Physics* **80**, 71–85.

CHRISTIANSEN, W.N. & MATHEWSON, D.S. 1959 The origin of the slowly-varying component. *Paris Symposium on Radio Astronomy* (ed. Bracewell). pp. 108–117. Stanford University, Stanford, California.

COVINGTON, A.E. 1947 Microwave solar noise observations during the partial eclipse of November 23, 1946. *Nature* **159**, 405–406.

COVINGTON, A.E. 1948 Solar noise observations on 10.7cm. *Proceedings of the IRE* **36**, 454–457.

DENISSE, J.F. 1948 Etude des émissions radioeléctriques solaires d'origine purement thermique. PhD. thesis, University of Paris, France.

ERSKINE, F.T., KUNDU, M.R. & RAO, A.P. 1983 6-cm observations of a solar active region and bursts with 6 seconds resolution. *Astronomy and Astrophysics* **83**, 256–260.

FELLI, M., POLETTO, G. & TOFANI, G. 1977 Magnetic and microwave structure in solar active regions. *Solar Physics* **51**, 65–81.

FELLI, M., LANG, K.R. & WILLSON, R.F. 1981 VLA observations of solar active regions 1, The slowly-varying component. *Astrophys. J.* **247**, 325–327.

GAIZAUSKAS, V. & TAPPING, K.F. 1980 Long-lived microwave pulsations observed in a complex solar active region. *Astrophys. J.* **241**, 804–810.

GAIZAUSKAS, V. & TAPPING, K.F. 1988 Compact sites of microwave emission at 2.8 centimeter wavelength inside solar active regions. *Astrophys. J.* **325**, 912–926.

HARVEY, K.L. 1992 Measurements of solar magnetic fields as an indicator of solar activity evolution. In *Proceedings of the Workshop on the Solar Electromagnetic Radiation Study for Solar Cycle 22* (ed. R.F. Donnelly). pp. 113–129. NOAA ERL SEL, Boulder, CO., USA.

KAKINUMA, T. & SWARUP, G. 1962 A model for the sources of the slowly-varying component of microwave solar radiation. *Astrophys. J.* **136**, 975–994.

KOUTCHMY, S., ZIRKER, J.B., STEINOLFSON, R.S. & ZHUGZDA, J.D. 1991 Coronal activity. In *The Solar Interior and Atmosphere* (ed. A.N. Cox, W.C. Livingston & M.S. Matthews). pp. 1044–1086. Univ. Arizona Press, Tucson, AZ, USA.

KRÜGER, A. 1979 *Introduction to Solar Radio Astronomy and Radio Physics.* Geophysics and Astrophysics Monographs, No. 6, Vol. 16, D. Riedel Hingham, Mass.

KRÜGER, A., HILDEBRANDT, J. & FURSTENBERG, F. 1985 A working model for the solar S-component radio emission. *Astronomy and Astrophysics* **143**, 72–76.

KUNDU, M.R. 1965 *Solar Radio Astronomy.* Wiley-Interscience, New York.

KUNDU, M.R., SCHMAHL, E.J. & RAO, A.P. 1981 VLA observations of solar active regions at 6-cm wavelength. *Astron. Ap.* **94**, 72–79.

LANTOS, P. 1968 A model for magnetic gyromagnetic radioemission from solar active regions. *Ann. Astrophys* **31**, 105–113.

LEHANEY, F.J. & YABSLEY, D.E. 1949 Solar radiation at 1200Mc/s, 600Mc/s and 200Mc/s. *Aust. J. Sci. Res., Ser. A* **2**, 48–62.

NICOLET, M. & BOSSEY, L. 1984 *Aeronomica Acta: "Solar fluxes as indices of solar activity".* published by the Belgisch Instituut voor Ruimte Aeronomie, **A-No. 282**.

PALLAVICINI, R., VAIANA, G.S., TOFANI, G. & FELLI, M. 1979 The coronal atmosphere above solar active regions: the comparison of high spatial resolution soft X-ray and centimetric observations. *Astrophys. J.* **229**, 375–386.

RABIN, D.M., DEVORE, C.R., SHEELEY, JR. N.R., HARVEY, K.R. & HOEKSEMA, J.T. 1991 The solar activity cycle. *The Solar Interior and Atmosphere* (ed. A.N. Cox, W.C. Livingston & M.S. Matthews). pp. 781–843. Univ. Arizona. Press, Tucson, AZ, USA.

SHEVGAONKAR, R.K. & KUNDU, M.R. 1985 Time variability of solar active regions at centimeter wavelengths. *Solar Physics* **98**, 119–126.

SHIBASAKI, K., CHIUDERI DRAGO, F., MELOZZI, M., SLOTTJE, C. & ANTONUCCI, E. 1983 Microwave, ultraviolet and soft X-ray observations of Hale Region 16898. *Solar Physics* **89**, 307–321.

SLOTTJE, C. 1978 Millisecond microwave spikes in a solar flare. *Nature* **257**, 520–521.

SLOTTJE, C. 1980 Fast, fine structure in solar microwave bursts. In *Radio Physics of the Sun* (ed. M.R. Kundu & T. Gergely). IAU Symposium **86**, 195–201.

SMERD, S.F. & WESTFOLD, K.C. 1949 The characteristics of radiofrequency radiation in an ionized gas, with applications to the transfer of radiation in the solar atmosphere. *Philos.*

Mag. Ser. 7 **40**, 831–848.

TAPPING, K.F., KUIJPERS, J., KAASTRA, J.S., VAN NIEUWKOOP J., GRAHAM, D. & SLOTTJE, C. 1983 VLBI of solar flares. *Astronomy and Astrophysics* **122**, 177–180.

TAPPING, K.F. 1987 Recent solar radio astronomy at centimeter wavelengths, the temporal variability of the 10.7cm flux. *J. Geophys. Res. No. D1* **92**, 829–838.

TAPPING, K.F. & DETRACEY, B. 1980 The origin of the 10.7cm flux. *Solar Physics* **127**, 321–332.

WALDMEIER, M. & MÜLLER, K. 1950 Die Sonnenstrahlung im Gebeit von $\lambda = 10$cm. *Z. für Astrophys.* **27**, 58–63.

WALDMEIER, M. 1959 A comparison between radioheliograms and optical observations of the solar corona. In *Paris Symposium on Radio Astronomy* (ed. Bracewell). pp. 118–122. Stanford University, Stanford, California.

WEBB, D.F., DAVIS, J.M., KUNDU, M.R. & VELUSAMY, T. 1983 X-ray and microwave observations of active regions *Solar Physics* **85**, 267–283.

ZHAO, R. & JIN, S. 1981 C.A.S.: Astronomy Circular #9. *Chinese Astron. Soc. Astron. Circ.* **9**, 3–7.

ZHELEZNAKOV, V.V. 1962 On the origin of the slowly-varying component of the solar radio emission. *Soviet Astronomy* **6**, 3–9.

ZLOTNIK, E.Y. 1968 On the theory of the slowly-varying component of the solar radio emission. *Astron. Zh.* **45**, 310–317.

ZWAAN, C. 1987 Elements and patterns in the solar magnetic field. *Annual Reviews of Astronomy and Astrophysics* **25**, 83–111.

Spatial Resolution of Solar Total Irradiance Variability: The *YOHKOH* White–Light Observations

By HUGH S. HUDSON

Institute for Astronomy, Honolulu, HI 96822, USA

The observations from the ACRIM instrument on board the *Solar Maximum Mission*, and other recent experiments, have allowed us to study the time variations of solar irradiance. These observations identified several mechanisms, ranging from sunspots to global oscillations produced by waves trapped in the solar interior (the p-modes) that cause variations in the total irradiance. In all cases the variations have small amplitudes in integrated light from the visible photosphere. The SXT instrument on board *Yohkoh* has provided a new data set that can resolve some of these variations spatially. Most of these data are in the form of whole-Sun images: rate about 5 per day; effective pixel size of 4.92 arc sec; passband 28Å centered at 4310Å; data interval beginning in September, 1991 and ending in October, 1992. These observations have several advantages over ground-based observations for the characterization of solar global variability, namely the lack of atmospheric effects and the long-term stability of the instrumentation. We present an assessment of these data in the context of the existing total irradiance data, and also discuss their application to determination of the figure of the Sun, particularly measurements of the variations of the solar radius.

1. Introduction

The ACRIM instrument on board *Solar Maximum Mission* greatly stimulated interest in the solar irradiance from two points of view: both as the defining forcing agent for climate change on Earth, and as a means of learning about the solar interior via inferences regarding the mechanisms of solar luminosity variability (*e.g.* Newkirk 1983). ACRIM and other total irradiance instruments detected variability on a wide range of time scales, up to the length of the present time series (for reviews see Hudson 1988; Fröhlich *et al.* 1991). Observations from the entire disk of the Sun confuse all of these mechanisms of solar variability. Thus time-resolved imaging observations, with high photometric accuracy, clearly would help in understanding the physics of at least some of the processes involved. Observations from space would be preferred for this purpose, although ground-based observations also hold a great deal of promise — for spatially resolved data, ground-based image photometry can be normalized to the whole Sun, and thus avoid many photometric problems.

The history of solar imaging at optical wavelengths from space is quite limited. About the only observations have been the balloon-borne telescopes (*e.g.*, Danielson 1961) and the time series of optical images from the SOUP experiment on *Spacelab-2* (*e.g.*, Title *et al.* 1989). This paper centers on the observations from the aspect camera of the SXT instrument on board *Yohkoh*, which has provided the first lengthy time series of solar white-light images from space. These data can be used in studies of the development of solar activity, the figure of the Sun, the limb distribution of surface brightness, g-modes, flares, solar rotation, etc. For a review of the *Yohkoh* X-ray observations, see Shibata (1994).

This paper tries to describe the properties of the data and the observing system, and in doing so it lists all the problems that we know about. The reader should not construe

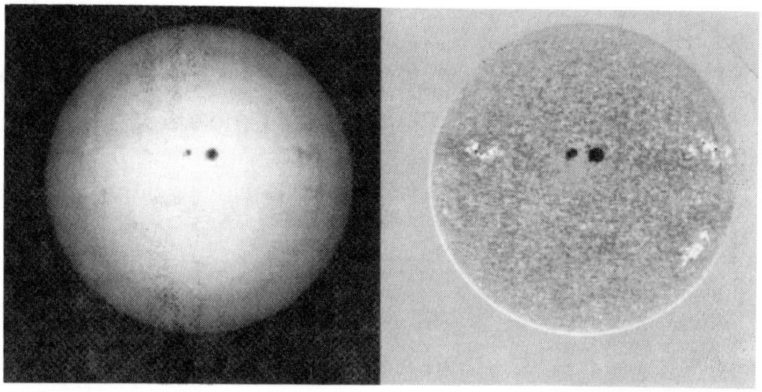

FIGURE 1. Image of the Sun obtained from the SXT instrument on board the *Yohkoh* spacecraft. Left, raw data; right, residuals as described in Section 4. This image (N top, E left, pixel size 4.92 arc s square, exposure time 0.339 ± 0.004 s) is from 03:27:36 UT, 18 August 1992. The large spot group is NOAA active region 7260, with a tabulated area of 940 millionths. The observations were made in the *Yohkoh* "narrow-band" filter, 28Å bandwidth centered at 4310Å.

this negatively. These data are unique and will be used in many important applications in the future. The CCD detector and optical system of the *Yohkoh* instrument clearly cannot produce the remarkable photometric stability displayed by ACRIM-type sensors, but the spatially resolved variations have much larger relative amplitudes.

2. The *Yohkoh* white-light data base

The SXT instrument on *Yohkoh* (see Tsuneta *et al.* 1991, for a general description) provided white-light observations of the solar disk as a byproduct of its primary function of soft X-ray imaging. The main purpose of these white-light images is to provide coalignment information for the X-ray observations via sunspot locations relative to the limb. The time series began in September 1991, and ended in October, 1992. The essential part of the database consists of about thirteen months of routine observations. The total number of images such as that of Figure 1 exceeeded 2,000 over this time interval, corresponding to a rate of approximately five per day.

The white-light images come from a 1024 × 1024 virtual-phase CCD illuminated by a 5-cm lens. The telescope has a fixed focus maintained by a graphite-epoxy metering structure, with the focus established by observations in Palo Alto, California, before launch. Most of the data reported here come from 2 × 2 pixel summations, resulting in an effective pixel size of 4.92″, and were synthesized from two separate exposures in order to extend the 8-bit telemetry to 12 bits. Limited amounts of full-resolution data are available, almost always during "flare mode." This resulted in stretches of data with approximately one image each 10-12 s, but over a restricted field of view of 64 × 64 pixels (about 2.62′ square). Note that the pixelization does not critically sample the spatial structure of the image ($1.22\ \lambda/d = 2.17''$), so that relatively large pixel-to-pixel variations may occur. The optical layout seems quite simple, consisting of the lens, blocking filters, and bandpass filters isolating one of two spectral bands, but the Appendix gives the flavor of some of the problems encountered in making it work properly. The data reported here come mainly from a filter with a nominal bandpass of 28Å centered at 4310Å. This is in

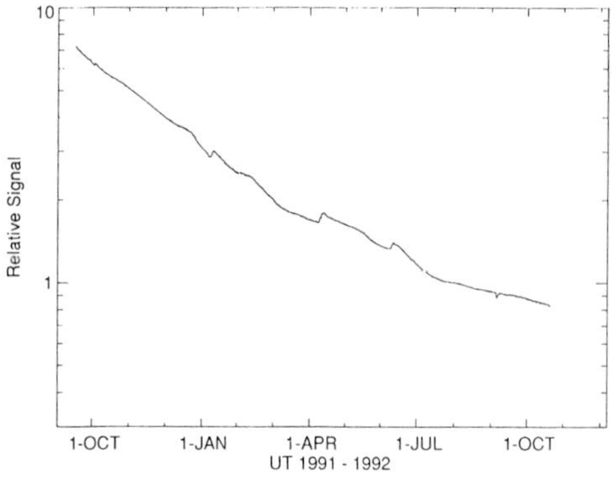

FIGURE 2. Relative signal level for *Yohkoh* white-light images, showing the steady decrease of signal level during the duration of the observations. The necessary exposure time grew from about 60 msec to about 948 msec during this period. The sudden increases occurred when the CCD was warmed up to drive off contaminants from its surface.

the Fraunhofer G band and was chosen for the aspect-sensing function of SXT in order to show the chromospheric network and faculae.

The aspect telescope performed well until the loss of an X-ray aperture filter in October 1992, which "blinded" the optical system with visible light reflected off the X-ray optics. In orbit the CCD slowly became contaminated, possibly as a result of the dehydration of the telescope's metering structure. This effect probably presents the most serious problem for precise analyses of these data, because the contamination has fine spatial structure of relatively large amplitude and also varies with time. Periodic heating of the CCD, at intervals of one to two months, drove some of the contamination off.

There was also a continual loss of sensitivity over the total span of the data, as shown in Figure 2. This quasi-exponential decay, which could be localized to the pre-filter/lens assembly, resulted in gradually increasing exposure times as the mission went on. Finally, we also observed a gradual degradation of the "flat-field" map, which was routinely observed using sunlight through a diffusing glass filter. None of these problems seriously affected the use of the data for aspect sensing, and it remains to be seen how much they will impact the scientific use of the data.

The images are nominally oriented to solar north by *Yohkoh*, but there are limited intervals during which this system was not maintained well and the N orientation wandered a small amount. There is also a one-year periodic error with an amplitude of approximately 0.45°, due to the on-board ephemeris for the Canopus reference sensor (T. Kosugi, personal communication).

Figure 3 compares "flat-field" maps derived from routine diffuse-light images, illustrating the changes in the spatial flatness of the response during the period of observation. This resulted from an unanticipated degradation of the CCD surface layers by the same soft X-radiation that SXT was designed for (fortunately, the degradation does not appear to have affected the X-ray response). The flat-field maps therefore reflect a kind of time integral of the detected X-ray flux, including some artifacts of individual flares that were not intercepted by the *Yohkoh* flare trigger and for which the X-ray exposure times were therefore not properly adjusted downwards by the automatic exposure-control logic. The detailed appearance of the flat-field maps can also be derived from the white-light data directly, by using the stability of the solar brightness distribution in the mean (A. McAllister, private communication). For the central region of the Sun, this may be the preferred technique for future studies of these data, since the flat-field irregularities shown in Figure 3 eventually became deeper than 20% near the limb positions, and because the diffuser flats do not have the same passband as the two analysis filters.

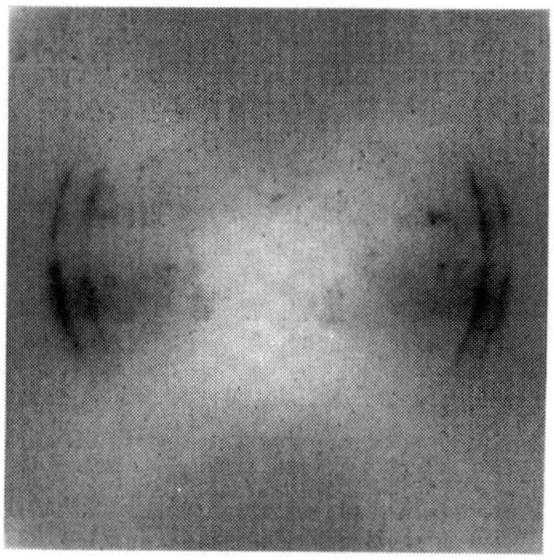
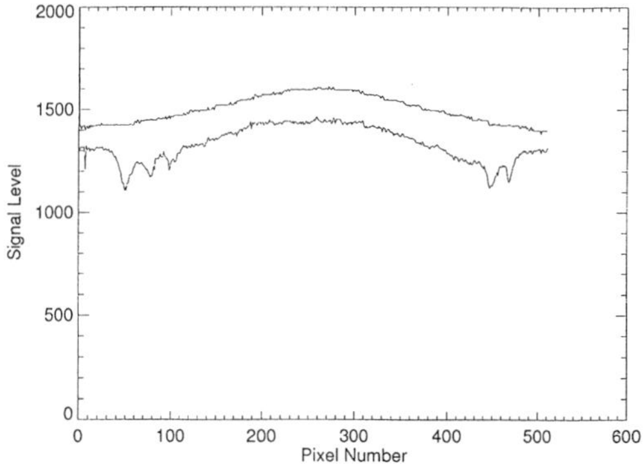

FIGURE 3. Flat-field image (N top, E left) obtained from near the end of the *Yohkoh* white-light data interval, 21 October 1992. The large-scale structure comes from the optics of the opal glass diffuser used to provide the illumination, and has an amplitude of a few percent; the small-scale structure comes from radiation damage to the CCD induced by the solar soft X-rays. The pattern represents a time integral over the whole mission, so that one can see the effects of pointing changes of the spacecraft, and even of individual bright flares. This damage has no appreciable effect on the X-ray sensitivity of the detector. The line plot on the right compares the flat-field structures from this image with one taken a year earlier, 31 October 1991 (plot shows the middle row of the 512 × 512 pixel images).

3. The solar radius

One of the initial engineering analyses involved measuring the apparent location of the solar limb. This was essential for SXT image registration. The results from this analysis are described in this section. The program (FIND_LIMB) used for this purpose is unsophisticated, consisting of an image differentiation based on the "Sobel" filter: $G(i,j) = \sqrt{(X^2 + Y^2)}$, in an approximation (from the IDL library) that defines X and Y as functions of the values in the eight pixels surrounding the (i,j)-th pixel. The differentiated image is subjected to various tests designed to screen out artifacts due to charged particles and to sunspots. The coordinates of the individual pixels of the differentiated image that pass these tests are then analyzed harmonically as a function of azimuth angle around the limb. FIND_LIMB returns the coordinates of the center of the image, the radius, a value for the scatter of the radius pixels around the best-fit limb trace, plus higher-order harmonic terms. The limb location defined in this way does not include effects due to the limb-darkening function at the extreme limb, and in this manner retains some direct sensitivity to low-level distortions of the limb-darkening function resulting from faculae. The use of the 4310Å filter for the bulk of these data probably enhances this effect, but it is also present in the SXT "broad-band" filter (184Å centered at 4616Å).

The current precision of the radius determination from a single image is about 0.1 arc s (rms) on short time scales, as determined by the scatter of the individual measurements. This compares favorably with the best ground-based observations (the H.A.O. observations of T. Brown) quoted by Ribes *et al.* (1991) to be ~0.14 arc s per day. The relatively short span of the *Yohkoh* data makes it difficult to compare these observations with longer runs of ground-based observations of the solar radius, at least in terms of the slow variations that have been reported by the CERGA group and others (Ribes *et al.*

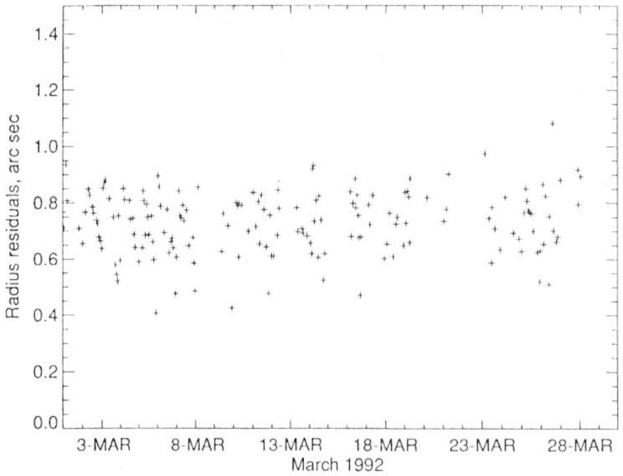

FIGURE 4. Apparent variation of solar radius determined from the SXT white-light images for one month in 1992.

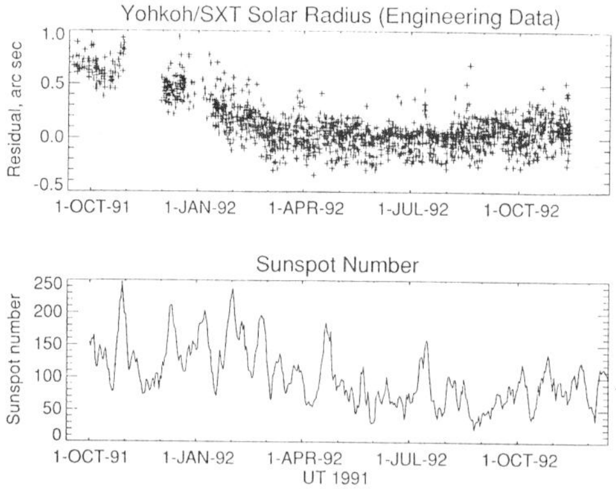

FIGURE 5. Apparent variation of solar radius determined from the SXT white-light images for the entire span of the data (upper). The lower panel shows the variation of sunspot number over the same time interval. The apparent solar radius decreased by about 0.7 arc s in coincidence with the decrease of solar activity in early 1992.

1991). Overlapping data exist but have not yet been jointly analyzed. The short-term scatter of these SXT radius determinations is considerably greater than theoretical limits imposed by pixelization or counting statistics (Buffington *et al.* 1991), but Figure 4 suggests that this scatter is approximately random.

On longer time scales the SXT radius determinations, made as described above, show a systematic variation (Figure 5). There is a clear variation of apparent solar radius with time, roughly correlated with the general level of solar activity as shown by the sunspot number. One possible cause of this observed variation, if solar in origin, could be confusion of facular brightness variations with true figure variations. Equally likely would be systematic errors of observation that coincidentally resemble the sunspot number variation. These data show a larger apparent radius variation than any present in the H.A.O. monthly means for 1981–1988. This paper cannot report a definitive conclusion here, and it remains possible that all or a part of the observed apparent radius change results from a true spatial effect. The stability of the measurements suggests that a more careful analysis of these data will provide better limits on true radius variations.

4. The brightness distribution

The main observational goal here consists of tracking the time variations of solar irradiance on a spatially resolved basis, for example to follow an individual active region across the disk and thus to determine the time dependence of the sunspot irradiance deficits and facular excesses (Willson *et al.* 1981). The discussion below describes the properties of the data relative to this goal.

The *Yohkoh* white-light images require dark subtraction and correction for flat-field imperfections. Routine operations of *Yohkoh* provide many dark and diffuser frames with which to make these corrections. Some pixels are "hot", with high dark-current rates;

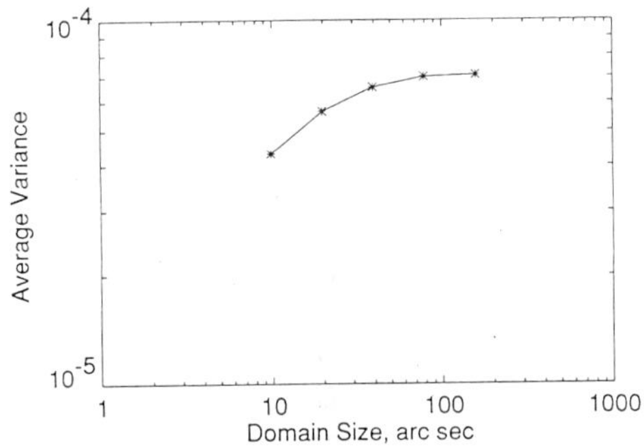

FIGURE 6. Statistics of image structure in a region 5.24′ × 5.24′ from the image in Figure 1, after flat-fielding and subtraction of a mean image free of sunspots. The plot shows the increase of pixel-to-pixel variance with increasing domain size.

the images are improved cosmetically by dark subtraction, but the noise fluctuations remain. Individual pixels or groups of pixels often show cosmic-ray tracks. Such effects of course cannot be removed except on a case-by-case basis, for example by filtering the image to eliminate pixels that look wrong. For the time of the image in Figure 1, the flat-field map had a total range of 34% and an rms fluctuation of 3.7% of the mean value.

Each of the images in the time *Yohkoh* series should be dark-subtracted, flat-fielded, and reduced to residuals according to the empirical limb-darkening function for this system (see Nishikawa 1994, for an example of such image preparation). Such a comprehensive data reduction is not available yet for the *Yohkoh* data, unfortunately. We have assessed the noise properties of the images in a preliminary manner by carrying out dark subtraction, flat-fielding, and subtraction of a mean image, for the whole-Sun images during the disk passage of AR 7260 (August 1992). The mean image was constructed by omitting the pixels containing sunspots from the summation. In a quiet region near disk center, these procedures resulted in a standard deviation of approximately 1% of the mean brightness for a single 4.92″ × 4.92″ pixel, and did not appreciably increase over a test area as large as 5.24′ × 5.24′ from the image in Figure 1. This level of variation is consistent with the level expected from p-modes and granulation. The statistics of this test region of Figure 1 are shown in Figure 6. The decrease of fractional variance for larger bin sizes indicates that the sources of the fluctuations remain largely unresolved, and that the image preparation has not left much larger-scale structure in the residuals.

For time-series analysis, the temporal fluctuations of the photospheric structures cause a basic limitation to precise photometry. Since the *Yohkoh* images discussed here have such a low cadence, much of this temporal variation will be aliased and not averaged down. Pointing jitter of the *Yohkoh* spacecraft also contributes in principle to the temporal variations by converting spatial patterns into time-series variations.

Thus far, the main application of time-series analysis of the *Yohkoh* white-light data are in the study of "white-light" flares (Hudson *et al.* 1992) and potentially "black light" flares as well (van Driel *et al.* 1994). Anwar *et al.* (1993) also used these flare-mode data to study rapid motions of sunspots associated with a major flare. The data show that all flares above a certain threshold have clearly detectable optical emission excesses as observed in the 4310Å filter. The contrasts are as large as about 50%, depending upon the flare. We have made a preliminary estimation of the ratio of the contrasts in this filter and near Hα for a flare simultaneously observed at Mees Solar Observatory (Canfield *et al.* 1992), and find that the 4310Å filter had a contrast ~2.2 times as great. The flares thus detected have a strong impulsive-phase component appearing to originate at the footpoints of coronal loops, and also may show longer-lasting emission.

The flare studies typically use difference images over a short time span as a means of

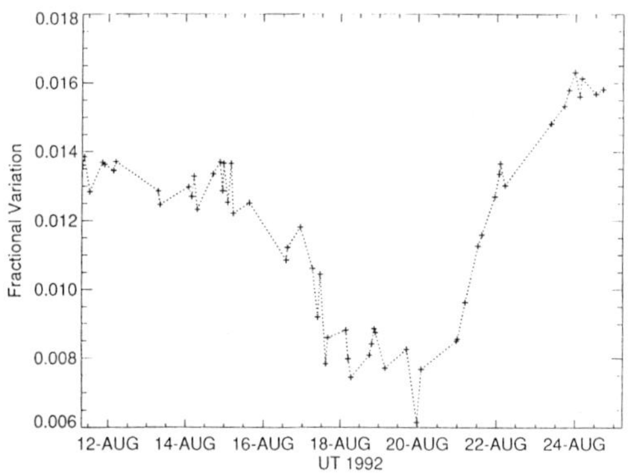

FIGURE 7. Solar brightness variations during the disk passage of NOAA active region 7260 in August, 1992. The plotted data represent the difference between the total signal in a N latitude strip (containing the sunspot group, see Figure 1), and an equivalent S latitude strip. The data are normalized to allow a comparison with the expected total irradiance variation caused by the spot group. The amplitude of the "dip" is larger than what would have been observed by ACRIM I for this sunspot area. This is consistent with the short wavelength (4310Å) of this observation, but detailed comparisons with ACRIM II data need to be made.

levelling the high-contrast features (sunspots) found in active regions. This works well on short time scales, but over the typical span of a "flare mode" observation (about ten minutes) the photospheric background features vary at the level expected from the statistics in Figure 6. Temporal fluctuations on the order of 1% amplitude are present and provide a limiting noise floor for the detection of white-light flare emissions.

We can use the whole-Sun images to study the sunspot "dips" discovered in the ACRIM total irradiance data (Willson *et al.* 1981). Figure 7 shows the effects of the disk passage of the sunspot group of NOAA region 7260. This is the analog of a "dip" in total irradiance as observed by ACRIM. What limits the precision of the photometry reported here? The scatter of individual data points in the time series of Figure 7 exceeds that expected from counting statistics by a large factor. Most of the additional variance must come from unresolved p-modes, granulation, or uncorrected jitter modulation of spatial structures. The latter may be somewhat improved by proper image preparation and rebinning, but the former cannot be escaped without adequate sampling. As Title *et al.* (1989) have shown, the aliasing of the p-modes can be suppressed effectively by spatio-temporal filtering of adequately sampled data. The message for future observations appears to be that the complete spatial resolution of the total irradiance variations will require observations of the whole solar disk at a sufficiently frequent cadence to be able to average down the high-frequency variations induced by waves and convection.

5. Conclusions

The *Yohkoh* data exhibit sufficient precision, stability, and sampling frequency for many of the problems of time-series image photometry needed for the "spatial resolution of the solar variability." To claim this might seem preposterous: the stability of the output of an individual CCD pixel cannot even be known from this data base (but see Buffington *et al.* 1990) very accurately at all, mainly because of the limited number of samples. However one does not even need one-percent photometry per pixel to see the effect of a sunspot as illustrated in Figure 7, much less the far better precision ACRIM demonstrated. I will conclude with a list, not necessarily comprehensive, of observational problems that could be tackled with these data at some future time:

(*a*) Are there novel types of irradiance variability that have not been recognized yet in the ACRIM-type data?

(*b*) Is there evidence for p- or g-modes in the limb figure or photometric data?

(*c*) Are there time-dependent effects in sunspot blocking (*e.g.*, a "thermal shadow" prior to sunspot formation)?

(*d*) Does the solar radius vary with time appreciably?

(*e*) Does the solar limb-darkening function vary with time appreciably?

(*f*) What is the photospheric signature of a coronal hole?

(*g*) What are the limits of detectability of continuum in white-light flares?

(*h*) Are there negative flares?

(*i*) Are flare-related sunspot distortions or motions a common phenomenon?

(*j*) Is there evidence for interplanetary material (*e.g.*, comets) striking the Sun?

These problems touch on several areas in solar physics, with some possibility of elucidating solar interior structure and dynamics. Because scientific studies with a telescope's aspect sensor obviously have secondary importance, it is a tribute to the *Yohkoh* scientific team that the data have such high quality. Even though we are just at the beginning of analysis of these data, I believe that they already demonstrate that spatially resolved irradiance observations can make major contributions. Future instrumentation with more direct optimization for this kind of study would be worthwhile. Based upon the *Yohkoh* experience, the key properties of such observations should include simplicity of optical design, large data throughput, careful control of systematic errors (telescope rotation around the solar direction, desirably), Sun-synchronous or deep-space orbit, careful calibration, and a field of view containing the whole solar disk.

Acknowledgments. I would like to acknowledge the help of many people associated with the white-light capability of SXT and *Yohkoh*: Loren Acton, Marilyn Bruner, Tom Cruz, Sam Freeland, Jim Lemen, Mons Morrison, Bill Rosenberg, Yoshiaki Ogawara, Saku Tsuneta, Bachtiar Anwar, Alan McAllister, Lidia van Driel-Gesztelyi, Dick Canfield and Jean-Pierre Wülser. NASA supported this work under contract NAS 8–37334.

Appendix A. The design of the SXT aspect telescope

The design parameters of the SXT aspect telescope are given in the summary description by Tsuneta *et al.* (1991). Much other material is contained in LPARL SXT Design Note #31 (Grillot & Cruz 1989). The following paragraphs borrows liberally from the latter source and attempts to give all of the relevant unpublished information about the materials and testing of the SXT aspect telescope.

A.1. *Attenuating filters*

The Aspect Telescope consists of an attenuating filter, a bandpass pre-filter, an imaging lens assembly, and four relevant filters in the SXT filter-wheel assembly near the focal plane of the system: an open position, two interference filters for optical wavelength bandpass determination, and an opal glass diffuser used for flat-field calibration of the CCD. See Tsuneta *et al.* (1991) for layout drawings.

The attenuating filter consists of a synthetic fused silica substrate, anti-reflection coated on the sunward surface, with an aluminum attenuating layer on the rear surface. The aluminum thickness is slightly over 500Å, and it has a dielectric overcoating layer for durability. Following this element, there is a bandpass-limiting filter (substrate Hoya CM-500, 6 mm thick). A near-IR coating on the sunward surface of this element was found to attenuate the region from $\sim$4800Å to $\sim$7000Å to 10^{-4} or better. The half-maxima were experimentally determined to be at approximately 3700Å and 4700Å. The transmission of the first elements together was measured to be approximately 1.5×10^{-4}

$\pm$ 5 $\times 10^{-5}$ from approximately 3600Å to 4650Å. The combined out-of-band rejection of the attenuator and bandpass filter together approach or surpass 10^{-8} absolute.

A.2. *Doublet lens*

The visible image is formed by a doublet lens, achromatic across the passband of the entrance filter assembly. The lens has a 50 mm clear aperture and forms a beam at approximately f/31. The doublet has a depth of focus of approximately $\pm$ 0.5 mm, with a fixed focal distance adjusted before flight and an Airy disk diameter about 34 microns.

The sunward element (element #1) of the doublet is a bi-convex lens composed of Schott BK7-G18, a borosilicate crown glass stabilized with cerium. Element #2 of the doublet is a concave-convex lens made from Schott LF5-G15 radiation-damage-resistant flint glass, also cerium-stabilized. The doublet elements are spaced 0.2 mm apart by a stainless steel ring. Both surfaces of both lenses are anti-reflection coated over the passband of the entrance filter. These lenses assist in attenuating incident light at wavelengths below 4000Å. The spectral half-maxima of the elements at the front end of the aspect telescope are at approximately 4000Å and 4700Å. Additional blocking is provided by the analysis filters.

A.3. *Analysis filters*

The Aspect Telescope package includes two interference filters in the filter wheel near the focal-plane end of the telescope. The "narrowband" filter has a center wavelength at 4310Å, bandpass (FWHM) of 28Å, and maximum transmission of 33%. It was chosen to show the chromospheric network as well as possible. The "broadband" filter is characterized by 4616Å, 184Å, and 2.2%. Both surfaces of each interference filter are anti-reflection coated at their respective center wavelengths. These filters, plus the spectral response of the CCD detector itself, determine the wavelength of observation. Figure 7 of Tsuneta (1991) shows these filter responses.

The in-flight "flat field" calibration of the CCD is done with an opal glass diffuser at one of the positions of the filter wheel. This provides approximately uniform illumination (few percent rms) with sunlight over the CCD area, in the spectral band defined by the pre-filter and lens assemblies. To match signal levels, this opal glass diffuser has an additional attenuator with $\sim$50% transmission.

A.4. *Radiation resistance*

Materials resistant to radiation damage were chosen for all optical components of the SXT aspect telescope where it was practical. The entrance filter used synthetic fused silica, because it has additional resistance to radiation damage from UV, neutrons, protons, electrons, or γ-rays as compared with crushed fused silica or fused quartz. The CM-500 substrate of the bandass filter needed further investigation because radiation damage information was not available before launch. Accordingly Grillot & Rosenberg (1990) conducted successful radiation-damage tests of this material and the Schott equivalent BG-39, with doses as high as 100$\times$the expected three-year dosage for the planned *Yohkoh* orbit. In the design of the SXT aspect telescope, the first element of the system shields the CM-500 element from electrons. The substrates of the analysis filters are cerium-stabilized BK7-G18 material.

A.5. *Control of ghost images*

The white-light attenuator and the bandpass filter elements 0.5-degree tilts in opposite directions to eliminate ghost images over a field of view larger than the Sun. Near the focal plane, however, the problem of suppressing the ghost images is more complex. The uncovered surface of the CCD reflects a portion of the incident light. Anti-reflection coatings were not possible because of concern about the soft X-ray quantum efficiency. The focal-plane filters are about 20% reflective and 10% absorbing. The reflection will thus be partially reflected back onto the CCD to form a secondary image. Measurements indicated that the CCD was approximately 25% reflective at 4310Å and at 4600Å, which would yield a 5% ghost. This ghost cannot be eliminated from the solar image by tilting the focal-plane filters relative to the CCD because of proximity. However, by incorporating an attenuating glass filter onto CCD-facing side of the focal-plane filters this ghost image intensity was reduced to approximately 1% or less of the primary image intensity. This was accomplished with 2.5 mm thick pieces of Schott UG-5 absorbing glass in each of the two focal-plane filters.

REFERENCES

ANWAR, B., ACTON, L.W., HUDSON, H.S., MAKITA, M., MCCLYMONT, A.N. & TSUNETA, S. 1993 Rapid sunspot motion during a solar flare. *Solar Phys.* **147**, 287–303.

BUFFINGTON, A., HUDSON, H.S. & BOOTH, C.H. 1990 A laboratory measurement of CCD photometric stability and dimensional stability. *Publ. Ast. Soc. Pacific* **102**, 688–697.

CANFIELD, R.C., HUDSON, H.S., LEKA, K.D., ACTON, L.W., STRONG, K.T., KOSUGI, T., SAKAO, T., TSUNETA, S., CULHANE, J.L., PHILLIPS, A. & FLUDRA, A. 1992 The X flare of 1991 November 15: coordinated Mees/*Yohkoh* observations. *Publ. Astr. Soc. Japan* **44**, L111–L115.

DANIELSON, J. 1961 The structure of sunspot penumbras. I. Observations. Ap. J. **134**, 275–288.

FRÖHLICH, C., FOUKAL, P., HICKEY, J., HUDSON, H. & WILLSON, R. C. 1991 Solar irradiance variability from modern measurements. In *The Sun in Time* (ed. C.P. Sonett, M.S. Giampapa & M.S. Mathews). pp. 11–29. Univ. Arizona Press, Tucson, AZ, USA.

GRILLOT, P. & CRUZ, T. 1989 Soft X-ray telescope aspect telescope design note # 31. (unpublished).

GRILLOT, P.N. & ROSENBERG, W.J. 1990 Proton radiation damage in optical filter glass. *Applied Optics* **28**, 4473–4477.

HUDSON, H.S. 1988 Observed variability of the solar luminosity. *Ann. Revs. Astron. Astrophys.* **26**, 473–501.

HUDSON, H.S., ACTON, L.W., HIRAYAMA, T. & UCHIDA, Y. 1992 White-light flares observed by *Yohkoh. Publ. Astr. Soc. Japan* **44**, L77–L81.

NEWKIRK, G. 1992 Variation of the solar luminosity. *Ann. Revs. Astron. Astrophys.* **21**, 429–467.

NISHIKAWA, J. 1994 On the cause of total irradiance variability observed by the CCD solar surface photometer., submitted to *Solar Phys.*.

RIBES, E., BEARDSLEY, B., BROWN, T.M., DELACHE, PH., LACLARE, F., KUHN, J.R. & LEISTER, N.V. 1991 The variability of the solar diameter. in *The Sun in Time* (ed. C.P. Sonett, M.S. Giampapa & M.S. Mathews), pp. 59–97. Univ. Arizona Press, Tucson, AZ, USA.

SHIBATA, K. 1994 The variable structure of the X-ray corona as viewed by *Yohkoh.* (ed. J.M. Pap, C. Fröhlich, H.S. Hudson & S.K. Solanki). Cambridge Univ. Press, in press.

TITLE, A., TARBELL, T.D., TOPKA, K.P., FERGUSON, S.H., SHINE, R.A. & THE SOUP TEAM 1989 Statistical properties of solar granulation observed by the SOUP instrument on Spacelab-2. *Ap. J.* **336**, 475–494.

TSUNETA, S., ACTON, L, BRUNER, M., LEMEN, J., BROWN, W., CARAVALHO, R., CATURA, R., FREELAND, S., JURCEVICH, B. MORRISON, M., OGAWARA, Y., HIRAYAMA, T. & OWENS, J. 1991 The soft X-ray telescope for the *Solar-A* mission. *Solar Phys.* **136**, 37.

VAN DRIEL, L. G., HUDSON, H.S., ANWAR, B., HIEI, E. & TSUNETA, S. 1994 A diagnostics of solar atmospheric models: Search for black flares. *Solar. Phys.*, in press.

WILLSON, R.C., GULKIS, S., JANSSEN, M., HUDSON, H.S. & CHAPMAN, G.A. 1981 Observations of solar irradiance variability. *Science* **211**, 700.

The Sun in Time: Detecting and Modelling Magnetic Inhomogeneities on Solar-Type Stars

By J. DAVID DORREN AND EDWARD F. GUINAN

Astronomy & Astrophysics Dept., Villanova Univ., Villanova, PA 19085, USA

We report on results of a program of coordinated multiwavelength observations of single G stars. While the spectral types are restricted to lie within a narrow range of the Sun's, they have ages ranging from $\sim$ 70 Myr to $\sim$ 9 Gyr, which makes them suitable proxies for the Sun at several stages of its life history from the ZAMS to the very late main-sequence phase.

---

1. Introduction

A detailed understanding of the probable evolution of solar magnetic activity throughout the Sun's lifetime cannot be obtained from evolutionary model calculations. It is necessary to study proxies for the Sun corresponding to different stages in its history. From several studies of stars of different ages (*e.g.,* Soderblom 1983) it can be concluded that the early Sun was rotating more rapidly than at present. Consequently its magnetic activity – its photospheric starspot and chromospheric plage activity, flares, coronal X-ray and radio emission, *etc.* – would have been enhanced over present levels, when magnetic braking has significantly reduced the strength of the solar dynamo. The correlation between age, rotation rate, and activity levels in single cool stars is well-established (*e.g.,* Skumanich 1972; Simon *et al.* 1985).

In 1990 we initiated a coordinated multiwavelength study which now includes ground-based *UBVRI* or *uvby* photoelectric photometry, *UV* (*IUE* and *EUVE*) and X-ray (*ROSAT*) observations of a number of nearby, single, solar-type stars which were selected as proxies for the Sun from an age of $\sim$ 70 Myr (the ZAMS Sun) to $\sim$ 9 Gyr (the terminal-age main-sequence or TAMS Sun). By *solar-type*, we mean main-sequence stars with masses, spectral types, and colors close to the Sun's: approximately $0.9M_\odot < M < 1.1M_\odot$, $F9/G0V < Sp.\ Type < G5V$, and $0.58 < B - -V < 0.70$. The group also includes the nearby G2 subgiant β Hyi as a proxy for the Sun at $\sim$ 9 Gyr (Dravins *et al.* 1993). However, the expression *solar-type* is frequently extended to the whole range of spectral types (F – M) to which stars showing evidence of solar-type magnetic activity belong. With the exception of β Hyi, our program stars in the G0 – G5 group will have similar convection-zone depths. On the other hand, there is no restriction on age ($\sim$ 70 Myr to $\sim$ 9 Gyr for the stars in our sample) and rotation periods ($P_{rot} \simeq 2.7d$ to $\sim 45d$). Ages for many of these stars are determined from moving-group membership. The program also contains stars selected for their close spectral similarity to the present Sun, with a view to determining whether the Sun is typical of stars of its age.

Up to the late 1970s it was by no means obvious that it would be possible to detect the presence of starspots on single dwarf G stars through observation of a rotational brightness modulation, as had already been done for the highly active RS CVn binaries. The solar analogy would suggest that the light variations might be only a fraction of one percent, beyond the limit of detectability by conventional ground-based photometry. Indeed, the solar irradiance variation due to rotational modulation has only recently been established by spacecraft measurements, and is up to 0.3% (Willson 1982; Foukal & Lean 1986). However, the Wilson (1978) survey of chromospheric Ca II HK emission revealed chromospheric variability to be common among solar-type stars. This program, continued by Baliunas and collaborators at Mt. Wilson, shows both rotational modulation of the Ca II HK flux and year-to-year variations in mean flux (Noyes *et al.* 1984; Radick *et al.* 1990).

Among the first photometric evidence of *light* variations in single solar-type stars that could be ascribed with some certainty to starspots were the observations by Dorren & Guinan (1982) of a small sample of the more active G – K stars in the Mt. Wilson Ca II HK program. Contemporaneous photometry and Ca II HK measures indicated that in the case of the K2 V star HD 149661 (12 Oph), the chromospheric emission was strongest at times of reduced stellar brightness, a clear suggestion that the bright chromospheric active areas (plages) are associated with dark photospheric inhomogeneities (starspots). In a follow-up project in 1983 at Kitt Peak National Solar Observatory (KPNO), we observed several single F – K stars, finding light variations of up to ~ 0.1 mag in several. In particular, we discovered HD 129333, a young G0 V solar-type star, to have the largest light variation of the G-stars (~ 0.06 mag) and a period of ~ 2.7d. This star is the youngest in our sample, with an age of ~ 70 Myr determined by its membership in the Pleiades Moving Group (Dorren & Guinan 1994a).

The advent of automatic photometric telescopes (APTs) has made long-term monitoring programs feasible, removing the drudgery of routine observation. Our program since 1988 has relied heavily on photometric data obtained by telescopes on Mt. Hopkins, AZ (the Phoenix-10; the Four College Consortium and Fairborn Observatory 0.8m APTs).

Other groups have investigated solar-type stars. A systematic program of high precision photometric monitoring of solar-type stars has been carried out since 1980, initially at Cloudcroft Observatory and subsequently at Lowell Observatory (Lockwood & Skiff 1988; Radick 1992). Their sample (including solar-type stars in the Hyades and Coma clusters, and 33 stars from the Mt. Wilson HK project) constitutes the largest set of solar-type stars which has been systematically monitored using concurrent photometry and Ca II HK observations. A number of important conclusions concerning the link between stellar rotation, age, and chromospheric activity have emerged from these programs (see Radick *et al.* 1990; Radick 1993; Lockwood 1993 and refs. therein).

Finally, direct measurements of magnetic fields on single stars have been made by Saar (1992 and refs. therein) using the Zeeman-broadening technique. For active G stars, the field strengths are typically 1,000 – 2,000 gauss, and filling factors 10 – 40%.

2. The Sun in time project

A list of the more intensively studied stars in the program is given in Table 1. Our own *UV* and X-ray observations have generally been supplemented by extracting all the available *IUE* archival spectra and existing X-ray observations. Most of the stars selected are nearby, bright, and have well-determined rotation periods from photometry or Ca II HK observations. Their ages are reasonably well-known from membership in moving groups or from isochronal fits and span the range from ZAMS to TAMS.

The principal aims include:

a) Investigating the evolution of the solar dynamo and solar magnetic activity using stars with nearly identical physical properties to the Sun (i.e., mass, radius, spectral type, temperature and depth of convective zone), but of different ages, and therefore different rotation rates. The prime concerns are the level of activity as a function of age, rotational modulation of optical brightness and *UV* and X-ray continuum and/or emission-line fluxes, and the search for activity cycles. The stars are not members of close binaries

STAR	Spectral Type	m_V	B-V	dist. (pc)	P_{rot} (days)	Age	Log L_x ergs s^{-1}	Log F_{CIV} ergs s^{-1}	Log F_{MgII} ergs s^{-1}	Light Range $< \Delta V >$
HD 129333 = EK Dra	G0 V	7.52	+0.61	31	2.75	~70 Myr	29.85	28.42	29.55	0.060
HD 72905 = π^1 UMa	G1.5 V	5.64	+0.62	14	4.68	300 Myr	29.20	27.81	29.40	0.034
HD 206860 = HN Peg	G0 V	5.94	0.59	15.1	4.86	(300 Myr)	28.85	27.67	29.22	0.040
HD 39587 = χ^1 Ori	G1 V	4.41	0.60	10.0	5.08	300 Myr	28.95	27.78	29.52	0.030
HD 1835 = BE Cet	G2 V	6.39	0.65	20.4	7.65	600 Myr	28.70	27.89	29.23	0.028
HD 28099 = VB 64	G2 V	8.12	0.66	46.6	8.7	600 Myr	29.41	27.89	29.19	0.021
HD 20630 = κ^1 Cet	G5 V	4.83	0.68	9.3	9.2	(1 Gyr)	28.78	27.63	29.30	0.020
HD 114710 = β Com	G0 V	4.26	0.58	8.1	12.4	(~2 Gyr)	28.30	27.23	29.08	0.011
HD 190406 = 15 Sge	G5 V	5.8	0.61	17.2	13.5	(~2 Gyr)	28.30	–	29.02	0.008
HD 128620 = α Cen A	G2 V	-0.01	0.71	1.33	~30	5-6 Gyr	27.11	26.78	28.93	< 0.005
HD 2151 = β Hyi	G2 IV	2.80	0.62	6.5	~45	9 Gyr	26.90	26.80	29.13	< 0.006
SUN	G2 V	-26.74	0.66	1AU	25.5	4.6 Gyr	27.48	26.70	28.95	0.0015

TABLE 1. Properties of solar proxies in Sun in Time Program

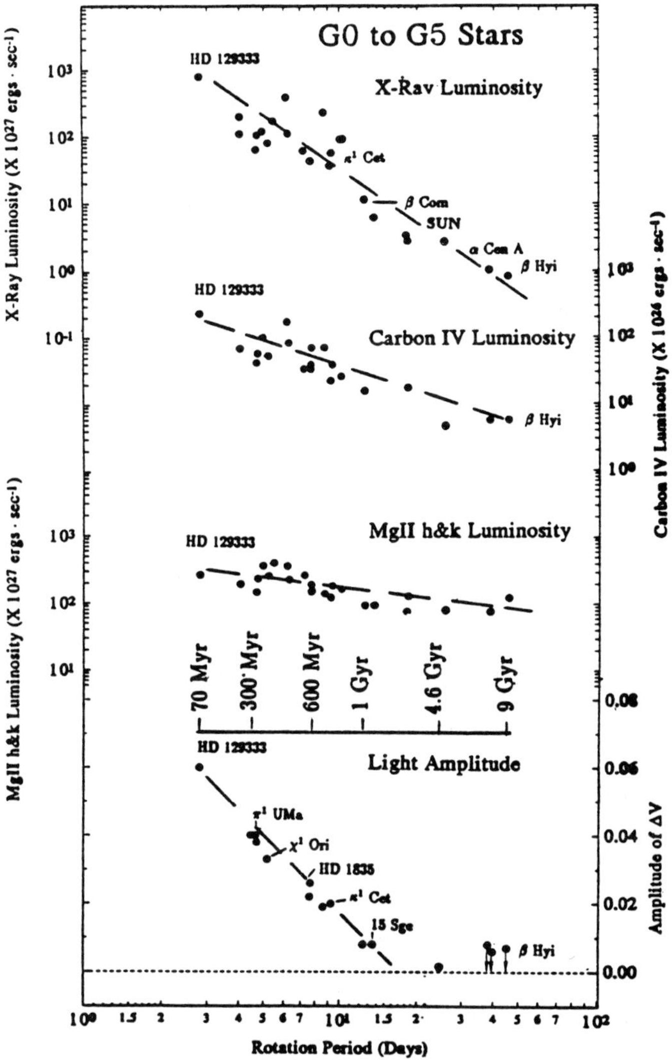

FIGURE 1. The dependence of magnetic activity on rotation for close solar analogs: dwarf G0 – G5 stars and the subgiant β Hydri. X-ray luminosity (coronal activity), C IV and Mg II h+k emission (TR and chromospheric activity) and maximum light curve amplitude (photospheric activity) all decline smoothly with increasing rotation period, or equivalently, age.

and thus are not spun-up or influenced by tidal effects or interaction as in the case of the RS CVn binaries and most BY Dra binaries. Figure 1 shows the decline of X-ray luminosity, C IV emission, and Mg II h+k emission with increasing rotation period (and age) for a sample of strictly solar-type (G0 – G5) stars, includings the stars of Table 1 (Dorren *et al.* 1994). The most dramatic decline occurs in the X-ray emission, which decreases by a factor of 1,000 between ZAMS and TAMS. An important consequence of limiting the range of spectral types (and hence masses) is the clear delineation of these activity-rotation-age relationships. Also shown is the dependence of the light variation on rotation period, to be discussed in the next section.

b) Establishing the extent to which the Sun is typical of stars of its age and spectral type. We have included several bright nearby stars which are close solar analogs. In particular, HD 44594 is a close spectral match for the Sun in the 3,000 – 8,000 Å range (Hardorp 1978). *IUE* observations which we obtained in 1991 and 1992 show that the spectral similarity persists into the *IUE* wavelength regime, but also that *there is evidence for variability of about 20-30% in the Mg* II *(λ 2800) emission.*

c) Obtaining estimates of the solar UV *and X-ray emission fluxes throughout the Sun's post-ZAMS history.* This has an important bearing on the chemical and dynamical evolution of primeval planetary atmospheres and on the dispersal of the remnants of the solar nebula (see *e.g.*, Giampapa & Imhoff 1985). As an example we have derived the *UV* spectral irradiance of the ZAMS sun using HD 129333 (Dorren & Guinan 1994a).

3. Methods and results

3.1. *Photosphere: starspots*

Light variations in active chromosphere stars are observed using multiwavelength photoelectric photometry, usually *UBVRI* or *uvby*, often supplemented by wide and narrow-band Hα filter pairs to obtain a measure of the chromospheric Hα emission. If the stellar photosphere has a sufficiently extensive starspot coverage (a few percent or more) there will be a detectable modulation of the stellar brightness as the star rotates. Multiwavelength observations are essential to obtain the starspot temperatures and an estimate of their areal extent and distribution. In most cases the area obtained is probably a lower limit; what is actually being observed is the contrast between maximally and minimally spotted hemispheres.

Details of the modelling procedure have been given by Dorren (1987), whose analytic solution to the problem of integrating over a distribution of circular spots is particularly transparent, and straightforward to apply. As an example of modelling a particular star in some detail, see the article by Dorren *et al.* (1981) on the highly active RS CVn binary, HR 1099. Similar techniques have been used by many authors to obtain spot areas and temperatures for a large number of active chromosphere stars (*e.g.*, Eaton & Hall 1979; Rodonó *et al.* 1986; Budding & Zeilik 1987; Strassmeier 1990). For reviews see Eaton (1992) and Guinan & Giménez (1993).

From observations of the solar-type stars we draw the following general conclusions:

- All the G0 – G5 V stars studied with periods less than 12-13d show definite rotationally-modulated light variations. At ∼ 13d, the range of the $\langle v \sin i \rangle$-band variation is $\Delta V \simeq 0.008$ mag, and at 2.8d, $\Delta V \simeq 0.06$ mag. No definite rotationally-modulated light variations greater than 0.004 mag (the limiting precision of our photometry) have yet been detected for stars with $P \geq 14$d. However, detectable light variations do occur in single dwarf K and M stars with rotation periods of up to ∼ 40d. Variability appears to be rare among F stars earlier than about F8. The values of brightness changes given in Table 1 for α Cen and β Hyi are *upper limits*, obtained from published visual magnitudes. Figure 2 shows the light variations in three of the program stars. The relation between the maximum light curve range (corresponding to the greatest observed contrast between stellar hemispheres, and taken to be an indication of the maximum spot coverage) and rotation period is shown at the bottom of Figure 1.

- Starspot modelling of the *UBVRI* and *uvby* light curves, which show a clear wavelength dependence, indicates that spots cover an area of up to 6% of the total stellar surface in the G stars with temperature differences between spot and photosphere of 300 – 600 °K. In most cases, two spots are needed to fit the shapes of the light curves.

• The light curves are observed to change on a timescale comparable to the rotation period. This is illustrated by the light curve of χ^1 Ori shown in Figure 3. This is probably due to differential rotation together with waxing and waning of starspot groups. The other active, fast-rotating stars in our sample (HD 129333, π^1 UMa, HD 1835) show similar behavior. These stars also show evidence for differential rotation in their Ca II HK measures (Donahue 1993). Probably the clearest example of stellar differential rotation is seen in the long-period RS CVn system λ And (G8 IV-III + ?; $P_{rot} \sim$ 54d), for which light curves obtained over several seasons display a systematic pattern of evolution which is most simply understood in terms of differential rotation of two spot groups (Dorren & Guinan 1984). Remarkably, a very similar pattern of light-curve evolution to that of λ And is found in the RS CVn binary HR 7275 (K1 III-IV + ?; P_{rot} = 28.6 d). This has recently been analyzed by Strassmeier *et al.* (1994), who also conclude that differential rotation is a major factor.

• Long term photometry of solar-type stars has indicated the presence of activity cycles with timescales typically of several years (Lockwood & Skiff 1988). We shall show evidence for a 12-yr cycle in HD 129333 later in the paper.

But a word of caution – the G5 V star HD 135262, an apparently undistinguished star which originally had been used as comparison star for the active solar-type star HD 134319, exhibits relatively large ($\sim$ 0.1 mag), usually smooth variations which are clearly not produced by spots. The star is not a known binary, and in any case there are no indications of an eclipse. At present the origin of its light variations is a puzzle.

3.2. *Chromosphere and transition region*

IUE observations of chromospherically active stars have provided direct measures of chromospheric, TR, and coronal activity through net emission-line strengths. For example, the Mg II (2800) chromospheric line emission, which is generally strong, is a reliable indicator of plage activity. Hotter lines, among the strongest of which are C IV (1550), Si IV (1400), He II (1640), and N $\langle v \sin i \rangle$ (1240) signal the strength of the concurrent activity in the TR and corona, and physical conditions such as electron densities from emission line ratios (Jordan & Linsky 1987; Linsky 1990).

Figure 4 shows the Mg II h and k emission lines for 3 solar-type stars of ages from $\sim$ 70 Myr to $\sim$ 9 Gyr together with the Sun, illustrating the decline of chromospheric emission with age. Rotational modulation of the Mg II emission has been observed in HD 129333 (Dorren & Guinan 1994a). Boesgaard & Simon (1984) observed a clear modulation of the C IV emission in χ^1 Ori due to rotation. For most of the other solar-type stars in our program the C IV emission shows variations but there are insufficient observations to establish rotational modulation with certainty. On the other hand, a large number of stars in the Mt. Wilson Ca II HK project show rotational modulation of the chromospheric emission, from which accurate rotational periods have been derived. Several stars also show activity cycles of typically 5-17 years (Baliunas *et al.* 1985; Donahue 1993). For example, the active, Pleiades-age G0 V star, HD 129333, has a well-defined activity of $\sim 12 - 13$ years (see Figure 5).

3.3. *Corona*

The PSPC instrument on the *ROSAT* satellite provides high resolution X-ray spectra in the 0.1 – 2.4 keV range from which coronal temperatures may be derived. For single

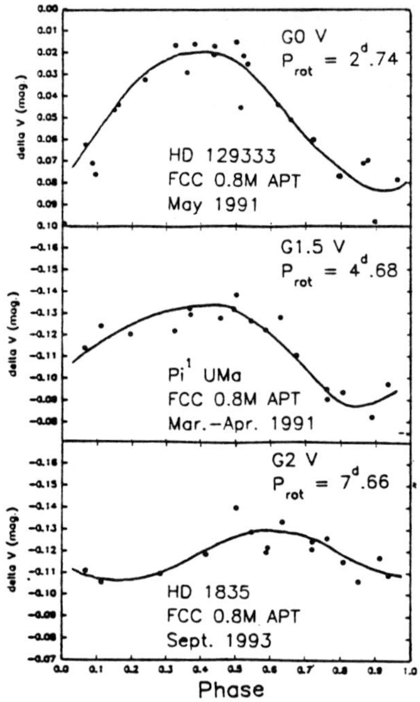

FIGURE 2. Light curves of solar proxies of different ages: HD 129333 (age ~ 70 Myr: the ZAMS Sun), π^1 UMa (~ 300 Myr), and HD 1835 (~ 600 Myr). The light variations are produced by the rotation of a spotted star. The ages of these stars are inferred from membership in moving star cluster groups.

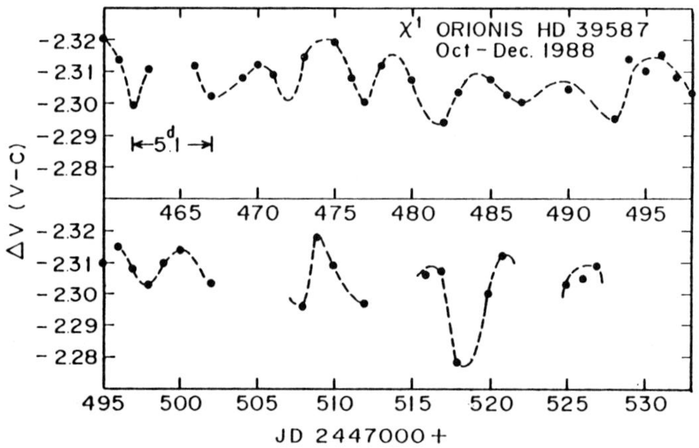

FIGURE 3. Long-term V-band light variations of χ^1 Ori (~ 300 Myr; P_{rot} = 5.1d) obtained with the Phoenix-10 APT. The changes in mean light level and light curve shape are indicative of variations in the spot coverage together with differential rotation.

solar-type stars, these temperatures are typically $10^6 - 10^7$ °K. The X-ray luminosities of single, solar-type stars are ~ $10^{27} - 10^{30}$ ergs s^{-1}, or up to 1,000 times the X-ray luminosity of the quiet Sun. For RS CVn systems, L_x is $10^{29} - 10^{31}$ ergs s^{-1}. In addition to the decrease in L_x with increasing rotation period (Figure 1), the *ROSAT* observations indicate a decrease in coronal temperature with increasing period. For example, the

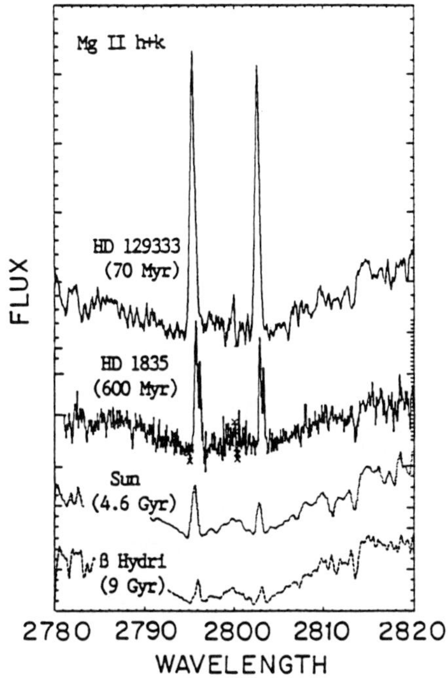

FIGURE 4. Mg II h and k emission lines of HD 129333 (age = 70 Myr), HD 1835 (600 Myr), the Sun (4.6 Gyr), and β Hydri (9 Gyr).

youngest star in our sample, HD 129333, has a coronal temperature of up to 10^7 °K; the oldest star, β Hyi, has a temperature of $1 - 2 \times 10^6$ °K (Dorren & Guinan 1994b).

In addition, radio (VLA or VLBI) observations provide information on non-thermal emission from coronal regions with temperatures of $10^7 - 10^8$ °K. Güdel (1993) recently detected HD 129333 as a radio source with a specific intensity of $\sim 1.8 \times 10^{14}$ ergs $s^{-1}Hz^{-1}$, or about 300 times stronger than the Sun. He has also reported a possible rotational modulation of the X-ray and radio flux from this star.

The recent launch of the *EUVE* satellite has opened a hitherto unexplored wavelength range (50-750 Å) to observation. *EUVE* is, moreover, the only instrument capable of providing *spectroscopy* of stellar coronae. Together, *IUE* and *EUVE* observations provide full spectroscopic coverage of the chromosphere, TR, and corona, *i.e.* over a temperature range from below 10^4 °K to above 10^7 °K. Altogether, the optical photometry and spectroscopy, ultraviolet spectroscopy, X-ray and radio observations produce a detailed picture of the relationship between active regions throughout the stellar atmosphere.

3.4. *Activity cycles*

Activity cycles have been observed for many of the F–K type stars in the Mt. Wilson HK project (Baliunas *et al.* 1985; Donahue 1993). Analysis of these data indicate no definite relation between the presence and length of an activity cycle and spectral type or rotation period (see Donahue 1993). We also note that a number of RS CVn-type binaries show long-term variations in their light curves that suggest the presence of a magnetic activity cycle (see Dorren & Guinan 1990).

In our program the best example of an activity cycle is seen in the long term variation of HD 129333 in mean light, mean Ca II HK index, and *UV* chromospheric and TR

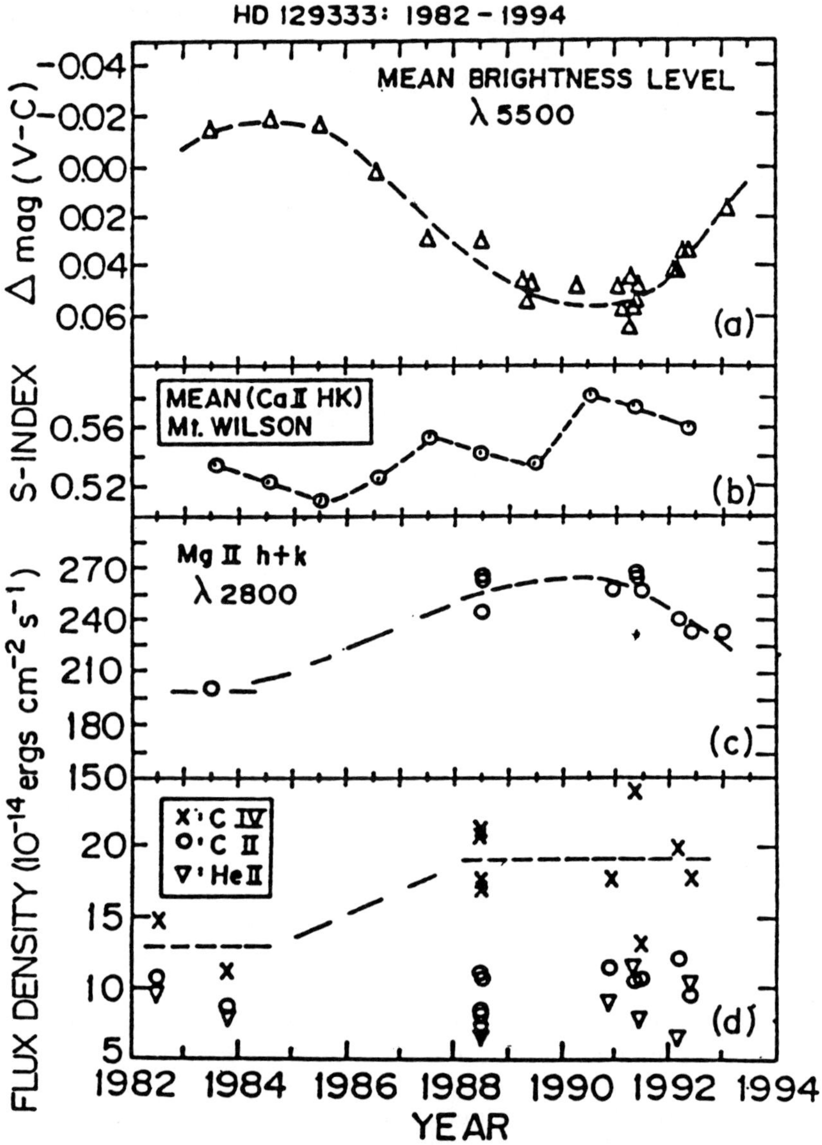

FIGURE 5. Long-term variations in HD 129333: mean V-band brightness, $\overline{\Delta V}$, referred to a comparison star; the Ca II HK emission line index (S); Mg II h+k emission line fluxes; C IV (1550Å), C II (1335Å), and He II (1640Å) emission line fluxes. An activity cycle of about 12-13 years' duration is suggested (from Dorren & Guinan 1994b.)

emission lines shown in Figure 5. There is clear evidence for an activity cycle lasting about 12 years.

4. The Sun — Normal or not? A final word

Whereas solar-type (G0-G5 V) stars with P_{rot} > 14 d show no measurable *photometric* variability, some K stars in this period range do have significant variations, most likely because of a greater convection zone depth. Some studies have suggested that the Sun may have smaller light variations than other solar-type stars (Radick 1994; Lockwood 1994), but this conclusion is based on a sample of stars that includes K stars as well as G stars and may therefore not be reliable. We reiterate the importance of restricting attention to *close* solar analogues.

Acknowledgments. This work has been supported in part by NASA *IUE* grants NAG 5-382, NAG 5-1703, *ROSAT* grant NAG 5-1662 and NSF grant (to EFG) AST-861362. We wish to thank Elizabeth Jewell for her editorial assistance and David Steelman and Anne McGhie for figure preparation.

REFERENCES

BALIUNAS, S.L., HORNE, J. J., PORTER, A., DUNCAN, D.K., FRAZER, J. CANNING, H., MISCH, A., MUELLER, J., NOYES, R., SOYUMER, D., VAUGHAN, A.H. & WOODARD, L. 1985 Time-series measurements of chromospheric Ca II H and K emission in cool stars and the search for differential rotation. *Ap.J.* **294**, 310–325.

BOESGAARD, A.M. & SIMON, T. 1984 Rotational modulation of the chromospheric activity in χ^1 Orionis (G0 V). *Ap.J.* **277**, 241–249.

BUDDING, E.A. & ZEILIK, M. 1987 An analysis of the light curves of short period RS Canum Venaticorum stars: Starspots and fundamental properties. *Ap.J.* **319**, 827–835.

DONAHUE, R. A. 1993 Surface Differential Rotation in a Sample of Cool Dwarf Stars. PhD thesis, New Mexico State University, Las Cruces, NM.

DORREN, J.D. 1987 A new formulation of the starspot model, and the consequences of starspot structure. *Ap.J.* **320**, 756–767.

DORREN, J.D. & GUINAN, E.F. 1982 Evidence for starspots on single, solar-like stars. *AJ* **87**, 1546–1557.

DORREN, J.D. & GUINAN, E.F. 1984 Starspots, differential rotation, and a possible six-year spot cycle on λ Andromedae. In *Lecture Notes in Physics No. 193* (ed. S. L. Baliunas & L. Hartmann). pp. 259–263. Springer-Verlag: N.Y.

DORREN, J.D. & GUINAN, E.F. 1990 The evolution of magnetic activity on V711 Tauri and evidence for a significant facular contribution. *Ap.J.* **348**, 703–711.

DORREN, J.D. & GUINAN, E.F. 1994a HD 129333 – The Sun in its infancy. *ApJ*, in press.

DORREN, J.D. & GUINAN, E.F. 1994b, in preparation.

DORREN, J.D., GUINAN, E.F. & DEWARF, L. 1994 The Sun in time: The decline of solar magnetic activity with age. In *Eight Cambridge Workshop: Cool Stars, Stellar Systems, and the Sun* (ed. J.-P. Caillault). Kluwer: Dordrecht, in press.

DORREN, J.D., SIAH, M.J., GUINAN, E.F. & MCCOOK, G.P. 1981 Starspots on V711 Tauri (HR 1099). *AJ* **86**, 572–582.

DRAVINS, D., LINDEGREN, L., NORDLUND, A. & VANDENBERG, D.A. 1993 The distant future of solar activity: A case study of β Hydri. I. Stellar evolution, lithium abundance, and photospheric structure. *Ap.J.* **403**, 385–395.

EATON, J.A. 1992 Models of starspots and results. In *Surface Inhomogeneities in Late-Type Stars* (ed. P.B. Byrne & D.J. Mullan). pp. 15–26. Springer-Verlag: NY.

EATON, J.A. & HALL, D.S. 1979 Starspots as the cause of the intrinsic light variations in RS Canum Venaticorum type stars. *Ap.J.* **227**, 907–922.

FOUKAL, P. & LEAN, J. 1986 The influence of faculae on total solar irradiance and luminosity. *Ap.J.* **302**, 826–835.

GIAMPAPA, M.S. & IMHOFF, C.L. 1985 The ambient radiation fields of young solar systems: Ultraviolet and X-ray emission from T Tauri stars. In *Protostars and Planets II* (ed. D. C. Black & M.S. Matthews). pp. 386–404. Univ. Arizonaz: Tucson, AZ, USA.

GÜDEL, M. 1993, private communication.

GUINAN, E.F. & GIMÉNEZ, A. 1993 Magnetic activity in close binaries. In *The Realm of Interacting Binary Stars* (ed. J. Sahade, Y. Kondo & G. McCluskey). pp. 51–110. Kluwer: Dordrecht.

HARDORP, J. 1978 The Sun among the stars: A search for solar spectral analogs. *A&A* **63**, 383–390.

JORDAN, C. & LINSKY, J.L. 1987 Chromospheres and transition regions. In *Exploring the Universe with the IUE Satellite* (ed. Y. Kondo *et al.*). pp. 259–293. Reidel: Dordrecht.

LINSKY, J.L. 1990 Modeling the coronae and chromospheres of RS CVn systems by the analysis of ultraviolet, X-ray and radio observations. In *Active Close Binaries* (ed. C. Ibanoğlu). pp. 747–759. Kluwer: Dordrecht.

LOCKWOOD, G.W. 1994 Irradiance variations of stars. In *The Sun as a Variable Star: Solar and Stellar Irradiance Variations* (ed. J.M. Pap, C. Fröhlich, H.S. Hudson & S.K. Solanki). Cambridge University Press, in press.

LOCKWOOD, G.W. & SKIFF, D.A. 1988 Luminosity variations of stars similar to the Sun. *Air Force Geophysics Lab No. AFGL-TR-88-0221.*

NOYES, R.W., HARTMANN, L.W., BALIUNAS, S.L., DUNCAN, D.K. & VAUGHAN, A.H. 1984 Rotation, convection, and magnetic activity in lower main sequence stars. *Ap.J.* **279**, 763–777.

RADICK, R.R. 1992 Report on luminosity variability of lower main-sequence stars. *Sacramento Peak Observatory*, preprint.

RADICK, R.R. 1994 Photometric variations of solar type stars. In *The Sun as a Variable Star: Solar and Stellar Irradiance Variations* (ed. J.M. Pap, C. Fröhlich, H.S. Hudson & S.K. Solanki). Cambridge University Press, in press.

RADICK, R.R., LOCKWOOD, G.W. & BALIUNAS, S.L. 1990 Stellar activity and brightness variations: A glimpse at the Sun's history. *Science* **247**, 39–44.

RODONÓ, M. *et al.* 1986 Rotational modulations and flares on RS CVn and BY Dra-type stars. *A&A* **165**, 135–156.

SAAR, S. 1992 Evidence for a complex distribution of magnetic field strengths on the flare star AD Leo. In *Cool Stars, Stellar Systems, and the Sun* (ed. M. Giampapa & J. Bookbinder). (ASP Conf. Ser. 26), pp. 252–254.

SIMON, T., HERBIG, S. & BOESGAARD, A.M. 1985 The evolution of chromospheric activity and the spin-down of solar-type stars. *Ap.J.* **293**, 551–574.

SKUMANICH, A. 1972 Time scales for Ca II emission decay, rotational braking, and lithium depletion. *ApJ* **171**, 565–567.

SODERBLOM, D.R. 1983 Rotational studies of late-type stars and the rotational history of the Sun. *ApJS* **53**, 1–15.

STRASSMEIER, K.G. 1990 Photometric and spectroscopic modeling of starspots on the RS Canum Venaticorum Binary HD 26337. *ApJ* **348**, 682–699.

STRASSMEIER, K.G., HALL, D.S. & HENRY, G.W. 1994 Time-series spot modeling II: Fifteen years of photometry of the bright RS CVn binary HR 7275. *A&A*, in press.

WILLSON, R.C. 1982 Solar irradiance variations and solar activity. *J. Geophys. Res.* **87**, 4319–4326.

WILSON, O.C. 1978 Chromospheric variations in main-sequence stars. *ApJ* **226**, 379–396.

Irradiance Models Based on Solar Magnetic Fields

By KAREN L. HARVEY

Solar Physics Research Corporation, Tucson, AZ 85718, USA

A method to separate the active region and quiet network components of the magnetic fields in the photosphere is described and compared with the corresponding measurements of the He I λ 10830 absorption. The relation between the total He I absorption and total magnetic flux in active regions is roughly linear and differs between cycles 21 and 22. There appears to no relation between these two quantities in areas outside of active regions. The total He I absorption in the quiet Sun (comprised of network, filaments, and coronal holes) exceeds that in active regions at all times during the cycle. As a whole, active regions of cycle 22 appear to be less complex than the active regions of cycle 21, hinting at one possible cause for a differing relation between spectral-irradiance variations and the underlying magnetic flux for these two cycles.

1. Introduction

Variations of the radiative output of the Sun are directly linked to the short- and long-term evolution of the magnetic fields during a solar cycle. Understanding the spatial and temporal character of the solar magnetic fields is necessary in order to understand the thermal structures that ultimately result in the observed solar irradiance variations.

Visual examination of high-resolution, full-disk magnetograms of the photosphere show a myriad of structures that can be roughly described by three components: (1) active regions, (2) stronger network elements that are the dispersing remnants of active regions, and (3) relatively weak network elements that are distributed in a cell-like pattern over the entire solar surface. Several investigations (Lean *et al.* 1982; Skumanich *et al.*. 1984; Foukal & Lean 1988; Livingston *et al.* 1988; Schatten 1988; Willson & Hudson 1988; Foukal *et al.* 1991) indicate that these components of activity play important, but different, roles in the observed solar-irradiance variations.

For this study, magnetic flux structures are partitioned into two components: active regions and the quiet Sun. The quiet Sun component includes both the remnants of active regions and the weak magnetic network elements. These two components are considered in relation to spectral-irradiance variations in the He I λ10830 absorption. Described in Section 2 of this paper is the method used to isolate active regions from the quiet Sun magnetic fields. One additional parameter also is investigated, that is, the complexity of the magnetic fields. In Section 3, the total magnetic flux and its complexity in and outside these so-defined active regions are compared with the corresponding total He I λ10830 Å absorption in these two components.

2. Data

The National Solar Observatory/Kitt Peak (NSO/KP) synoptic rotation maps of the magnetic field and He I λ10830 spectroheliograms are used in this analysis. These rotation maps are constructed from the daily NSO/KP full-disk magnetograms; a description of the method used can be found in Gaizauskas *et al.* (1983) and Harvey (1992). The rotation maps shown in Figure 1 depict the distribution of magnetic flux and He I λ10830 structures during one complete rotation of the Sun, in this case Carrington Rotation 1712 (at the peak of magnetic activity in cycle 21, August 1981). Each equal-area pixel in the

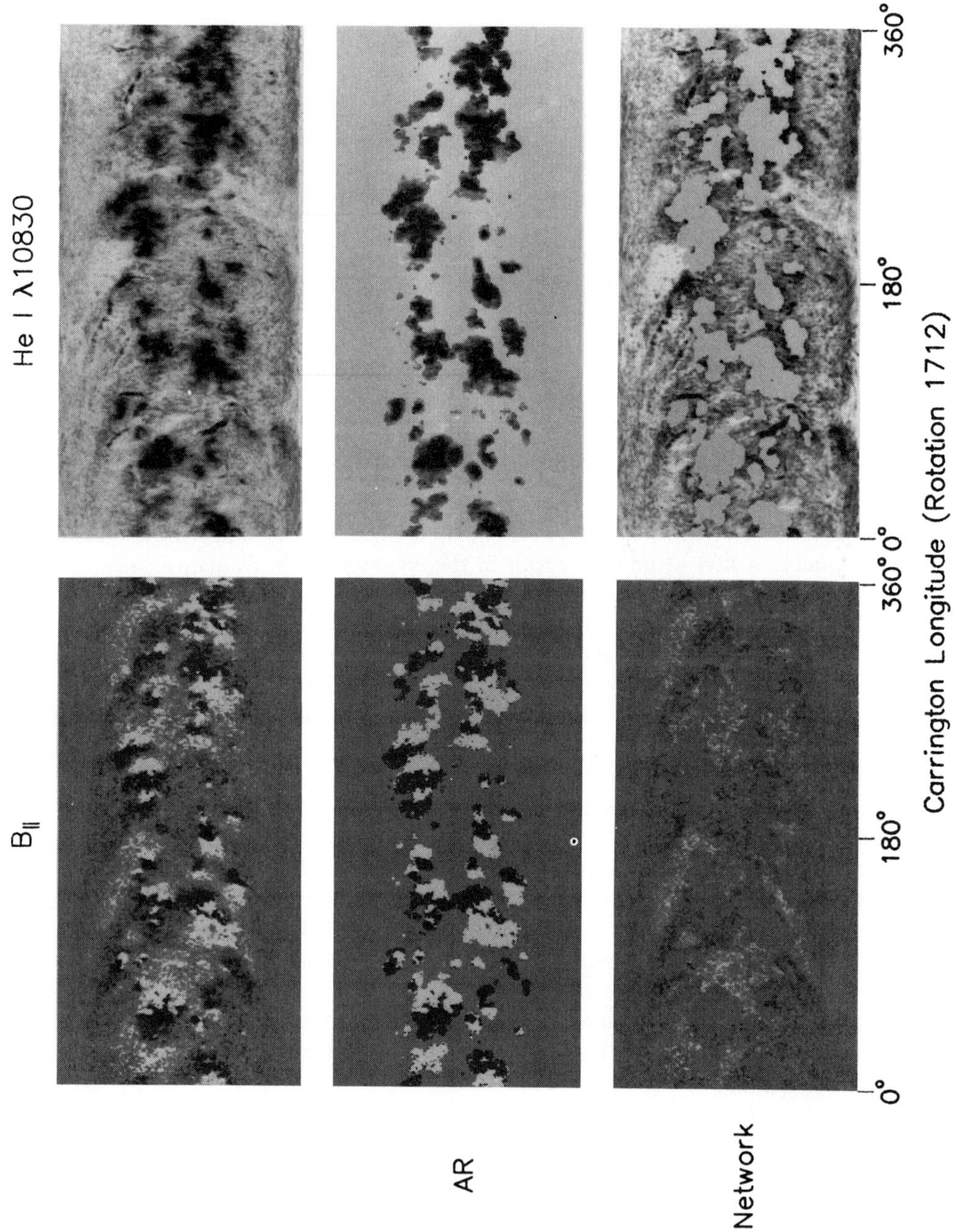

FIGURE 1. Synoptic rotation maps of the magnetic field *(left)* and He I λ10830 Å *(right)* for Carrington Rotation 1712. The *top* maps are separated into their active region *(middle)* and quiet network *(bottom)* components using the technique described in Section 2.1.

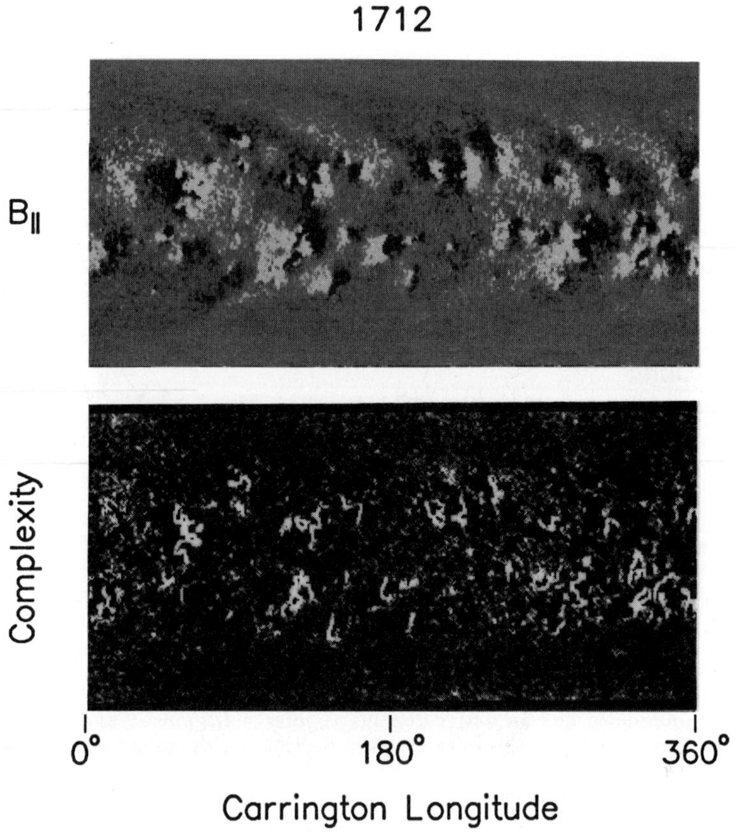

FIGURE 2. Synoptic magnetic map for Carrington Rotation 1712 *(top)* and the corresponding complexity map *(bottom)*.

maps is 1° in longitude and 0.011 in sine latitude; the noise level in the magnetic field data is estimated to be 0.3 Gauss.

2.1. *Technique for isolating active regions*

The first attempt to isolate active regions in the synoptic rotation maps used simply a threshold of 25 Gauss, above which a pixel was considered to be in an active region and below in the 'quiet' Sun (Harvey 1992). In such a procedure, we find that, based on the value of the magnetic field alone, there is an overlap between these two components of magnetic structures. To minimize this effect, a second technique was developed. Active regions were identified in the synoptic rotation magnetic maps by examining the amplitude of the variation of the field within a sub-array of 5×5 pixels, i.e. within an area 5° in longitude and 0.055 in sine latitude. Within active regions, it is not expected that over this size area the magnetic flux will be uniform. For each of the 25 pixels in the sub-array, the absolute value of the magnetic field is differenced against the absolute value of the center pixel, and the square of the differences is summed. The square root of the sum

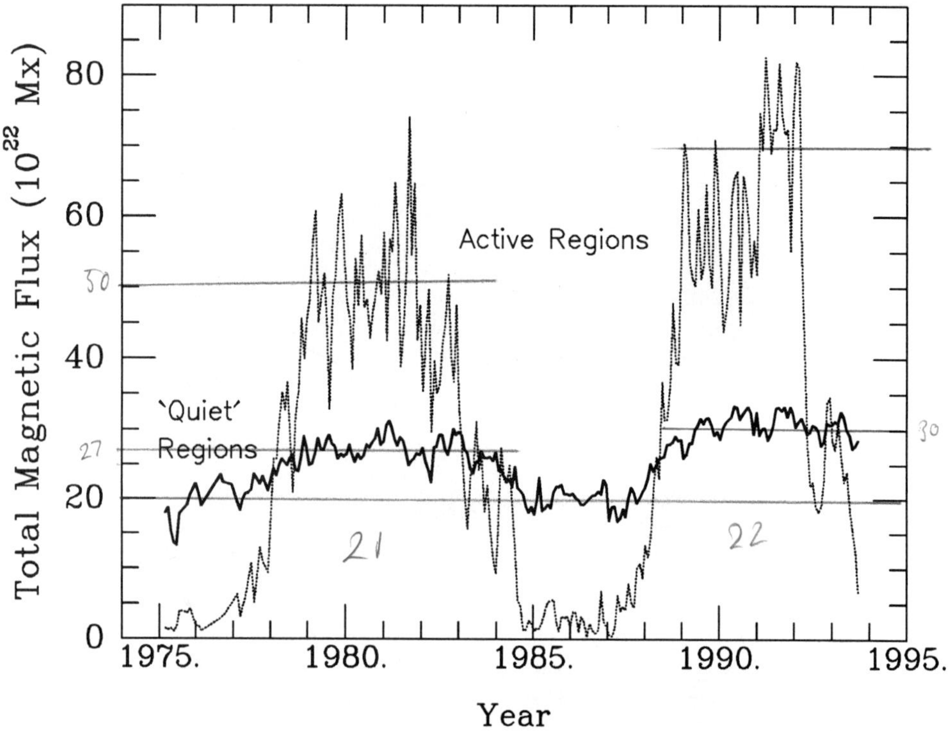

FIGURE 3. Total magnetic flux in active regions *(lighter solid curve)* and in the quiet Sun network *(heavier curve)*, using method described in Section 2.1 to isolate active regions.

over the sub-array, i.e. the root-mean-square-difference, is determined for every pixel in the rotation map. The resulting map of root-mean-square-differences is smoothed to eliminate patchiness and is then thresholded to objectively define the location of active regions. The two separated components of magnetic structures, active regions and the quiet Sun network, are shown in the bottom panels at the left of Figure 1.

Compared to thresholding the data at 25 Gauss, this method to isolate active regions results in less patchy selection of pixels considered to be in an active region, as well as an active region boundary that is expanded slightly outside the isogauss contour of 25 Gauss. There is little magnetic flux in the pixels excluded by thresholding at 25 Gauss, but included by the root-mean-square difference technique, while the area, is larger. In Carrington Rotation 1712, for example, the total magnetic flux in active regions is 3.3% higher using the root-mean-square-difference compared to thresholding at 25 Gauss (71.8×10^{22} Mx to 74.2×10^{22} Mx), while the area within active regions is 60% larger (0.82×10^{22} cm^2 to 1.31×10^{22} cm^2).

2.2. *Technique for defining magnetic complexity*

A measure of the complexity of the magnetic field is determined in a similar way as described above. The sub-array is 3×3 pixels. Within this sub-array, the polarity of a pixel is compared with that of the center pixel. If the pixel is of opposite polarity, the difference between the magnetic field of the pixel and that of the center pixel is squared and summed. This quantity identifies magnetic inversion lines in the rotation maps and

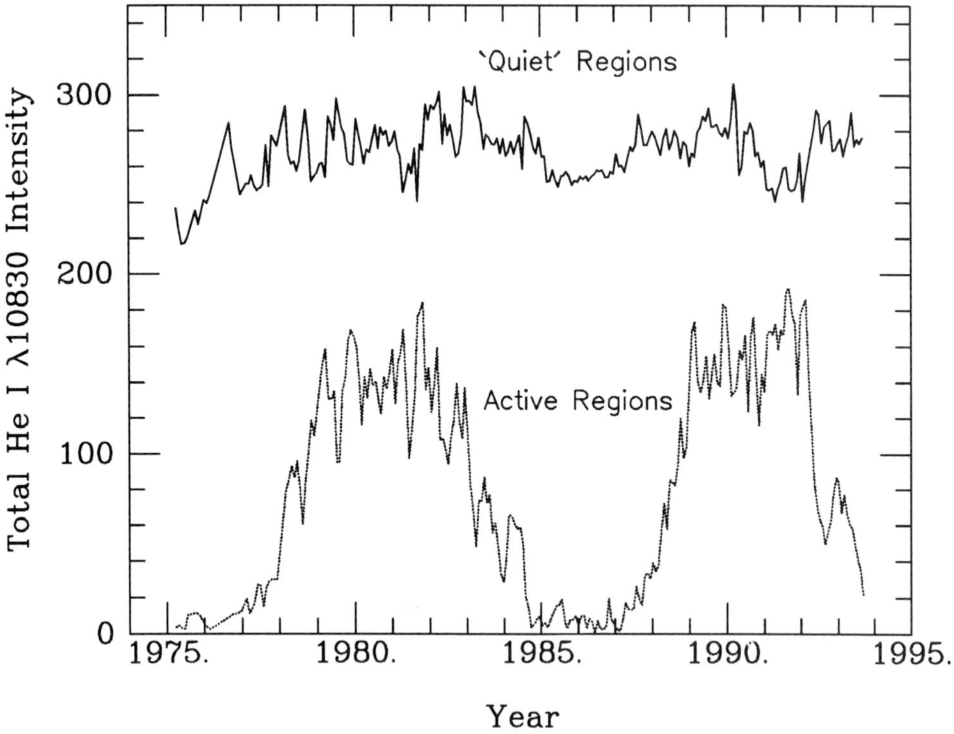

FIGURE 4. He I λ10830 in active regions *(solid line)* and in the quiet Sun network *(dotted line)*, using the method described in Section 2.1 to isolate active regions.

is a measure of the gradients of the field across the inversion line. The 'complexity' map for Carrington Rotation 1712 is shown in Figure 2.

A complexity index defined in this way is sensitive to the noise and the completeness in the daily coverage of magnetograms used to contruct the synoptic maps. If data are missing over extended periods of time, i.e., for more than 4 to 5 days, the information at Carrington Longitudes not observed at their Central Meridian passage is provided by measurements at large heliographic longitudes, where the data are affected by geometrical effects. The poorer coverage (and noisier data) will primarily affect the determination of the complexity index in the quiet sun, leading to generally higher (20 to 50%) values of this index.

3. The active region and the quiet Sun components

3.1. *Total magnetic flux*

Figure 3 shows the variation of the total magnetic flux (the sum of the absolute values of the positive and negative flux) in the two defined components: active regions and the quiet Sun network. These curves are differ little from the total flux determined by thresholding the data at 25 Gauss to separate active regions from the quiet network (Harvey 1992). However, the area included in active region and quiet components does differ significantly using these two methods to define active regions (see section 2.1).

The total magnetic flux in active regions and in the quiet Sun network varies in phase

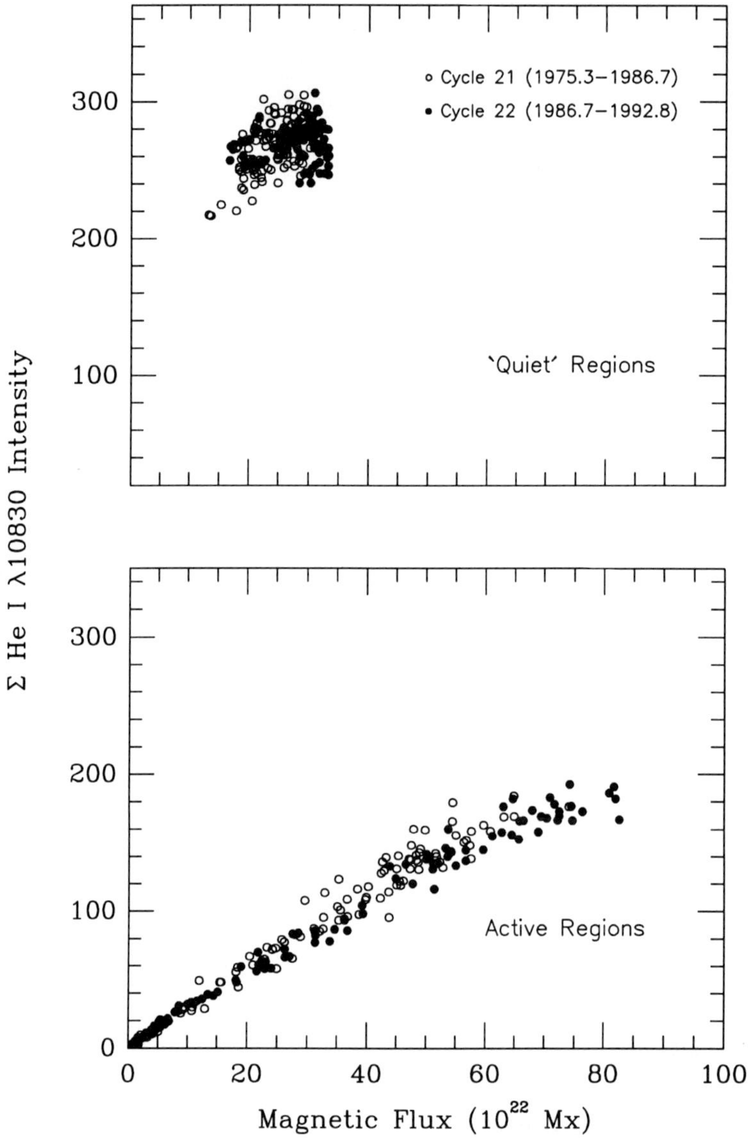

FIGURE 5. Plot of the total He I λ10830 compared to the total magnetic flux in active regions *(bottom)* and in the quiet Sun network *(top)*. Data for cycle 21 is indicated by ∘ and for cycle 22 by •.

during the cycle. However, the total magnetic flux in the quiet Sun network varies from the minimum to maximum phases of a cycle, by no more than a factor of 2, a significantly lower factor than observed for active regions. For approximately half of a solar cycle, the total flux in the quiet Sun network exceeds that in active regions.

3.2. *He I λ10830 absorption*

The active region mask was applied to He I λ10830 rotation maps to determine the corresponding components of He I absorption related to active regions and the quiet Sun

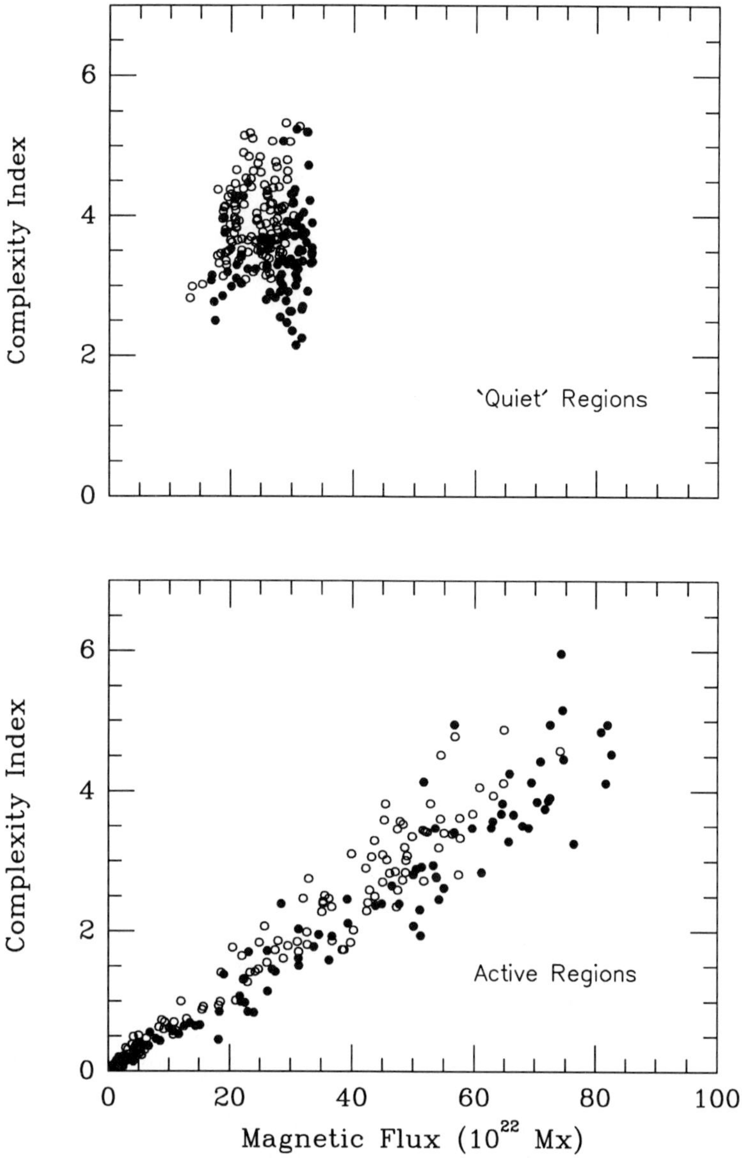

FIGURE 6. Plot of the magnetic complexity index against the total magnetic flux in active regions *(bottom)* and in the quiet Sun network *(top)*. Data for cycle 21 is indicated by ○ and for cycle 22 by •.

network. The results are shown in the two bottom panels at the right of Figure 1. The variation with time of the total He I λ10830 absorption (summing the He I intensity over the area designated as active regions and as the quiet Sun) is shown in Figure 4.

The total absorption in He I λ10830 observed in the quiet Sun exceeds that observed in active regions over the entire interval of the observations. The quiet Sun component varies little during the cycle. In active regions, on the other hand, the total He I λ10830 shows a strong cycle variation, varying nearly as the total area of active regions. That is, *per unit area* the average level of He I λ10830 in active regions is roughly constant, and is a factor of 2 to 2.5 higher than the He I λ10830 *per unit area* in quiet sun structures.

It should be noted that the He I λ10830 structures in the quiet Sun include network, associated with magnetic network elements, coronal holes, and filaments (see Figure 1). No attempt has yet been made to further separate these individual features in the He I λ10830 data, as is done by Harvey and Livingston (1993).

In Figure 5, the total He I λ10830 absorption is plotted against the total magnetic flux for each rotation, separately for active regions and for the quiet Sun. In the quiet Sun, there appears to be no relation of He I absorption with the magnetic flux. In active regions, however, they are strongly correlated. This is in part a result of their similar cycle variation. The He I λ10830 absorption in active regions increases roughly linearly with increasing total magnetic flux, but differs between cycles 21 and 22. In cycle 22, their relation has a slightly lower slope than is seen in cycle 21. Such variations between cycles also are seen in other spectral irradiance measurements (e.g., Rottman 1988; Donnelly 1991; Harvey 1992).

3.3. *Magnetic field complexity*

The complexity of the magnetic field was investigated in an attempt to understand the differing relation between He I λ10830 and the total magnetic flux between cycles. The premise is, that given the same amount of total magnetic flux, more complex magnetic field structures (activity complexes, active region nests) are associated with higher emission (or absorption in the case of He I λ10830) than a simple bipolar active region. The complexity index defined in Section 2.2 measures, in effect, the linear extent and gradient of the magnetic inversion lines (see Figure 2). Active regions with a complicated magnetic structure have more inversion lines and, therefore, a higher complexity index, than a simple bipolar region.

In Figure 6, the complexity index for each Carrington rotation is compared with the total magnetic flux, separately for active regions and for the quiet Sun network. The complexity of the quiet sun appears to be poorly correlated with the total magnetic flux.

In active regions, for increasing levels of the total magnetic flux (i.e. more active regions on the Sun), there is a corresponding increase in the complexity index of the ensemble of active regions. In cycle 22, however, this increase is less than in cycle 21, suggesting that as a whole, active regions in cycle 22 are less complex than active regions in cycle 21. The sense of the relation between active region magnetic complexity and total magnetic flux between cycles is similar to that seen in the relation of He I λ10830 absorption and magnetic flux. The less complex active regions, overall, in cycle 22, if real, offers one possible explanation of the differing relations of He I λ10830 and other measurements of spectral irradiances between different cycles.

It is important in this interpretation to verify, one objective of future analyses, that there are no long-term instrumental effects that may contribute to the cycle difference in the complexity index either in the quiet sun or active regions.

4. Conclusions

In this analysis, a new objective method to isolate active regions and measure their magnetic complexity is described. Using this method on the synoptic rotation magnetograms, the quite different cycle variation of the total magnetic flux in active regions and the quiet Sun network found by Harvey (1992) is verified. The magnetic flux in the quiet Sun varies in phase with but by a smaller factor (2) than that in active regions (>15). The separation of the active and quiet regions of the Sun was applied to the He I λ10830 synoptic maps. The total He I λ10830 absorption in the quiet Sun exceeds that observed in active regions and varies little with the cycle, while that associated with active regions shows a strong cycle variation. The relation between He I λ10830 absorption and the magnetic flux is poor in the quiet Sun, but is strong in active regions. This relation differs between cycles 21 and 22, with the absorption in He I λ10830 being less in cycle 22. Active regions in cycle 22 appear to be as a whole less complex than in cycle 21, offering one explanation for the lower spectral irradiances observed in cycle 22 despite higher levels of total magnetic flux.

Acknowledgments. My thanks to the referee for the suggested improvements in this paper. This work was supported under NASA contract NASW-4721 and was done while the author was a visitor at the National Solar Observatory (National Optical Astronomy Observatories, operated by the Association of Universities for Research in Astronomy, Inc. (AURA), in cooperative agreement with the National Science Foundation). The NSO/Kitt Peak data used in this analysis are produced cooperatively by NSF/NOAO, NASA/GSFC, and NOAA/SEL.

REFERENCES

Donnelly, R. F. 1991 Solar UV spectral irradiance variations. *J. Geomagn. Geoelect.* **43**, Suppl. Ser., 835–842.

Foukal, P. & Lean, J. 1988 Magnetic modulation of solar luminosity by photospheric activity. *Astrophys J.* **328**, 347–357.

Foukal, P., Harvey, K. L. & Hill, F. 1991 Do changes in the photospheric magnetic network cause the 11–year variation of total solar irradiance? *Astrophys J.* **383**, L89–L92.

Gaizauskas, V., Harvey, K. L., Harvey, J. W. & Zwaan, C. 1983 Large scale patterns in solar activity during the ascending phase of cycle 21 *Astrophys. J.* **265**, 1056–1065.

Harvey, J. W. & Livingston, W. C. 1993 Variability of the solar He I 10830 Å line. In IAU Symposium 154 *Infrared Solar Physics* (ed. D.M. Rabin, J.T. Jefferies & C. Lindsey). pp. 59–64.

Harvey, K. L. 1992 Measurements of solar magnetic fields as an indicator of solar activity evolution. In *Proceedings of the Workshop on the Solar Electromagnetic Radiation Study for Solar Cycle 22* (ed. R. F. Donnelly). pp. 113–129. NOAA/ERL/SEL, Boulder, CO.

Lean, J. L., White, O. R., Livingston, W. C., Heath, D. F., Donnelly, R. F. & Skumanich, A. 1982 A three–component model of the variability of the solar ultraviolet flux: 145–200 nM. *J. Geophys. Res.* **87**, 10307–10317.

Livingston, W., Wallace, L. & White, O. 1988 Spectrum line intensity as a surrogate for solar irradiance variations. *Science* **240**, 1765–1767.

Rottman, G. J. 1988 Observations of solar UV and EUV variability. *Adv. Space Res.* **8**: (7), 53–66.

Schatten, K. 1988 A model for solar constant secular changes. *Geophys. Res. Lett.* **15**, 121–124.

Skumanich, A., Lean, J. L., White, O. R. & Livingston, W. 1984 The Sun as a star: three component analysis of chromospheric variability in the calcium K line. *Astrophys J.* **282**, 776–783.

Willson, R. C. & Hudson, H. S. 1988 Solar luminosity variations in solar cycle 21. *Nature* **332**, 810–813.

Irradiance Effects of Small-Scale Magnetic Fields on the Sun

By SAMI K. SOLANKI

Institute of Astronomy, ETH-Zentrum, CH-8092 Zürich, Switzerland

Small-scale magnetic fields affect the solar luminosity mainly on long time scales. To understand their contribution to solar luminosity variations we must know and understand the contribution of a typical small-scale magnetic feature. In this review I briefly outline our theoretical understanding of the processes leading to the enhancement (or reduction) of the brightness of flux tubes. I also present a brief overview of our observational knowledge.

1. Introduction

Observations show that whereas sunspots dominate solar luminosity variations on a time-scale of days to months it is the bright faculae which dictate the variation of the global solar luminosity on time scales approaching the period of the solar cycle (Willson & Hudson 1988; Foukal & Lean 1988; Fröhlich 1994). It is equally well established that faculae are composed of small-scale magnetic features (see the reviews by Stenflo 1989; Solanki 1993). Consequently, in order to understand long term solar irradiance variations we must study and understand the brightness and energy balance of small-scale magnetic features. For the purposes of brightness variations small-scale magnetic features may be divided into two classes.

a) *Flux tubes*: These are magnetic flux concentrations having a field strength of a kG or more in the middle and lower photosphere (e.g.; Stenflo 1973; Rabin 1992; Rüedi *et al.* 1992). The magnetic energy density of such features is sufficiently high to dominate over the thermal and kinetic energy of the internal gas at and above the visible solar surface. This implies that the magnetic field significantly affects the way that radiation is channelled through the solar surface. The most successful description of these strong-field elements is provided by the flux-tube model (briefly described in Sect. 2) and in the following we will refer to such features simply as flux tubes. Note that a dominant fraction of the non-sunspot magnetic flux seen in the daily Kitt Peak magnetograms is in the form of small flux tubes.

b) *Weak field features*: They are composed of less concentrated fields, with field strengths below one kG throughout the photosphere (Rüedi *et al.* 1992). The thermal and kinetic energy densities are larger than or equal to the magnetic energy density. Therefore we expect such magnetic fields to affect the solar luminosity only slightly, at most. The necessary measurements have, however, never been carried out.

In the following I shall consider only flux tubes, since we know too little about the brightness or the irradiance effects of weak field features.

2. Luminosity changes due to small-scale magnetic features: Theory

A considerable number of MHD models of magnetic flux tubes and sheets exist in the literature. Of interest here are those which include an energy equation that adequately takes into account at least some of the important effects. Such models have been constructed by, e.g., Spruit (1976, 1977); Deinzer *et al.* (1984a,b); Herbold *et al.* (1984); Kalkofen *et al.* (1986); Knölker *et al.* (1988); Knölker & Schüssler (1988); Hasan (1988);

Grossmann-Doerth *et al.* (1989)i; Nordlund & Stein (1989); Steiner (1990); Pizzo *et al.* (1993a,b) and Fabiani Bendicho *et al.* (1993). Unfortunately, no model exists which adequately includes all the relevant energetics.

There are four main effects which determine the final luminosity of a flux tube:

1. The magnetic field inhibits convection, so that heat transport from below is reduced. Consequently the flux tube is cooled.
2. The magnetic pressure causes a gas pressure deficit within the flux tube, so that its $\tau = 1$ level is lowered (Wilson depression). Radiation flows in through the walls of the flux tube above the Wilson depression, which heats the flux tube near its $\tau = 1$ level and cools its surroundings (due to radiation losses into the flux tube).
3. Radiation from the 'hot walls' and the bottom of the flux tube heats the middle and upper photosphere within the tube, and partly also in its near surroundings.
4. Mechanical energy, e.g. in the form of waves, is transported upwards along the field lines to heat the upper photosphere and chromosphere. The waves are thought to be excited by the buffeting of the flux tubes through granules.

The importance of effects 2, 3, and 4 decreases as the size of the flux tube increases. The important quantity determining the brightness of magnetic features is the ratio of horizontal size to the Wilson depression. For larger horizontal size more magnetic flux is blocked, while for a larger Wilson depression the surface area of the flux tube and its temperature are increased, allowing more radiation to escape. Since the Wilson depression changes only slowly with changing flux tube size, it is mainly the horizontal extent of flux tubes which determines their brightness.

Theory predicts that small flux tubes (smaller than 200–300 km in diameter) appear bright all over the solar disc, in both the continuum and in spectral lines. They are surrounded by a dark ring. Somewhat larger tubes are predicted to appear dark in the continuum at disc centre, but bright near the limb. They may appear bright in spectral lines as well. Such tubes are also surrounded by a dark ring. Very large flux tubes (pores, sunspots) are predicted to be dark everywhere on the solar disc, in both the continuum and photospheric spectral lines.

3. Luminosity changes due to small-scale magnetic features: Observations

Basically two types of observations are of interest:

a) Contrast measurements: These are narrow- or broad-band measurements of the intensity of magnetic features relative to the quiet Sun intensity.

b) Spectropolarimetric measurements: Stokes V profiles (spectral profiles in net circularly polarized light) are measured and the temperature stratification of the magnetic features is determined therefrom.

3.1. *Contrast measurements*

The basic idea is to measure the intensity of the 'magnetic feature', I_m, and of the average quiet Sun, I_q, and therewith to construct the contrast δ:

$$\delta = I_m/I_q - 1$$

One major advantage of this approach is its simplicity. Thus a large body of data has been collected (for reviews see: Akimov *et al.* 1987; Solanki 1993). Another advantage

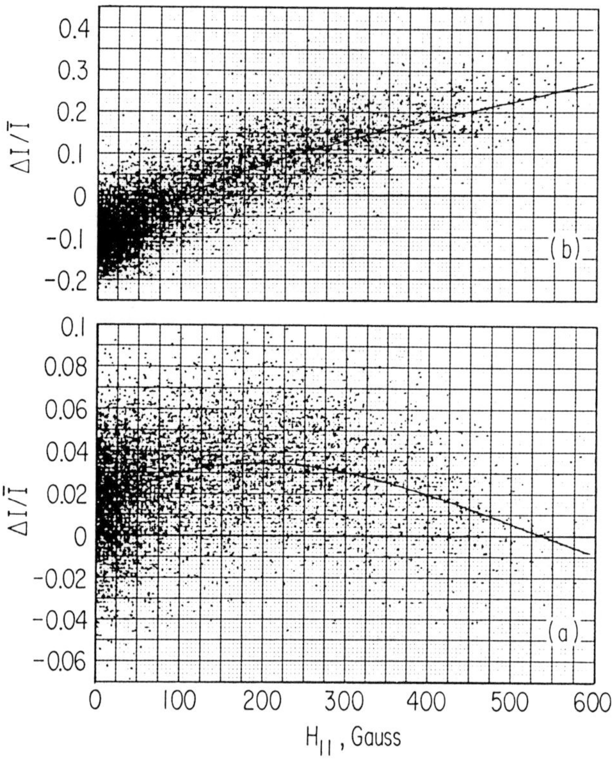

FIGURE 1. Contrast $\Delta I/\langle I\rangle$ vs. spatially averaged field strength, $\langle B\rangle$, where $\langle I\rangle$ is the average brightness of the whole observed image and differs somewhat from I_q. a) Continuum, b) line core of Fe I 5250Å. The solid curves are polynomial least-squares fits (adapted from Frazier 1971).

is that in this manner the effect of the magnetic feature on its surroundings, for example the production of a dark ring around the flux tube, is automatically taken into account.

The main problem is that the measured δ depends on many factors, which makes its interpretation and the comparison between different observations difficult. There are four main parameters on which δ depends.

a) *Spatially averaged field strength* $\langle B\rangle$: δ initially increases with increasing $\langle B\rangle$, before decreasing again and finally turning negative. This happens at much smaller $\langle B\rangle$ for the continuum than for spectral lines (Figure 1). Sheeley & Engvold (1970); Frazier (1971, 1978); Simon & Zirker (1974); Mehltretter (1974); Frazier & Stenflo (1978); Foukal & Fowler (1984); Title *et al.* (1990, 1992); Moran et al. (1992); Topka et al. (1993) give further examples.

b) *Limb distance*: The continuum contrast δ_c increases rapidly towards the limb (e.g. Waldmeier 1949; Rogerson 1961; Schmahl 1967; Livshits 1968; Chapman 1970; Frazier 1971; Muller 1975; Ingersoll & Chapman 1975; Hirayama 1978; Chapman & Klabunde 1982; Libbrecht & Kuhn 1984, 1985; Akimov *et al.* 1982, 1987; Lawrence 1988; Auffret & Muller 1991), while δ in line cores varies much less rapidly. Figure 2 illustrates the centre to limb variation of δ at different λ (from Foukal *et al.* 1991).

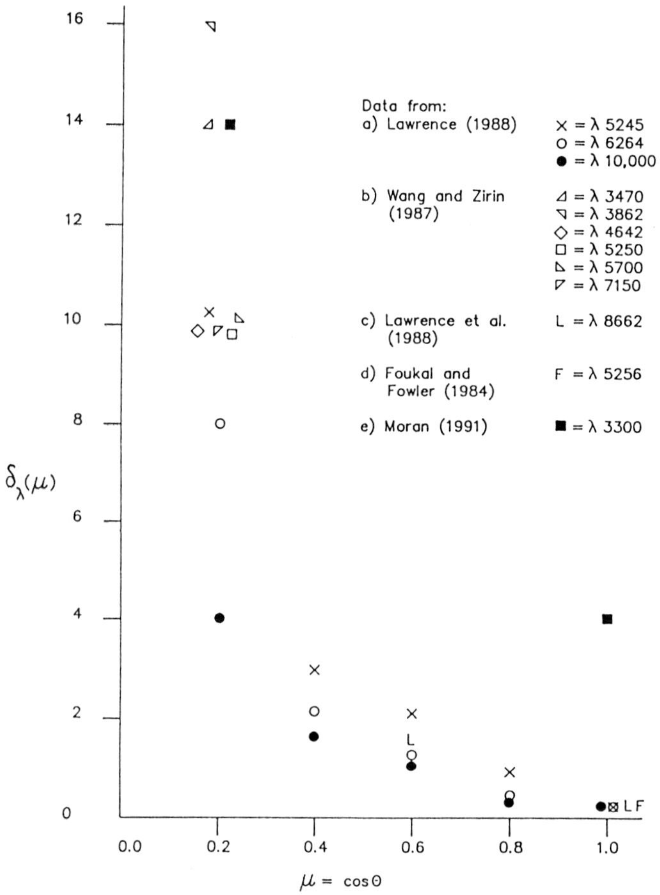

FIGURE 2. Measured values of the broad-band contrast, δ (in %), at different wavelengths vs. $\mu = \cos\vartheta$, where ϑ is the heliocentric angle. The data are from various sources (adapted from Foukal *et al.* 1991).

c) *Wavelength* λ: δ decreases strongly from the UV to the IR (cf. Figure 2) due to two effects. Firstly, the Planck function has a larger temperature sensitivity in the UV, so that the intensity reacts more strongly to a given temperature perturbation. Secondly, the density of spectral lines decreases with λ and spectral lines show larger contrasts than the continuum. Consequently, the exact wavelength and width of the passband may strongly affect the results. The wavelength dependence of δ has been widely studied (e.g. Frazier 1978; Chapman & McGuire 1977; Wang & Zirin 1987; Moran *et al.* 1992).

d) *Spatial resolution*: Higher spatial resolution usually leads to higher $|\delta|$ values (e.g., Mehltretter 1974; Koutchmy 1977; Muller & Keil 1983; Auffret & Muller 1991; Keller 1992). Thus Auffret & Muller (1991) find δ values (around 5750 Å) up to 25% at disc centre ($\mu = 1$). Keller (1992) even finds values ranging up to 30% in the true continuum near 5250 Å. Compare this with Figure 2. Note, however, that the highest δ values (1.8–2.0) have often been obtained in bandpasses containing strong, temperature sensitive lines and do not reflect white-light contrasts.

The above list on the one hand illustrates the wide variety of brightness signatures produced by small flux tubes. On the other hand it also demonstrates that different contrast measurements often cannot be directly compared to each other. A consistent set of measurements under good seeing conditions at many λ and μ values is required. It is important to measure $\langle B \rangle$ simultaneously.

3.2. *Spectropolarimetric measurements*

The procedure for determining contrasts from spectropolarimetric measurements is as follows: 1.) A number of spectral lines are recorded in polarized light. 2.) Clean diagnostics (e.g. line ratios) are identified. 3.) Models are constructed which ideally reproduce all diagnostics. The end products are the simultaneously determined magnetic, velocity and thermal stratifications within the magnetic feature. Using these, continuum or line contrasts may be derived.

We may count among the advantages of this approach that its results are (to first order) spatial resolution and wavelength independent. In addition, $\langle B \rangle$ is automatically determined with high accuracy. Disadvantages are that some model dependence always remains and considerable effort must generally be invested to obtain good models, which limits the number of relevant investigations. Also, current diagnostics mainly give information on the magnetic features alone and not so much on their surroundings which are also important for the net radiation budget. Here I concentrate on two results of spectropolarimetric observations.

Dependence of temperature on $\langle B \rangle$: The temperature in the lower photosphere, and consequently the continuum intensity I_c, depends strongly on $\langle B \rangle$ (e.g. Frazier 1978; Hirayama 1978; Solanki & Stenflo 1984, 1985; Solanki 1986; Zayer *et al.* 1990; Solanki & Brigljević 1992).

$$
\begin{aligned}
\text{Small } \langle B \rangle \lesssim 100 \text{ G} \quad &\rightarrow \quad \delta_c\,(5000\text{Å}) \approx 5\% \text{ to } 30\%\ , \\
\text{Larger } \langle B \rangle \approx 200 - 350 \text{ G} \quad &\rightarrow \quad \delta_c\,(5000\text{Å}) \approx -5\% \text{ to } -20\%\ .
\end{aligned}
$$

Also, as $\langle B \rangle$ increases, so does the average diameter d of the magnetic features. The dependence on $\langle B \rangle$ is most probably a result of this dependence of d on $\langle B \rangle$ (e.g. Keller 1992; Grossmann-Doerth *et al.* 1994). $\langle B \rangle$ is, however, the more convenient quantity to measure. The δ_c inferred from spectropolarimetric observations is at least in good qualitative agreement with direct δ_c measurements.

Height dependence of the temperature: In the deeper layers the temperature in the flux tube is lower than in the surroundings at the same height, while in the upper photosphere and the chromosphere the opposite is the case (Keller *et al.* 1990; Bruls & Solanki 1992). At equal optical depth the flux tubes may be hotter *or* cooler than their surroundings near $\tau_c = 1$, but in the middle and upper photosphere the flux tubes are invariably hotter (e.g. Stenflo 1975; Chapman 1979; Solanki 1986; Walton 1987; Keller *et al.* 1990; Solanki *et al.* 1991). In particular, the chromospheric temperature rise starts 200 – 300 km deeper within small-scale magnetic features than in the quiet Sun (Bruls & Solanki 1992; cf. Ayers *et al.* 1986; Chapman 1981). This leads to considerably enhanced Ca II H and K emission. The temperature stratification of small flux tubes is plotted in Figure 3.

4. Conclusions

The basic parameter determining the brightness of small-scale magnetic features is their size. Since the average size of the magnetic features increases with increasing $\langle B \rangle$ or

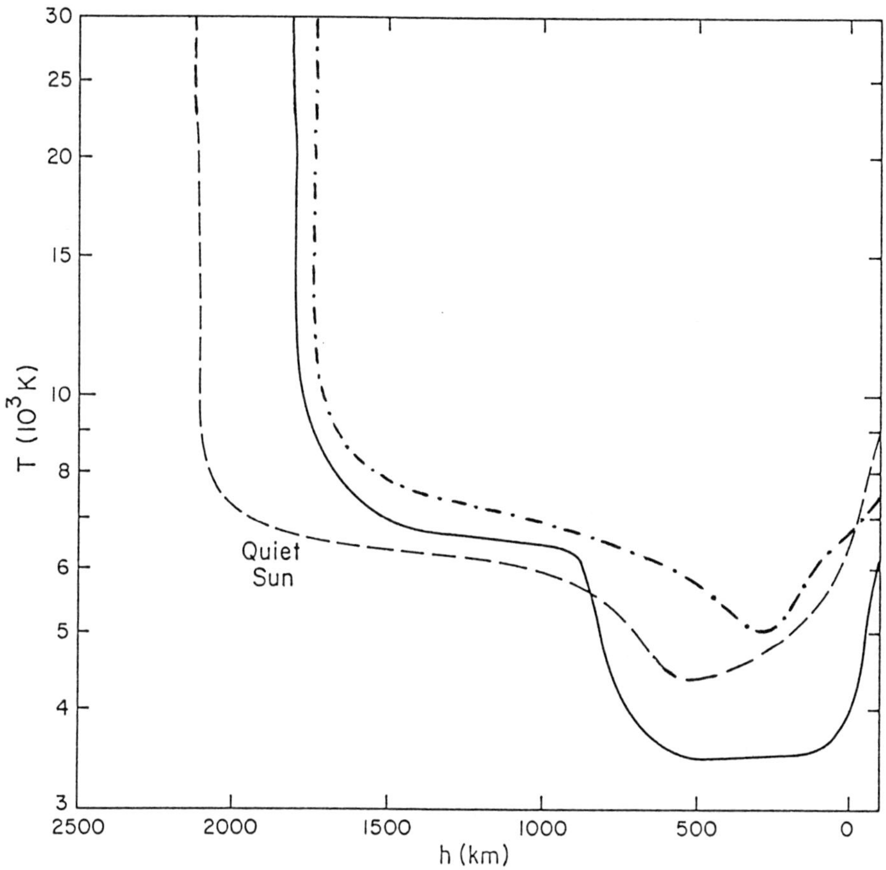

FIGURE 3. Temperature T vs. height Z of a small flux tube (dot-dashed), of the quiet Sun (dashed) and a sunspot umbra (solid) (adapted from Avrett 1985).

magnetic filling factor, their average luminosity decreases with increasing $\langle B \rangle$. It follows that it is not sufficient to know the magnetic field averaged over the Sun to predict the solar irradiance variation. We must know the detailed distribution of the field. Thus solar irradiance need not peak simultaneously with the disc-integrated magnetic flux, even if the magnetic field is responsible for solar irradiance variations. For example, the B-field is more uniformly distributed in the falling phase of the solar cycle, so that at that time there may be more field in regions of small $\langle B \rangle$, which show a larger δ_c.

In the subphotospheric layers small magnetic features receive an excess of energy from their surroundings through radiation, through the excitation of internal motions (e.g. waves), and partly through storing energy in more complex magnetic geometries. This excess energy is:

- partly radiated away immediately at $\tau_c \approx 1$. This gives rise to a limb continuum enhancement from the flux-tube walls and, for small tubes, to a disc-centre continuum enhancement from the bottom of the flux tube.
- partly transported upwards to heat the middle and upper photosphere, which weakens

most spectral lines, thus enhancing line emission. In addition, it further enhances continuum emission from flux tubes at the extreme limb.

- to a minor part transported still higher to heat the chromosphere and corona, enhancing Ca II H and K emission, X-ray and 11 cm radio flux and He I 10830 Å equivalent width.

Most of these proxies to solar irradiance are thus produced by channelling into the upper atmosphere of a minor amount of the excess energy flowing into magnetic features. The proxies thus have basically a common origin to the long-term irradiance variations themselves which are caused by the dominant part of this excess energy flux. The heating of the upper atmosphere, however, is a non-linear function of the magnetic flux, so that it is not a priori clear which particular indicator should be a good proxy of solar luminosity.

In addition to simply possessing a higher radiation flux than the quiet Sun, small flux tubes also redistribute the energy vertically in the atmosphere, transporting it to greater heights before allowing it to be radiated away. Therefore, the contribution of spectral lines to the total irradiance variations should not be neglected.

In summary, small-scale solar magnetic features affect the solar luminosity and even appear to dominate its evolution on the time scale of the solar activity cycle. Nevertheless there are sizable gaps in our understanding of how small-scale magnetic features influence the global luminosity. In order to understand the complex influence of the small-scale fields we must know both, the evolution of their distribution on the solar surface and the luminosity of individual features as a function of their position on the solar disc and of the local magnetic filling factor. The distribution and large-scale evolution of magnetic features is well documented by the daily Kitt Peak magnetograms. Our knowledge of individual magnetic features has also progressed substantially, but hardly any attempt has so far been made to combine these two subjects of study. I expect that a number of open questions regarding the mechanism by which small-scale magnetic features affect the global solar luminosity can be answered by carrying out this synthesis.

REFERENCES

AKIMOV, L.A., BELKINA, I.L. & DYATEL, N.P. 1982 Brightness of the photosphere and faculae at the limb based on eclipse observations. *Sov. Astron.* **26**, 334–340.

AKIMOV, L.A., BELKINA, I.L., DYATEL, N.P. & MARCHENKO, G.P. 1987 Contrast of faculae near the solar limb. *Sov. Astron.* **31**, 64–68.

AUFFRET, H. & MULLER, R. 1991 Center-to-limb variation of the network bright points. *Astron. Astrophys.* **246**, 264–279.

AVRETT, E.H. 1981 Reference model atmosphere calculations: The Sunspot sunspot model. In *The Physics of Sunspots* (ed. L.E. Cram & J.H. Thomas). pp. 235–255. National Solar Obs., Sunspot, NM, USA.

AYRES, T.R., TESTERMAN, L. & BRAULT, J.W. 1986 Fourier transform spectrometer observations of solar carbon monoxide. II. Simultaneous cospatial measurements of the fundamental and first-overtone bands, and Ca II K, in quiet and active regions. *Astrophys. J.* **304**, 542–559.

BRULS, J.H.M.J. & SOLANKI, S.K. 1993 The chromospheric temperature rise in solar magnetic flux tubes. *Astron. Astrophys.* **273**, 293–303.

CHAPMAN, G.A. 1970 On the physical conditions in the photospheric network: An improved model of solar faculae. *Solar Phys.* **14**, 315–327.

CHAPMAN, G.A. 1979 New models of solar faculae. *Astrophys. J.* **232**, 923–928.

CHAPMAN, G.A. 1981 Active regions from the photosphere to the chromosphere. In *Solar Active Regions* (ed. F.Q. Orrall). pp. 43–82. Colorado University Press, Boulder, CO, USA.

CHAPMAN, G.A. & KLABUNDE, D.P. 1982 Measurements of the limb darkening of faculae near the solar limb. *Astrophys. J.* **261**, 387–395.

CHAPMAN, G.A. & MCGUIRE, T.E. 1977 The wavelength dependence of the facular excess brightness. *Astrophys. J.* **217**, 657–660.

DEINZER, W., HENSLER, G., SCHÜSSLER, M. & WEISSHAAR, E. 1984 Model calculations of magnetic flux tubes I. Equations and method. *Astron. Astrophys.* **139**, 426–434.

DEINZER, W., HENSLER, G., SCHÜSSLER, M. & WEISSHAAR, E. 1984 Model calculations of magnetic flux tubes II. Stationary results for solar magnetic elements. *Astron. Astrophys.* **139**, 435–449.

FABIANI BENDICHO, P., KNEER, F. & TRUJILLO BUENO, J. 1992 On the photospheric temperature in small-scale magnetic flux concentrations. *Astron. Astrophys.* **264**, 229–235.

FOUKAL, P. & FOWLER, L. 1984 A Photometric study of heat flow at the solar photosphere. *Astrophys. J.* **281**, 442–454.

FOUKAL, P. & LEAN, J. 1988 Magnetic modulation of solar luminosity by photospheric activity. *Astrophys. J.* **328**, 347–357.

FOUKAL, P., DUVALL, T. JR. & GILLESPIE, B. 1981 Detection of a temperature deficit in magnetic faculae at the solar photosphere. *Astrophys. J.* **249**, 394–398.

FOUKAL, P., HARVEY, K. & HILL, F. 1991 Do changes in the photospheric magnetic network cause the 11 year variation of total solar irradiance? *Astrophys. J.* **383**, L89–L92.

FRAZIER, E.N. 1971 Multi-channel magnetograph observations. III: faculae. *Solar Phys.* **21**, 42–53.

FRAZIER, E.N. 1978 Line profile families of faculae and pores. *Astron. Astrophys.* **64**, 351–358.

FRAZIER, E.N. & STENFLO, J.O. 1978 Magnetic, velocity and brightness structures in solar faculae. *Astron. Astrophys.* **70**, 789–799.

FRÖHLICH, C. 1994 Irradiance observations of the Sun. In *The Sun as a Variable Star: Solar and Stellar Irradiance Variations* (ed. J.M. Pap, C. Fröhlich, H.S. Hudson & S.K. Solanki). Cambridge Univ. Press, in press.

GROSSMANN-DOERTH, U., KNÖLKER, M., SCHÜSSLER, M. & WEISSHAAR, E. 1989 Models of magnetic flux sheets. In *Solar and Stellar Granulationx* (ed. R.J. Rutten & G. Severino). pp. 481–492. Kluwer, Dordrecht.

GROSSMANN-DOERTH, U., KNÖLKER, M., SCHÜSSLER, M. & SOLANKI, S.K. 1993 The deep layers of solar magnetic elements. *Astron. Astrophys.* in press.

HASAN, S.S. 1988 Energy transport in intense flux tubes on the Sun. I. Equilibrium atmosphere. *Astrophys. J.* **332**, 499–513.

HERBOLD, G., ULMSCHNEIDER, P., SPRUIT, H.C. & ROSNER, R. 1985 Propagation of nonlinear radiatively damped longitudinal waves along magnetic flux tubes in the solar atmosphere. *Astron. Astrophys.* **145**, 157–169.

HIRAYAMA, T. 1978 A model of solar faculae and their lifetime. *Publ. Astron. Soc. Japan* **30**, 337–352.

INGERSOLL, A.P. & CHAPMAN, G.A. 1975 Temperature variation with latitude in the upper solar photosphere: relevance to solar oblateness measurements and facular models. *Solar Phys.* **42**, 279–288.

KALKOFEN, W., ROSNER, R., FERRARI, A. & MASSAGLIA, S. 1986 The equilibrium structure of thin magnetic flux tubes. II. *Astrophys. J.* **304**, 519–525.

KELLER, C.U. 1992 Resolution of magnetic flux tubes on the Sun. *Nature* **359**, 307–308.

KELLER, C.U., SOLANKI, S.K., STEINER, O. & STENFLO, J.O. 1990 Structure of solar magnetic fluxtubes from the inversion of stokes spectra at disk center. *Astron. Astrophys.* **233**, 583–597.

KNÖLKER, M. & SCHÜSSLER, M. 1988 Model calculations of flux tubes. IV. Convective energy transport and the nature of intermediate size flux concentrations. *Astron. Astrophys.* **202**, 275–283.

KNÖLKER, M., SCHÜSSLER, M. & WEISSHAAR, E. 1988 Model calculations of magnetic flux tubes III. Properties of solar magnetic elements. *Astron. Astrophys.* **194**, 257–267.

KOUTCHMY, S. 1977 Photospheric faculae: the contrasts at the center of the solar disk using filigree pictures. *Astron. Astrophys.* **61**, 397–404.

LAWRENCE, J.K. 1988 Multi-color photometric observations of facular contrasts. *Solar Phys.* **116**, 17–32.

LAWRENCE, J.K., CHAPMAN, G.A. & HERZOG, A.D. 1988 Photometric determination of facular contrasts near the solar disk center. *Astrophys. J.* **324**, 1184–1193.

LIBBRECHT, K.G. & KUHN, J.R. 1984 A new measurement of the facular contrast near the solar limb. *Astrophys. J.* **277**, 889–896.

LIBBRECHT, K.G. & KUHN, J.R. 1985 On the facular contrast near the solar limb. *Astrophys. J.* **299**, 1047–1050.

LIVSHITS, M.A. 1968 *Soln. Akt.* **3**, 78.

MEHLTRETTER, J.P. 1974 Observations of photospheric faculae at the center of the solar disk. *Solar Phys.* **38**, 43–57.

MORAN, T., FOUKAL, P. & RABIN, D. 1992 A photometric study of faculae and sunspots between 1.2 and 1.6 μm. *Solar Phys.* **142**, 35–46.

MULLER, R. 1975 A model of photospheric faculae deduced from white light high resolution pictures. *Solar Phys.* **45**, 105–114.

MULLER, R. & KEIL, S.L. 1983 The characteristic size and brightness of facular points in the quiet photosphere. *Solar Phys.* **87**, 243–250.

NORDLUND, Å& STEIN, R.F. 1989 Simulating magnetoconvection. In *Solar and Stellar Granulation* (ed. R.J. Rutten & G. Severino). pp. 453–470. Kluwer, Dordrecht.

PIZZO, V.J., MCGREGOR, K.B. & KUNASZ, P.B. 1993a A numerical simulation of two-dimensional radiative equilibrium in magnetostatic flux tubes. I. The model. *Astrophys. J.* **404**, 788–798.

PIZZO, V.J., MCGREGOR, K.B. & KUNASZ, P.B. 1993b A numerical simulation of two-dimensional radiative equilibrium in magnetostatic flux tubes. II. Computational results. *Astrophys. J.* **413**, 764–777.

RABIN, D. 1992 Spatially extended measurements of magnetic field strength in solar plages. *Astrophys. J.* **391**, 832–844.

ROGERSON, J.B. 1961 On photospheric faculae. *Astrophys. J.* **134**, 331–336.

RÜEDI, I., SOLANKI, S.K., LIVINGSTON, W. & STENFLO, J.O. 1992 Infrared lines as probes of solar magnetic features. III. Strong and weak magnetic fields in solar plages. *Astron. Astrophys.* **263**, 323–338.

SCHMAHL, G. 1967 Zum Zustand der photosphärischen Schichten von Sonnenfackeln. *Z. Astrophys.* **66**, 81–117.

SHEELEY, N.R.,JR. & ENGVOLD, O. 1970 Simultaneous measurements of magnetic fields and brightness fields using a 4-image spectroheliograph. *Solar Phys.* **12**, 69–83.

SIMON, G.W. & ZIRKER, J.B. 1974 A search for the footprints of solar magnetic fields. *Solar Phys.* **35**, 331–342.

SOLANKI, S.K. 1986 Velocities in solar magnetic fluxtubes. *Astron. Astrophys.* **168**, 311–329.

SOLANKI, S.K. 1993 Small-scale solar magnetic fields: An overview. *Space Sci. Rev.* **61**, 1–188.

SOLANKI, S.K. & BRIGLJEVIĆ, V. 1992 Continuum brightness of solar magnetic elements. *Astron. Astrophys.* **262**, L29–L32.

SOLANKI, S.K. & STENFLO, J.O. 1984 Properties of solar magnetic fluxtubes as revealed by Fe I lines. *Astron. Astrophys.* **140**, 185–198.

SOLANKI, S.K. & STENFLO, J.O. 1985 Models of solar magnetic fluxtubes: Constraints imposed by Fe I and II lines. *Astron. Astrophys.* **148**, 123–132.

SOLANKI, S.K., STEINER, O. & UITENBROEK, H. 1991 Two-dimensional models of the solar chromosphere. I. The Ca II K line as a diagnostic: 1.5-D radiative transfer. *Astron. Astrophys.* **250**, 220–234.

SPRUIT, H.C. 1976 Pressure equilibrium and energy balance of small photospheric fluxtubes.

Solar Phys. **50**, 269–295.

SPRUIT, H.C. 1977 Heat flow near obstacles in the solar convection zone. *Solar Phys.* **55**, 3–34.

STEINER, O. 1990 Model calculations of solar magnetic fluxtubes and radiative transfer. *Ph.D. thesis*, No. 9292, ETH-Zürich.

STENFLO, J.O. 1973 Magnetic-field structure of the photospheric network. *Solar Phys.* **32**, 41–63.

STENFLO, J.O. 1975 A model of the supergranulation network and of active-region plages. *Solar Phys.* **42**, 79–105.

STENFLO, J.O. 1989 Small scale magnetic structures on the Sun. *Astron. Astrophys. Rev.* **1**, 3–48.

TITLE, A.M., TARBELL, T.D., TOPKA, K.P., CAUFFMAN, D., BALKE, C. & SCHARMER, G. 1990 Magnetic flux tubes and their relation to continuum and photospheric features. In *Phys. of Magn. Flux Ropes* (ed. C.T. Russell, E.R. Priest & L.C. Lee). pp. 171–179. Geophysical Monograph 58, American Geophys. Union, Washington, DC, USA.

TITLE, A.M., TOPKA, K.P., TARBELL, T.D., SCHMIDT, W., BALKE, C. & SCHARMER, G. 1992 On the differences between plage and quiet Sun in the solar photosphere. *Astrophys. J.* **393**, 782–794.

TOPKA, K.P., TARBELL, T.D. & TITLE, A.M. 1992 Properties of the smallest solar magnetic elements. I. Facular contrast near Sun center. *Astrophys. J.* **396**, 351–363.

WALDMEIER, M. 1949 Die Sichtbarkeitsfunktion der Sonnenfackeln. *Z. Astrophys.* **26**, 147–157.

WALTON, S.R. 1987 Flux tube models of solar plages. *Astrophys. J.* **312**, 909–929.

WANG, H. & ZIRIN, H. 1987 The contrast of faculae near the solar limb. *Solar Phys.* **110**, 281–293.

WILLSON, R.C. & HUDSON, H.S. 1988 Solar luminosity variations in solar cycle 21. *Nature* **332**, 810–812.

ZAYER, I., SOLANKI, S.K., STENFLO, J.O. & KELLER, C.U. 1990 Dependence of the properties of solar magnetic flux tubes on filling factor. II. Results of an inversion approach. *Astron. Astrophys.* **239**, 356–366.

Estimating Solar Forcing of Climate Change during the Maunder Minimum

By JUDITH LEAN[1], ANDREW SKUMANICH[2], ORAN R. WHITE[2] AND DAVID RIND[3]

[1]Naval Research Laboratory, Washington DC, 20375, USA

[2]High Altitude Observatory, National Center for Atmospheric Research, Boulder, CO 80303, USA

[3]Goddard Institute for Space Studies, New York, NY 10025, USA

Solar variability models which account for contemporary irradiance variations in terms of modulation by dark sunspots and bright faculae on the solar disk are used to estimate solar radiative output during the Maunder Minimum by postulating the absence of such sources, and additional reduction in the basal solar emission. When the resultant irradiance reduction of 0.25% relative to contemporary mean levels is input to the GISS GCM, the global mean temperature is reduced by 0.46°C. However, as a result of differential heating of the land and oceans, some regions may cool, and others warm, by as much 1°C, in response to the reduced solar irradiance.

1. Introduction

Evidence for variations in the activity level of the Sun, illustrated by the irradiance and sunspot data in Figure 1, exists on time scales from days to centuries, and raises the possibility of solar forcing of past and future climate change. During the period from about 1645 to 1715, called the Maunder Minimum, sunspots were absent from the solar disk for long periods (Eddy 1976) whereas numerous sunspots occur at contemporary activity maxima, and are rarely absent even at activity minima. The Maunder Minimum in the sunspot record is but the most recent of recurrent epochs of depressed solar activity evident in both the ^{10}Be and ^{14}C cosmogenic solar activity proxies (McHargue & Damon 1991). Since the Maunder Minimum, solar activity has risen steadily to the current Modern Maximum and, according to the ^{10}Be proxy data, has now reached levels as high as those last seen in the 12th century. Space-based measurements with sufficient precision to identify real changes in solar radiative have been made near the peak of the Modern Maximum but for barely more than one 11-year solar activity cycle (Willson & Hudson 1991; Hoyt *et al.* 1992).

The variations in solar radiative output recorded by the space-based monitoring are associated with eruptions of magnetic activity into the solar atmosphere, producing regions where the local solar emission may be depleted (in sunspots) or enhanced (in faculae). As the mix of the bright and dark active regions changes throughout the 11-year activity cycle, so too does the integrated radiative output from the solar disk (e.g., Foukal & Lean 1988). Numerical estimates of solar radiative output during times beyond the limited observational data base are possible using variability models in which the relative roles of sunspot blocking and faculae enhancement are derived from the extant radiometry. With this approach we investigate the effect of various scenarios of depleted bright magnetic elements on the full disk solar emission, estimate the probable reduction of the total solar irradiance during the Maunder Minimum, and use the Goddard Institute for Space Studies (GISS) General Circulation Model (GCM) to calculate the resultant geographical temperature changes that might be expected from this reduction.

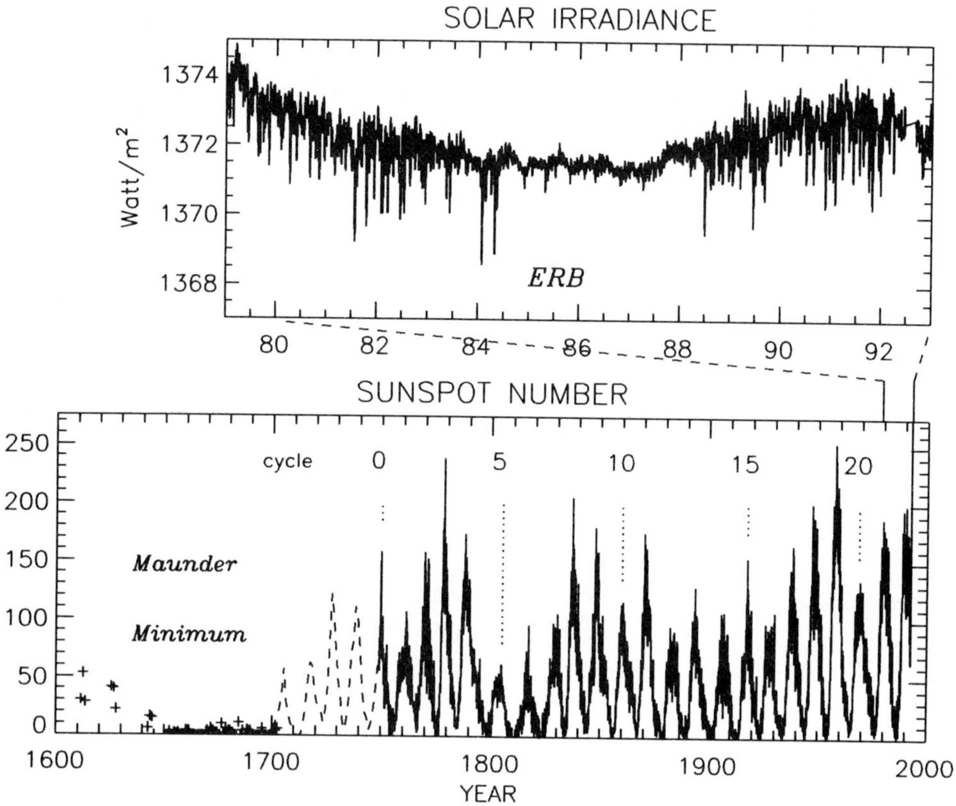

FIGURE 1. Changes in solar activity over daily, decadal and century time scales, as evidenced in (upper panel) total solar radiative output (Hoyt *et al.* 1992) and (lower panel) the sunspot number.

2. Solar irradiance observations and models

Satellite radiometry during the past decade has established that the solar radiative output varies continuously, on all time scales and at all wavelengths measured thus far (Lean 1991). Shown in Figure 2 are variations during solar cycles 21 and 22 in the total solar irradiance, S, measured by the ACRIM radiometer (Willson & Hudson 1991) and in the ultraviolet spectral irradiance of the H I Lyman α line at 121.57 nm, measured by the Solar Mesosphere Explorer (Rottman 1988). Reductions of as much as 0.3% in total solar irradiance can occur during the Sun's 27-day rotation as sunspots, which block solar radiation, rotate across the face of the Sun's disk seen at the Earth. Over longer time scales, from the minimum to the maximum of the 11-year solar cycle, an irradiance increase of 0.1% results from the build-up of enhanced emission in bright regions, which more than compensates for the reduction in the dark sunspots. Solar variability at ultraviolet wavelengths exceeds that of the total irradiance. Variations over the solar cycle are of the order of 5-10% at 200 nm and 50% at H I Lyman α (Rottman 1988; Lean 1991). In the spectral region 120 to 250 nm, the variations exhibit very similar temporal structure (Donnelly 1988), responding alike to the presence on the solar disk of bright magnetic regions, the dominant source of variability over both the 27-day and 11-year

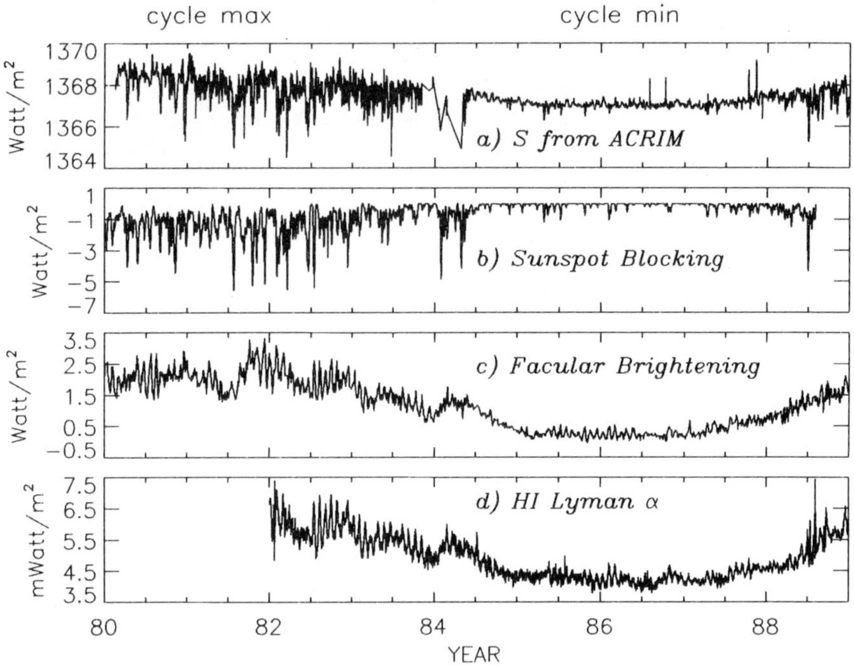

FIGURE 2. Variations throughout the 11-year solar activity cycle in a) total solar radiative ouput, b) sunspot blocking, c) facular brightening and d) the UV flux at HI Lyman α.

cycles. In comparison, the total solar radiative output has a different temporal structure that reflects the competing effects of both bright faculae and dark sunspots.

Also shown in Figure 2 are the variations in the sunspot blocking and the facular brightening associated with magnetic modulation of the *quiet* Sun irradiance, S_Q, during solar cycles 21 and 22. These quantities parameterize, respectively, the dark and bright regions whose competing effects determine the overall solar radiative output. The sunspot blocking was calculated according to Foukal (1981). The facular brightening was estimated from ground-based measurements of solar Ca II K emission, linearly regressed against $S-S_Q(1+P_s)$ (Livingston *et al.* 1988), where $S_Q P_s$ is a measure of the sunspot blocking. As expected, and evident in Figure 2, the H I Lyman α variations, which reflect changes in enhanced emission from bright faculae, track closely the calculations of facular brightening derived from Ca II K data (White *et al.* 1990), as well as determined directly from the ACRIM radiometry (Foukal & Lean 1990).

Models of solar irradiance variability constructed from the ground-based sunspot blocking and facular brightening time series in Figure 2 are generally able to reproduce both the 27-day and 11-year variations measured by the satellite radiometers during the past decade (Foukal & Lean 1990; Willson & Hudson 1991; Lean 1991). However, the radiometric database extends over barely one 11-year solar activity cycle and some discrepancies between the models and measurements remain unexplained (Foukal & Lean 1988; Kuhn & Libbrecht 1991; Willson & Hudson, 1991). Variations that may be occurring on time scales longer than the 11-year cycle have yet to be recorded by the radiometers, or included in these types of variability models (Fröhlich 1994). As well, refinements to the sunspot blocking function are in progress, incorporating the dependence of sunspot contrast on spot area (Brandt *et al.* 1992).

3. Solar and stellar CA II flux observations

The generally good agreement between the contemporary solar irradiance measurements and reconstructions from sunspot blocking and facular brightening time series provides strong evidence that variations in total solar irradiance arise from magnetic modulation of a *quiet* signal, at least for the recent 11-year solar cycle. With knowledge of the character and evolution of solar magnetic active regions, inferences can thus be made about irradiance variations on longer time scales. Observations of the Sun and stars made at the wavelength of ionized calcium have proven particularly helpful in this regard because Ca II emission is enhanced in magnetic active regions. On the Sun, these regions of bright Ca emission are spatially coherent with bright photospheric faculae, and thus provide proxy data for the brightness source of irradiance variations.

Spatially resolved observations of the solar disk in the K Fraunhofer line of Ca II indicate that large active regions are a significant component of disk-integrated brightness variations. However, studies comparing resolved Ca II K images with independent measurements of the disk-integrated flux indicate that the large active regions are not the only component. To explain the 11-year cycle increase in the disk-integrated Ca II K flux data, Skumanich *et al.* (1984) postulated that in addition to increased emission from large active regions, the area of the solar disk covered by chromospheric network was larger at solar maximum than the 39% coverage determined by Skumanich *et al.* (1975) for the quiet Sun. This additional brightness source, postulated to arise in *active* network, may explain why the bolometric facular brightening is sufficient to exceed the sunspot blocking over the 11-year activity cycle, giving rise to total radiative output variations that are in phase with solar magnetic activity. Global variations in solar structure, rather than magnetic activity alone, may also provide an explanation for this (Kuhn & Libbrecht 1991).

A broader perspective of the range of temporal variability possible for the Sun is afforded from comparative observations of solar-like stars (Lockwood 1994). This is illustrated in Figure 3 which compares contemporary solar Ca II emission with that from 13 solar-like stars. Values for both the solar Ca II K index and the stellar Ca II HK index are indicated, using the conversion HK = 0.04 + 1.53 °K (White *et al.* 1992).

The Sun's Ca K emission is seen to be typical of Ca emission levels from the brightest one-third of those solar-like stars that exhibit cyclic patterns similar to the Sun's 11-year cycle (HK > 0.15). This is consistent with the contemporary solar Ca II data having been collected during an epoch in which overall levels are higher than in the past few hundred years, as indicated by the sunspot and cosmogenic isotope data. In stars for which long-term cycles are absent (HK < 0.15), the Ca II emission is reduced even below the minimum levels seen in the cycling stars. One explanation for the reduced Ca emission in some solar-like stars is that the fraction of their disk covered by bright Ca emission is less than the 39% coverage of the contemporary minimum Sun. White *et al.* (1992) have demonstrated that this hypothesis can explain the range of Ca II emission in the cycling stars (the upper portion of the stellar distribution in Figure 3). However, reduction of the basal solar emission (i.e., the emission from the centers of the chromospheric cells around which the active network forms) to a value as low as that emitted from only 15% of the contemporary Sun must also be invoked to account for Ca II emission levels of the non-cycling stars (the lower part of the distribution in Figure 3).

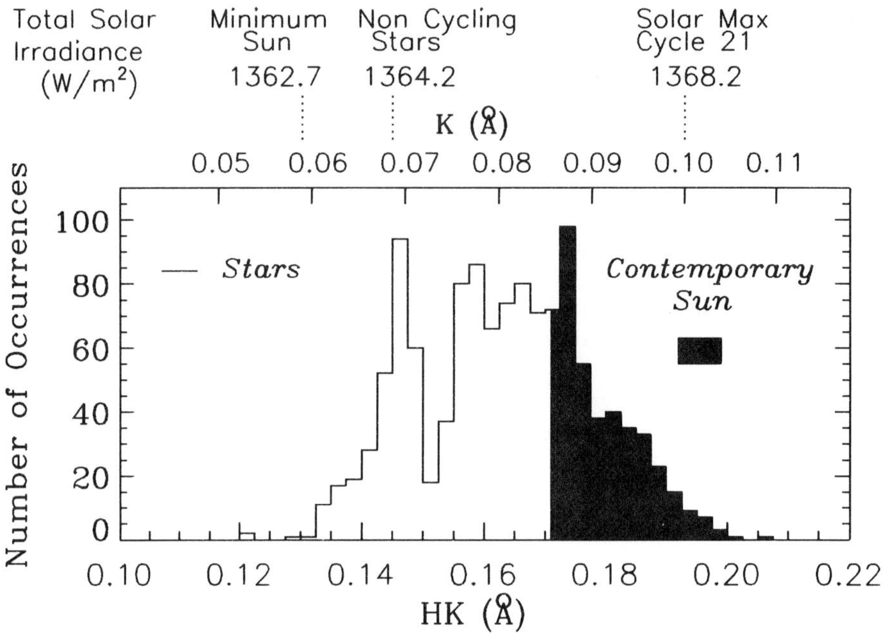

FIGURE 3. Distribution of Ca emission from 13 solar-like stars, compared with the emission from the contemporary Sun. Stars with HK > 0.15 exhibit cycles similar to the 11-year solar activity cycle.

4. Inferences for the Maunder Minimum Sun

During the Maunder Minimum, the absence of sunspots (Eddy 1976), indicates zero sunspot blocking, $P_s = 0$. The level of facular brightening can be inferred from solar and stellar Ca emission data. Assuming that the distribution of solar-like stars in Figure 3 is pertinent to the Sun, Baliunas & Jastrow (1990) postulated that the non-cycling stars (HK < 0.15) are in phases of inactivity analogous to the Maunder Minimum Sun. In this case, the solar Ca II K level for the Maunder Minimum is 0.0686, the peak of the distribution of the non-cycling stars. This value is significantly less than the lowest level of K=0.085 seen in the contemporary minimum Sun.

The level of solar facular emission corresponding to the Maunder Minimum Ca II K value is obtained from the correlation of contemporary Ca II K flux with direct solar radiometery, as indicated in Figure 4. The values of $S - S_Q(1+P_s)$ were calculated using S from the ACRIM radiometry, with $S_Q = 1366.9$ Watt/m^2 and P_s determined from ground-based sunspot data following Foukal (1981), using a mean spot contrast of 0.67. From Figure 4, a Ca II K value of 0.0686 (non-cycling stars) implies a reduction in facular emission of 2.7 Watt/m^2 relative to contemporary solar minimum values. Thus, the results in Figure 4 imply that the total solar irradiance was reduced to 1364.2 Watt/m^2 during the Maunder Minimum.

5. Solar forcing of climate during the Maunder Minimum

The Maunder Minimum coincided with the coldest period of the Little Ice Age. Since then, the mean global temperature of the Earth has increased by about 1°C, although

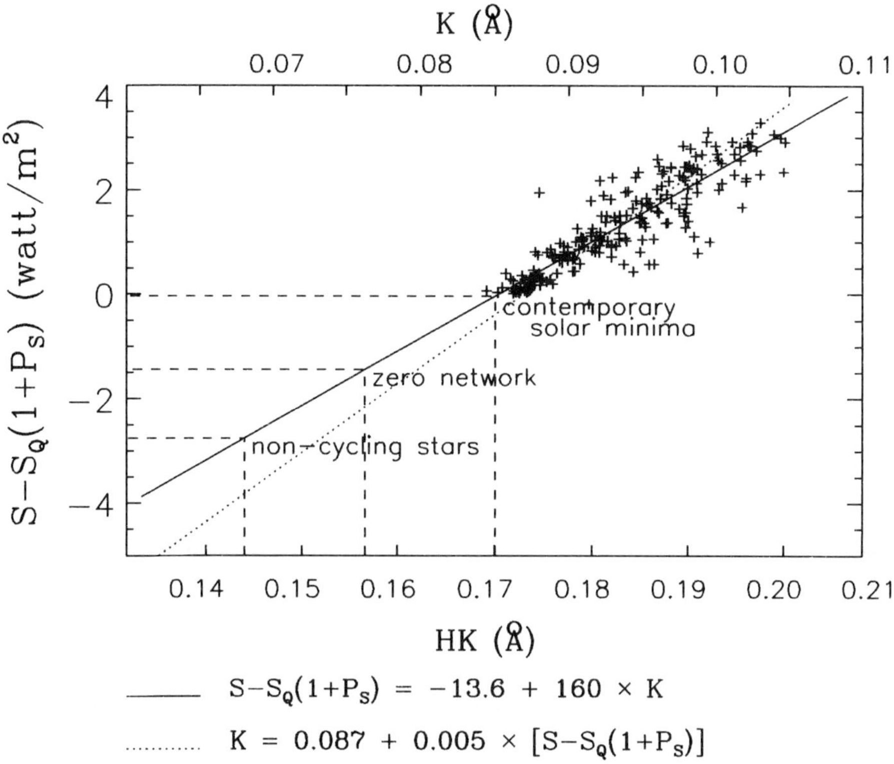

FIGURE 4. Correlation between solar Ca emission (abscissa) and total irradiance brightness (ordinate), from Lean *et al.* (1992). The dotted line is the fit obtained when the abscissa and ordinate are interchanged, and provides an estimate of the uncertainty of the regression.

the exact magnitude and geographic extent remain uncertain. Mechanisms for climate change over decadal to century time scales may potentially be found in combinations of changing atmospheric trace gas concentrations, North Atlantic Deep Water production, volcanic aerosols, solar radiative output and insolation, and natural variability (Rind & Overpeck 1993). Shown in Figure 5 are calculations by the GISS GCM of the geographical distribution of surface temperature changes corresponding to a reduction of 0.25% in solar total irradiance. The global reduction is 0.46°C. There is significant cooling (0.5°C to 1.5°C) at tropical latitudes, but not all regions on the Earth's surface are predicted to cool. Rather, because of differential heating between the land and the oceans, some geographical regions may have been warmer (by as much as 1°C) during the Maunder Minimum.

6. Discussion

Our estimates of solar radiative output variations and their impact on the global surface temperature during the Maunder Minimum suggest that improved knowledge of solar forcing will be needed to help reduce uncertainties in interpreting both paleo- and future climate change. Determining whether solar radiative output variations of as much as 0.2% - 0.3% are indeed occurring over decadal time scales will require continuous, high-precision solar monitoring for many decades. The first evidence for such long term

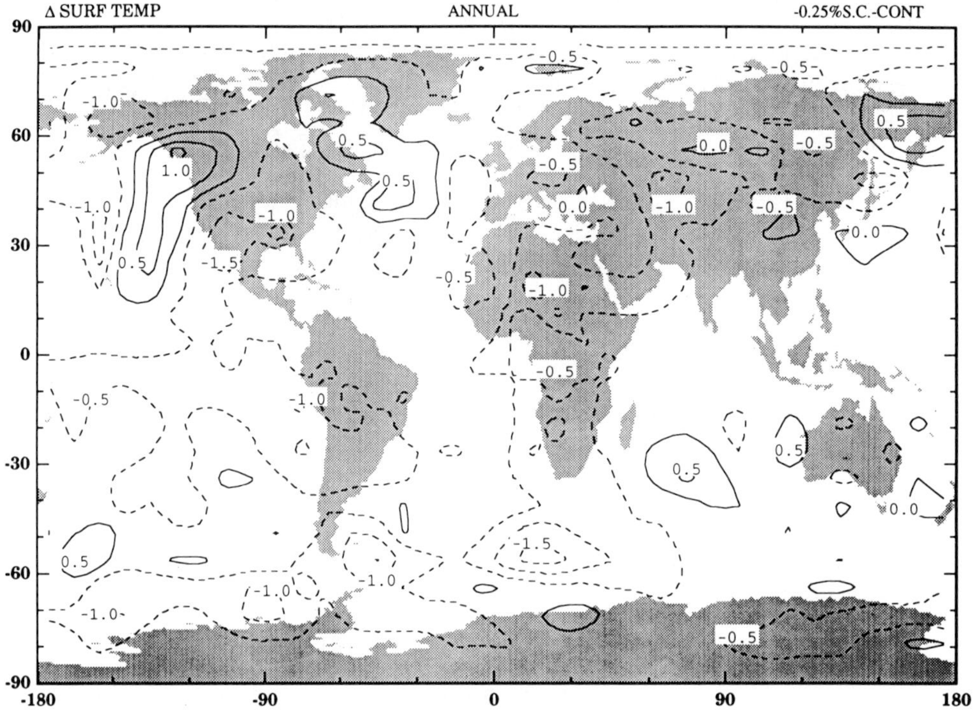

FIGURE 5. Calculations by the GISS GCM of the geographical distribution of the temperature change arising from a spectrally uniform reduction in solar radiative output of 0.25%.

changes might come from comparison of successive minima in the 11-year activity cycle, since the mechanism for the longer term variations, if they are occurring, is expected to affect the overall background level. However, assuming that solar activity is presently at the peak of the Modern Maximum, only very small changes can be expected until a period of rapid change ensues.

Determining the exact nature of the terrestrial response to solar forcing will also require additional effort. The calculations used to produce the results in Figure 5 assumed a flat spectral irradiance reduction. However the solar emission at ultraviolet wavelengths is also expected to be reduced during the Maunder Minimum (Lean *et al.* 1994, manuscript in preparation). Changes in ultraviolet irradiance at wavelengths shortward of 300 nm can also impact the surface temperature. To the extent that the irradiance variations occur at wavelengths that are deposited higher in the Earth's atmosphere (rather than penetrating to the Earth's surface), the direct solar forcing of the Maunder Minimum temperature would be reduced. However, atmospheric ozone concentrations would be less because of the diminished ultraviolet radiation, with the corresponding radiative forcing dependent on the altitude profile of the depletion. To investigate this, further studies with the GISS GCM are planned.

Acknowledgments. The solar Ca K data are measured at the National Solar Observatory at Kitt Peak by W. Livingston.

REFERENCES

BALIUNAS, S. & JASTROW, R. 1990 Evidence for long term brightness changes of solar-type stars. *Nature* **348**, 520–523.

BRANDT, P. N., SCHMIDT, W. & STEINEGGER, M. 1992 Photometry of sunspots observed at Tenerife. In *Proceedings of the Workshop on the Solar Electromagnetic Radiation Study for Solar Cycle 22* (ed. R. F. Donnelly). pp. 130–134. NOAA ERL SEL, Boulder, CO, USA.

DONNELLY, R. F. 1988 Uniformity in solar flux variations important to the stratosphere. *Annales Geophys.* **6**, 417–424.

EDDY, J. A. 1976 The Maunder Minimum. *Science* **192**, 1189–1202.

FOUKAL, P. 1981 Sunspots and changes in the global output of the Sun. In *Physics of Sunspots* (ed. L. E. Cram & J. H. Thomas). pp. 391. Sacramento Peak Observatory, Sunspot, NM, USA.

FOUKAL, P. & LEAN, J. 1988 Magnetic modulation of solar luminosity by photospheric activity. *Astrophys. J.* **328**, 347–357.

FOUKAL, P. & LEAN, J. 1990 An empirical model of total solar irradiance variation between 1874 and 1988. *Science* **247**, 505–604.

FRÖHLICH, C. 1994 Irradiance observations of the Sun. In *The Sun as a Variable Star: Solar and Stellar Irradiance Variations* (ed. J.M. Pap, C. Fröhlich, H.S. Hudson & S.K. Solanki). Cambridge University Press, in press.

HOYT, D. V., KYLE, H.L., HICKEY, J. & MASCHHOFF, R. 1992 The Nimbus 7 total solar irradiance: A new algorithm for its derivation. *J. Geophys. Res.* **97**, 51–63.

KUHN, J. R. & LIBBRECHT, K. G. 1991 Nonfacular solar luminosity variations. *Astrophys. J.*, L35–L37.

LEAN, J. 1991 Variations in the Sun's radiative output. *Rev. Geophys.* **29**, 505–535.

LEAN, J., SKUMANICH, A. & WHITE, O. 1992 Estimating the Sun's radiative output during the Maunder Minimum. *Geophys. Res. Lett.* **19**, 1591–1594.

LIVINGSTON, W. C., WALLACE, L. & WHITE, O. R. 1988 Spectrum line intensity as a surrogate for solar irradiance variations. *Science*, **240**, 1765–1767.

LOCKWOOD, W. 1993 Irradiance variations of stars. In *The Sun as a Variable Star: Solar and Stellar Irradiance Variations* (ed. J.M. Pap, C. Fröhlich, H.S. Hudson & S.K. Solanki). Cambridge University Press, in press.

MCHARGUE, L. R. & DAMON, P.E. 1991 The global Beryllium 10 cycle. *Rev. Geophys.* **29**, 141–158.

RIND, D. & OVERPECK, J. 1993 Hypothesized causes of decadal-to-century climate variability: climage model results. *Quat. Sci. Rev.*, in press.

ROTTMAN, G. J. 1988 Observations of solar UV and EUV variability. *Adv. Space Res.* (7)53–(7)66.

SKUMANICH, A., SMYTHE, C. & FRAZIER, E. N. 1975 On the statistical description of inhomogeneities in the quiet solar atmosphere. I. Linear regression analysis of chromospheric variability in the Calcium K line. *Astrophys. J.* **200**, 747–764.

SKUMANICH, A., LEAN, J. L., WHITE, O. R. & LIVINGSTON, W. C. 1984 The Sun as a star: Three-component analysis of chromospheric variability in the Calcium K line. *Astrophys. J.* **282**, 776–783.

WHITE, O. R., ROTTMAN, G. J. & LIVINGSTON, W. C. 1990 Estimation of the solar Lyman alpha flux from ground based measurements of the CaII K line. *Geophys. Res. Lett.*, **17**, 575–578.

WHITE, O. R., SKUMANICH, A., LEAN, J., LIVINGSTON W. C. & KEIL, S. L. 1992 The Sun in a non-cycling state. *Publ. Astronom. Soc. Pac.* **104**, 1139–1143.

WILLSON, R. C. & HUDSON, H. S. 1991 The Sun's luminosity over a complete solar cycle. *Nature* **351**, 42–44.

Solar Rotation, Irradiance Changes and Climate

By ELIZABETH NESME-RIBES[1], DMITRY SOKOLOFF[1,2] AND ROBERT SADOURNY[3]

[1]URA 326, Observatoire de Paris, 5 Place Janssen, 92195 Meudon, France

[2] Department of Physics, Moscow State University, 119899, Moscow, Russia

[3] Laboratoire de Météorologie Dynamique, Ecole Normale Supérieure, 75005-Paris, France

Magnetic activity cycles for solar-type stars are believed to originate from non-uniform internal rotation. To determine this depthwise angular velocity distribution, helioseismology is a valuable source of information. Surface rotation, as traced by sunspot motion, is a well-observed parameter with data going back to the beginning of the telescopic era. This long sunspot series can be used in understanding the behaviour of the Sun's surface rotation, the connection with its internal rotation, and thereby its magnetic activity. Apparent solar diameter is another important parameter. This is related to the structure of the convective envelope and how it reacts to the presence of magnetic fields. Both these parameters are related to the solar output, and can provide a surrogate for total solar irradiance, by way of a theoretical modeling of the response of the convective zone to the emergence of periodic magnetic fields. The impact of solar variability on the terrestrial climate is also addressed.

1. Introduction

Direct observation of solar rotation is available for the surface only. It is possible, however, to obtain information on the internal rotation of the Sun from surface oscillations, provided we have a theory explaining the 5 min oscillations. This is known as an inverse problem.

Magnetic tracers (sunspots, faculae, network, filaments) can be used to reconstruct solar rotation. This is also an inverse problem, for similar reasons, because a rotation curve deduced from tracers provides information only about those layers where the tracers originate. We believe that sunspots provide information on the distribution of azimuthal large-scale magnetic fields, and we adopt the idea that the azimuthal field is generated by the α-ω dynamo mechanism. These concepts are not as yet unanimously accepted. However, if the rotation curve deduced from seismic observations is compatible with that derived from magnetic tracers, this is a clear indication that observations and theories are compatible. So we shall adopt the framework of the α-ω dynamo theory and relate the properties of solar dynamics (meridional circulation and internal rotation) to the emergence of magnetic fields, and thereby to the strength of solar cycles. Then applying a simple model, we shall use the dynamics as a surrogate for solar output. An estimate of the solar output variation will become available and will be introduced in atmospheric circulation models to study the climatic response to solar forcing.

2. Parker migratory dynamo and the solar rotation curve

The distribution of solar activity through the solar cycle can be interpreted as a dynamo wave propagating from pole to equator (Parker 1955). From observations, this wave starts at about 70° latitude (Leroy & Noens 1982) and moves equatorwards in approximately 11 years.

According to Parker's migratory dynamo, the wave generation is controlled by two ingredients. The first is a radial shear in the angular velocity, $\frac{\partial\omega}{\partial r}$. The second ingredient is the helicity of the flow α. According to the usual explanation of helicity (due to the influence of Coriolis forces on cyclonic-type motions in a stratified medium), helicity is positive in the northern hemisphere and negative in the southern. So a dynamo wave moving equatorward, as shown in the butterfly diagram of sunspot activity, implies that $\frac{\partial\omega}{\partial r}$ should be negative.

It becomes clear that Parker's scenario should be adapted to this new picture of solar activity. There are two basic alternative ways to produce the observed distribution of solar activity of Figure 1, in the framework of the α-ω dynamo. One is to change the sign of helicity (as suggested by P. Gilman). Several sophisticated (compared with the standard explanation) ways exist to create this α effect, with an anisotropic tensor structure for α. The other alternative is to understand the complex solar activity distribution through the solar cycle as a result of the combination of two waves, as shown in Figure 1. In other words, the modification of the Parker migratory dynamo will depend on the description of the solar activity. Either we consider that the branches B_1 and A_1 are connected so as to make a continuous equatorwards wave (similarly, B_2 and A_2 are a continuous poleward wave) or we concentrate essentially on the A_2 and B_2 branches and leave the interpretation of the secondary A_1 and A_2 branches for further study. In the next section, we discuss these two possibilities for interpreting the new solar activity diagram, in the framework of the Parker dynamo.

Parker's dynamo is not completely successful, as it gives rise to a number of predictions inconsistent with observational data (Parker 1989). For example, the predicted period for the solar cycle should be controlled by the Sun's rotation period, i.e. 30 days, while the cycle lasts about 10 years. The role of helicity also has to be considered. However, if both helicity and radial shear are located in the same layers, it is difficult to extend the solar cycle to more than, say 10 rotation periods. Several ways have been suggested to overcome this difficulty, the most reasonable being to say that the helicity belongs to convective layers other than those subject to radial shear. The two regions of interest would be connected by turbulent diffusion, and the slow diffusion time-scale would lead to a long cycle of activity (Parker 1993). This view is supported by recent seismic observations, which seem to place the existence of a radial shear of the angular velocity at the bottom of the convective zone or even below it.

Another difficulty has come from seismic observations, with the sign of the observed angular velocity shear, $\frac{\partial\omega}{\partial r}$, and the direction of propagation of the dynamo wave in Parker migratory dynamo. However, some recent observations of solar activity provide new insight into the subject. It has been noticed (Harvey 1992) that a few active regions do not follow the Hale polarity rule: namely the leading sunspot has a magnetic polarity that is opposite to what is expected in a given cycle, for a given hemisphere. These active regions appear close to the equator shortly after sunspot maxima, and move poleward (up to 20°). This is suggestive of another activity wave of opposite direction and of weak amplitude (Mouradian & Soru-Escaut 1991). Furthermore, the large-scale magnetic field as traced by H_α filaments do show a poleward migration from sunspot minimum to maximum (Makarov 1984). On the other hand, the small ephemeral regions that appear in each hemisphere near the latitudes of 70° (Martin & Harvey 1979) move equatorwards. So there seems to exist two waves, one poleward and the other equatorward. The two corresponding branches of solar activity are represented schematically in Figure 1. The solid lines in Figure 1 correspond to the well-established migration of sunspots, the so-called sunspot butterfly diagram (equatorward branch, B_1). The dotted lines describe the migration of the small bipolar active regions A_1 (Martin & Harvey 1979). The dashed lines represent the migration of sunspots violating the Hale polarity rule A_2. The dotted-dashed lines trace the migration of large-scale magnetic fields (poleward branch B_2).

3. Possible scenario for normal activity cycles

3.1. *Two intersecting dynamo waves*

From seismic observations, the radial shear has a negative sign in the latitude range between 90° and 40°. So the dynamo number will be negative there, with the conventional sign for helicity (positive in the northern hemisphere). As a result, a wave propagating equatorward can be generated there as long as the (negative) dynamo number exceeds some critical value. Because helicity is antisymmetric with respect to the equator, it is natural to expect the dynamo number to be maximum somewhere around ± 40°, which corresponds precisely to the latitudes where sunspots appear (branch A_1).

To account for the additional dynamo wave generated near the equator and moving polewards, the sign of the radial shear should be positive; this is the case for the latitudes ranging between the Equator and 30°. So a poleward wave will be generated at around 5° to 10° (branch A_2).

What occurs when these two waves collide, which is somewhere near $\theta \sim 30°$? The answer probably lies in our understanding of the nonlinear influence of the dynamo wave on the α-coefficient. In a nonlinear regime, α is the sum of both hydrodynamic and magnetic helicity. If we accept such an idea, the magnetic helicity can exceed the hydrodynamic, so the value of $\alpha\frac{\partial\omega}{\partial r}$ varies within the dynamo wave. This means that, after collision, the poleward and equatorward waves may continue on their way in a nonlinear regime through regions which would not be permitted in the kinematic approach. As a result of such propagation, one obtains the branches B_2 and B_1 respectively. Though these waves probably become unstable, it is natural to believe that their lifetime is of the order of the solar cycle. Two intersecting waves can be produced in the framework of α-ω dynamo theory.

3.2. *Two distinct dynamo waves*

There is another way to produce the schematic butterfly diagram of Figure 1. We assume that helicity does not have the conventional sign, i.e. it is negative in the northern hemisphere. In this case, the poleward wave (branch B_2) would occupy the latitude domain $\theta > 40°$, and the equatorwards wave (branch B_1), the region $\theta < 30°$. There would be a region without sunspots around $\theta \sim 35°$. One reasonable test for this type of model would be the existence of an empty region on the butterfly diagram. This seems to be the case, as a gap around $\theta \sim 30°$ is visible on both the Meudon and Greenwich sunspot data, approximately one year before sunspot maximum. However, this gap is not observed during the whole cycle. If we accept this explanation for the butterfly diagram, it would be necessary to explain how and why this gap disappears in the more intensive phase of solar activity and what is the origin of the phenomena, referred previously to as branches A_1 and A_2.

4. Characteristics of the Maunder Minimum

There are times when solar activity cycles are very low. Such episodes have possibly occurred several times over the past millenia (Stuiver & Braziunas 1993). We have at our disposal only one such long-lived minimum that is well-documented: the "Maunder Minimum". This minimum can be described as follows (Ribes & Nesme-Ribes 1993).

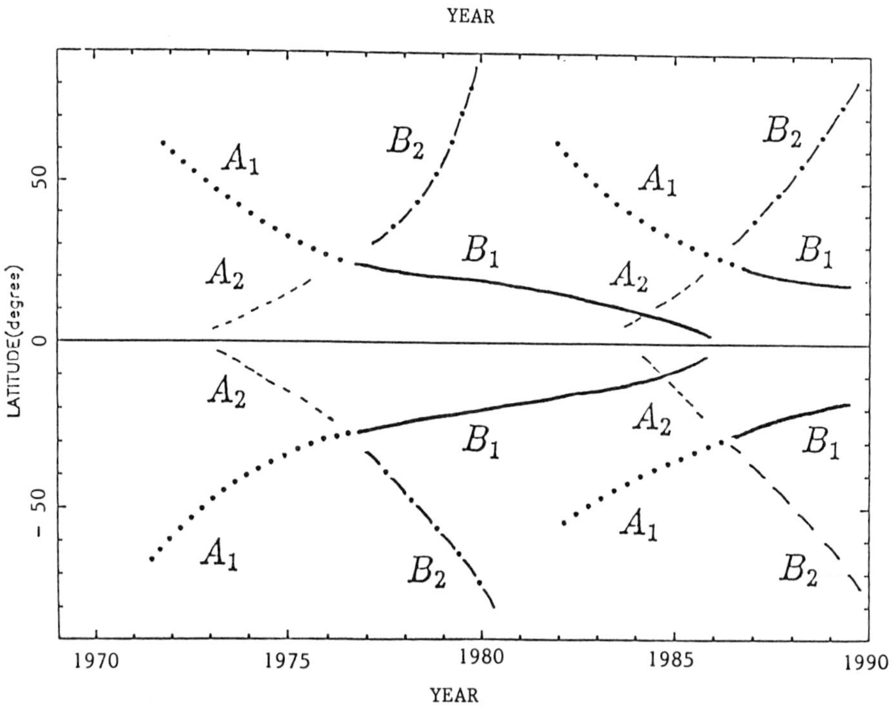

FIGURE 1. Schematic butterfly diagram of solar activity: two dynamo waves (equatorward and polarward) are presented during each 11-year cycle. The dotted lines describe the migration of the small ephemeral regions, the solid lines correspond to the well-known sunspot butterfly diagram. The dashed lines represent the migration of sunspots violating the Hale magnetic polarity rule, the dotted-dashed lines correspond to the migration of large-scale magnetic field as traced by H_α filaments.

The sunspot number was extremely low for at least 50 years (from 1660 to 1710), and was restricted to the southern hemisphere only. Activity resumed in the northern hemisphere starting in 1715, though the cycle was nonetheless still weak. The sunspots then remained close to the Equator, as happens during the decreasing phase of the solar cycle (after sunspot maximum). The sunspot rotation rate was low and the variation of rotation (with respect to latitude) was more differential. This is in conformity with what is observed during the minimum of a modern 11-year cycle, when the rotation is more rigid at the sunspot maximum and more differential near the minimum (Nesme-Ribes *et al.* 1993b). The difference, however, is that the differential rotation was more pronounced during the Maunder Minimum than at present. Lastly, the solar diameter appeared larger than at present, and showed a strong 10-year modulation, thereby suggesting that the cycle was still at work (Ribes *et al.* 1989). It should be noted that, in spite of difficulties in correctly interpreting the observed variations of the solar diameter, the modulation (larger at the time of sunspot minimum and smaller at the time of high activity) was in conformity with the present variation of the solar diameter (Ribes *et al.* 1992). Furthermore, the abundance of cosmogenic isotopes during the Maunder Minimum (Stuiver & Braziunas 1993) coincides with a 10-year modulation in phase with the sunspot number and the apparent solar diameter (Ribes *et al.* 1989).

There are also some more empirical indications that a 10-year cycle was at work.

The state of the corona was observed at various times, during solar eclipses (Ribes & Nesme-Ribes 1993), showing periods of alternating quiet and activity.

5. Possible Maunder Minimum scenario

It seems likely that the (rigid) rotation of the radiative zone has remained unchanged over the last three centuries. Assuming sunspots are an effective tracer of the convective zone rotation, it becomes clear that this rotation changed during the Maunder Minimum, for the latitude between $\pm$ 30°, as shown in Figure 1 of (Nesme-Ribes *et al.* 1993a). Adopting a conventional (positive) sign for helicity in the northern hemisphere, the low-latitude poleward branch of activity was suppressed, as both helicity near the Equator and positive shear were weak. Above 30°, no tracers were available during the 17th century to determine the rotation rate at high latitudes. It is possible, however, that the negative shear was still large, allowing an equatorward dynamo wave. It is also natural to believe that the latitudinal distribution of $\alpha\frac{\partial\omega}{\partial r}$ was much sharper near the Equator during Maunder Minimum than during a normal cycle. It can be shown that the solar dynamo is then nearer the bifurcation point between non-oscillatory and oscillatory regimes. Under such conditions, mixed parity solutions are expected (Brandenburg *et al.* 1989). A corresponding mixed parity solution has been obtained by Jennings (1991) in numerical simulations: the dynamo wave was suppressed in one hemisphere and confined to the equatorial region in the other (see more about computer simulations of mixed parity in solar dynamo Brandenburg *et al.* 1990). Azimuthal magnetic fields (that of active regions) in this mixed parity solution are produced from the combination of a dipole and a quadrupole field of comparable strengths (Sokoloff & Nesme-Ribes 1993).

This scenario of Maunder Minimum solar activity assumes a standard sign for helicity. As a result, one obtains a single wave A_1 in the southern hemisphere and no wave in the northern.

6. Meaning in terms of solar output variation and climatic impact

We must first address the problem of relating the strength of the solar cycles to the solar output. To do so, a self-consistent theoretical model has to be developed to solve the dynamics and the thermodynamics. This is far from being done. However, an approximation has been made by relating the change in the kinetic energy throughout the solar cycle empirically to the change of total solar irradiance (Nesme-Ribes & Mangeney 1992). The total solar irradiance has been reconstructed from the properties of solar dynamics as observed during the 17th century. Furthermore, a certain amount of structural change in the solar envelope (apparent diameter and luminosity) (Spruit 1992) also results from the magnetic field emergence. So the diameter constitutes another surrogate for deriving the solar output. Systematic observations of the apparent diameter were made during the Maunder Minimum, which gives us two independent ways of reconstructing the solar output (Nesme-Ribes *et al.* 1993a).

One interesting feature was a 10-year modulation of the solar output, over the period 1666 to 1719. The decrease ranged from .1% to 1% with respect to a reference value, which was the total solar irradiance measured by radiometers at the sunspot maximum in the year 1989 (Wilson & Hudson 1991). To address the question of climatic response to such solar forcing, equilibrium climate models were used. We adopt a constant decrease of the total solar irradiance, i.e. 0.4% over approximately 50 years. Assuming a planetary albedo of 30%, a 0.4% decrease of the solar output would correspond, on the average over the spherical surface of the Earth, to a variation of approximately -1W m^{-2} of the absorbed solar energy. For such a perturbation equilibrium climate simulations are expected to produce an average cooling of 0.3 to 1.2 °K, depending on the atmospheric general circulation model used (Cess *et al.* 1990). These values are compatible with those found in the literature of the Little Ice Age reconstructions, although most data there concern Europe only (e.g. Pfister 1992, among others). On the other hand, the equilibrium climate assumption is not very realistic, as it neglects the effect of slow oceanic circulation transients; other external perturbations like volcanic dust might have played a significant role in the Little Ice Age event.

The global structure of equilibrium climate change associated with a -0.4% perturbation of solar luminosity has been described in Nesme-Ribes *et al.* (1993a) for a relatively sensitive version of the Laboratoire de Météorologie Dynamique atmospheric general circulation model (referred to as LMD agc model, Sadourny & Laval 1984). The response is found to be very similar to the case of a greenhouse gas perturbation of equivalent radiative magnitude, which points to the major role of internal feedback processes in climatic change; this would also mean that, if the Little Ice Age was indeed due to the weakening of solar output during the Maunder Minimum, the difference between the present climate and that of the XVIIth century would be a good analogue to the present anthropogenic greenhouse warming (at least to the second order, as we must take into account the strong nonlinearity of the response, which, starting from the present state, produce an asymmetry between warming and cooling). Some of the main characteristics of the response (likely to be equally valid for a solar or long-wave radiative perturbation) have been discussed by Nesme-Ribes *et al.* (1993a). One prominent feature of the climatic modeling is the decreased weight of latent energy compared to enthalpy in the global energetics of the atmosphere (including the mean Hadley-Walker circulation and middle-latitude transients) associated with a decrease of the solar constant, which follows from the strong concavity of the Clausius-Clapeyron law relating saturation water vapor mixing ratio to temperature. The response of the Hadley circulation has been analyzed in detail by Sadourny (1993). It is shown that the Hadley circulation is likely to increase in strength as the solar constant decreases, due to the combined effects of cooling and the increased vertical static stability following tropical drying in the lower layers. In addition, the meridional extent of the Hadley circulation would have decreased at the Maunder Minimum, due to the decrease in the meridional temperature gradient in the upper troposphere. This is a direct consequence of angular momentum dynamics theory (Held & Hou 1980).

The analysis of the simulated decrease of Monsoon rainfall is at first sight, more elusive. In the LMD agc model, the decrease of the water cycle strength overcomes the increase of the divergent Hadley-Walker circulation mass flux. This again might be a reliable result: a decrease of the solar constant should induce a lesser continental energy excess in the summer season, and thus weaker energy transport, reflected primarily in the rainfall, again because of the sensitivity of the Clausius-Clapeyron relation.

7. Conclusion

In the framework of the α-ω dynamo, we have described qualitatively how to relate the changes in the internal rotation with the strength of the solar cycles and the solar output. As there is compelling evidence that the dynamics (thereby the strength of the 11-year cycles) changed significantly during the Maunder Minimum, we have been able to estimate crudely the solar output decrease from the available surrogates, the apparent solar diameter and the sunspot rotation rate. A mean 0.4 % decrease in the solar constant is likely to have lasted at least 50 years.

Such a variation in the solar output (which is equivalent to a solar perturbation of 1 W/m^2 at the top of the Earth's atmosphere) is sufficient to induce climatic fluctuations which are similar to those expected during a little ice age. The main climatic characteristics will be a reduction of the mean surface temperature and a decrease of water precipitations and soil moisture. However, the energetics of the Hadley circulation is complex; and as a result, we should expect less Monsoon rainfall in spite of a more vigorous Hadley circulation.

In conclusion, we mention two problems of importance for climatic predictions. The first is that solar activity should be monitored if we are to predict episodes of the Maunder Minimum type, insofar they coincide with periods of lower total solar irradiance. Along this line of thought, it would be helpful to study the onset of the Maunder Minimum and the corresponding characteristics of solar rotation. The second problem is the pronounced solar activity asymmetries which seem to be typical of solar activity minima. This could be the case for the Dalton Minimum detected in the abundance of cosmogenic isotopes (cycles V and VI of the early ninenteenth century). Such minima seem to influence solar irradiance significantly.

Acknowledgments. This work was supported by a D.R.E.T. contract n° 9220011/A and RFFR grant n° 93-02-3638.

REFERENCES

BRANDENBURG A., MEINEL R., MOSS D. & TUOMINEN I. 1990 Variation of even and odd parity in the solar dynamo. In *Solar Photosphere: Structure, Convection and Magnetic Fields* (ed. J.O. Stenflo). pp. 379–382. Kluwer. Dordrecht.

BRANDENBURG A., TUOMINEN I. & MOSS, D. 1989 On the nonlinear stability of dynamo modes. *Geophys. Astrophys. Fluid Dynam.* **49**, 129–141.

CESS R.D., POTTER G.L., BLANCHET J.P., BOER G.J., DEL GENIO A.D., DÉQUÉ M., DYMNIKOV V., GALIN V., GATES W.L., GHAN S.J., KIEHL J.T., LACIS A.A., LE TREUT H., LIANG Z.-H., MCAVANEY B.J., MELESHKO V.P., MITCHELL J.F.B., MORERETTE J.-J., RANDALL D.A., RIKUS L., ROECKNER E., ROYER J.F., SCHLESSE U., SHEININ D.A., SHINGO A., SOKOLOV A.P., TAYLOR K.E., WASHINGTON W.M., WETHERALD R.T., YAGAI I. & ZHANG M.-H. 1990 Intercomparison and interpretation of climate feedback processes in nineteen atmospheric general circulation models. *J. Geophys. Res.* **95**, 166501–16615.

HARVEY, K. 1992 The cyclic behaviour of solar activity. In *The Solar Cycle* (ed. K. Harvey). pp. 335–337. Astron. Soc. Pacif. San Francisco.

HELD I.M. & HOU A.Y., 1980 Nonlinear axially symmetric circulations in a nearly inviscid atmosphere. *J. Atmosp. Sci.* **37**, 515–533.

JENNINGS R.L. 1991 Symmetry breaking in stellar dynamos. *Month. Not. R. astr. Soc.* **252**, 249–260.

LEROY, J.L. & NOENS, J.C. 1983 Does the solar activity extend over more than an 11-year period? *Astron. Astrophys.* **120**, 1–2.

LE TREUT H. & LI Z.X. 1991 Sensitivity of an atmospheric general circulation model to prescribed SST changes: feedback effects associated with the simulation of cloud optical properties. *Climate Dynamics* **5**, 175–187.

MAKAROV, V.I. 1984 Do prominences move equatorwards? *Solar Phys.* **93**, 59–79.

MARTIN, S.F. & HARVEY, K.L. 1979 Ephemeral active regions during solar minimum. *Solar Phys.* **64**, 93–108.

MOURADIAN Z., SORU-ESCAUT I. 1991 On the dynamics of the large-scale magnetic fields of the Sun and sunspot cycle. *Astron. Astropys.* **251**, 649–654.

NESME-RIBES E., MANGENEY A. 1992 On a plausible mechanism connecting the Maunder Minimum to the Little Ice Age. *Radiocarbon* **32**, 63–270.

NESME-RIBES E., FERREIRA E. N., SADOURNY R., LE TREUT H. & LI L. 1993 a Solar Dynamics and its impact on solar irradiance and the terrestrial climate. *J. Geophys. Res.* **98**, 18923–18935.

NESME-RIBES E., FERREIRA E. N. & MEIN P. 1993 b Solar dynamics over solar cycle 21 using sunspots as tracers. I. Sunspot rotation. *Astron. Astrophys.* **274**, 563–570.

PARKER E.N. 1955 Hydromagnetic dynamo models. *Astrophys. J.* **122**, 293–314.

PARKER E.N. 1989 Solar and stellar magnetic fields and atmospheric structures:theory. *Solar Phys.* **121**, 271–288.

PARKER E.N. 1993 A solar dynamo surface wave at the interface between convection and nonuniform rotation. *Astrophys. J.* **408**, 707–719.

PFISTER C. 1992 *Monthly Temperatures and precipitation in central Europe 1525-1979: quantifying documentary evidence on weather and its effects.* Climate since A.D. 1500. London.

RIBES E., RIBES J.-C., MERLIN PH. & BARTHALOT R. 1989 Absolute periodicities in the solar diameter derived from historical and modern data. *Annales Geophysicae* **7**, 321–329.

RIBES J.C. & NESME-RIBES E. 1993 The solar sunspot cycle in the Maunder Minimum AD1645 to AD1715. *Astron. Astrophys.* **276**, 549–563.

RIBES E., BEARDSLEY B., BROWN T. M., DELACHE PH., LACLARE F., KUHN J. & LEISTER N. V. 1992 The variability of the solar diameter. In *The Sun in Time* (ed. C.P. Sonett, M.S. Giampapa & M.S. Matthews) pp. 59–97. Arizona Univ. Tucson.

SADOURNY R. 1993 Climate sensitivity, Maunder Minimum and the Little Ice Age. ESF, Bologna.

SADOURNY R. & LAVAL, K. 1984 January and July performances of the LMD general circulation model. *New Perspectives in Climate Modelling.* Elsevier.

SCHWEINGRUBER F.H., FRITTS O., BRÄKER U., DREW L.G. & SCHAAR E. 1978 The X-ray technique as applied to dendrochronology. *Tree Ring Bulletin* **38**, 61–91.

SOKOLOFF D. & NESME-RIBES E. 1993 The Maunder Minimum, a mixed-parity dynamo mode? *Astron. Astrophys.*, in press.

SPRUIT H. 1992 Theory of luminosity and radius variations. In *The Sun in Time* (ed. C.P. Sonett, M.S. Giampapa & M.S. Matthews), pp. 118–159. Arizona Univ. Tucson.

STUIVER M. & BRAZIUNAS T.F. 1993 Sun, Ocean, Climate and Atmospheric $^{14}CO_2$ variation: an evaluation of causal and spectral relationships. *The Holocene* **3**, in press.

WILLSON R. C. & HUDSON H. 1991 The Sun's luminosity over a complete solar cycle. *Nature* **351**, 42–44.

What can we Learn from Oscillation Studies about Irradiance and Radius Changes?

By DOUGLAS GOUGH

Institute of Astronomy and Department of Applied Mathematics and Theoretical Physics, University of Cambridge, UK and Joint Institute for Laboratory Astrophysics, University of Colorado and National Institute of Standards and Technology, Boulder, CO 80303, USA

Dynamical and thermal modulation of the internal structure of the Sun can be manifest at the surface as changes in irradiance and radius. The relative magnitudes of these changes could provide a diagnostic of at least the location of the primary modulation, if only the mechanism were known. Variations of the frequencies of solar oscillations offer additional potentially valuable diagnostics, but unfortunately at present the mechanisms causing those variations and their relation to the structural and irradiance changes are not yet understood. In this lecture I shall review some of the theoretical conjectures that have been put forward to explain the observations.

1. Introduction

Associated with the solar cycle are variations in irradiance and in the apparent radius of the photosphere. How do these arise? I do not know the answer. However, recent helioseismic data have enabled us to eliminate some notions. In this review I shall try to explain how. To set the scene, let me first recall Kuhn's presentation at this colloquium, in which he analysed the latitudinal variation of the brightness of the solar surface and deduced that the luminosity of the Sun also varies with the cycle; it is almost in step with the irradiance variation and has essentially the same amplitude. This is an important result, for it implies that the spherically averaged thermal structure of the Sun varies with the cycle. There is also an aspherical component, associated with which must be large-scale material circulation, advecting heat, magnetic fields and angular momentum. Without an understanding of these processes we cannot understand the solar cycle, and without some means of measuring them, we cannnot test our theories adequately. Therefore I shall discuss also what has recently been learnt about variations of the Sun's angular velocity.

2. The luminosity and radius variations

Even without a precise model of solar variation, some information can be gleaned from the ratio W of the relative variations in the photospheric radius R and luminosity L: $W=\Delta \ln R/\Delta \ln L$. This was a quantity of some interest a decade or so ago, for its magnitude provides some indication of where in the Sun the process directly responsible for changing the structure is located. On the basis of a few particular calculations, it has been conjectured (Gough 1981) that if the radius and luminosity variations have the same origin, at some radius r_0, then on the whole W is a positive decreasing function of r_0, and virtually vanishes as $r_0 \to R$. In particular, $W \simeq 0.5$ when $r_0 = 0$, and $W \simeq 0.2$ when the perturbing process is situated near the base of the convection zone; a modification to the upper superadiabatic convective boundary layer yields a very much smaller value: $W \lesssim 10^{-3}$.

The reason for this dependence on radius is not difficult to appreciate. First, let us remind ourselves that the characteristic internal thermal readjustment time for the convection zone is given by $\tau_a = \int_{r_c}^{R} w^{-1} dr \simeq 1$y, where w is the rms vertical velocity of the convective motion and $r = r_c$ locates the base of the convection zone, whereas the thermal cooling time for the zone as a whole is $\tau_c \simeq 10^5$y, and, because it is much longer than τ_a, is determined only by the heat content of the zone and the radiative properties of the photosphere (Gough 1981; Spruit 1990). Thus, on a timescale of 11y, the convection zone can be considered to be almost in internal thermal balance, providing a large reservoir to supply or absorb heat associated with any perturbation to the luminosity without a significant concomitant change in its hydrostatic stratification. Any modification to the thermal impedance of the surface layers generally changes both L and the stratification of the superadiabatic boundary layer by similar amounts, but does not modify the adiabat beneath. Consequently if b is the characteristic thickness of the boundary layer, $\delta \ln L$ and $\delta \ln b$ are of the same order, and $|W| = \mathrm{O}(b/r) \ll 1$. If, on the other hand, the perturbation is beneath the convection zone, the reaction of the zone involves an overall hydrostatic change, which typically entails comparable changes in $\ln R$ and $\ln L$.

It is important to realize that this argument is specific to timescales τ satisfying $\tau_a \ll \tau \ll \tau_c$. As τ increases towards τ_c the thermal reservoir responds more, and W increases; and once τ becomes comparable with the thermal diffusion time of the radiative interior, $\tau_{KH} \simeq 3 \times 10^7$y, the entire star responds. Däppen (1983) has shown how this trend can be studied. In particular, when $\tau \gg \tau_{KH}$, increasing the efficacy of convection (whose immediate effect is to modify the stratification of the thin boundary layer) causes the entire star to shrink to a new state of thermal balance, and the value of W becomes quite large and negative (cf. Ulrich 1986; Gough & Novotny 1993). Now the luminosity is determined not solely by the thermal properties of the photosphere, but also by the augmented rates of the thermonuclear reactions under the increased weight of the more compact star.

The solar-cycle variation of irradiance (Willson & Hudson 1988) is now well established (e.g. Kyle *et al.* 1994). However, positive reports of radius variations are controversial, partly because the observations are difficult to interpret. Brown (1987), for example, finds no significant signal; his upper bound implies $|W| \lesssim 5 \times 10^{-2}$, apparently implying that the thermal variation substantially modifies the structure of the Sun only in its outer layers. On the other hand, Delache *et al.* (1993) take the CERGA astrolobe data literally as measurements of R, from which one would deduce $W \simeq -0.3$. If correct, that implies the existence of a deeply seated process modifying the stratification of the Sun. However, it is difficult to reconcile so large a negative W on an 11-year timescale with a single process. If the processes that regulate R and L were different, though presumably connected by some solar-cycle mechanism, then W would no longer be a robust indicator of the whereabouts of those processes, even if they occurred in the same place.

3. Lifting of p-mode degeneracy

The publication by Duvall *et al.* (1986) of an even component to the frequency-averaged splitting of p-mode multiplet frequencies, observed near sunspot maximum in winter 1981-82, demonstrated that the hydrostatic structure of the Sun was not spherically symmetric. The asphericity was a factor 10 greater than the oblateness caused by the centrifugal acceleration from the angular velocity Ω (Ω is measured by the odd component of the splitting) and of the opposite sign; acoustically, the Sun was prolate.

In attempting to explain the phenomenon, Gough & Thompson (1988) observed that the dependence of the splitting on m/Λ, where m is azimuthal order and $\Lambda^2 = \ell(\ell+1)$, ℓ being the degree of the mode, though not well resolved, implied a latitudinal variation

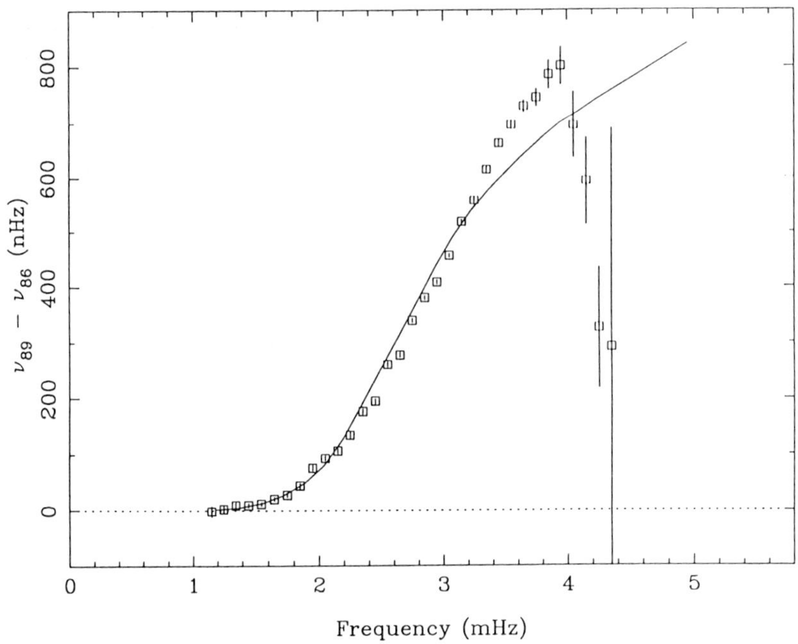

FIGURE 1. Frequency changes ν_{89} - ν_{86} from 1986 to 1989 measured at the Big Bear Solar Observatory, averaged in 100 μHz bins. The continuous line is proportional to the inertia of low-degree modes, normalized at a height of 430 km above the photosphere (after Woodard & Libbrecht 1993b).

of the hydrostatic stratification not dissimilar to that of the sunspots. They supposed that the frequency perturbations were due to magnetic activity in the form of fibril fields, which augmented the wave propagation speed and thereby increased the frequencies of the acoustic modes. The variation from pole to equator of fibril intensity required to reproduce the observations was not obviously at variance with observation, so the conjecture was not implausible. By considering the dependence of the degeneracy splitting on Λ, the variation of the fibril perturbation with depth can in principle be ascertained. Those early data were quite noisy. Nonetheless, they indicated that the perturbation was somewhat concentrated near the surface of the Sun, but that it was not completely confined there. Indeed, by fitting the data to a power law in Λ, a dependence on depth $R-r$ of the form $(R-r)^{0.1}$ was inferred.

Subsequent observations showed that the frequency splitting diminished towards solar minimum. The data were considered by Kuhn (1988a, 1988b) who compared them with surface brightness temperature measurements (Kuhn *et al.* 1988). Kuhn pointed out that if one were to assume simply that there were a depth-independent (at least down to $r = 0.95R$ and possibly as deep as $r = 0.2R$) relative sound-speed variation of the same magnitude as that implied by the apparent surface temperature variation, with no other modification to the Sun's structure, then the oscillation data could be reproduced. As will become evident in the next section, it appears that that idea cannot describe what is actually the case, because, as with the latitudinal variation, the temporal acoustical changes are concentrated very near the surface of the Sun. However, the coincidence is striking, and it would be interesting to find out whether it is a signature of an important physical constraint, or is merely fortuitous.

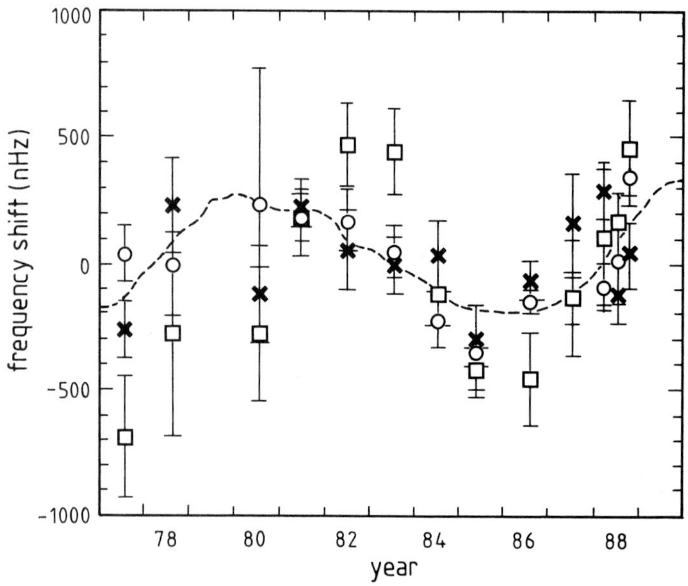

FIGURE 2. Frequency shifts of low-degree p modes, obtained from the whole-disk Doppler measurements from the Birmingham Solar Network (after Elsworth *et al.* 1990). The dashed line represents sunspot number, and is plotted on a scale chosen to fit the frequency shifts.

4. Solar-cycle frequency variations

A substantial step forward was made in 1990 with two announcements of temporal frequency changes. One was by Libbrecht & Woodard (1990), who reported changes from 1986 to 1988 both in multiplet frequencies (the mean frequencies of modes of like order n and degree ℓ, averaged over all admissible values of m; they depend on the spherically averaged structure of the star) and in the even contribution to degeneracy splitting (expressed, essentially as also had Duvall, Harvey and Pomerantz and subsequent observers, as coefficients α_i in a Legendre series expansion in the variable m/Λ). For the first time, the frequency dependence of the data was presented, and it was shown that the changes in both the multiplet frequencies (Figure 1) and the even splitting coefficients $\alpha_i(\omega, \ell)$ were roughly inversely proportional to the inertia I of the mode. The other announcement, by Elsworth *et al.* (1990), was of a variation over an 11-year interval of low-degree frequencies (averaged over n) of Doppler measurements in spatially integrated light (Figure 2), which demonstrated a connexion with the solar cycle. Like the luminosity, oscillation frequencies are greatest at solar maximum, the total variation being about 0.46μHz.

The importance of the I^{-1} proportionality can be appreciated from the formula $\delta\nu \simeq \nu^{-1}I^{-1}\int(\mathcal{K} - \nu^2\mathcal{I})dV$ for the small difference between the cyclic frequencies ν of corresponding modes of two similar states of the Sun; the integrands $\mathcal{K}$ and $\mathcal{I}$ are sums of terms which are bilinear in the components, or their spatial derivatives, of the displacement eigenfunction ξ of one of the states of the Sun (the reference), and which are linear in quantities defining the difference between the two states, and they are integrated over the volume V of the reference state. The inertia I is given by $\int \rho\xi\cdot\xi dV$, which is also bilinear in ξ, so the arbitrary amplitude of ξ cancels. For convenience, suppose the eigenfunctions to be normalized such that the average of $|\xi|$ over the surface of the Sun is unity. Then it turns out that I is a rapidly decreasing function of ν at low ν; roughly, $I \propto \nu^{-\alpha}$ with $\alpha \simeq 7$, the precise value of α depending on where in the solar atmosphere ξ is normalized. Thus, to within low powers of ν, it is evident that $\delta\nu$ is approximately proportional to I^{-1} if $\int \mathcal{K} dV$ and $\int \mathcal{I} dV$ are constants, independent of which mode is being considered; that is achieved if $\mathcal{K}$ and $\mathcal{I}$ differ from zero only in the very surface layers near where ξ is normalized. Thus the data of Libbrecht and Woodard convincingly demonstrated that both the mean and the latitudinally varying change in the solar structure occurs predominantly in the outer layers. With this in mind, one could then demonstrate that the low-degree data were similarly explained; indeed, the frequency dependence of frequency differences between sunspot maximum and sunpot minimum determined by P. L. Pallé, C. Régulo and T. Roca Cortés, though noisy, were not inconsistent with the I^{-1} scaling (Gough 1990a). I should point out that because I is so rapidly varying a function of ν at low ν, it is not easy to tell whether $\delta\nu$ is proportional strictly to I^{-1} or to some low-degree polynomial in (or slowly varying function of) ν multiplied by I^{-1}. Indeed, one would expect the latter, the details depending on precisely in what manner the outer layers of the Sun vary. For example, it has been argued (Gough 1990b) that the relative frequency variation is likely to be of the form $(\nu^2 I)^{-1}Q(\nu)$, where Q is a low-degree polynomial.

5. How can the frequency variations be explained?

I have already mentioned that it has been suggested (Gough & Thompson 1988) that a latitudinal variation of fibril magnetic field in the outer layers of the Sun is responsible for the degeneracy splitting. It is natural to adopt a unified conjecture, and to suppose that temporal variations in the same field are responsible for the solar-cycle changes too. This conjecture has been taken up by Goldreich *et al.* (1991), who discuss the influence on the oscillations in more detail. They also found that the magnitude of the field variation required is not obviously at variance with observation. Strong evidence for a unified cause of both temporal multiplet frequency variations and degeneracy splitting has been provided recently by Woodard *et al.* (1991), Woodard & Libbrecht (1993a), and Backman & Brown (1993), who demonstrate that there is a unique relationship between oscillation frequency changes and surface activity, irrespective of the characteristic timescale of the variation.

It is perhaps more natural to suppose that the frequency shifts result from a thermal variation modifying the sound speed, brought about, for example, by a modification to the efficacy of convection by a magnetic field. This has been discussed in various guises by Kuhn (1988a, 1988b, 1989a, 1989b) following ideas developed by Parker (1987). He basically postulates thermal disturbances rising from regions of magnetic activity at the base of the convection zone. The numerical experiment he discusses in these proceedings indicates that the relative temperature fluctuation increases near the top of the convection zone, and thereby modifies the sound speed predominantly in the outer layers of the Sun, as implied by Libbrecht and Woodard's data. However, ideas of this kind are difficult to reconcile with the observed luminosity variations. Indeed, Gough & Thompson (1988) had already considered the possibility that such thermal disturbances are responsible for degeneracy splitting, as an alternative to the direct dynamical effect of a fibril magnetic field. They modified the efficacy of convection by changing the mixing-length parameter in a model of the solar envelope that matches onto an unperturbed radiative interior. Although it was possible to reproduce the splitting data, it required a radiative flux some 15 per cent smaller at the equator than at the poles. This is contradicted by the surface brightness measurements published by Kuhn *et al.* (1985, 1988); not only were the theoretical flux variations too large by a factor of about 50, but they also had the wrong sign. A similar conclusion was drawn subsequently by Goldreich *et al.* (1991) and Balmforth *et al.* (1993) when considering solar cycle variations of multiplet frequencies.

The reason why purely thermal readjustments lead to the wrong sign of the frequency perturbation relative to the luminosity change is easy to explain (Goldreich *et al.* 1991; Balmforth *et al.* 1993). It is adequate to consider spherical perturbations, such as can result from increasing the efficacy of convection, which, on a timescale of 11y diminishes the superadiabatic gradient $\Delta\nabla \equiv d\ln T/d\ln p - (\partial \ln T/\partial \ln p)_{\mathrm{ad}}$ in the boundary layer without modifying the structure of the adiabatically stratified interior. (The entropy deep in the interior is essentially unchanged on timescales much less than τ_c, and therefore the pressure p, and consequently temperature T and density ρ, are also unchanged, because a slight expansion or contraction of the outer layers hardly changes their weight.) The dominant change is to the gradient $d\ln T/d\ln p$, since $(\partial \ln T/\partial \ln p)_{\mathrm{ad}}$ is a thermodynamic quantity and is more robust; therefore the temperature of a given material layer in the vicinity of the photosphere increases. Gravity hardly varies through the thin outer layers, and pressure, which depends on the weight of the overlying material, is essentially a Lagrangian variable (being directly related to the weight, and therefore the mass of the overlying material). Thus, the photospheric temperature increases, and so does the radiative flux. Roughly speaking, the oscillation frequencies ν are proportional to $(\int c^{-1}dr)^{-1}$, which can be considered to be both proportional to sound speed $c \propto T^{1/2}$ and inversely proportional to the characteristic length scale λ of the cavity within which the modes are confined. If one disregards the movement of the outer boundary of the cavity with respect to Lagrangian coordinates (which is a good first approximation), then, since the Lagrangian pressure perturbation δp vanishes, $\delta\lambda/\lambda = -\delta\rho/\rho = \delta T/T$, whereas $\delta c/c = \frac{1}{2}\delta T/T$. Thus $\delta\nu/\nu = \delta c/c - \delta\lambda/\lambda = -\frac{1}{2}\delta T/T$. The relative increase in the cavity size resulting from heating the layer is twice the relative increase in sound speed, and consequently the frequencies of oscillation are reduced. To show that the magnitude of the frequency change is much too small it is hardly necessary to be subtle. It suffices to estimate $\delta\nu/\nu$ by $-\frac{1}{2}\int T^{-1}\delta T d\tau / \int d\tau$ where $d\tau = c^{-1}dr$. For low-degree modes the integrals are taken over the entire Sun, whose acoustical radius is $\int d\tau \simeq 3500$ s. Taking $\delta T/T$ to decline linearly with respect to τ, from its photospheric value of 2.5×10^{-4} implied by a 10^{-3} relative variation in L to zero near the bottom of the boundary layer, at say where hydrogen is 50 per cent ionized (at an acoustical depth of about 200 s), therefore leads to a frequency variation of 0.011 μHz for a 3 mHz mode. This is smaller than the value reported by Elsworth *et al.* (1990) by a factor of more than 40.

Because this result deviates by so much from observation it seems inevitable that a very different mechanism must be responsible for the frequency variations. Neglected Reynolds stresses or slow large-scale circulation in the convection zone is hardly likely to alter the argument materially. Moreover, Balmforth *et al.* (1993), who also performed a time-dependent calculation when perturbing the star, have confirmed that nonadiabatic processes also play only a minor role. Aside possibly from horizontal inhomogeneity, the direct influence of Lorentz forces on the acoustic oscillations appears to be the most likely remaining candidate for causing the variation. This could be of the form of a fibril field in and beneath the photosphere, as I have already discussed. Alternatively, it has been suggested that instead it might be a relatively smooth horizontal field in the chromosphere (Campbell & Roberts (1983); Evans & Roberts (1990, 1991); Wright & Thompson 1992). Whatever is the case, it is unfortunate that there is no robust theory to link the variations of the magnetic field to those in irradiance.

6. Variations in the angular velocity

One expects horizontal variations in structure to be intimately linked with material circulation and consequent redistribution of angular momentum. Therefore angular velocity variations represent an integral aspect of solar variability. The original inversion of rotational splitting of sectoral modes (Duvall *et al.* 1984) revealed a radial variation of angular velocity Ω in the vicinity of the equatorial plane that immediately suggested temporal variation (Gough 1985). Goode *et al.* (1991) claimed that subsequent helioseismic data sets indeed provided direct evidence for such a variation, particularly in the radiative interior. However, the significance of the claim has since been brought into doubt (Schou 1991; Goode & Dziembowski 1993; Gough & Stark 1993a; Woodard & Libbrecht 1993a).

Significance of inferences from helioseismic data is difficult to assess, because the reliability of the means by which uncertainties in the data are estimated is hard to judge. Of special concern are systematic errors. When looking for temporal trends it is always safer to work with homogeneous data sets, obtained with the same instrument and analysed in the same way; in subtracting them one expects that many of the systematic errors will cancel. For that reason I shall limit the subject of my discussion to the data collected from 1986 to 1990 at the Big Bear Solar Observatory (BBSO), presented by Libbrecht & Woodard (1990) and Woodard & Libbrecht (1993a, 1993b). However, one must be warned of a particularly insidious problem apparently associated with the filtering by the observing window function, which has most power at $1\mathrm{d}^{-1} = 11.6\mu\mathrm{Hz}$, with the signal from modes of adjacent degree whose multiplet frequencies differ by a similar amount. Backmann *et al.* (1993) report a contribution to their splitting data which might arise from such filtering. Unfortunately, it is not invariant in time, and even changes sign. It is also unfortunate that, as Schou (1993) has pointed out, modes of different order that suffer from this coincidence all penetrate to the same depth, to about $r = 0.8R$, so they experience essentially identical averages of Ω, at least if Ω varies on a length scale greater than the characteristic wavelength of the oscillations. Unlike many other errors, such a systematic error cannot therefore be revealed from the inversions as an obvious inconsistency in the data.

The BBSO rotational splitting data were originally represented by the first three odd coefficients a_i in an expansion in Legendre polynomials in m/Λ, from which inversions, evidently poorly resolved in latitude, have been carried out (Gough *et al.* 1993). Some variation between the years is evident, particularly in the polar regions, though it is noticeable that much of that occurs near or above a radius of $0.8R$. The variation has been related to large-scale magnetic-field patterns by Komm *et al.* (1993). Unfortunately, the measurements are separated by too great a time interval to determine whether any of the features propagate, as is predicted by many dynamo theories. There is really no firm evidence for vertical propagation, for example, which is consistent with the finding of Dziembowski & Goode (1993) that the angular momenta of spherical shells appear not to vary in time. Furthermore, the spherically averaged angular momentum density h varies with r continuously across the base of the convection zone (Dziembowski & Goode 1993), despite a near discontinuity in angular velocity Goode & Dziembowski (1993), in agreement with previous evidence presented by Brown *et al.* (1989) and earlier anticipations (Gough 1985). It implies that there is probably no substantial net torque between the convection zone and the radiative interior.

A statistical analysis of the BBSO rotational splitting data, represented by the three odd coefficients a_i, does indicate some significant temporal variation (Gough & Stark 1993a, 1993b). In particular, certain radial averages $\bar{\Omega}(\theta, t)$ of Ω appear to have increased from 1986 to 1988 near latitude 20°, the location of maximum sunspot activity, and near latitudes 50°-60°. A change in $\bar{\Omega}$ in the vicinity of the sunspots is hardly surprising, because it is presumably a region of dynamical activity, though one must be aware that that activity might also have biassed the data. A marked change in Ω at high latitude is also not unexpected, because the moment of inertia is smaller there. Indeed, it is particularly likely in view of the temporal invariance of h, which suggests that transport of angular momentum is principally on spherical surfaces, even in the convection zone. However, it must be borne in mind that the inferences at the pole are less significant because the rotation (considered as a function of latitude) has a lesser influence on the degeneracy splitting. As time progressed beyond 1988, the increase in $\bar{\Omega}$ was found to become significant at all latitudes, except near 30° where the magnitude of the θ-dependent contribution to $\bar{\Omega}$ is relatively insensitive to α_3 and α_5. In addition, a radial trend emerged, augmenting both the negative mean radial gradient in Ω at latitudes above about 30°, and the positive gradient near the equator.

In a recent paper, Woodard & Libbrecht (1993b) have reported an extension of the Legendre expansion of the rotational splitting to six terms, and discuss the variation of $\bar{\Omega}$, now more highly resolved with respect of θ than previous estimates. They too report the greatest mean angular velocity change to be at high latitudes, near 60°; in the equatorial regions the greater latitudinal resolution revealed a rather more complicated pattern of temporal change than was discernible in the three-term expansion. No evidence of latitudinal propagation of $\bar{\Omega}$, as one might expect from the torsional oscillations of LaBonte & Howard (1982), was detected. It will be interesting to see what is revealed by full inversions of the data.

7. Concluding remarks

I conclude with a few further brief remarks about several phenomena raised at this meeting that await explanation. The first is to point out that the frequency shifts depicted in Figure 1 exhibit a rapid decline in the vicinity of the theoretical chromospheric resonance frequency, which is probably a little above 4 mHz. Such a sharp drop was a property of the shifts computed by Gough & Thompson (see Gough 1990a), and resulted from the fact that the structure of the model atmosphere did not change substantially as the subphotospheric layers varied; the frequency shift of chromospheric modes was therefore very small. However, the resonance was sharp, and just above the resonant frequency the shift continued along the original curve. This behaviour does not seem to agree with the rather smoother decline in the observations. Goldreich *et al.* (1991), on the other hand, assume that both chromospheric temperature and fibril fields increase towards solar maximum; modes resonating with the chromospheric cavity are influenced more by the chromospheric temperature rise, and therefore suffer a decline in frequency. The decline is gentler than that found by Gough and Thompson, and can be made to fit the observations very well. Once again, the shift returns to being positive at frequencies above resonance. Recently, Jain & Roberts (1993) have investigated the combined effect of a horizontal chromospheric magnetic field and a chromospheric temperature rise in a simple model. This produces two terms in the polynomial Q in the formula for the frequency shift quoted in Section 4: a linear term from the magnetic field and a cubic term from the temperature change. Thus, as ν increases, dominance switches from changes in the Lorentz forces to changes in sound speed and hydrostatic structure, and the frequency perturbation declines and becomes negative. However, because both mechanisms take place in the same location, there is no return to positive values at higher frequency.

Whilst on the subject of the frequency shifts, I wish to draw attention to the hint in

Figure 1 of an oscillatory deviation of the observations from the smooth curve. This feature was pointed out by Goldreich *et al.* (1991), and is probably the signature of a changing property of the solar structure that is localized in radius and situated at an acoustical depth of about $(2P)^{-1}$ beneath the photosphere (more strictly speaking, beneath the upper turning points of the modes), where P is the "period" (with respect to ν) of the oscillation (Gough 1990a). It is evident from the figure that $P \simeq 680\ \mu$Hz, which corresponds to a region of variation situated about $0.03R$ beneath the photosphere, located in the He II ionization zone where about 60 percent of the helium is completely ionized. This immediately raises the issue of the influence of the magnetic field both on the thermodynamic properties of the gas, principally through its influence on the adiabatic exponent γ via ionization which modifies the variation of sound speed in the ionization zone, and the hydrostatic stratification through its influence on the adiabatic gradient and on the balance of forces, which both modifies the ionization zone and moves it relative to the upper turning points of the acoustic modes. However, in seeking an explanation of the phenomenon, one must be aware that the oscillation might just be a product of the manner in which the data were binned. I must also point out that although most of the variation in the frequencies of the low-degree modes, illustrated in Figures 1 and 2, can be accounted for by the surface phenomena responsible for the shifts observed by Libbrecht and Woodard, it is by no means clear that the whole of the variation can be so explained. If there remains an additional contribution, that would be indicative of a much more deeply seated variation associated with the solar cycle, possibly in the solar core.

Finally, I return to Kuhn's interesting discussion at this meeting. His numerical convection experiments were very instructive, but were not, if I understood correctly, carried out consistently over the entire convection zone. To do so would still be a daunting task, which is why stellar modellers often continue to revert to mixing-length models. And I conclude by doing so myself. In particular, I revisit the point raised by Kuhn that convection transports heat perturbations vertically more readily than it does horizontally, which is why a magnetically induced thermal anomaly starting out at the base of the convection zone is not severely washed out by the time it reaches the photosphere. How is that explained? Let us estimate the fate of a lateral temperature discontinuity. I shall at first temporarily assume that the usual small-eddy-size approximation is valid, even though that is certainly not the case in the lower regions of the zone. Nonetheless, if it were, then the anomaly would rise through the convection zone with the rms convective velocity w (Gough 1981), and suffer lateral diffusion that tends to smooth out the discontinuity. Assuming that, as the anomaly rises, the lateral turbulent diffusion coefficient $\kappa = \beta w l$, where l is the mixing length and β is a dimensionless correlation parameter of order unity, varies slowly compared with the evolution of the discontinuity, one can estimate the scale width of the smoothed discontinuity (at its mid-point) at time t to be $(\pi \int \kappa dt)^{1/2}$. In this simplified discussion, I ignore the dynamical influence of the lateral temperature variation on the convection. Since the disturbance is travelling upwards with speed w, the scale width at the photosphere is approximately $\sigma = (\pi \int \kappa w^{-1} dr)^{\frac{1}{2}}$, where the integral extends from the bottom to the top of the convection zone. If one now sets $l = \alpha H$, where H is the local pressure scale height and α is another constant of order unity, and adopts the approximation $H \simeq 0.4z$, where $z = R - r$ is depth beneath the photosphere, one readily deduces that $\sigma \simeq (0.2\pi\alpha\beta)^{\frac{1}{2}} d$, where $d \simeq 0.3R$ is the depth of the convection zone. Setting $\alpha \simeq 2$ and $\beta \simeq \frac{1}{2}$, say, yields $\sigma \simeq 0.24R$, which subtends an angle of approximately 14° on the solar surface (about the centre of the Sun). Of course this estimate exaggerates the diffusion in any particular realization, because in reality the eddies deep in the convection zone are larger than the characteristic lateral scale of variation of the anomaly, so part of the influence of convection would be more akin to laminar advection than diffusion. In that case an initial near discontinuity would emerge at the surface on a scale less than 14°, but would be moved and distorted within that characteristic angle. If the convection were truly random, the 14° would then represent the scale of a frequency (probability) distribution of realizations. In any case, it is evident from this estimate that a thermal (density) anomaly whose lateral scale exceeds 14° would not be obliterated as it rises to the photosphere. I must emphasize that this estimate relates only to density anomalies, which produce buoyancy and contribute to the dynamics of the convection. A passive scalar diffuses both horizontally and vertically, and would suffer rather more smoothing. Evidently, one of the tasks of the helioseismologists is to test Kuhn's ideas by analysing the even splitting coefficients in search for corrugations in the base of the convection zone. I tried that once using the early data provided by Libbrecht & Woodard (1990), but was unsuccessful. Perhaps the new data will bear fruit.

REFERENCES

BACHMANN, K. T. & BROWN, T. M. 1993 P-mode frequency variation in relation to global solar activity. *Astrophys. J. Lett.* **411**, L45–L48.

BACHMANN, K. T., SCHOU, J. & BROWN, T. M. 1993 Observations of intermediate-degree solar oscillations: 1989 April-June. *Astrophys. J.* **412**, 870–879.

BALMFORTH, N. J., GOUGH, D. O. & MERRYFIELD, W. J. 1993 Structural changes of the Sun during the solar cycle. *Mon. Not. R. astr. Soc.*, submitted.

BROWN, T. M. 1987 Ground-based observations relating to the global state of the Sun. *Solar Radiative Output Variation* (ed. P. Foukal). pp. 176–188. NCAR, Boulder, CO, USA.

BROWN, T. M., CHRISTENSEN-DALSGAARD, J., DZIEMBOWSKI, W. A., GOODE, P. R., GOUGH, D. O. & MORROW, C. A. 1989 Inferring the Sun's internal angular velocity from observed p-mode frequency splittings. *Astrophys. J.* **343**, 526–546.

CAMPBELL, W. R. & ROBERTS, B. 1989 The influence of a chromospheric magnetic field on the solar p and f modes. *Astrophys. J.* **338**, 538–556.

DÄPPEN, W. 1983 Hydrostatic reaction of the Sun to local disturbances. *Astron. Astrophys.* **124**, 11–22.

DELACHE, PH., GAVRYUSEV, V., GAVRYUSEVA, E., LACLARE, F. RÉGULO, C. & ROCA CORTÉS, T. 1993 Time correlation between solar structural parameters: p-mode frequencies, radius, and neutrino flux. *Astrophys. J.* **407**, 801–805.

DUVALL JR, T. L., DZIEMBOWSKI, W. A., GOODE, P. R., GOUGH, D. O., HARVEY, J. W. & LEIBACHER, J. W. 1984 The internal rotation of the Sun. *Nature* **310**, 22–25.

DUVALL JR, T. L., HARVEY, J. W. & POMERANTZ, M. A. 1986 Latitude and depth variation of the solar rotation. *Nature* **321**, 500–501.

DZIEMBOWSKI, W. A. & GOODE, P. 1993 The Sun's internal angular velocity from seismology. *GONG 1992: Seismic Investigation of the Sun and Stars* (ed. T. M. Brown). Astr. Soc. Pac. Conf. Ser. **Vol. 42**, pp. 225–227.

ELSWORTH, Y., HOWE, R., ISAAK, G. R., MCLEOD, C. P. & NEW, R. 1990 Variation of low-order acoustic solar oscillations over the solar cycle. *Nature* **345**, 322–324.

EVANS, D. J. & ROBERTS, B. 1990 The influence of a chromospheric magnetic field on the solar p and f modes, II: uniform chromospheric field *Astrophys. J.* **356**, 704–719.

EVANS, D. J. & ROBERTS, B. 1991 The sensitivity of chromospherically induced p- and f-mode frequency shifts to the height of the magnetic canopy. *Astrophys. J.* **371**, 387–395.

EVANS, D. J. & ROBERTS, B. 1992 Interpretation of solar-cycle variability in high-degree p-mode frequencies. *Nature* **355**, 230–232.

GOLDREICH, P. MURRAY, N. WILLETTE, G. & KUMAR, P. 1991 Implications of solar p-mode frequency shifts. *Astrophys. J.* **370**, 752–762.

GOODE, P. R. & DZIEMBOWSKI, W. A. 1990 Solar-cycle dependence of the Sun's deep internal rotation shown by helioseismology. *Nature* **349**, 223–225.

GOODE, P. R. & DZIEMBOWSKI, W. A. 1993 The Sun's internal rotation during and after the 1986 activity minimum. *GONG 1992: Seismic investigation of the Sun and Stars* (ed. T. M. Brown). Astr. Soc. Pac. Conf. Ser. vol. 42, 217–219.

GOODE, P. R., DZIEMBOWSKI, W. A. KORZENNIK, S. G. & RHODES JR, E. J. 1991 What we know about the Sun's internal rotation from solar oscillations. *Astrophys. J.* **367**, 649–657.

GOODE, P. R. & KUHN, J. R. 1990 Consistency in trends in helioseismic data and photospheric temperature data through the solar cycle. *Astrophys. J.* **356**, 310–314.

GOUGH, D. O. 1981 On the seat of the solar cycle. In *Variations of the Solar Constant* (ed. S. Sofia). NASA Conf. Publ. **2191**, 185–206.

GOUGH, D. O. 1985 Immediate and long-term prospects for helioseismology. *Adv. Sp. Res.* **4**, 85–102.

GOUGH, D. O. 1990 Shaky clues to solar activity. *Nature* **345**, 768–769.

GOUGH, D. O. 1990 Comments on helioseismic inference. In *Progress in Seismology of the Sun and Stars* (ed. Y. Osaki & H. Shibahashi). pp. 283–318. Springer, Heidelberg.

GOUGH, D. O., KOSOVICHEV, A.G., SEKII, T., LIBBRECHT, K. G. & WOODARD, M. F. 1993 Seismic evidence of modulation of the structure and angular velocity of the Sun associated with the solar cycle. In *Inside the Stars* (ed. W. W. Weiss & A. Baglin). Astr. Soc. Pac. Conf. Ser. **Vol. 40**, 93–96.

GOUGH, D. O. & NOVOTNY, E. 1993 On the asteroseismic calibration of a solar-type star. In *Inside the Stars* (ed. W. W. Weiss & A. Baglin). Publ. Astr. Soc. Pac. Conf. Ser. **Vol. 40**, pp. 550–553.

GOUGH, D. O. & STARK, P. B. 1993a Are the 1986-1988 changes on solar free-oscillation frequency splitting significant? *Astrophys. J.* **415**, 376–382.

GOUGH, D. O. & STARK, P.B. 1993b The significance of changes in solar free-oscillation splitting from 1986 to 1990. In *GONG 1992: Seismic Investigation of the Sun and Stars* (ed. T. M. Brown). Astr. Soc. Pac. Conf. Ser. **Vol. 42**, 221–224.

GOUGH, D. O. & THOMPSON, M. J. 1988 Magnetic perturbations to stellar oscillation eigenfrequencies. In *Advances in Helio- and Asteroseismology* (ed. J. Christensen-Dalsgaard & S. Frandsen). pp. 175–180. Reidel, Dordrecht.

JAIN, R. & ROBERTS, B. 1993 Do p-mode frequency shifts suggest a hotter chromosphere at solar maximum? *Astrophys. J.* **414**, 898–907.

KOMM, R. W., HARVEY, J. W. & HOWARD, R. F. 1993 Torsional oscillations and internal rotation. In *GONG 1992: Seismic Investigation of the Sun and Stars* (ed. T. M. Brown). Astr. Soc. Pac. Conf. Ser. **Vol. 42**, pp. 269–272.

KUHN, J. R. 1988a Radial and temporal structure of the internal solar asphericity. In *Seismology of the Sun and Sun-Like Stars* (ed. V. Domingo & E. J. Rolfe). pp. 87–90 ESA SP-286, Noordwijk.

KUHN, J. R. 1988b Helioseismological splitting measurements and the nonspherical solar temperature structure. *Astrophys. J. Lett.* **331**, L131–L134.

KUHN, J. R. 1989a Helioseismic observations of the solar cycle. *Astrophys. J. Lett.* **339**, L45–L47.

KUHN, J. R. 1989b Calculating the internal solar asphericity from frequency splitting measurements. *Solar Phys.* **123**, 1–5.

KUHN, J. R. 1990 Measuring solar structure variations from helioseismic and photometric observations. In *Seismology of the Sun and Stars* (ed. Y. Osaki and H. Shibahashi). pp. 157–162. Springer, Heidelberg.

KUHN, J. R., LIBBRECHT, K. G & DICKE, R. H. 1985 Observations of a solar latitude-dependent limb brightness variation. *Astrophys. J.* **290** 758–764 (erratum **319**, 1010 (1987)

KUHN, J. R., LIBBRECHT, K. G. & DICKE, R. H. 1988 The surface temperature of the Sun and changes in the solar constant. *Science* **242**, 908–911.

KYLE, H. L., HOYT, D. V. & HICKEY, J. R. 1994 A review of the Nimbus-7 ERB solar dataset.

Solar Phys., in press.

LaBonte, B. & Howard, R. F. 1982 Are the high-latitude torsional oscillations of the Sun real? *Solar Phys.* **80**, 373–378.

Libbrecht, K. G. & Woodard, M. F. 1990 Solar-cycle effects on solar oscillation frequencies. *Nature* **345**, 779–782.

Parker, E. N. 1987 Magnetic activity complexes, thermal relaxation oscillations, and the dynamics of the azimuthal magnetic field of a star. In *The Internal Solar Angular Velocity* (ed. B. R. Durney & S. Sofia). pp. 289–301. Reidel, Dordrecht.

Schou, J. 1991 An inversion for the rotation rate of the solar interior. In *Challenges to Theories of the Structure of Moderate-Mass Stars* (ed. D. O. Gough & J. Toomre). pp. 81–86. Springer, Heidelberg.

Schou, J. 1993 On the analysis of helioseismic data. Ph.D thesis, University of Aarhus.

Spruit, H. 1990 Theory of luminosity and radius variations. In *The Sun in Time* (ed. C.P. Sonett, M.S. Giampapa & M.S. Matthews). pp. 118–158. Univ. Arizona Press, Tucson, AZ, USA.

Ulrich, R. K. 1986 Determination of stellar ages from asteroseismology. *Astrophys. J. Lett.* **306**, L37–L40

Willson, R.C. & Hudson, H.S. 1988 Solar luminosity variations in solar cycle 21. *Nature*, **332**, 810–812

Woodard, M. F. & Libbrecht, K. G. 1993a Solar activity and oscillation frequency splittings. *Astrophys. J. Lett.* **402**, L77–L80

Woodard, M. F. & Libbrecht, K. G. 1993b Observations of time variation in the Sun's rotation. *Science* **260**, 1178–1181

Woodard, M. F., Kuhn, J. R., Murray, N. & Libbrecht, K. B. 1991 Short-term changes in solar oscillation frequencies and solar activity. *Astrophys. J. Lett.* **373**, L81–L84

Wright, A. N. & Thompson, M. J. 1992 On the effects of chromospheric magnetic perturbations on solar oscillation frequencies. *Astron. Astrophys.* **264**, 701–707.

Theoretical Interpretation of Magnetic Activity

By EUGENE N. PARKER

Enrico Fermi Institute and Departments of Physics and Astronomy, University of Chicago, Chicago, Illinois 60637, USA

Magnetic fields generated and driven by thermal convection are the primary cause of solar activity. There are many facets of the activity, such as plages, flares, sunspots, coronal heating, and the variation of solar luminosity or irradiance whose nature and cause are understood only partially or not at all, although detailed superficial observational descriptions are available. It is suggested that the inferred 10^5 gauss azimuthal field bundles may be a direct result of the emergence of Ω–loops to form bipolar magnetic regions on the surface in association with an increase in solar irradiance.

1. Solar activity

The surface of the Sun exhibits a variety of suprathermal phenomena collectively called *activity*. Close examination shows that the activity is mostly magnetic in character, driven by the interaction of convection with magnetic fields that are generated somewhere beneath the visible surface. It is evident, then, that the theory of magnetic activity begins with the perplexing problem of the solar dynamo, generating azimuthal magnetic field with a 22 year periodicity somewhere in or below the convective zone.

The magnetic suprathermal effects observed at the surface of the Sun include faculae, plages, prominences, coronal mass ejections, and flares of all sizes. Then there is the subthermal magnetic sunspot to give variety. The chromosphere, and particularly the corona, are the outward manifestations of the activity. In regions where the mean magnetic field is relatively weak (< 10 gauss) the coronal gas is heated to a temperature of $1 - 1.5 \times 10^6$K and continually expands outward into space, gradually accelerating to hundreds of km/sec to produce the solar wind. The expansion carries the weak field along with it, sweeping out through the solar system to produce the heliosphere whose outer boundary, the heliopause, is estimated to lie at 10^2 AU or beyond. So in the large the activity of the Sun dictates the conditions throughout the planetary system. Even the powerful galactic cosmic ray gas is swept away to some degree by the outward rush of magnetic field in the solar wind.

The radiative output of the Sun is modulated in important ways by the magnetic activity. It has been known for many decades that the electromagnetic radiation from the Sun, which has its peak intensity in yellow light, has a UV excess that extends into X-rays as a consequence of the superheated transition layer and corona. Such radiation is the product of the suprathermal activity, and its intensity varies widely with the changing level of magnetic activity. The X-ray intensity at 10 keV fluctuates by several decades over the magnetic cycle, with transient excursions up to a factor of 10^{10} at the time of a large flare (cf. Dennis 1985). The UV varies by several percent with the magnetic cycle.

The most remarkable aspect of the magnetic activity of the Sun is the variation $\Delta L_\odot$ of the total luminosity $L_\odot$ (the irradiance, as observed near the equatorial plane of the Sun) in proportion to the general level of activity. Direct observation by NIMBUS 7 shows a decrease by 2 parts in 10^3 between 1979 and 1986 (Hoyt *et al.* 1992). The work of Radick *et al.* (1990) and Zhang *et al.* (1993) shows that there is a universal quantitative ΔL – activity relation for middle aged solar–type stars. Foukal & Lean (1986) show that a large part of $\Delta L_{\odot}$ results from the competition between plages and sunspots, with the plages dominating so that the mean $L_{\odot}$ increases with the activity.

Now there are two different aspects to the problem of solar activity that should be clearly understood. First of all, although the magnetic activity has been studied for many decades, and has been described at length with an extensive vocabulary of technical terms, there is only a limited understanding of the physics of the various aspects of the activity. For instance, there is a variety of ideas on the form of the solar dynamo. The nature of the plage, the general fibril state of the magnetic field, the clustering of fibrils to form sunspots, and the basic reasons for $\Delta L_{\odot}$ are not clearly understood. That is to say, one cannot yet establish from the principles of physics why a star like the Sun is obliged to produce these phenomena. The heat sources that create the active X-ray corona and the quiet corona and the coronal hole have not been established for lack of the crucial observations of the small-scale behavior of the fibril magnetic field. The scientific program is seriously handicapped by the seeing limitations of the ground based telescope, and by the curious reluctance of NASA to fly a 1m diffraction limited telescope either in orbit or with a balloon, to resolve the critical aspects of the activity at the Sun.

Second, there is enough known of the variation of the solar output, from IR to X-rays, to recognize the broad impact on the terrestrial atmosphere. The chemistry and general state of the upper atmosphere and the ionosphere are driven and controlled by the varying UV and X-rays. The ozone-UV problem is widely known and is an example of the complicated chemical interaction of anthropogenic pollutants in the variable UV irradiance. More recently the observational determination of a substantial $\Delta L_{\odot}$ indicates that the recorded century by century terrestrial forays into hot and cold climate in the northern temperate zone are evidently connected with the changing activity level and the corresponding change in the luminosity of the Sun. The human consequences of the climatic extremes are well recorded (Eddy 1976; 1977; Wigley & Kelly 1990; Friis–Christensen & Lassen 1991) from the agricultural disasters in northern climes during the cold 15th and 17th centuries to the devastating dessication of southwestern United States in the warm 12th century. The scientific study of the effects should, therefore, be recognized as the central concern of the worldwide space science program.

It must be appreciated that the study of the variable activity and luminosity of the Sun extends across diverse fields of scientific inquiry. The oceanic and atmospheric sciences are directly involved, of course, at all levels through the upper atmosphere. Aeronomy and ionospheric physics and chemistry play a central role, which also involves magnetospheric convection and activity. The system is driven by changes across the solar spectrum, and in some degree by the solar wind and momentarily by transient bursts of energetic particles from the Sun. Therefore it poses a central problem for solar physics to quantify and explain the variability of the thermal, suprathermal and particle emissions from the Sun while the climatic consequences are studied by the geophysical and atmospheric sciences. The scientific community should be the first to recognize the many dimensions and the central priority of the *solar variability – terrestrial environment* problem, now that it is known to be a major player in the climate in both the upper and lower levels of the terrestrial atmosphere.

The remainder of this paper is directed to recent questions about the solar dynamo and the requirement of flux bundles of 10^5 gauss in the overshoot layer below the convective zone in order to account for the location and orientation of the active bipolar magnetic regions at the surface of the Sun. The inquiry leads to a curious barometric effect in the buoyant rise of Ω–loops of azimuthal field to the surface, and suggests that the increased luminosity may be an automatic consequence of the emergence of the active regions.

2. Origins of solar activity

The magnetic field of the Sun, created and driven by the subsurface convection, is the basis for solar activity. It is a matter of conjecture whether the fibril state of the magnetic field observed at the surface implies a fibril state throughout the entire convective zone. (cf. Parker 1984; Zwaan 1985; see also Solanki 1993). The origin of the magnetic field appears to be an $\alpha\omega$ dynamo (Parker 1955, 1957). Helioseismology indicates that the principal shear (nonuniform rotation, $\partial\omega/\partial r$) is confined to a thin layer immediately below the base of the convective zone. Therefore it is presumed that the azimuthal field resides mainly in that layer. Dynamical studies of azimuthal field in the sheared subadiabatic temperature gradient below the convective zone (Spruit & Van Ballegooijen 1982; Fisher *et al.* 1991; Moreno-Insertis *et al.* 1993) indicate that azimuthal fields up to about 10^5 gauss can be stably confined, but that, when the field strength exceeds some such magnitude, it becomes unstable to the formation of Ω–loops. It is presumably these Ω–loops that emerge through the surface of the Sun to provide the observed bipolar magnetic active regions (Parker 1955). It is interesting to note that recent calculations by D'Silva & Choudhuri (1993) and Fan *et al.* (1993, 1994) on the dynamics of buoyant flux bundles show that Ω loops emerge through the surface at the proper latitude with the observed inclination to the east–west direction only if the emerging Ω–loops are formed from fields of $0.5 - 1 \times 10^5$ gauss. Evidently, then, the conditions for the onset of buoyant instability in the shear layer determine the observed location and orientation of the bipolar magnetic regions at the surface.

Now the field intensity of 5×10^4 gauss does not follow from the available solar dynamo effects. To be precise, a nonuniform rotation of the order of 3×10^4 cm/sec across a shear layer of thickness $h = 3 \times 10^9$ cm below the convective zone yields

$$\begin{aligned} G &\equiv dv/dr \\ &\cong 10^{-5}/\text{sec.} \end{aligned}$$

The rate of generation of azimuthal field B_φ from a vertical poloidal field B_p is

$$\frac{\partial B_\varphi}{\partial t} = GB_p$$

so that if B_φ is to reach 5×10^4 gauss in 3 years it follows that $B_p = 50$ gauss. The Maxwell stress $B_\varphi B_p/4\pi \cong 2 \times 10^5$ dynes/cm^2 would spread out the differential rotation from h to $2h$ in just half a year. Evidently, then, the $\alpha\omega$–dynamo of the Sun generates a mean azimuthal field of the order of a couple of kilogauss (Parker 1993), necessary to account for the 10^{23} Maxwells that appear in large bipolar magnetic regions at the surface (Gaizauskas *et al.* 1983). Some other effect concentrates the mean field into widely separated intense fibrils of $0.5 - 1 \times 10^5$ gauss.

We suggest that the concentration of the field is an automatic byproduct of the formation of Ω-loops and their emergence through the visible surface of the Sun. It appears, too, that the formation of Ω–loops may increase the vertical mixing of gas and entropy sufficiently as to account for the increase in the solar luminosity observed to accompany the magnetic activity of emerging Ω–loops. The effect is a simple one. A buoyant azimuthal flux bundle gives rise to an Ω–loop, driving a convective updraft all the way to the surface of the Sun. The essential point for concentrating the field is that the temperature within the rising Ω–loop increases by the amount $\Delta T(z)$ above the ambient $T(z)$ at the same height z by the amount

$$\Delta T(z) = \int_0^z dz \Delta \frac{dT}{dz} \tag{2.1}$$

where $\Delta dT/dz$ is the excess of the ambient temperature gradient $|dT/dz|$ over the nearly adiabatic temperature gradient of the rising fluid. The effect is, then, for the hotter gas within the field of the Ω–loop to billow upward within the flux bundle, drawing gas up behind it from the feet of the Ω–loop and hence from the azimuthal flux bundle in which it is formed. This partial evacuation of the azimuthal flux bundle shrinks the flux bundle and intensifies the field in the bundle. The limiting value of the field B_{max} is determined by the limiting value of the pressure reduction

$$\Delta p(0) \cong p(0) \int_0^z \frac{dz}{\Lambda(z)} \frac{\Delta T(z)}{T(z} \tag{2.2}$$

in the base of the Ω–loop, where $\Lambda(z)$ is the pressure scale height kT(z)/Mg.

To deduce this result formally, consider the momentum equation for a vertical flux bundle extending upward from the base $z = 0$ of the convective zone. Then the fluid velocity v along the bundle satisfies

$$\frac{dv}{dt} = -\frac{1}{\rho}\frac{\partial p}{\partial z} - g \tag{2.3}$$

where we may as well approximate the gravitational acceleration by a constant value of about 3×10^4 cm/sec^2. Multiply by M/kT and integrate upward from $z = 0$. The result can be written

$$p(z) = p(0) \exp\left[-\int_0^z \frac{ds}{\Lambda(s)} - \int_0^z \frac{ds}{\Lambda(s)} \frac{1}{g}\frac{dv}{dt}\right]. \tag{2.4}$$

The relation can be applied to the gas within the vertical flux bundle, where $dv/dt \neq 0$ and where the gas temperature is $T(z) + \Delta T(z)$, and it may be applied to the ambient gas outside $(dv/dt \cong 0)$ which has a temperature $T(z)$. Thus, for the ambient pressure outside,

$$p(z) = p(0) \exp\left[-\int_0^z \frac{ds}{\Lambda(s)}\right]. \tag{2.5}$$

Inside the flux bundle write $\Lambda(z) + \Delta\Lambda(z)$ for the scale height, with $\Delta\Lambda(z) = k\Delta T(z)/Mg$. Note, then, that $\Delta\Lambda/\Lambda$ and $g^{-1}dv/dt$ are both small compared to one, so that to first order, the pressure $p_i(z)$ inside the bundle is

$$p_i(z) \cong \frac{p_i(0)}{p(0)} p(z) \left[1 + \int_0^z dz \frac{\Delta\Lambda}{\Lambda^2} - \frac{1}{g}\int_0^z \frac{dt}{\Lambda}\frac{dv}{dt}\right]. \tag{2.6}$$

This equation may now be used to compute

$$\Delta p(z) = p(z) - p_i(z)$$

in terms of $\Delta p(0)$, yielding

$$\frac{\Delta p(0)}{p(0)} = \frac{\Delta p(z)}{p(z)} + \int_0^z \frac{dz \Delta\Lambda(z)}{\Lambda^2(z)} - \frac{1}{g}\int_0^z \frac{dz}{\Lambda(z)}\frac{dv}{dt} \tag{2.7}$$

upon neglecting terms second order in $\Delta p/p$.

The condition for local transverse equilibrium of the slender flux bundle is $B^2(z)/8\pi = \Delta p(z)$, of course, so that

$$\frac{B^2(0)}{8\pi} = p(0)\left\{\frac{B^2(z)}{8\pi p(z)} + \int_0^z \frac{dz\Delta\Lambda(z)}{\Lambda(z)^2} - \frac{1}{g}\int_0^z \frac{dz dv/dt}{\Lambda(z)}\right\}. \tag{2.8}$$

In the simplest case $dv/dt = 0$ and the field at the apex $z = \lambda$ of the Ω–loop expands so much that the first term in braces can be neglected. There remains the basic barometric term

$$\frac{B^2(0)}{8\pi} = p(0) \int_0^\lambda \frac{dz}{\Lambda(z)} \frac{\Delta T}{T(z)}. \tag{2.9}$$

A standard convective zone model (Spruit 1974) gives $\Lambda(0) = 7 \times 10^9$ cm at the bottom of the convective zone ($z = 0$) where $p(0) = 6 \times 10^{13}$ dynes/cm^2 and $T(0) = 2 \times 10^6$K. At the midlevel ($z = 10^{10}$ cm) $p = 6 \times 10^{12}$ dynes/cm^2, $T \cong 10^6$K, and $\Lambda \cong 4 \times 10^9$ cm, with $\Delta T \cong 1K$. At the three quarter level ($z = 1.5 \times 10^{10}$cm) $p = 10^{12}$ dynes/cm^2 $T = 4 \times 10^5$, and $\Delta T \cong$ 6K. We find at the three quarter level that the integral of $\Delta\Lambda/\Lambda^2$ reaches a value of 10^{-5} and increases rapidly upward from there with rapidly increasing ΔT and decreasing Λ. Thus the limiting pressure reduction is $\Delta p(0) \cong 10^{-5} p(0) \cong 6 \times 10^8$ dynes/cm^2, equivalent to $B(0) \cong 1.2 \times 10^5$ gauss. So the Ω–pumping is capable of limiting fields B_{max} of the required order of 10^5 gauss.

One imagines that the process begins with azimuthal fields of a few kilogauss at middle latitudes where the nonuniform rotation $d\omega/dr$ is not very strong. The convective overshoot may initiate an Ω–loop in spite of the stability of the mean field. The Ω–pumping begins, eventually concentrating many, if not all, of the azimuthal flux bundles into intense separated magnetic fibrils of 0.5×10^5 gauss or more.

The process is certainly more complicated than this simple idealized example. For instance the term $B^2(s)/8\pi p(s)$ on the right hand side of equation (8) can only increase the effect, perhaps substantially. On the other hand, the third term on the right hand side represents inertial effects and reduces the effect where dv/dt is positive (upward). In fact dv/dt may have either sign, but there is no reason to think that the effect is large either way. For instance, the upward motion v of an Ω–loop is presumably of the order of 2×10^4 cm/sec, covering the thickness 2×10^{10} cm of the convective zone in a time $\tau = 10^6$ sec. The characteristic mean upward acceleration is, then, $2v/\tau = 4 \times 10^{-2}$ cm/sec^2. Hence $2v/g\tau = 1.3 \times 10^{-6}$ and is integrated over only a few scale heights, so that the integral may approach 10^{-5}. Remembering that the vertical motion decelerates to zero at the surface, where there are many scale heights, the net effect of the inertia is not only small but may actually contribute to the net effect as the Ω–loop approaches the surface.

3. Summary

The emergence of Ω–loops from the azimuthal field to provide the magnetic activity at the surface of the Sun poses a problem as to what effect can convert the mean kilogauss azimuthal field in the overshoot-shear layer into a fibril field with individual intensities of the order of 10^5 gauss. It looks as though it may be the emergence of the Ω–loops themselves that is responsible, through the Ω–pumping effect. Moreno-Insertis & Schüssler (1990) have pointed out an intensification of the vertical legs of Ω–loops in their numerical experiments on the dynamics of Ω–loop, resembling the Ω–pumping described above.

As a final point, the vigorous buoyant rise of Ω–loops, pushing all the way through from the bottom to the top of the convective zone, may be expected to enhance the total upward convective heat transport by substantial amounts. It may be an important factor in producing the enhanced solar luminosity during the years of high activity and clearly needs quantitative consideration.

Acknowledgments. This work was supported in part by the National Aeronautics and Space Administration under NASA grant NAGW-2122.

REFERENCES

DENNIS, B.R. 1985 Solar hard X–ray bursts. *Solar Phys.* **100**, 465–490.

D'SILVA, S. & CHOUDHURI, A.R. 1993 A theoretical model for tilts of bipolar magnetic regions. *Astron. Astrophys.* **272**, 621–633.

EDDY, J.A. 1976 The Maunder Minimum. *Science* **192**, 1189–1202.

EDDY, J.A. 1977 Historical evidence for the existence of the solar cycle. In *Solar Output and its Variation* (ed. O.R. White). pp. 51–77. Colorado Association University Press.

FAN, Y., FISHER, G.H. & DELUCA, E.E. 1993 The origin of morphological asymmetries in bipolar active regions. *Astrophys. J.* **405**, 390–401.

FAN, Y., FISHER, G.H. & MCCLYMONT, A.N. 1994 *Astrophys. J.*, submitted.

FISHER, G.H., MCCLYMONT, A.N. & CHOU, D.Y. 1991 The strerching of magnetic flux tubes in the convective overshoot region. *Astrophys. J.* **374**, 766–772.

FOUKAL, P. & LEAN, J. 1986 The influence of faculae on total solar irradiance and luminosity. *Astrophys. J.* **302**, 826–835.

FRIIS–CHRISTENSON, E. & LASSEN, K. 1991 Length of the solar cycle: An indicator of solar activity closely associated with climate. *Science* **254**, 698–700.

GAIZAUSKAS, V., HARVEY, K.L., HARVEY, J.W. & ZWAAN, C. 1983 Large-scale patterns formed by solar active regions during the ascending phase of cycle 21. *Astrophys. J* **265**, 1056–1065.

HOYT, D.V., KYLE, H.L., HICKEY, J.R. & MASCHNOFF, R.H. 1992 The Nimbus 7 total irradiance: A new algorithm for its deviation. *J. Geophys. Res.* **97**, 51–63.

MORENO–INSERTIS, F., SCHÜSSLER M., & FERIZ–MAS, A. 1984 Private communication.

MORENO–INSERTIS, F., SCHÜSSLER, M. & FERIZ–MAS, A. 1993 *Astron. Astrophys.*, submitted.

PARKER, E.N. 1955 The formation of sunspots from the solar toroidal field. *Astrophys. J.* **121**, 491-507.

PARKER, E.N. 1984 Stellar fibril systems: I. Reduced energy state. *Astrophys. J.* **283**, 343–348.

PARKER, E.N. 1993 A solar dynamo surface wave at the interface between convection and nonuniform rotation. *Astrophys. J.* **408**, 707–719.

RADICK, R.R., LOCKWOOD, G.W. & BALIUNAS, S.L. 1990 Solar activity and brightness variations: A glimpse at the Sun's history. *Science* **24**, 39–44.

SOLANKI, S.K. 1993 Small-scale solar magnetic fields: An overview. *Space Sci. Rev.* **63**, 1–186.

SPRUIT, H.C. 1974 A model of the solar convective zone. *Solar Phys.* **34**, 277–290.

SPRUIT, H.C. & VAN BALLEGOOIJEN, A.A. 1982 Stability of toroidal flux tubes in stars. *Astron. Astrophys.* **113**, 350.

WIGLEY, T.M.L. & KELLY, P.M. 1990 Holocene climatic change, ^{14}C wiggles and variations in solar irradiance. *Phil. Trans. Roy. Soc. London* **A330**, pp. 547–560.

ZHANG, Q., SOON, W.H., BALIUNAS, S.L., LOCKWOOD, G.W., SKIFF, B.A. & RADICK, R.R. 1993 Possible method for determining brightness variations of the Sun in past centuries from solar type stars *Astrophys. J. Letters*, submitted.

ZWAAN, C. 1985 The emergence of magnetic flux. *Solar Phys.* **100**, 397–414.

Theoretical Interpretation of Solar and Stellar Irradiance Variations

By HENK C. SPRUIT

Max-Planck-Institut für Astrophysik, Postfach 1523, D-85740 Garching bei München, Germany

The main cause of variability of solar type stars are their varying magnetic fields. To compute irradiance variations one has to compute the magnetic field (the dynamo problem), and from this the irradiance effects. The second problem is considered here. The theoretical work of the past decade has shown that the dominant effect of magnetic fields is a surface effect: a change of effective emissivity of the magnetic parts of the surface while the nonmagnetic part of the surface contributes very little to the irradiance variation on almost all time scales. No other processes have yet been found that would cause variations exceeding (at the current level of magnetic activity) the observed 0.1% irradiance fluctuation of the Sun. This implies that a knowledge of the surface magnetic fields [separated into its bright small scale (faculae, network) and dark large scale (spots) components] is sufficient for pre- or postdicting the solar irradiance. It is hypothesized that the discrepancy remaining between the measured irradiance variations and values reconstructed from proxies is due to the difficulty of finding a proxy that accurately correlates with the continuum contrast of a dispersed small scale magnetic field. Stellar structure theory predicts that the variations in the solar radius associated with magnetic activity are quite small. For stars, color and brightness variations should primarily be interpreted in terms of variations in the fraction of the surface covered by magnetic patches. Their (long term) displacement from the main sequence is not very large.

1. Introduction

The solar luminosity varies on a wide range of time scales due to several processes. It increases as Hydrogen burns into Helium in the core. This increase is climatologically relevant ($\sim$ 30% over the past $4.6\,10^9$ yr), but does not cause changes on human time scales (see review by Kasting & Grinspoon 1991). The Sun oscillates in a large number of sound modes, but their combined effect on irradiance is small ($\sim 10^{-5}$ level, Fröhlich 1990; Fröhlich *et al* 1991) and confined to short time scales (minutes). The random pattern of the granulation at the surface causes statistical noise at an amplitude of a few times 10^{-5} and similar time scales. Larger scale convective flows are known to exist and have longer time scales, but their effect on irradiance probably does not exceed that of the granulation. This leaves magnetically related variations as the main known cause of variability on all time scales from hours to millions of years.

1.1. *Luminosity modulation by magnetic fields*

To make a successful theory of luminosity variations by magnetic fields one needs two ingredients. A theory is needed that tells us how much magnetic field is present and where, and given the distribution of magnetic fields, a theory to compute the resulting luminosity variations, during the Sun's magnetic cycle. The first of these questions is the *dynamo problem.* Dynamo theories have a long tradition and very extensive literature, but unfortunately have not progressed yet to the stage that the location and strength of the magnetic field in a star can be predicted. Hence one is forced to rely mostly on observations instead of theory. The second question turns out to be very much simpler. Once the magnetic fields are known, the problem of luminosity variations reduces to a calculation of thermal perturbations in which the sources are known. The theory of stellar structure is sufficiently well developed and tractable to produce reliable answers for this problem. These answers, and the underlying physics are the subject of the following sections.

Conceptually, it is easiest to separate the thermal effects of magnetic fields into two classes:

1. 'Sources and sinks'. The generation of a magnetic field involves the conversion of kinetic energy of motion into magnetic energy. Since the motions in the solar envelope are thermally driven, this ultimately means conversion of thermal into magnetic energy: building up a magnetic field produces a thermal sink somewhere. The opposite happens when the field decays. These thermal effects exist only during *changes* in the magnetic energy content of the envelope.

2. 'Changes in thermal transport coefficients'. Magnetic fields interfere with convection, causing a reduction in the efficiency of heat transport in the envelope. In contrast to (1), these changes last as long as the magnetic field itself is present.

Both these sources and sinks and changes in the transport coefficient cause thermal perturbations, varying with magnetic activity, which propagate through the envelope and cause variations in surface energy flux. Related to the second class of perturbations are the effects of magnetic fields at the surface of the star. Starspots, being dark, radiate less than the surrounding photosphere, while the small elements that make up plage and the network have an excess emission.

In the following I will refer mainly to the Sun, questions specifically relating to stars are discussed in Section 5.

2. Time scales

The response of the Sun to thermal perturbations is not governed by a single time scale, but by a wide range of time scales. The longest of these is the thermal time scale of the Sun as a whole, called the Kelvin-Helmholtz time scale, about 10^7 years. If the central heat source of the Sun were switched off, the internal structure and the luminosity would start to change on this time scale. More generally, we can define the thermal time τ_t scale as a function of depth:

$$\tau_t \equiv U(z)/L(z) \approx \frac{1}{L}\int_{R-z}^{R} 4\pi r^2 u dr, \tag{2.1}$$

where L is the luminosity at depth z, U the thermal energy of the envelope down to a depth z, and u the thermal energy per unit volume, approximately (ideal gas of constant γ) given by $u = P/(\gamma - 1)$. Some rough values for this quantity are $\tau_t \sim 10^5$yr at $z = 2\,10^5$km (depth of the convection zone), 10 yr at 20 000km (the size of a supergranule), 20 hrs at 2000km (size of a granule). This shows that the thermal time scale is a very sensitive function of depth in the Sun. As a result, the thermal response of the Sun also depends critically on the location of the disturbance. A fairly good approximation to the stratification of the convective envelope of the Sun is a polytrope of index $n = 2$ (this is better than the standard 'convective' value n=1.5 because $\gamma < 5/3$ due to partial ionization). Hence the gas pressure and thermal time scale vary roughly as $P \sim z^3$, $\tau_t \sim z^4$, where the depth z is counted from a level 3 scale heights above the surface $\tau = 1$.

A second kind of time scale involved in thermal readjustments is the *diffusive* time scale. In the mixing length approximation transport in the convection zone can be computed with a turbulent diffision coefficient $\kappa_t \approx \frac{1}{3} l_c v_c$ where l_c and v_c are the convective length scale and velocity. The time scale on which thermal inhomogeneities (more precisely:

entropy perturbations) are smoothed on a length scale d by turbulent diffusion is

$$\tau_d = d^2/\kappa_t. \tag{2.2}$$

For $d = 2\,10^5$km this is about 1 yr, for $d = 2000$km about 1 hr. Comparing τ_t and τ_d, we see that they are of similar magnitude close to the surface (to be precise: in the surface boundary layer where convection is not efficient enough to keep the stratification close to adiabatic). In deeper layers, the thermal time scale is much longer than the diffusive time scale, by a factor of up to 10^5. The two time scales measure different types of thermal adjustment process. These same processes appear in the thermal behavior of, say, a spacecraft. Let me approximate a satellite as a chunk of aluminum with a heater in it. The thermal time scale is the time scale on which its temperature adjusts to a change in the heat input, and is determined by the heat capacity [U in eq. (2.1)] and the power level (L). The time scale on which different parts of the chunk equilibrate to the same temperature is governed by a different process, namely the thermal conduction (the equivalent of the turbulent diffusion in the Sun). The diffusion (conduction) time scale is much shorter than the thermal time scale, because of the large conductivity of Al. In the Sun, it is the very large turbulent diffusivity in the bulk of the convection zone that causes the very short diffusive time scale compared with the thermal time scale.

How do these different time scales come into play when the convection zone is thermally perturbed by, say, the storage of energy in a growing magnetic field? Such perturbations can be computed in detail, either by numerical methods (Endal *et al.* 1985) or more analytically. We can, for example, consider the initial value problem in which a perturbation is allowed to evolve in time by thermal transport in the convection zone. In general this evolution has components on all the time scales of the problem, including the very long thermal time scale. Formal aspects of this problem are discussed briefly in the next subsection (for a more complete analysis, see Spruit 1982a,b, 1991; Arendt 1992). For quantitative results, skip to Section (2.2).

2.1. *Modes of thermal relaxation*

The problem can be studied in detail within the mixing length approximation for convective energy transport:

$$\mathbf{F} = -\kappa\rho T\nabla S, \tag{2.3}$$

where $\mathbf{F}$ is the heat flux, κ the turbulent diffusivity, ρ the density, T the temperature and S the entropy. The energy equation, in the absence of sources, is given in terms of the entropy by

$$\rho T\frac{\mathrm{d}S}{\mathrm{d}t} = -\mathrm{div}\mathbf{F}. \tag{2.4}$$

As an example, consider the one dimensional problem in which all quantities depend on depth z only. In this case, combining (2.3) and (2.4), and neglecting flows we get

$$\frac{\partial S}{\partial t} = \frac{F_z}{\rho TH} + \kappa\frac{\partial S}{\partial z^2}, \tag{2.5}$$

where $H = \partial\ln\rho T/\partial z^{-1}$ is the pressure scale height. For perturbations in which the left hand side is balanced by the first term on the right, the time scale is of the order $F/(c_p\rho TH)$, which is just a thermal time scale (cf. 2.1). For modes in which the left hand side is balanced by the second term, we evidently have an ordinary diffusion problem, hence these modes evolve on diffusive time scales (cf. 2.2). The first term appeared as a consequence of the stratification of pressure, ie. the inhomogeneity of the convection zone. In a homogeneous medium (as in a chunk of Al), the thermal time scale comes in only through the radiating surface boundary condition (which one could also consider as a form of inhomogeneity). If T_s is the surface temperature, this condition is

$$\sigma T_s^4 = F_z. \tag{2.6}$$

Since the perturbations in entropy outside spots and the magnetic elements making up the small scale field are small, the reponse of the convection zone can be computed by linearizing eqs. (2.3),(2.4),(2.6). An arbitrary initial perturbation may then be decomposed in the standard way in terms of the eigenmodes of the problem. These are found by setting $\delta S = e^{\eta t} f(\mathbf{r})$, and solving for $f(\mathbf{r})$ with the decay rate η as eigenvalue (the problem thus defined has only decaying solutions). Specializing to the one dimensional case ($f(z)$) one finds a series of modes ordered by the number of nodes n of f. The fundamental ($n = 0$) has a decay rate of the order $\eta_0 \sim \tau_t^{-1}$ that is, this mode decays on the thermal time scale. The higher modes decay with rates of the order $\eta_n \sim n^2 \tau_d^{-1}$, they decay on diffusive time scales.

3. Effect of subsurface magnetic fields on luminosity

In this section I consider the effect on luminosity of magnetic fields in the convection zone which vary on the time scale $t_c \approx 10$yr of the solar cycle (for a discussion of more general perturbations and for details of the calculations, see Spruit 1982a,b 1991). The best current guess is that most of the flux, of the order $10^{23} - 10^{24}$ (Zwaan 1978) Gcm^2 is located near the base of the convection zone, and has a strength of the order 50–100 kG (D'Silva & Choudhuri 1993; D'Silva & Howard 1993). These quantities are somewhat model dependent.

To reduce the dependence on these models, let me keep the depth z_b at which the field is located as a free parameter. To obtain upper limits to the luminosity effects, assume that the field at z_b covers the entire horizontal surface, and covers a depth range of one pressure scale height H (this is sufficiently general for the present purpose, since $H(z_b)$ is a significant fraction, of the order 1/3, of z_b itself). The convection zone, being nearly isentropically stratified, offers no restoring force to the motion of the magnetic field. If the field is in temperature equilibrium with its surroundings, the gas density in it is less, causing magnetic *buoyancy*. The speed at which the field starts rising under the influence of this force is of the order of the Alfvén speed. If the field is to remain inside the convection zone for a significant fraction of the cycle time, this yields the so-called buoyancy limit on the field of the order of 100G (Parker 1984). Buoyancy can be avoided by creating the field with a slight temperature deficit, in such a way that it is neutrally buoyant. Such fields are found to be unstable however, on an Alfvén travel time scale (Parker 1979, 1984; Spruit & van Ballegooijen 1982). This again limits the field to the buoyancy limit. This restrictive limit can be relaxed a bit by noting that convective downflows exist that are larger than the Alfvén speed at 100G. In principle, these flows can be used to prevent a field from rising through the convection zone. Making this assumption, the upper limit to the field becomes equal to the *equipartition* value, at which the energy density in the field equals the kinetic energy of the convective flows. This value increases with depth, to a maximum of about 10 000G near the base of the convection zone.

With these values for the strength, extent and location of the field, the predicted upper limits to fluctuations in luminosity can be estimated using the formalism of Section (2.1). This is shown in Figure 1. The results depend on the nature of the perturbations. The result of conversion between thermal and magnetic energy is called a 'β-perturbation', that caused by a reduction of the convective energy transport efficiency an 'α-perturbation'

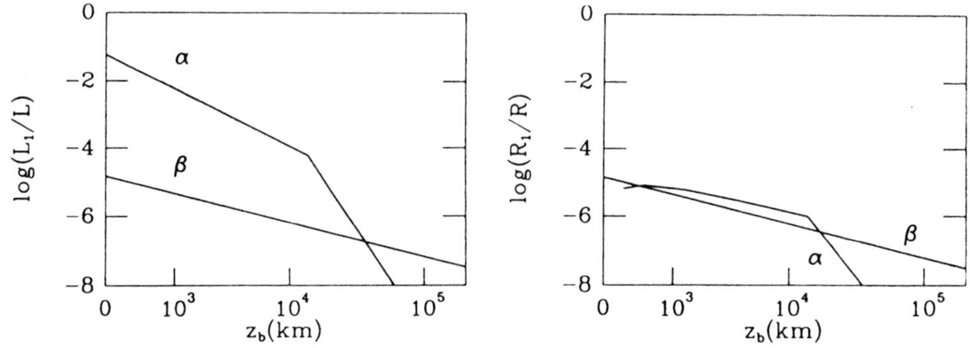

FIGURE 1. Upper limit on luminosity and radius variations on a solar cycle time scale caused by magnetic fields in the convection zone, as a function of their assumed depth z, for fields limited by equipartition with convection

(Endal *et al.* 1982). In computing the α-perturbations, it is assumed that the reduction of convective efficiency by the field is of the order unity if the field strength is equal to the equipartition value (Spiegel & Weiss 1980). From Figure 1 it is clear that α-perturbations are the more effective ones (Gilliland 1988) except near the base of the convection zone. This has to do with the fact that the field can interfere with the convective flow as long as it exists, whereas the β-effect is limited by the magnetic energy content of the field. [Near the base the relative effectiveness of the two is reversed because convection there is so efficient that a reduction by the field has little effect]. The results in Figure 1 agree qualitatively with results obtained from perturbed stellar structure models (e.g. Endal *et al.* 1982; Gilliland 1988).

Figure 1 shows that luminosity variations on the order of the observed 0.1% irradiance variation can not be caused by subsurface magnetic fields unless they are located at a depth less than a few thousand km. This is so close to the surface that observations probably give a better estimate of the magnetic field at this depth than the theoretical limits used above. The surface field is nearly vertical, 1–3kG, and generally has a filling factor of the order 1% or less. Because of the low filling factor (compared with the 100% filling factor assumed so far), such fields have much smaller effect than the fields assumed above (see Section 4).

At the base of the convection zone, the upper limit on the field strength based on the buoyancy argument is not valid, since the radiative core is quite stably stratified, and acts as a nearly solid surface towards motions in the convection zone. Fields close to this surface could plausibly be higher than the equipartion value. Ignoring the buoyancy limit, the next most plausible upper limit to the field strength is obtained by assuming that the entire luminosity of the Sun is spent in building up a magnetic field near the base of the convection zone during the ascending phase of the solar cycle, and this same amount released again during the descending phase. One finds that the luminosity effect in this case is $\delta L/L < 10^{-5}$ (for β-perturbations, which are the most effective ones in this case). This perhaps surprising result has to do with the very large heat capacity of the convection zone, related to its thermal time scale of 10^5yr.

The variations in solar radius predicted are small. The maximum values in Figure 1 may be observable, but if we take the current estimate (see above) that most of the field is located near the base of the convection zone, a plausible estimate is $\delta R/R \sim 10^{-7}$. The vertical magnetic fields resulting from eruption of active regions have an additional effect on δR, but this is even smaller. These radius variations are defined as changes of the *nonmagnetic* part of the solar surface. In spots and small magnetic elements, the $\tau = 1$ surface is depressed (Wilson depression). If one were to compute a mean geometric radius including these depressions, radius changes up to 10^{-5} (a reduction at solar maximum) would result. This mean radius is not directly observably however, and the radius of the nonmagnetic photosphere is, also observationally, the more meaningful quantitity.

4. Effect of surface magnetic fields

The strongest effects on luminosity are obtained from fields at the radiating surface itself (cf. Figure 1). This is mostly a reflection of the steep depth dependence of the thermal inertia of the convection zone. The surface effects are caused by changes in the efficiency of energy transport associated with the field. A reduction of the convective energy flux due to the resistance of magnetic field lines against bending causes sunspots to have a (bolometric) surface emissivity of about 20-25% of that of the normal solar surface. In very small magnetic concentrations (less than about three pressure scale heights across) this is more than compensated by a net *excess* emissivity. Due to the lower internal pressure such concentrations are small 'pits' in the surface. Through the sides of these depressions, an extra energy flux is emitted, which is visible mainly when viewing them at an angle from the vertical i.e. when seen near the limb (Spruit 1976, 1977). Most of the magnetic field on the solar surface is in the form of these small concentrations, and it appears from the irradiance measurements of the last cycle that their effect dominates over the reduced emission of sunspots (this may not have been the case in all cycles, however, see Foukal & Lean 1990). Theoretical prediction is difficult, since the excess emission from a facular area is a sensitive function of the poorly known size distribution of the concentrations (see however Spruit & Zwaan 1981). For a given size, the effect can be computed in principle through detailed numerical simulation of the energy transport in the concentration and its surroundings. Such calculations have already been undertaken by Knölker *et al.* (1991); Schüssler (1991).

Changes in surface emissivity are much more effective than magnetic energy storage or release in the same structures. Consider for example a small concentration with a contrast ϕ (averaged over its surface cross section). The life time of such elements can be estimated from the decay time of an active region, on the order of a few months. During this time τ, the element emits, per unit of magnetic flux, an excess energy $E_e = \phi\tau F_\odot/B_s$, where B_s is the surface magnetic field strength. If the field lines of the element reach down to base $z = D$ of the convection zone, their magnetic energy content per unit magnetic flux is $E_B = DB/8\pi$. Putting in $B \sim 2$kG, one finds that E_e is more than a factor 10 larger than E_B. This is only a simple energetic argument. In practice, the magnetic energy content of a magnetic structure is not so simply available to be emitted at the surface. In fact, there is no known way of transporting the energy in the deeper layer of the magnetic structure to the surface on the required time scale, even if all of it were available. This contrasts with the effect of the field on the surface emissivity. This effect develops as soon as the magnetic structure appears at the surface, because of the very short thermal time scale of the surface layers.

We conclude from the discussion so far that changes of the surface emissivity are predicted to be the main cause of irradiance variations. Since the brightness of sunspots and small scale fields can be measured, an irradiance variation can be calculated by just multiplying contrasts and areas and adding up. Is the situation as simple as this? What if some of the heat flux blocked by sunspots for example were reemitted elsewhere on the surface in diluted form? Calculations based on the formalism mentioned in Section (2.1) predict that, in fact, this reemission 'at a distance' does not take place (Foukal et al. 1983; Spruit 1982a,b; Chiang & Foukal 1984), on time scales of human interest. On time scales of the order $\tau_t \sim 10^5$ yr, the blocked flux does reappear. It does so in distributed form, nearly uniformly over the surface. Thus, the current nonmagnetic surface of the Sun radiates the blocked flux of all the spots on its surface avaraged over the past 10^5 yr (and similarly for the small scale fields). Due to this averaging effect, the nonmagnetic surface does not change on observable time scales.

Since the contrasts and numbers of magnetic elements of various sizes at the surface are in principle measurable at any point in time, the prediction that the brightness nonmagnetic surface does not take part in the cycle can in principle be checked. In practice this is rather difficult, since the largest effect comes from the small scale field. The predicted net emission of these small scale structures is a very sensitive function of their size (Spruit 1977; Spruit & Zwaaan 1981), and the largest effect is predicted from structures that are near or below the current spatial resolution. To account for the small scale field meaningfully, an intensive study of the effective brightness of the small scale field, and its magnetic flux has to be made, as a function of its evolutionary state (since it is known that the typical size of magnetic elements is larger in young active regions than in old ones, for example). The current proxy indicators do not represent the relevant quantity (photospheric brightness) accurately enough. The prediction that the smallest elements give the largest contribution would mean that the contribution of the most dispersed small scale fields, which are hard to observe, may have been strongly underestimated. The results of Lean et al. (1992, 1994) point in this direction.

4.1. *The 'blocked heat flux problem' of sunspots*

Suppose at $t = 0$ a spot appears instantly on the surface, reducing the heat flux. At least a part of the blocked heat flux will find its way 'around' this obstacle, to show up as a 'bright ring'. How long does it take for this to happen, and what fraction of the blocked heat flux is reemitted elsewhere? To see this, consider a terrestrial analogy, consisting of an aluminum brick heated from below, and losing heat to its environments from the top. On this I put, again at $t = 0$, a thermally insulating patch. Does a 'hot ring' develop around the patch? Is all of the heat flux blocked by the patch radiated from some other part of the surface? The answer is no. The high thermal conductivity of aluminum does not allow strong inhomogeneities like a hot ring. By conduction an even temperature is maintained, and the blocked heat flux, instead of reappearing somehwere on the surface, appears in the energy balance as a slow heating up of the brick. A new equilibrium is reached on the thermal time scale of the brick, which is long on account of its high heat capacity. This situation also applies to a convective stellar envelope, with only a small additional complication. In a convective envelope both the thermal (turbulent) conductivity and the heat capacity are strong functions of depth. The comparison with aluminum applies only at depths where the convective stratification is close to adiabatic. Above this, convection is less efficient, and the heat capacity small. For this reason, it matters how deep into the envelope the blocking object extends. For the Sun, one finds (Spruit 1982a; Foukal *et al.* 1983) that less than 5% of the blocked flux reappears at the surface, on time scales of observational interest ($< 10^5$yr), if the blocking extends to a depth greater than about 1000km. The part that does return to the surface does so very close to the spot or facular element.

The prediction is thus that in practice the solar surface outside magnetic regions *does not take part* in irradiance variations associated with the solar cycle. For dissenting views on the spot blocking problem and related matters, see Chapman *et al.* (1984), Sofia *et al.* (1982), Schatten & Mayr (1985), Fox *et al.* (1991), and Kuhn (1994).

In numerical simulations of spot blocking (Foukal *et al.* 1983; Fox *et al.* 1991), the very large range of time scales causes problems. This is one of the reasons why the depth of the convection zone is reduced artificially in such calculations. As a result, the thermal time scale in such calculations is also much shorter than in the real convection zone. In the simulation by Fox *et al.* (1991) with a depth of only 2000km the thermal time scale is on the order of a few hours instead of 10^5yr for the real convection zone. The rapid return to a steady heat flux observed by Fox *et al.* is a consequence of this shallow depth. Secondly, the depth of their 'spot', 150 km, is much shallower than a real spot. The Wilson depression alone of a spot is about four times as deep as this. The strength of the sunspot field, 3000 G, is large enough to interfere strongly with the convective heat flow for several thousand km below the surface. Whereas a convective diffusion model predicts that less than 10% of the blocked heat flux returns to the surface if the blocking effect extends 1000 km or more (Spruit 1982a), it predicts that 60% returns if the blocking is only 150 km deep. This is the consequence of the very steep increase of the turbulent thermal conductivity with depth (roughly proportional to ρ). Finally the upper boundary condition used by Fox *et al.* , which sets the temperature fluctuations to zero at the top surface instead of a proper radiative boundary condition, overestimates the fraction of the blocked heat flux that returns to the surface (Spruit 1976, Section 4.2). In other words the results by Fox *et al.* are in qualitative agreement with the diffusion model, and do not say much about sunspots.

For small magnetic concentrations, in which a lateral influx of heat enters into the tube rather close below the solar surface, a significant fraction of the influx of heat may derive from the immediate surroundings (forming a 'dark ring') around it. The extent to which this happens depends on the precise nature of the energy carrying flow near the tube, which is the subject of current numerical study. Since the concentrations that show a net excess are so small, the measured excess is usually an average over the concentration (see however Keller 1992). This average can still be positive, because part of the lateral influx derives from the interior of the convection zone (this is a consequence of the rapid increase of the thermal conductivity with depth). In a simple diffusion model for the heat transport near the tube, an average excess emissivity of up to 100% was found (Spruit 1977, Figures 16,17, see also Spruit & Zwaan 1981).

5. Stars

Since main sequence stars of type F and later are known to have spots and chromosheric emission like the Sun (for reviews see, e.g. Byrne & Mullan 1992), we may endeavor to extrapolate the conclusions in the preceding sections to these stars. One predicts then that the nonmagnetic part of the stellar surface does not change in brightness or color on observable time scales, except perhaps in early F-stars in which the convection zone is much shallower than in the Sun. Variations in brightness and colors of these stars, whether due to rotational modulation or to changing magnetic fields, can then be interpreted directly in terms of the color properties of the magnetic structures involved and their distribution on the surface. Even secular changes in brightness on time scales of centuries do not involve the nonmagnetic surface. Since the umbrae of spots are rather dark in the visual range, they contribute relatively little to color changes; more effective are the *penumbrae* (cf. Spruit & Weiss 1986). This depends on the size of starspot penumbrae however, about which little is known.

For the position of an active star in the Hertzsprung–Russel Diagram (HRD), changes in the nonmagnetic part of the surface *do* play a role, since the age of the star is much longer than the thermal time scale of the convection zone. The brightness of the non-magnetic surface in these stars reflects the average of the magnetic activity over the last thermal time scale of its convection zone. Changes in position in the HRD, taking this into account, have been computed by Spruit & Weiss (1986), assuming spots with properties similar to sunspots. The stars were found to remain close to the main sequence, even for spots covering a significant farction of the stellar surface, because the displacement in the HRD is nearly parallel to the main sequence.

6. Conclusions

The theory of structure and evolution of stellar envelopes gives a rather simple picture for the effect of magnetic fields on luminosity (irradiance). This is because it predicts that for irradiance variations on time scales short compared with 10^5 yr the non-magnetic surface† participates only at a very low level in the variations due to magnetic fields. Taking account of effects that are ignored in the calculations, such as systematic flows and uncertainties in the mixing length model used, may certainly change the contribution by factors of order unity. The main conclusion is robust with respect to such changes because this contribution is so small in the first place. Most of the uncertainty in making comparisons with observations comes from our poor knowledge of the intrinsic brightness and magnetic flux in the small scale magnetic field, especially in its most dispersed form, since this field contributes disproportionately to the net irradiance. Its brightness for a given size and shape of the magnetic elements can in principle be obtained from numerical simulations (Knölker *et al.* 1991). The spectrum of scales in the small scale field however depends on aspects of the solar dynamo process that are intrinsically difficult and have hardly been adressed theoretically.

Acknowledgments. I thank Dr. Sami Solanki and an anonymous referee for corrections and useful comments.

REFERENCES

ARENDT, S. 1992 Large-scale flows and solar luminosity variations. *Astrophys. J* **389**, 421–427.

BYRNE, P. B. & MULLAN, D. J. 1992 Influence of starspots on internal stellar structure. In *Lecture Notes in Physics* **397**, 78–84. Springer-Verlag.

CHAPMAN, G. A., HERZOG, A. D., LAWRENCE, J. K. & SHELTON, J. C. 1984 Solar luminosity fluctuations and active region photometry. *Astrophys. J.* **282**, L99–L101.

CHIANG, W. H. & FOUKAL, P. V. 1984 The influence of facuale on sunspot heat blocking. *Solar Phys.* **97**, 9–20.

D'SILVA, S. & CHOUDHURI, A. R. 1993 A theoretical model for tilts of bipolar magnetic regions. *Astron. Astrophys.* **272**, 621–633.

D'SILVA & HOWARD, R. A. 1993 Limits on the magnetic field strength at the base of the solar convective zone. *Solar Phys.* **148**, 1–9.
bitem[]ENDAL, A.S., SOFIA, S. & TWIGG, L.W. 1982 Changes of solar luminosity and radius following secular perturbations in the convective envelope. *Astrophys. J.* **290**, 748–757.

FOUKAL, P. V., & LEAN, J. 1990 An empirical model of total solar irradiance variations between 1874 and 1988. *Science* **247**, 556–558.

† To be precise: the surface outside about 1 granule diameter from a magnetic element, since the magnetic field has some influence on the convective flow in its neighborhood.

FOUKAL, P., FOWLER, L.A. & LIVSHITS, M. 1983 A thermal model of sunspot influence on solar luminosity. *Astrophys. J.* **267**, 863–871.

FOX, P., SOFIA, S., CHAN, K. L. 1991 Convective flows around sunspot-like objects. *Solar Phys.* **135**, 15–42.

FRÖHLICH, C. 1990 Irradiance variability of the Sun. In *Climate Impact of Solar Variability* (ed. K. H. Schatten & A. Arking) NASA CP 3086, pp. 269–278.

FRÖHLICH, C., FOUKAL, P. V., HICKEY, J. R. & HUDSON, H. S., WILLSON, R. C. 1991 Solar irradiance variability from modern measurements. In *The Sun in Time* (ed. C. Sonett, M. Giampapa & M. Matthews). pp. 11–29. University of Arizona, Tucson, AZ, USA.

GILLILAND, R. L. 1988 In *Solar radiative output variation*, Cambridge Research & Instrumentation Inc., P. Foukal, ed., 21 Erie st. Cambridge, MA 02139, pp. 239.

KASTING, J.F., & GRINSPOON, D.H. 1991 The faint young Sun problem. In *The Sun in Time* (ed. C. Sonnett, M. Giampapa & M. Matthews). University of Arizona, Tucson, pp. 447–462.

KELLER, C.U. 1992 Resolution of magnetic flux tubes on the Sun *Nature* **359**, 307–308.

KNÖLKER, M. GROSSMANN-DOERTH, U., SCHÜSSLER, M., WEISSHAAR, E. 1991 *Adv. Sp. Res.* **11**, 285.

KUHN, J. 1994 Brightness observations of the Sun. In *The Sun as a Variable Star: Solar and Stellar Irradiance Variations* (ed. J.M. Pap, C. Fröhlich, H.S. Hudson & S.K. Solanki). Cambridge University Press, in press.

LEAN, J., SKUMANICH, A. & WHITE, O.R. 1992 Estimating the Sun's radiative output during the Maunder Minimum. *Geophys. Res. Lett.* **19**, 1595–1598.

LEAN, J., SKUMANICH, A., WHITE, O.R. & RIND, D. 1994 Estimating solar forcing of climate change during the Maunder Minimum. In *The Sun as a Variable Star: Solar and Stellar Irradiance Variations* (ed. J.M. Pap, C. Fröhlich, H.S. Hudson & S.K. Solanki). Cambridge University Press, in press.

PARKER, E. N. 1979 *Cosmical Magnetic Fields* Oxford, Clarendon Press.

PARKER, E. N. 1984 Magnetic buoyancy and the escape of magnetic fields from stars. *Astrophys. J.* **281**, 839–845.

SCHATTEN, K. H. & MAYR, H. G. 1985 On the maintenance of sunspots – An ion hurricane mechanism. *Astrophys. J.* **299**, 1051–1062.

SCHÜSSLER, M. 1991 *Geophys. Astrophys. Fluid Dyn.* **62**, 271.

SOFIA, S. SCHATTEN, K.H. & OSTER, L. 1982 Solar irradiance modulation by active regions during 1980. *Solar Phys.* **80**, 87–98.

SPIEGEL, E.A. & WEISS, N.O. 1980 Magnetic activity and variations in solar luminosity. *Nature* **287**, 616–617.

SPRUIT, H.C. 1976 Pressure equilibrium and energy balance of small photospheric flux tubes. *Solar Phys.* **50**, 269–295.

SPRUIT, H.C. 1977 Heat flow near obstacles in the solar convective zone. *Solar Phys.* **55**, 3–34.

SPRUIT, H.C. 1982a Effect of spots on a star's radius and luminosity. *Astron. Astrophys.* **108**, 348–355.

SPRUIT, H.C. 1982b The flow of heat near a spot. *Astron. Astrophys.* **108**, 356–360.

SPRUIT, H.C. 1991 In *The Sun in Time* (ed. C. Sonett, M. Giampapa & M.S. Matthews). pp. 118–158. University of Arizona Press, Tucson, AZ, USA.

SPRUIT, H.C. & ZWAAN, C. 1981 The size dependence of contrasts and numbers of small magnetic flux tubes in an active region. *Solar Phys.* **70**, 207–228.

SPRUIT, H.C. & VAN BALLEGOOIJEN, A.A. 1982 Stability of toroidal flux tubes in stars. *Astron. Astrophys.* **106**, 58–66, *Erratum* in **113**, 350.

SPRUIT, H.C. & WEISS, A. 1986 Colors and luminosities of stars. *Astron. Astrophys.* **166**, 167–176.

ZWAAN, C. 1978 On the appearance of magnetic flux in the solar photosphere. *Solar Phys.* **60**, 213–240.

Convection and Irradiance Variations

By PETER A. FOX[1] AND SABATINO SOFIA[2]

[1]High Altitude Observatory, National Center for Atmospheric Research, Boulder CO 80307, USA

[2] Center for Solar and Space Research, Yale University New Haven, CT 06511, USA

In the outer layers of the Sun ($\approx 30\%$ by radius), energy is transported by convection. The nature of the highly stratified and compressible convective flow is determined from the components of the energy flux (internal, kinetic, viscous, magnetic and radiative). Local suppressions or enhancements of any of these components may give rise to measurable changes in the emergent radiation.

On the solar surface there is direct evidence for modulation of the emerging heat flux covering a large range in spatial and temporal scales, particularly associated with concentrated magnetic fields (e.g. sunspots, plages). Associated with these surface features is the observation that the characteristics of convective motions are also modified. In the deeper layers, the interaction of convection and magnetic fields will play an important role in readjusting the local emerging heat flux and thus should contribute to the modulation of the total solar irradiance.

The task of calculating the response of the convection zone structure to developing active regions, and the solar activity cycle in general is difficult and complex due to the highly non-linear nature of the interaction of convection and magnetic fields. Theoretical work has ranged from empirical and global structure models, all the way to fine scale compressible convection simulations. This paper will highlight some recent theoretical advances that may have a direct bearing on the understanding of solar luminosity and irradiance variations and outline the important problems that must be addressed and what observational constraints may be used.

1. Introduction

When the Sun is viewed in terms of its main sequence evolution (timescales on the order of billions of years), processes in the convective envelope zone can be viewed as instantaneous. However, if the timescale of interest is reduced to centuries or decades, many of the physical processes become distinct and exhibit their own characteristic time variation. Examples of these timescales are those of radiative relaxation, thermal diffusion, convective turnover, and hydrostatic readjustment. The study of solar variability amounts to determining the response of the solar convection zone to disturbances. In general the resulting variability depends on the location, size, strength, and nature of the perturbation.

To date, the primary solar variability that is observed and can be largely explained, is that associated with short term surface modulations, (sunspots, faculae, and most likely the magnetic network) which have timescales ranging from hours to days, to months. The next established timescale is that of the solar cycle (decadal) variation of the solar irradiance (from Nimbus 7 and ACRIM I/II, for example). Nevertheless a complete theoretical explanation for the measurements is still lacking. Beyond these observed timescales, there is an inferred secular variation (on the order of centuries, or longer) which arises from statistical studies of global solar and geophysical parameters. It is reasonable to assume that these changes are associated with long term internal modulation of the solar heat flux, and thus the convective energy transport.

Convection is also likely to play a role in the conversion between the solar luminosity and irradiance; which involves a knowledge of the radiance of solar surface structures (sunspots, plage, etc). These issues will not be addressed in this paper.

As added motivation, there now exists increasing observational evidence for changes in convective energy transport within the Sun. In regions of enhanced magnetic field there is a noticeable change in both the surface pattern and the lifetime of granulation which is the dominant energy carrying scale in the surface layers. More quantitatively, the changes in solar absorption line bisectors both over the solar cycle and in/out of plage (Livingston 1990), gives some measure of the change in the convective flows. Even though these observations relate only to the surface layers, it seems reasonable to assume that convection motions in the deeper interior are also affected by concentrations of magnetic flux. More global variations in the convective heat transport are suggested by observations of limb temperature excesses (Kuhn 1989) whose location varies with the solar cycle.

An even more abrupt change in the Sun has been observed to occur around 2–3 years after solar cycle maximum (McIntosh 1994; White 1994), where all proxies of solar activity suddenly decrease and do not return to the previously high levels. Accompanying this effect in the most recent cycle (January 1992), is the tantalizing hint that a thermal shadow effect may be present in the ERBS irradiance measurements (corrected for sunspots) exactly at this time (White 1994). An observational limit for ground–based photometry of shadows, or bright rings localized around active regions is 0.1 to 0.3% (Fowler *et al.* 1983), whereas the irradiance measure is weighted across the solar disk. If confirmed, or established, this implies a submergence of active regions on a large scale, and thus a modification of the solar luminosity. To date all observational support for changes in the convective nature of the Sun is limited to timescales up to decades.

Even though this paper will only focus on recent advances in convection theory there is an existing body of numerical work on global solar perturbations (see Endal *et al.* 1985) in the solar convection zone which provides a strong incentive to advance our understanding of convective energy transport.

2. Convective energy transport

2.1. *Equations*

The basic equations which govern magnetized convective energy transport in a star like the Sun can be written:

$$\frac{\partial\rho}{\partial t} = -\nabla\cdot\mathbf{M}, \quad \frac{\partial\mathbf{M}}{\partial t} = -\nabla\cdot\mathbf{P}+\rho\mathbf{g}, \quad \frac{\partial\mathbf{B}}{\partial t} = -\nabla\times\mathbf{E}, \quad \frac{\partial\varepsilon}{\partial t} = -\nabla\cdot\mathbf{F}_{\text{total}},$$

where ρ is the density, ε is the total energy $\mathbf{M}$ is the momentum vector, $\mathbf{P}$ is the momentum flux tensor, $\mathbf{g}$ represents the gravitational field, $\mathbf{F}_{\text{total}}$ is the total heat flux vector, $\mathbf{B}$ is the magnetic field, and $\mathbf{E}$ is the electric field. The ancillary relations, such as the equation of state of the gas, the Ohm's law, etc., are not relevant to the following discussion, and rotational effects have been neglected at present although are certainly important over longer timescales.

Consider a particular layer in the Sun where the incident heat flux is uniform in the statistical sense, i.e. time averaged (in reality the boundary between the convection zone and radiation zone is unlikely to have uniform flux at a given instant). In the absence of any perturbations, that incident heat flux will be propagated through the convective layer until it reaches the surface. In doing so, it will be distributed according to the following components of the heat flux:

$$\mathbf{F}_{\text{total}} = \mathbf{F}_{\text{ep}} + \mathbf{F}_{\text{ke}} + \mathbf{F}_{\text{rad}} + \mathbf{F}_{\text{B}} + \mathbf{F}_{\text{visc}} + \mathbf{F}_{\text{turb}}, \tag{2.1}$$

which are the enthalpy flux, kinetic energy flux, radiative flux, Poynting flux, viscous flux and the turbulent flux, respectively. One of the problems then, is to determine how convection redistributes the emerging solar heat flux in response to magnetic (or other) variations, and over what timescale. Another key question is to determine the role of horizontal fluxes of energy which are not usually considered in mean structure models.

2.2. *Standard stellar structure*

Models of the spherically symmetric solar structure utilize an expression for the total heat flux to determine the radial thermodynamic profiles, and thus the effective temperature (luminosity). For example in Mixing Length Theory (MLT; Vitense 1953), the expression is $F_{\rm total} = F_{\rm conv} + F_{\rm rad}$, and radiation is modeled with the diffusion approximation. MLT provides an expression for $F_{\rm conv}$ in terms of local quantities and a number of adjustable parameters which are usually all expressed in terms of the mixing length to pressure scale height ratio, α. The primary drawback with this free parameter (which, to first order, is selected so that the model has the correct solar radius) becomes apparent when studying solar variability, i.e. relative changes in radius and luminosity.

An essential ingredient of solar structure models is that the physics be as realistic as possible, i.e equation of state, opacity, chemical composition, atmosphere model, etc. In any case, for the purpose of study solar variability, these models are viewed as instantaneous representations of the state of the convection zone of a star. When timescale of the internal changes become comparable to the characteristic times for convection, the assumption of instant adjustments needs to be examined carefully.

2.3. *Numerical simulations*

A long standing concern for simplified models of convective energy transport is how well they would capture the features of time dependent, three dimensional motions that actually take place within a star such as the Sun (the only star whose surface motions we can observe in detail, and whose structure we can reliably probe with helioseismology). Because of the highly turbulent nature of the solar convection zone, numerical simulations are not expected to accurately represent the real Sun either, however they do give a guide to the features that must be present and what the essential physics is. In this section we will highlight some recent results of numerical simulations that have a direct bearing on understanding solar variability.

Since the physical conditions in the solar convection zone vary considerably throughout the zone (that of a highly compressible and stratified gas, with the effects of partial ionization and radiation, etc.), numerical simulations/experiments have focused primarily on three distinct regions: the shallow layers – which are highly superadiabatic (with a radiative layer on top), the deeper layers – which are marginally superadiabatic or adiabatic, and the lower convection zone–radiation zone interface (also known as the overshoot region).

2.3.1. *Topology*

One of the very important results of turbulent compressible convection calculations is that the topology of the flow is quite unlike the cellular patterns reminiscent of Benard-type convection. The upflows are broad and extend over a scale height or two, whereas the downflow regions are narrow and often extend over multiple scale heights (Chan & Sofia 1986; Chan & Sofia 1989; Stein & Nordlund 1989; Cattaneo *et al.* 1991) with velocities comparable to the local sound speed.

These results strongly suggest that the energy carrying vertical scale of convection is that on the order of a pressure scale height whereas the surface horizontal scale naturally divides itself into granular, mesogranular and supergranular domains.

In addition to these general features, Chan & Sofia (1989) found that the vertical velocity autocorrelation function (whose FWHM is a measure of the so-called mixing length) scaled with the local pressure scale height and not the density scale height in agreement with standard MLT models. They also found that the mean vertical velocity, $\bar{v}$, was non-zero over most of the layer, in contrast with the MLT assumption.

2.3.2. *Energy fluxes*

Another extremely important result arises from studying the horizontally averaged energy flux components (in eq. (2.1)). The work of Chan & Sofia (1989; Figure 13) and Cattaneo *et al.* (1991) indicates the dominance of enthalpy and kinetic energy fluxes and the very minor role played by diffusive/turbulent fluxes. This result has also been confirmed by many authors and is now regarded as a robust feature of these flows.

Further, Chan & Sofia (1989) found that, although the creation can be described in terms of local quantities, the dissipation of kinetic energy flux is a non-local process.

2.3.3. *The role of downflows*

In the analysis of mean structure models of convection, the horizontal averaging provides the net fluxes. Since there is a marked asymmetry between upflows and downflows, an important question arises: can mean structure models be constructed with a net flux (i.e. one component) or should they allow for two streams (one upflow and one downflow component)? Alternately stated: how well does the horizontal averaging capture the high spatial variation in the convective patterns?

Cattaneo *et al.* (1991) examined the difference between energy fluxes in regions of upflow and downflow for both laminar and turbulent regimes (as measured by the Rayleigh number). They found that in the downflows, in the case of turbulent flows, the enthalpy and kinetic energy fluxes nearly balanced giving a net zero flux. This very striking result was not true for laminar flows where the enthalpy flux dominated. Thus, the overall role of intense downdrafts may be less important in the overall heat transport.

2.4. *Improved structure models*

An obvious question which arises from the analysis of numerical simulations is: do the mean properties of the simulations (as presented in Chan & Sofia (1989), for example) allow one to construct a (one dimensional) model of the Sun? This question has recently been addressed in a series of papers by Lydon *et al.* (1992, 1993a, 1993b) by using an expression for the total heat flux, $F_{\text{total}} = F_{\text{ep}} + F_{\text{ke}} + F_{\text{rad}}$, which includes the kinetic energy flux. Furthermore, the formulation does not include the adjustable "α" from the MLT and is thus more attractive for solar variability studies. The distribution of energy fluxes in a solar convection zone structure model using this formulation is shown in Figure 1.

The new formulation makes the resulting models very sensitive to the input physics (Lydon *et al.* 1992) but a solar model can be constructed within the observation bounds (radius, luminosity, p-mode oscillation frequencies, etc.) and furthermore, models have been constructed for the binary system α-Centauri A and B (Lydon *et al.* 1993b). Despite these advances, there is still much work to be done on the formulation of convective energy transport. In particular, since all three models stars are similar to each other, it is necessary to ascertain what needs to be added/modified to models elsewhere in the H–R diagram.

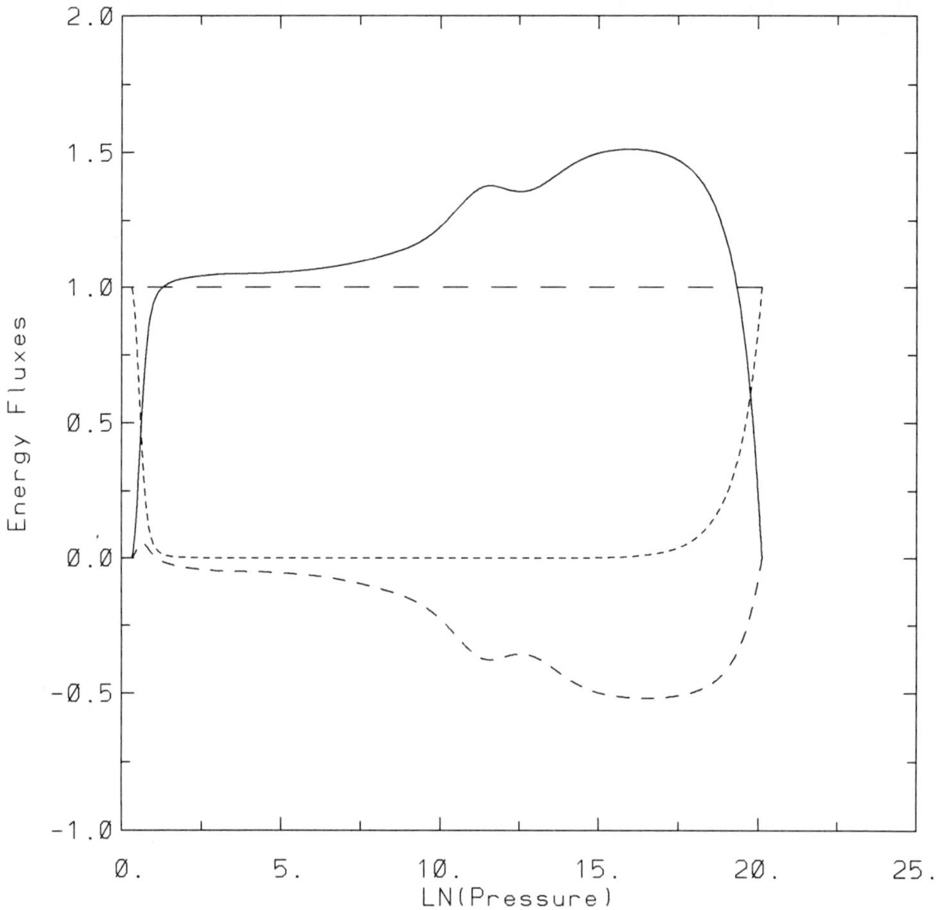

FIGURE 1. Schematic of vertical energy fluxes versus ln(pressure) for a solar model: enthalpy flux (solid), kinetic energy flux (medium dash), radiative flux (short dash), and the total heat flux (long dash).

2.5. *Magnetic fields*

So far in the discussion we have not dealt with the direct influence of magnetic fields on convective energy transport. A primary reason for this is that in the realm of three dimensional numerical simulations, this topic has received little attention until recently.

The general notion that magnetic fields inhibit convection seems to be an oversimplification. A more accurate, but less precise description, is that magnetic fields modify the convective flows depending on the scales and topology of the features and the magnetic field strength. A magnetic flux concentration can store and release heat (through Poynting Flux and radiative exchange; Fox *et al.* 1991) and thus modulates the total heat flux. There is also the possibility of generation of MHD waves which can amount to some fraction of the energy flux (perhaps up to 1%).

Perhaps most importantly for mean structure models, the presence of a magnetic field, either diffuse or concentrated, will alter correlations between flow and thermodynamic quantities, particularly ones like $C[v_z, T']$ which are important for obtaining expressions for the fluxes in eq. (2.1).

The discrete nature of the solar magnetic field also poses the question of how well can these features be represented in the horizontally averaged sense (i.e. the same question that was posed for concentrated downflows in the unmagnetized case)? Unfortunately this is still an open question.

2.6. *Energy blocking models*

The earliest models of this type were carried out by Spruit (1976, 1977) where time dependence of the problem as well as systematic flows were neglected, and by Foukal *et al.* (1983) and Chiang & Foukal (1984), where time dependence was considered, but still no flows were allowed. The results of these investigations indicated that, because of the much stronger thermal contact between the perturbed region and the deep convection zone than between the perturbed region and the surface, the large majority of the sunspot-blocked energy would be diffused into the entire solar convective region to be released on a thermal timescale of this region (i.e. about 10^5 years). More recently, Nye *et al.* (1988) considered this same problem, but included both energy and mass flow in the *linear* approximation. They found that significant mass flows were generated by the thermal energy blocked by the spot, but that these flows were not accurately in agreement with observations.

With the results mentioned in the previous sections in mind (i.e. importance of kinetic energy fluxes, etc.), we now turn to two simulations which deal with the response of a convective region to a disturbance in the energy flux. These simulations are the only ones available that are solved in the fully *non-linear compressible* regime and include dynamical effects self consistently. The two "experiments" in question are distinctly different and still far from addressing the actual conditions in the solar convection zone, but nevertheless they are both attempts to include detailed convective transport processes in the study of energy blocking.

2.6.1. *Deep seated perturbations*

Kuhn (1991) designed a numerical experiment which consisted of two boxes of convectively unstable gas, side by side in the conventional manner but with a differential heat flux ($1.01 \times F_{\text{total}}$ and $0.99 \times F_{\text{total}}$) imposed on either side. After a short relaxation time, the depth dependence of the average temperature difference between the hot and cold sides of the box shows a peak in the superadiabatic surface region, i.e. at the top of the layer. The temperature difference was about 6 times larger than that induced at the lower boundary.

A conclusion of these results was that a MLT–like representation for the heat transport (i.e. diffusive) is not accurate and that the disturbances were not distributed over the entire convective region.

2.6.2. *Surface layer perturbations*

Fox *et al.* (1991) considered the problem of calculating the flow around a sunspot-like object at various depths in the solar convection zone to study the energy balance in both the immediate area and far from the object, as a function of time. Although magnetic fields were not *explicitly* included, the sunspot-like object was modeled as a region prohibiting convective flow but allowing diffusive heat exchange due to radiation (and conduction). The geometry was two dimensional, in spherical coordinates. The width of the domain was 24.4 Mm and the depth was 2.0 Mm (approximately 6.65 pressure scale heights).

There were a number of results arising from these simulation but only one aspect will be discussed here, namely comparing the modification of the heat flux components due to the sunspot-like object compared to an otherwise identical control case where no object was present. The pore–sized object (of size 1.2 Mm wide and 0.132 Mm deep) was placed in the top, center of the domain. Further details are contained in Fox *et al.* 1991.

Once the object is in place, there is a short time of dynamic readjustment of the layer surrounding the object to accommodate the altered heat carrying capacity of the object. A closer examination of the region around the object reveals an enhanced horizontal kinetic energy flux which is distributed in both directions, indicating that there is a mix of small scale motions either to the left or right. Thus, once again, kinetic energy flux is an important factor in the heat transport.

To quantify this result, we examine the volume integrated components of the vertical enthalpy (Figure 2) and kinetic energy (Figure 3) fluxes in two regions (close to the object which represents $\approx$ 10% of the domain;the upper panel of Figs 2 & 3, and far from the object, which is the remainder of the domain; on the lower panel of Figs 2 & 3) as a function of time after the object was introduced. The dash line represents the disturbed case and the solid line represents the undisturbed case.

There are two distinctive effects, the first is the time delay over which the heat flux is modified in response to the object. This is evident by noting that the two cases always begin to differ for the "close" integration (about 10-20 minutes) before they do for the "far" (1-2 hours) integration. The time delay is slightly different for the enthalpy and the kinetic energy fluxes and also for the vertical and horizontal components (not shown here).

The next effect is in the relative changes of the enthalpy flux and kinetic energy flux components and how these can be interpreted as a global indication of modified heat transport.

One very important factor is displayed in Figure 3, which shows that the kinetic energy flux close to the object changes sign compared to the undisturbed case. This is due to the fact that there is a net downward, or near zero, flux close to the object, which is associated with downflows beside and away from the object, as distinct from the broad upflows that occur in the absence of the object. Further from the object both radial and kinetic energy fluxes have larger variations than the undisturbed case: essentially they are responding to the excess heat flux being diverted from the region close to the object.

Despite these horizontal diversions of heat flux, not all of the blocked heat was stored in the convective region. In fact a good fraction of it appeared at the surface (Fox *et al.* 1991). The extent to which horizontal diversion, storage in fluid circulations, or appearance at the surface could occur depends on the size and position of the object, which was limited to a pore–sized object.

Given the importance of the kinetic energy in the redistribution of the heat flux, it is not surprising that previous studies found markedly different behavior. Most of the previous work relied only on a diffusion treatment of the heat flow (e.g. Spruit 1977; Chiang & Foukal 1984), or used linearized equations (Nye *et al.* 1988), and thus excluded the process that is suggested to be important in these calculations. As a result, the timescales of heat redistribution can be markedly different compared to completely diffusive processes.

There are a number of limitations to these calculations: the sub-surface geometry of the sunspot-like objects (their small size and shallow depth extent) and the treatment of radiative losses within the blocking region (diffusive only). For a surface object (the ones that are directly observed) the depth extent may be an important factor. As the depth of the lower boundary of the object increases, so does the local convective efficiency, indicating that convection should deal even more efficiently in redistributing the heat flux and that the conduction of heat into the object would decrease and its timescale for diffusion would increase. The important factor now in the redistribution is the width to

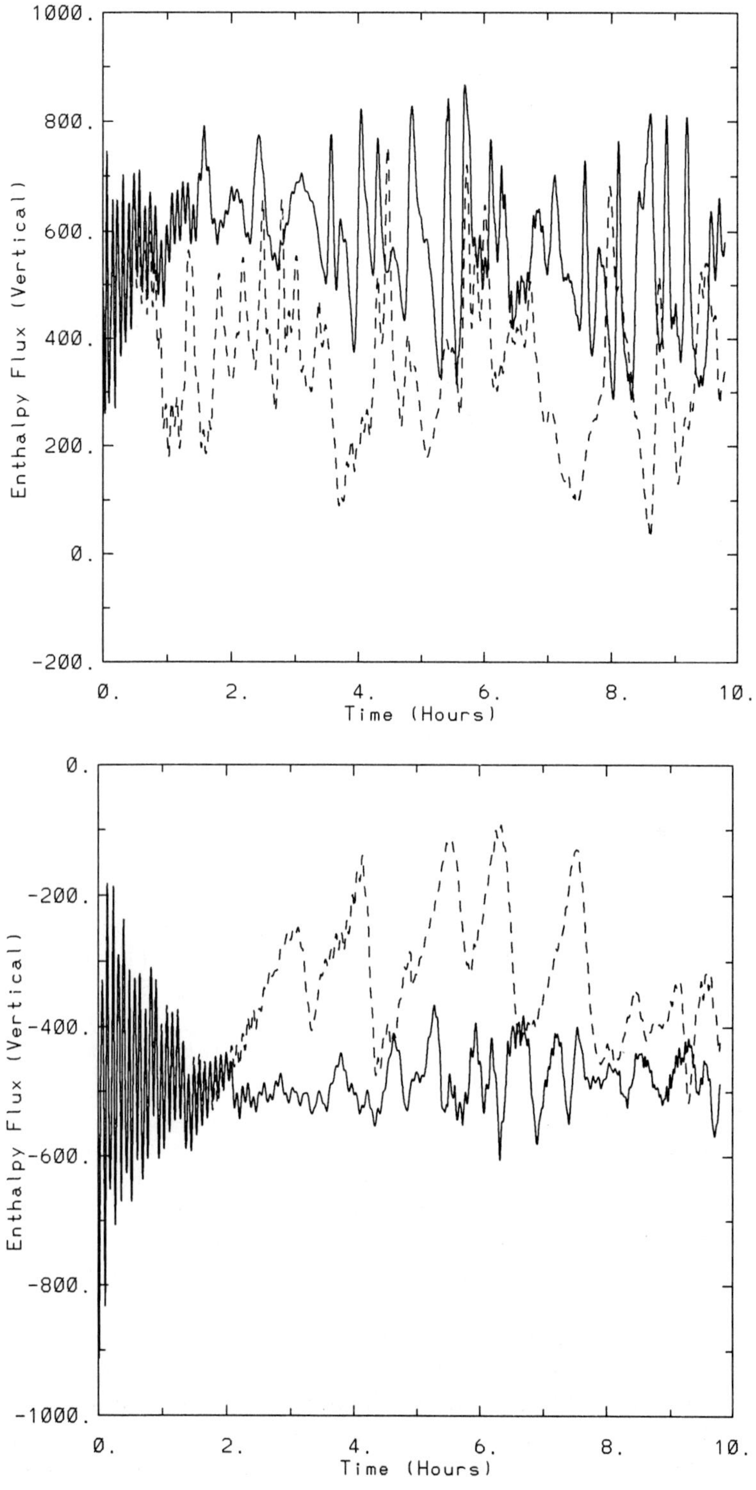

FIGURE 2. Vertical component of the enthalpy flux integrated over the domain close to (upper) and far from (lower) the object versus time after placement of the object. The solid line is the comparison case without the object, the dash line shows the response to the object.

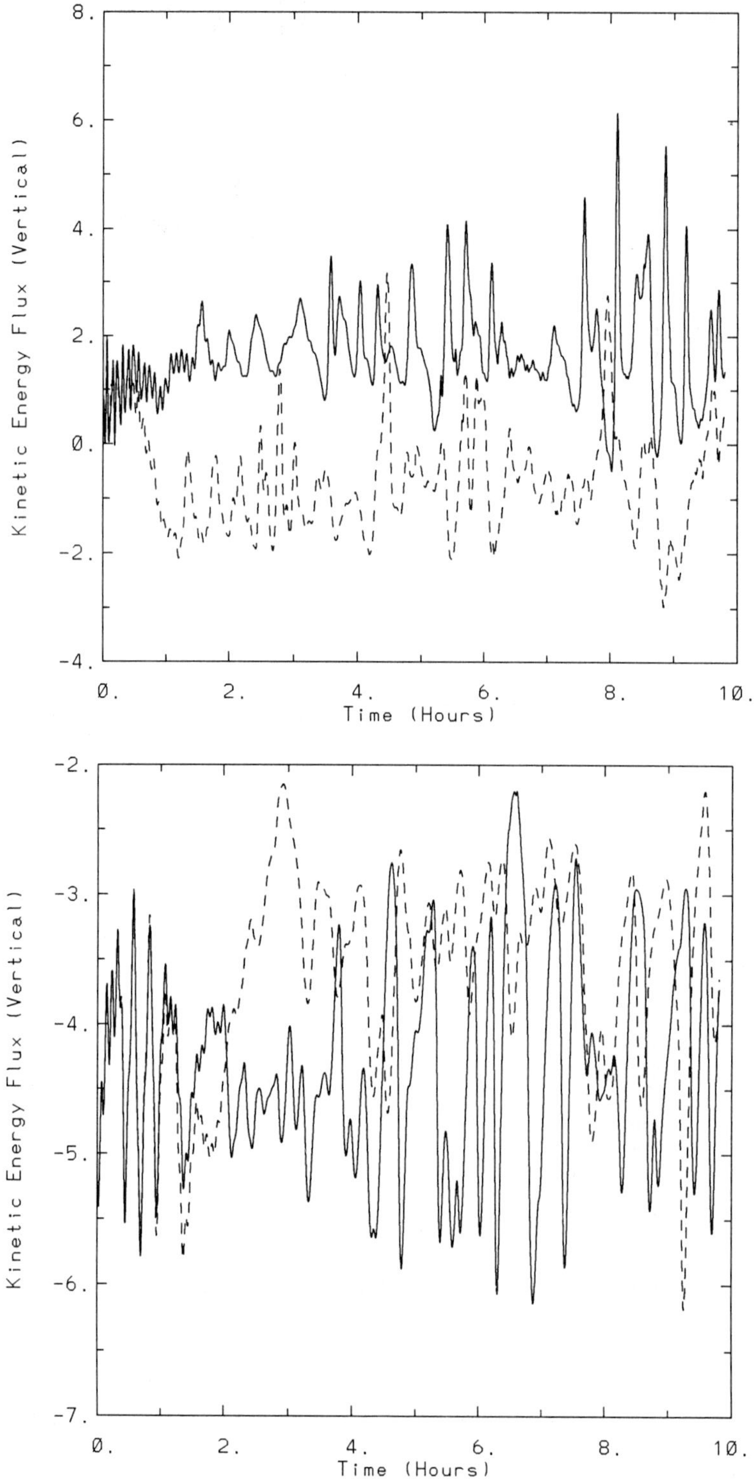

FIGURE 3. Vertical component of the kinetic energy flux integrated over the domain close to (upper) and far from (lower) the object versus time after placement of the object. The solid line is the comparison case without the object, the dash line shows the response to the object.

depth ratio, i.e. how long it takes the flow to go around. Also, because the pressure scale height is also increasing, the object may be "felt" at a larger distance, which would alter the way heat is transported. These considerations could be very important for sunspots.

3. Discussion

All of the preceding discussion highlights an increasing understanding of the physical nature of convective energy transport. The following general conclusions are relevant to solar variability studies:

(*a*) Kinetic energy flux is important in the heat budget in the interior
(*b*) Downflows may not be important in the overall heat transport
(*c*) Localized changes in heat transport can occur quickly
(*d*) Deep thermal perturbations may have a larger amplitude near the surface
(*e*) Heat can be stored in regions of intense magnetic field
(*f*) Diffusive, or mixing length descriptions can be misleading

Clearly, all simulation efforts, whether they deal with general characteristics of the compressible convection, or specifically with energy blocking and redistribution, must include the relevant physical effects. Because active regions have strong magnetic fields, closely surrounded by convective flow, it is likely that there will be some conversion between internal, kinetic and magnetic energy (Fox *et al.* 1991). The current assumption of the passive existence of a heat blocking object is only somewhat valid based on surface observations; however the lack of any knowledge of subsurface fields can be seen as an obvious drawback. For the energy blocking models, it will also be necessary to extend the existing calculations to larger length scales and longer timescales, to address the question of measurable radiance or irradiance variations. Further numerical experiments of the type performed by Kuhn are also essential in understanding how deep seated magnetic fields perturb the layers closer to the surface.

There are also a number of important questions to be addressed, and perhaps the most pressing is: what is the role of the solar activity cycle (on the decadal timescale) in longer term changes in the Sun? Further, can a global change in the Sun be effected in about 11 years? Most correlation studies between solar and terrestrial climate quantities point to timescales longer than a decade (usually 80-100 years), whereas other terrestrial indices apparently correlate with the length of a particular solar cycle (Friis-Christensen & Lassen 1991) and this is far from understood.

Another important area for study is at what scale do localized changes in convective transport alter the solar luminosity? Thus the question of a site for variability, much as solar physicists seek a site for the solar dynamo, becomes prominent.

Numerical simulations of convection and magnetic fields should soon address how the energy flux components are distributed under the influence of interior perturbations (of many varieties) and what their timescale for change is.

Finally, observations must play a crucial role in constraining the theoretical models. Already, helioseismology can probe the mean interior structure and, with improved databases (from global networks and satellites), should start to provide information on flows and latitudinal variations in the solar interior (such as those induced by large scale magnetic fields). The emergence of sunspot seismology also promises some insight into the subsurface structure of active regions, specifically spots, which would be essential in more detailed modeling of energy blocking.

Acknowledgments. The National Center for Atmospheric Research is sponsored by the National Science Foundation. S. Sofia thanks for AFOSR and NASA for support.

REFERENCES

CHAN, K. L. & SOFIA, S. 1986 Turbulent compressible convection in a deep atmosphere III: tests on the validity and limitations of the numerical approach. *Astrophys. J.* **307**, 222–241.

CHAN, K. L. & SOFIA, S. 1989 Turbulent compressible convection in a deep atmosphere IV: results of three-dimensional computations. *Astrophys. J.* **336**, 1022–1040.

CATTANEO, F., BRUMMELL, N. H., TOOMRE, J., MALAGOLI, A. & HURLBURT, N. E. 1991 Turbulent compressible convection. *Astrophys. J.* **370**, 282–294.

CHIANG, W. H. & FOUKAL, P. 1984 The influence of faculae on sunspot heat blocking. *Solar Phys.* **97**, 9–20.

ENDAL, A. S., SOFIA, S. & TWIGG, L. 1985 Changes in solar luminosity and radius following secular perturbations in the convective envelope. *Astrophys. J.* **290**, 748–757.

FOUKAL, P., FOWLER, L. A. & LIVSHITS, M. 1983 A thermal model of sunspot influence on solar luminosity. *Astrophys. J.* **267**, 863–871.

FOWLER, L. A., FOUKAL, P. & DUVALL, T. 1983 Sunspot bright rings and the thermal diffusivity of solar convection. *Solar Phys.* **84**, 33–44.

FOX, P. A., SOFIA, S. & CHAN, K. L. 1991 Convective flows around sunspot-like objects. *Solar Phys.* **135**, 15–42.

FOX, P. A., THEOBALD, M. L. & SOFIA, S. 1991 Compressible magnetic convection: formulation and two-dimensional models. *Astrophys. J.* **383**, 860–881.

FRIIS-CHRISTENEN, E. & LASSEN, K. 1991 Length of the solar cycle: an indicator of solar activity closely associated with climate. *Science* **254**, 698–700.

KUHN, J. R. 1989 Helioseismic observations of the solar cycle. *Astrophys. J.* **339**, L45–L47.

KUHN, J. R. 1991 Inferring solar structure variations from photometric and helioseismic observations., preprint.

LIVINGSTON, W. C. 1990 Sun-as-a-star. its convective signature and the activity cycle.. In *The Sun and Cool Stars: Activity, Magnetism and Dynamos* (ed. I. Tuominen, D. Moss & G. Rüdiger). Lecture Notes in Physics, vol. 380, pp. 246–251. Springer-Verlag.

LYDON, T. J., FOX, P. A. & SOFIA, S. 1992 A formulation of convection for stellar structure and evolution calculations without the MLT approximations I. application to the Sun. *Astrophys. J.* **397**, 701–716.

LYDON, T. J., FOX, P. A. & SOFIA, S. 1993a Improved solar models constructed with a formulation of convection for stellar structure and evolution calculations without the MLT approximations. *Astrophys. J.* **403**, L79–L82.

LYDON, T. J., FOX, P. A. & SOFIA, S. 1993b A formulation of convection for stellar structure and evolution calculations without the MLT approximations II. application to α Centauri A and B. *Astrophys. J.* **413**, 390–400.

MCINTOSH, P. S. 1994, in preparation.

NYE, A., BRUNING, D. & LABONTE, B. J. 1988 Mass and energy flow near sunspots: II. Linear numerical models of moat flow. *Solar Phys.* **115**, 251–268.

SPRUIT, H.-C. 1976 Pressure equilibrium and energy balance of small photospheric flux tubes. *Solar Phys.* **50**, 269–295.

SPRUIT, H.-C. 1977 Heat flow near obstacles in the solar convection zone. *Solar Phys.* **55**, 3–34.

STEIN, R. F. & NORDLUND, Å. 1989 Toplogy of convection beneath the solar surface. *Astrophys. J.* **342**, L95–L98.

WHITE, O. R. 1994 The solar spectral irradiances from X-ray to radio wavelengths. In *The Sun as a Variable Star: Solar and Stellar Irradiance Variations* (ed. J.M. Pap, C. Fröhlich, H.S. Hudson & S.K. Solanki). Cambridge University Press, in press.

VITENSE, E. 1953 Die Wasserstoffkonvektionszone. *Z. F. Astrophys.* **32**, 135–164.

Solar Variability Traced by Cosmogenic Isotopes

By JÜRG BEER[1], STEPHAN T. BAUMGARTNER[1], BEATE DITTRICH-HANNEN[2], JÜRG HAUENSTEIN[3], PETER KUBIK[4], CHRISTIAN LUKASCZYK[1], WERNER MENDE[5], RITA STELLMACHER[5] AND MARTIN SUTER[2]

[1]Swiss Federal Institute for Environmental Science and Technology (EAWAG), CH-8600 Dübendorf, Switzerland

[2]Institute of Particle Physics, ETH-Hönggerberg, CH-8093 Zürich, Switzerland

[3]Physics Institute, University of Bern, CH-3012 Bern, Switzerland

[4]Paul Scherrer Institute, c/o Institute of Particle Physics, ETH-Hönggerberg, CH-8093 Zürich, Switzerland

[5]Institute of Meteorology, Freie Universität Berlin, Germany

^{10}Be has been measured in a 300 m long ice core from Greenland. After a general analysis of the ^{10}Be data the potential and the limitations of ^{10}Be records in ice to reconstruct solar variability are discussed. Special emphasis is given to periods of low solar activity (Maunder Minimum etc.). The 'clock problem' is addressed, also the question whether the Schwabe cycle maintains its phase. Finally the question of what ^{10}Be can tell us about changes of the solar irradiance in the past is explored.

1. Introduction

Although the existence of sunspots has been known for a long time, it was not before 1843 that the cyclicity of sunspot occurences was recognised by Heinrich Schwabe. His discovery of the 11-year or better Schwabe cycle 150 years ago was the beginning of continuous systematic observations of solar variability. Other parameters like changes in the solar diameter or terrestrial occurences like auroras and magnetic disturbances at antipodal stations (aa index) are all related to the magneto-hydrodynamic processes taking place in the convective zone of the Sun. In recent years the Sun has been continuously monitored by Earth and satellite based instruments. Probably the most important result of the last 15 years is the observation of a positive correlation between the solar irradiance and the sunspot number (solar activity). The irradiance exhibits variations of about 0.1% during one cycle (Willson & Hudson 1988). According to the standard solar model the energy production by fusion is extremely stable over periods of centuries or millennia. Therefore, variations of solar irradiance should only be attributed to fluctuations of the energy transport within the convective zone and to sunspot blocking effects in the photosphere which have to be compensated after some time. Since the Sun represents the engine driving the climate system of the Earth, any variation of the solar irradiance is directly reflected in a change of solar forcing which in turn influences climate. Although solar variability over the last decade was rather small, a fundamental question arises: How large was the solar variability in the past and how large will it be in the future? Based on the level of variability observed for Sun-like stars the potential of the Sun seems to be rather large (Baliunas & Jastrow 1990).

When studying solar variability the main problem is that in spite of an almost complete set of direct observations of all emitted radiation and particles this information is restricted to a very short time period of decades. This time scale is by far not sufficient to investigate the long-term solar variability. The longest record of continuous, direct observations is the sunspot record. It goes back, with decreasing reliability, to about 1700 A.D.. Before the Maunder Minimum period (1645 – 1715 A.D.) there is a lack of consistent records of directly observed solar variability. Several attempts have therefore been made to develop indirect methods with which the history of solar variability can be reconstructed. The search for solar variability in various natural archives such as tree rings, sediments and ice cores turned out to be a rather difficult task due to the smallness of the effects induced by solar variability.

Probably the most successful indirect tool so far for establishing the history of solar variability is the use of cosmogenic radioisotopes (Raisbeck & Yiou 1980). Since their production rate is modulated by the interaction of the solar wind with the primary cosmic ray particles, they provide information which can be attributed to solar variability. Records of counters monitoring secondary cosmic ray particles since 1937 show a clear anticorrelation between the cosmic ray flux impinging on the Earth's atmosphere and the solar activity expressed in sunspot numbers (Beer *et al.* 1991). The main two cosmogenic isotopes used so far are ^{14}C and ^{10}Be. Due to their different geochemical behaviour they are somewhat complementary. ^{14}C oxidises to $^{14}CO_2$ after production and then exchanges between the main carbon reservoirs atmosphere, ocean and biosphere. Due to the large size of these reservoirs short-term variations are strongly dampened and delayed. On the other hand the atmospheric residence time is longer than the corresponding mixing time leading to a global mean value.

The main archive for ^{14}C is tree rings which have been analysed back to $\sim$ 11,000 years B.P. with a resolution of 20 years and for some periods even with one year (Stuiver *et al.* 1991). The absolute time scale is obtained by counting the rings.

^{10}Be attaches itself to aerosols and is removed from the atmosphere mainly by wet deposition within about 1.5 years after its production. As a consequence, the deposition flux at a specific site does not necessarily represent the global mean value. It contains also a local component which can be related to atmospheric processes (e.g. exchange stratosphere - troposphere, precipitation rate).

On the other hand, due to its short atmospheric residence time ^{10}Be is also sensitive to short-term production variations, such as expressed in the Schwabe cycle. Based on model calculations production variations of 30 - 50% are to be expected during a cycle depending on the mean level of solar activity. In the case of ^{14}C these variations are attenuated by about a factor of 100 (Siegenthaler & Beer 1988).

The best archive for ^{10}Be measurements are ice sheets which directly store the atmospheric fall-out. However, 1 kg of ice contains only about 10^7 atoms ^{10}Be which makes detection by decay counting impossible. With the development of accelerator mass spectrometry such small numbers of atoms became measurable.

Newkirk (1984) discussed the potential of this technique in a paper entitled: "What can accelerator mass spectrometry do for solar physics?". Some of the questions he raised were: 1. Does the 11-year cycle continue through the Maunder, Spoerer, and other past minima in the envelope of solar activity? 2. Does the cycle maintain phase through such minima? 3. Does solar activity follow an 80-90 year variation as suggested by the more limited optical and auroral records? 4. Is the modern character of the 11-year cycle typical of eras of high solar activity in the past? 5. Are climatological variations in any way induced or affected by solar variations? 6. What is the power spectrum of solar activity?

In search for answers to the above questions, we now discuss a continuous annual ^{10}Be record covering the last 563 years.

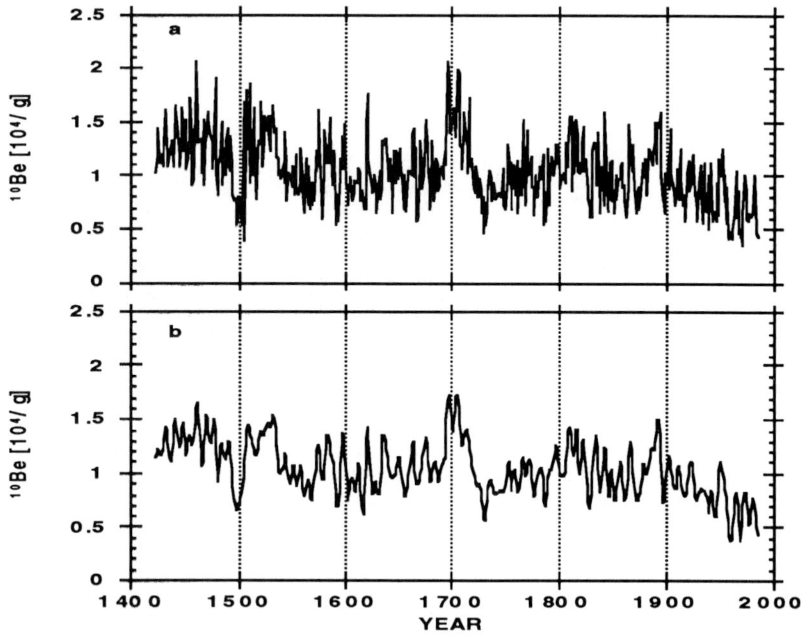

FIGURE 1. (*a*) Annual ^{10}Be concentration in 10^4 atoms per gram ice after interpolation of the measured data using cubic splines. The relative precision (1σ) of the measurements is 5-8%. (*b*) ^{10}Be concentration of the same data after applying a low-pass filter with a cut-off of 6 years.

2. Experimental

The analysed ice core was drilled at Dye 3 (65.2° N; 43.8° W), Greenland. It has a length of 300 m corresponding to 563 years. The ^{10}Be results of the upper part dating as far back as 1783 A.D. have been published earlier (Beer *et al.* 1990). Dating of the upper part is based on determinations of annual layer thicknesses by measuring H_2O_2 which shows strong seasonality. Before 1783 the core was cut into 50 cm and further down into 40 cm long pieces which is close to the annual accumulation. The time scale was established afterwards using δ^{18}O variations on the nearby deep-drilling core (Clausen, personal communication) assuming that the annual layers do not differ significantly between the two cores. Each sample of about 2 kg of ice was prepared following standard procedures and was analysed at the AMS facility of ETH/PSI Zürich (Suter 1990). ^{36}Cl measurements corresponding to the ^{10}Be data presented here are currently being carried out.

3. Results and discussion

3.1. *General analysis of the data*

The measured data which cover the period from 1423 AD. to 1985 AD. have been interpolated using natural cubic splines to obtain equidistant annual data (Figure 1(*a*)). To remove the rather large short-term fluctuations we applied a low-pass filter with a cut-off frequency of 0.164 y^{-1}. (Figure 1(*b*)). The data between 1783 and 1985 have been discussed elsewhere (Beer *et al.* 1990). Figure 1(*a*) shows that annual variations of ^{10}Be concentrations are rather large. We attribute this to fluctuations of transport

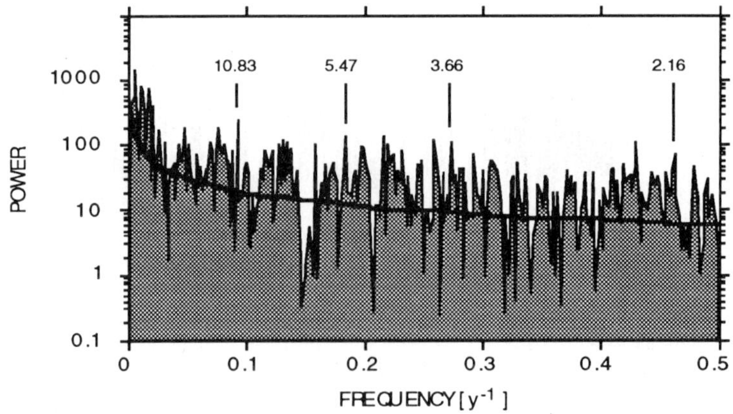

FIGURE 2. DFT-Fourier spectrum of the data of Figure 1(*a*). The solid line represents a hyperbolic model spectrum. The most prominent peak below 45 y corresponds to the 10.83 y Schwabe cycle. Its second, third and fifth harmonic can also be seen.

and deposition processes, because annual variations of the production rate are smaller ($< 50\%$) (Blinov 1988). These short-term variations are superimposed on a long term-trend with maxima near 1500, 1700 and 1800. The 20th century is characterised by a decrease of the ^{10}Be concentration. The filtration in Figure 1(*b*) reveals strong, nearly decadal oscillations which we relate to the solar Schwabe cycle.

Figure 2 displays the results of a DFT-Fourier analysis of the 563 data points shown in Figure 1(*a*) using a rectangular window. One of the most prominent peaks in the spectrum corresponds to a period of 10.83 years. It raises 9 dB above the fitted hyperbolic model spectrum. The second, third and fifth harmonics of this peak are also clearly visible (5.47, 3.66, 2.16 years). The deviations from the theoretical harmonic frequencies are small (-0.98%, -1.33% and 0.29%) and the phases are coherent indicating a common source of the signal. Precise determination of long periods is difficult due to the limited length of the record. Therefore, the transformation window has been extended by means of spline embedding. The Fourier spectrum of this extended series shows among others (60.86, 46.92 and 20.97 years) a spectral peak at 87.4 years which can be attributed to the Gleissberg-cycle.

The coherence of the record was tested by calculating an average spectrum. A window of 256 years was moved in steps of 30 years across the data interval yielding 12 Fourier spectra. Averaging of these spectra reveals again a peak at 10.85 years (including the second, third and fifth harmonic) in good agreement with the single spectrum of the whole ^{10}Be record.

3.2. ^{10}*Be and solar variability*

There are strong, but complex couplings between the ^{10}Be production rate and the magnetic properties of the solar wind which itself is related to various solar processes and their manifestations as observed from the Earth. The most famous phenomenon is the sunspot cycle. However, it seems that there are periods when this cycle almost disappears (Maunder Minimum). Other more recently observed phenomena related to the Schwabe cycle are fluctuations in the number of flares, plages, variations of the solar diameter and changes in the emission of solar wind. When the solar wind interacts with the Earth it causes fluctuations of the geomagnetic field strength (aa index), induces ionisation

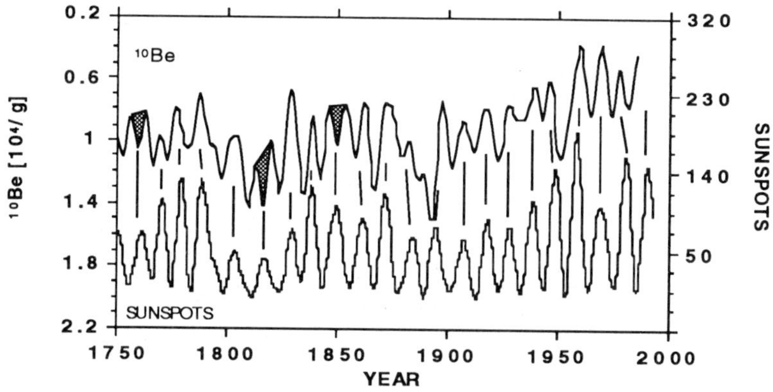

FIGURE 3. Comparison of the filtered ^{10}Be data of Figure 1(b) with the sunspot record. The correspondence between sunspot maxima and ^{10}Be minima is indicated by lines. Three sunspot maxima coincide with local ^{10}Be maxima (shaded).

processes in the upper atmosphere leading to auroras, and modulates the production rate of cosmogenic radioisotopes like ^{10}Be and ^{14}C. Finally, satellite measurements show changes in spectral and the total solar irradiance depending on the sunspot numbers and the intensity of the magnetic network (Lean 1991). Although all these effects are in the end related to the magneto-hydrodynamical processes taking place in the Sun's convective zone, their relationship is not yet well-known. It seems that they are not linearly coupled as far as amplitudes and phases are concerned. The situation is complicated by the fact that some of the parameters show only very small fluctuations and are strongly linked (solar diameter, total irradiance) or are influenced by additional parameters (cosmogenic isotopes). The longest direct record of solar variability, so far, is the sunspot record. Systematic observations began in 1826. The period from approximately 1720 - 1826 has been reconstructed from various individual observations. Going further back in time data become sparse and more and more unreliable. Nevertheless, the 17th century was obviously a very quiet period regarding sunspots (Eddy 1976). Since all other records of solar activity are much shorter (aa index, solar diameter and irradiance) or less reliable (aurora) we concentrate here on a comparison between ^{10}Be and the sunspot data.

A discussion of the modulation of the production rate of ^{10}Be is given by Lal & Peters (1967).

In Figure 3 the filtered ^{10}Be record from Figure 1(b) is compared with the sunspot record for the period 1750 – 1985. Since large sunspot numbers correspond to high solar activity leading to a reduction of the ^{10}Be production rate, the ^{10}Be scale is inverted. As indicated by the lines in Figure 3, 12 peaks are synchronous within one year but some are shifted up to 4 years. All the sunspot maxima can be related to corresponding ^{10}Be minima except at three points (1760, 1820, 1850) where ^{10}Be shows an additional maximum not represented in the sunspot record. These peaks can be interpreted either as real in the sense that they are the result of solar modulation not visible in the sunspot numbers or as a consequence of terrestrial transport and deposition processes. In the first case this would indicate that there are solar processes which lead to changes in the solar wind but do not influence the sunspot number. This seems rather unlikely, although during the Maunder Minimum we have a situation where we observe ^{10}Be variations in spite of the missing sunspots. In the second case we need additional information to separate production and transport effects. One possibility is to combine the ^{10}Be record

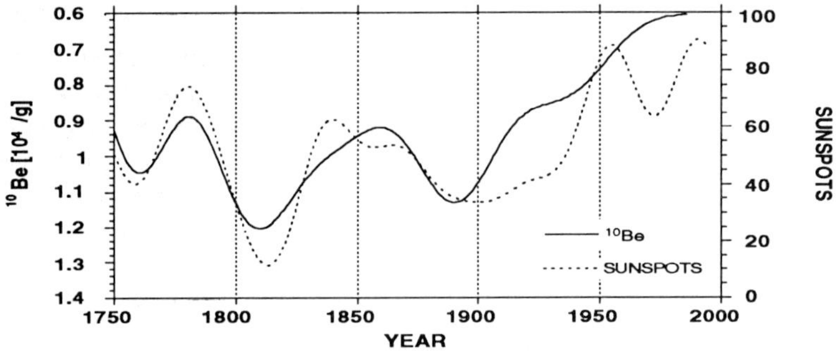

FIGURE 4. Long-term trends of the ^{10}Be concentration and the sunspot number after applying spectral low-pass filters (41 y for ^{10}Be, 31 y for sunspots).

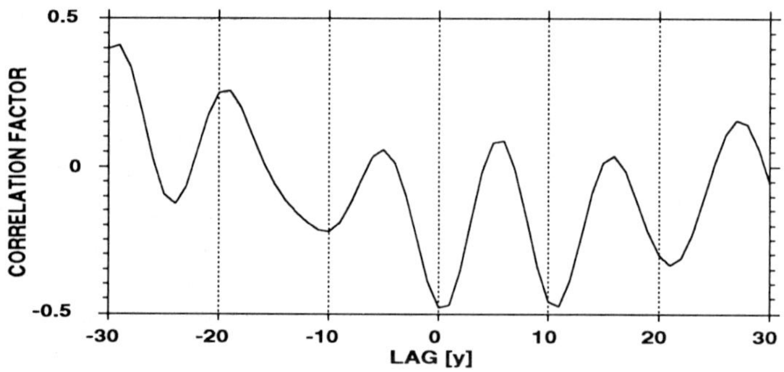

FIGURE 5. Cross correlation function between the ^{10}Be concentration and the sunspot number shown in Figure 3. The highest correlation of -0.46 corresponds to a ^{10}Be lag of 0.5 y.

with the ^{36}Cl record of the same samples. Due to the different geochemical behaviour of ^{10}Be and ^{36}Cl, transport and deposition effects should at least be partly removable. The second possibility is to combine the ^{10}Be record with a second ^{10}Be record from another site. This procedure has been successfully applied earlier (Beer *et al.* 1991).

Figure 4 shows the long-term trends obtained by low-pass filters (41 years for ^{10}Be, 31 years for sunspots). Both curves show the same features, namely the Dalton solar minimum (1795–1830 A.D.) and an increasing long-term trend during the 20th century. The cross-correlation of the data of Figure 3 is displayed in Figure 5. The largest correlation of -0.46 is obtained for a ^{10}Be lag of 0.5 years which is shorter than the 1.5 years found for the period 1783 – 1985 (Beer *et al.* 1990). The oscillations of the cross correlation function reflect the Schwabe cycle with minima about every 11 years.

Summarising the data discussed so far we come to the following conclusions:

1. The ^{10}Be concentration reflects the sunspot cycle for the period 1750 - 1985 quite well as far as the short-term (10-year scale) variations and the long-term (100-year scale) trend are concerned. However, there are three events (indicated in Figure 3) where an additional ^{10}Be peak appears.

2. The Fourier spectrum of the whole ^{10}Be record reveals a clear and sharp peak at 10.83 years with the second, third and fifth harmonic. This is in excellent agreement with the sunspot record which reveals a periodicity of 10.9 years for the same interval 1750–1985. The sharpness of the peak clearly indicates that in spite of the varying lengths

of individual Schwabe cycles the mean cycle length seems to be precisely defined (see discussion of clock problem).

3.3. *Solar minima*

Periods of very low sunspot numbers are especially interesting. The most famous one is the Maunder Minimum period (1645 – 1715 A.D.) when almost no sunspots have been observed (Eddy 1976). The Spörer Minimum (1415-1535 A.D.) is only partly covered by our data. Its evidence is based mainly on ^{14}C measurements on tree rings (Stuiver & Quay 1980).

During the Dalton Minimum (1795–1830 A.D.) sunspots were not completely absent but the numbers were relatively low.

^{14}C measurements on tree rings (Stuiver & Quay 1980) and earlier ^{10}Be measurements on ice cores (Beer *et al.* 1991) revealed higher concentrations of these isotopes during the Maunder and Spörer minima which are in agreement with reduced solar activity. However, the strong damping in the carbon system of short-term variations and the limited time resolution of the ^{10}Be records made a detailed investigation of the Schwabe cycle during these periods impossible. The present ^{10}Be data clearly show that cyclic processes of the same frequency as the Schwabe cycle were operating during the minima (Figure 1(*b*)). There are indications that the amplitude of such processes were even larger than during epochs with high sunspot activity. This result is consistent with the observation that during the Maunder Minimum period auroras did not disappear.

3.4. *The "clock" problem*

Based on an analysis of the Schwabe cycle, Dicke (1978) raised the question whether the sunspot cycle is driven by a precise clock deeply buried inside the Sun meaning that a constant frequency is being phase modulated by a variable lag-function. Depending on the propagation time of each stroke to the solar surface the observable cycle length is changing. Nevertheless, according to this model the frequency averaged over many cycles is kept constant. The eruption model assumes that, if an eruption should be late, the lateness will be passed on to all the following cycles leading to a random walk in phase. One way to look at this question is to calculate the "melody", the momentary frequency for all times in a record. This has been done for the ^{10}Be record based on band-passed data (6-20 years). The results are shown in Figure 6.

The mean frequency of the melody is 0.0928 $year^{-1}$ corresponding to a periodicity of 10.78 years in excellent agreement with the power spectrum mean frequency of the Schwabe peak. Although the frequency shows fluctuations caused by changes in the phase, the mean phase shift is in agreement with the clock model. This result is consistent with the sharpness of the Schwabe cycle peak in the power spectrum (Figure 2). However, the record is still too short for a final answer. We cannot exclude that during earlier periods the phase shows a random walk. Other records also show remarkably sharp lines with the same frequencies supporting the hypothesis that the Schwabe cycle is a phase modulated oscillation (Mende, in preparation).

3.5. *Changes of solar irradiance and climate*

Satellite observations since 1978 show that the solar irradiance varies during a solar cycle due to fluctuations of the energy transport within the convective zone. However, the measured variations over one solar cycle are very small (~ 0.1 %). Therefore, for a discussion of the relevance to the climate four basic questions have to be answered:
1) What is the potential variability of the solar irradiance? 2) How much did the irradi-

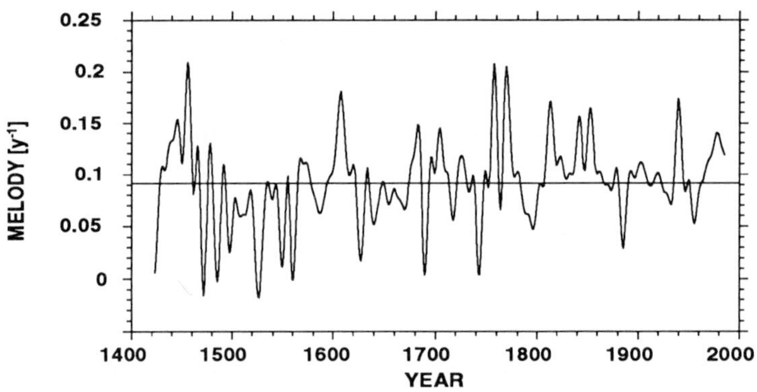

FIGURE 6. ^{10}Be melody derived from the ^{10}Be concentration after applying a band filter (6-20 y). The mean frequency corresponds to a periodicity of 10.78 y in excellent agreement with the 10.83 y peak in the Fourier spectrum (Figure 2).

ance change in the past? 3) How can we reconstruct these changes in the past? 4) What is the response of the climate system to such changes?

Based on a 25 year observation period of 33 Sun-like stars it has been concluded that the potential for brightness variability of the Sun is much larger than observed in the measurements since 1978 (Baliunas & Jastrow 1990; Lockwood *et al.* 1992). Compared to the energy content of the convective zone of approximately 10^{45} ergs, a change in the solar irradiance by 0.5% for one century corresponds to a change of 10^{-4} of the total energy storage. The potential of solar variability seems to be much larger than executed during the last millennium.

A complete discussion of these topics is beyond the scope of this paper. We concentrate therefore mainly on question three. A large number of publications (Stephenson 1988) deals with the question how proxies of solar variability can be used to reconstruct changes in the irradiance. Since there are no long records of direct observations available, the only possible approach is to compare proxies with climate records, preferably the mean global surface temperature. However, this approach is hampered by several unknowns:

The quality of the older climate data is limited, because they are increasingly based on proxies. The response of the climate system to long-term trends has to be estimated with models. The measured relationship between solar variability and irradiance is only based on the very short observation period since 1978 and the underlying physical processes are not yet understood. The climate system is also strongly influenced by the increase in greenhouse gases and by changes of the atmospheric aerosol content.

In spite of all these difficulties many studies (Friis-Christensen & Lassen 1991; Kelly & Wigley 1992; Lean *et al.* 1992; Schlesinger & Ramankutty 1992) have shown partially striking similarities between climate records and solar proxies.

In Figure 7 the inverted ^{10}Be record (Figure 1(*b*)) is compared to a combined temperature record of the northern hemisphere (Groveman & Landsberg 1979 and Jones *et al.* 1986). There is generally good agreement between the long-term trends. Also, some of the most prominent features such as the temperature minimum at 1815 and the subsequent sudden increase are well represented in the ^{10}Be record. During the 20th century the temperature increases more steeply than the ^{10}Be record. This could be interpreted as a result of greenhouse warming. On the basis of this simple correlation there is good reason to believe that the solar irradiance was changing in the past and that these changes are reflected in climate records. But for a quantitative discussion

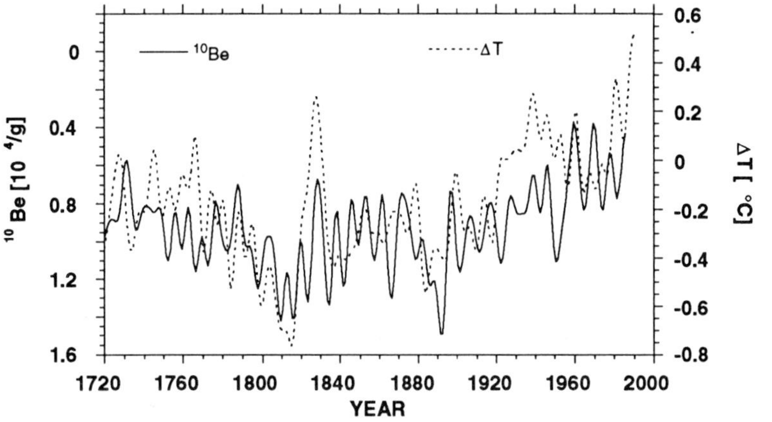

FIGURE 7. Comparison of the ^{10}Be record (Figure 1(*b*)) with deviations from the mean temperature of the northern hemisphere (Groveman & Landsberg 1979; Jones *et al.* 1986).

additional parameters like solar cycle lengths and atmospheric aerosol content have to be included.

4. Summary and conclusions

To date the Dye 3 ^{10}Be record provides the most impressive record of historical solar activity. It will be extended back by several thousand years in the near future. The analysis of the annual ^{10}Be concentration data covering the period 1423 to 1985 A.D. reveals interesting results concerning the solar-terrestrial relationship and provides the following answers to the questions raised by Newkirk (1984): The 11-year cycle does continue through the Maunder and most probably the Spörer Minimum. Although the cycle length varies deviations from the mean cycle length of 10.83 years seem to be compensated. The 87-year Gleissberg cycle is also present in the ^{10}Be record.

There are clear indications that at least in part climate changes are related to changes in solar activity. An important question to be studied in more detail is the coupling between the solar activity and the Earth system. The relatively large variations of the UV part of the solar spectrum during a Schwabe cycle leads to large effects in the upper atmosphere which may be coupled in some, yet unknown way, to the lower atmosphere.

Acknowledgments. The authors thank all their colleagues involved in the core drilling, the sample preparation and the AMS measurements. J. Beer is especially indebted to Gordon Newkirk for his encouragement to work in this field. This work was supported by the Swiss National Science Foundation.

REFERENCES

BALIUNAS, S. & JASTROW, R. 1990 Evidence for long-term brightness changes of solar-type stars. *Nature* **348**, 520–523.

BEER, J., BLINOV, A., BONANI, G., FINKEL, R. C., HOFMANN, H. J., LEHMANN, B., OESCHGER, H., SIGG, A., SCHWANDER, J., STAFFELBACH, T., STAUFFER, B., SUTER, M. & WOELFLI, W. 1990 Use of ^{10}Be in polar ice to trace the 11-year cycle of solar activity. *Nature* **347**, 164–166.

BEER, J., RAISBECK, G. M. & YIOU, F. 1991 Time variations of ^{10}Be and solar activity. In *The Sun in Time* (ed. C. P. Sonett, M. S. Giampapa & M. S. Matthews). pp. 343–359. Univ. of Arizona Press.

BLINOV, A. 1988 The dependence of cosmogenic isotope production rate on solar activitiy and geomagnetic field variations. In *Secular Solar and Geomagnetic Variations in the last 10,000 Years* (ed. F. R. Stephenson & A. W. Wolfendale). pp. 329–340. Kluwer Academic Publishers.

DICKE, R. H. 1978 Is there a chronometer hidden deep in the Sun? *Nature* **276**, 676–680.

EDDY, J. A. 1976 The Maunder Minimum. *Science* **192**, 1189–1201.

FRIIS-CHRISTENSEN, E. & LASSEN, K. 1991 Length of the solar cycle: An indicator of solar activity closely associated with climate. *Nature* **254**, 698–700.

GROVEMAN, B. S. & LANDSBERG H. E. 1979 Reconstruction of Northern hemisphere temperature. 1579 - 1880. University of Maryland, College Park, MD: Publication 79-181.

JONES, P. D., RAPER, S. C. B., BRADLEY, R. S., DIAZ, H. F., KELLY, P. M. & WIGLEY, T. M. L. 1986 Northern Hemisphere surface air temperature variations, 1851-1984. *J. Clim. Appl. Met.* **25**, 161–179.

KELLY, P. M. & WIGLEY, T. M. L. 1992 Solar cycle length, greenhouse forcing and global climate. *Nature* **360**, 328–330.

LAL, D. & PETERS, B. 1967 Cosmic ray produced radioactivity on the Earth. In *Handbuch für Physik, 46/2* (ed. S. Flügge). pp. 551–612.

LEAN, J. 1991 Variations in the Sun's radiative output. *Reviews of Geophysics* **29/4**, 505–535.

LEAN, J., HULBURT, E. O., SKUMANICH, A. & WHITE, O. 1992 Estimating the Sun's radiative output during the Maunder Minimum. *GRL* **19**, 1591–1594.

LOCKWOOD, G.W., SKIFF, B. A., BALIUNAS, S. L. & RADICK, R. R. 1992 Long-term solar brightness changes estimated from a survey of Sun-like stars. *Nature* **360**, 653–655.

NEWKIRK, G. JR. 1984 What accelerator mass spectrometry can do for solar physics. *Nucl. Instrum. Meth.* **B5**, 404–410.

RAISBECK, G. M. & YIOU, F. 1980 ^{10}Be in polar ice cores as a record of solar activity. In *Proc. Conf. on Ancient Sun* (ed. R.O. Pepin, J.A. Eddy & R.B. Merrill). pp. 185–190. Pergamon Press.

SCHLESINGER, M. E. & RAMANKUTTY, N. 1992 Implications for global warming of intercycle solar irradiance variations. *Nature* **360**, 330–333.

SIEGENTHALER, U. & BEER, J. 1988 Model comparison of ^{14}C and ^{10}Be isotope records. In *Secular Solar and Geomagnetic Variations in the last 10,000 Years* (ed. F. R. Stephenson & A. W. Wolfendale). pp. 315–328. Kluwer Academic Publishers.

STEPHENSON, F. R. 1988 Solar variability from historical records. In *Secular Solar and Geomagnetic Variations in the last 10,000 Years* (ed. F. R. Stephenson & A. W. Wolfendale). pp. 109–129. Kluwer Academic Publishers.

STUIVER, M., BRAZUNIAS, T.F., BECKER,B. & KROMER, B. 1991 Climatic, solar, oceanic and geomagnetic influences on Late-Glacial and Holocene atmospheric 14C/12C change. *Quat. Res.* **35**, 1–24.

STUIVER, M. & QUAY, P. D. 1980 Changes in atmospheric Carbon-14 attributed to a variable Sun. *Science* **207**, 11–19.

SUTER, M. 1990 Accelerator mass spectrometry: State of the art in 1990. *Nucl. Instrum. Meth.* **B52**, 211–223.

WILLSON, R.C. & HUDSON, H. S. 1988 Solar luminosity variations in solar cycle 21. *Nature* **332**, 810–812.

Solar Forcing of Global Climate Change

By PAUL E. DAMON AND JOHN L. JIRIKOWIC

Department of Geosciences, University of Arizona, Tucson, AZ 87521, USA

Using present global warming and paleoclimatic records from climatically sensitive regions as a frame of reference, we infer that global temperature changes did not exceed ±0.5°C during the current interglacial or ±2.0°C during the last glacial period. In order to completely explain such fluctuations by solar irradiance changes, all solar and terrestrial factors must be optimized. Since this is unlikely, we conclude that solar forcing of pre-anthropogenic climate change is a significant and perhaps dominant factor but other processes must also be significant. Solar irradiance changes alone cannot result in the temperature decrease required to change from interglacial to glacial as suggested by Opik (1965).

1. Introduction

1.1. *Glacial and interglacial climate*

During the Pleistocene (1.8 Ma to present), the Earth's climate has varied between two states, glacial and interglacial. This can be seen in the oxygen isotope stratigraphy in Figure 1. The meteorological cycle has been likened to a multiple-stage still with 6 to 7 theoretical distillation plates (Epstein 1958). The amount of fractionation in the hemispheric distillation process is a function of altitude, latitude and temperature gradient from equatorial to polar regions. The variations of the $^{18}O/^{16}O$ ratio during the hydrologic cycle are measured in ‰ relative to mean seawater. With reference to the Byrd core, Antarctica, it can be seen that $\delta^{18}O$ fluctuates around −40‰ from about 110 ka to 18 ka, gradually increasing after 18 ka to fluctuate around a new steady-state, −34‰, from about 10 ka to present. The increase in $\delta^{18}O$ from −40‰ to −34‰ is the result of global warming and coincident with rising sea-level as the continental ice masses melted (Fairbanks 1989). The Camp Century, Greenland core shows $\delta^{18}O$ variation around the same two steady states but during the transition form 18 ka to 10 ka, the warming is temporarily interrupted by a return to glacial conditions during the Younger Dryas from about 12,900 BP to 11,600 BP. Thus, the hydrologic cycle is responding to both regional and global climate changes. The North Atlantic is more sensitive to climate change during the glacial period as can be seen by the larger fluctuations of $\delta^{18}O$ in Greenland ice cores when compared to Antarctic ice cores but the two states are global in nature. In general, global climate during the Pleistocene can be modeled as a set of systems fluctuating around the glacial and interglacial steady states separated by transitional episodes. $\delta^{18}O$ fluctuations around steady state are three to four times less during the interglacial compared to the glacial epoch.

1.2. *Regional vs. global climate change*

According to estimates from the CLIMAP project (CLIMAP Project Members, 1981), the Laurentides ice mass reached a thickness of 3800 m and the northwestern European ice mass reached a thickness of 2500 m, whereas the Greenland ice cap reached an intermediate thickness of about 3200 m. Using an empirical equation from Dansgaard (1964), we can estimate the glacial age temperature at Camp Century to have been $\approx -38°C$. The Northern Hemisphere continental areas covered by ice must have been at similar temperatures, warming to their present annual temperatures. As a robust

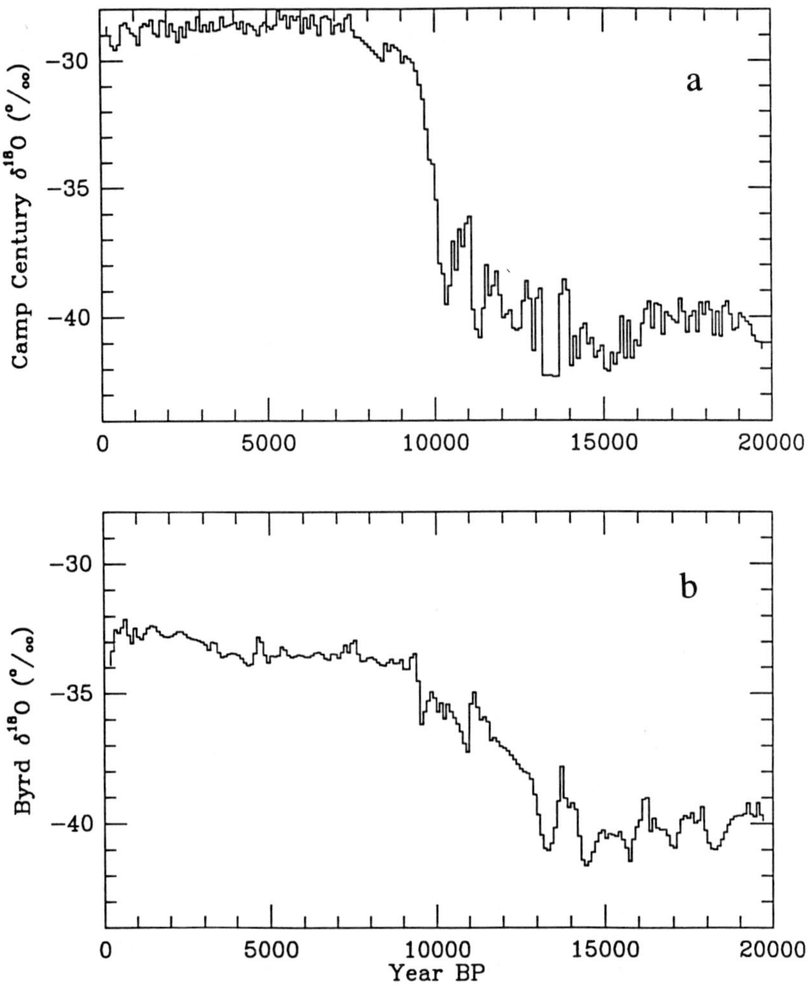

FIGURE 1. Ice core δ^{18}O (Camp Century, Greenland, after Dansgaard *et al.* 1982 (1a); Byrd Station, Antarctica, after Johnsen *et al.* 1972, (1b)). Note that both isotopic records show distinct modes between glacial and interglacial states. For the glacial state (20 ka-13 ka), δ^{18}O varies by an amplitude of 2‰ on a century time-scale about an average value of ≈ -40‰. During the interglacial (9 ka-present), δ^{18}O varies by an amplitude of <0.5‰ around an average of ≈ -29‰ for Camp Century and ≈ -33‰ for Byrd Station. The decreased δ^{18}O variability suggests a decrease in sensitivity to climate forcings. Camp Century shows greater glacial to interglacial contrast due the influence of surface ice and snow expanse during the glacial period. A sharp increase in the Camp Century record contrasts with the more gradual increase in the Byrd Station record. This may reflect the ameliorating influence of the Southern Ocean. The pronounced Younger Dryas negative deflection ($\approx$10.5 ka) present in the Camp Century may correlate with a much smaller feature ($\approx$10 ka) in the Byrd Station record (calibrated ages are about 1700 years older as in the text). The Younger Dryas event may have affected North Atlantic circulation greatest and thus adjacent continental indicators have more profound signatures. Thus, δ^{18}O records contain both global and region climate signals.

Region	ΔT (°C)	Area of Globe (%)	Contribution to Global Mean weighted by area (°C)
Presently Glaciated	−10	3	−0.3
Not glaciated at present	−40	5.7	−2.3
Other continental areas	−6	20.5	−1.2
The world oceans	−2.3	70.8	−1.6
Hansen *et al.*, 1984			$\Sigma = -5.4$
GCM Model:			$\Sigma = -5.3$

TABLE 1. Temperature changes (ΔT) from present to 18 ka.

but not rigorous estimate of the temperature increase for those areas, we suggest about 40°C. For the areas still covered by polar ice, we estimate a change of about 10°C. From a paleoclimatology summary by Crowley & North (1991), we take the average change at mid-latitudes to be between 5–8°C and 4–5°C for tropical continent and areas under the influence of maritime air masses. Let us assume an average of ≈6°C for continental areas not covered by ice sheets. CLIMAP Project (McIntyre *et al.* 1976) estimated an average change for the global oceans of only 2.3°C but with great variation between different ocean basins. For example, in the Northern Hemisphere, they estimated temperature changes varied from 0.8°C for the Indian Ocean to 3.8°C for the North Atlantic. We summarize and weight these changes by areas in Table 1, arriving at a global change of −5.4°C from the glacial to the present state. This compares favorably with more rigorous GCM (global circulation model) modeling by Hansen *et al.* (1984) and serves to illustrate several pertinent points. First, the global oceans serve as a great source and sink of heat moderating climate. Second, 8.7% of the area of the Earth accounts for ≈50% of the global warming during the transition from glacial to interglacial states. Third, there were large regional differences in temperature changes around the global mean.

As previously noted, oxygen isotope stratigraphy demonstrates that the hemispheric temperature fluctuations during the past 10,000 years (the Holocene) are relatively small. The linear empirical relationship between δ^{18}O and temperature (Bradley 1985) suggests that these fluctuations do not exceed ±0.6°C (or 1.2°C peak-to-peak). If this is correct, global warming during the past 100 years of 0.5°C based on instrumental records (Folland *et al.* 1990) should be a very significant event relative to previous events during the Holocene. Of course, regional temperature variations of sub-hemispheric scale have been both lesser and greater. For example, there was an insignificant increase in temperature during the last century for the contiguous United States comprising roughly 1.6% of global area (Hanson *et al.* 1989). However, from data in Trends, 1991 (Boden *et al.* 1991), we calculate an increase of (1.2±0.2)°C(s) from solar cycle 14 (1901-1913) to solar cycle 21 (1976-1987) for the southern deserts which is one of 23 regions of the U.S.A. reported in Trends, 1991. These sorts of regional variations of climate change together with conceptual and dating problems make it difficult to evaluate past changes in Holocene global temperatures. The instrumental record coupled with GCM studies can be helpful in resolving this problem. Specifically, in order to evaluate the feasibility of solar forcing of global climate, we need to know the magnitude of global Holocene temperature changes to be forced. As noted above, oxygen isotope studies of polar ice suggest global changes equal to or less than ±0.6°C during the Holocene. This suggestion needs further justification.

GCM studies suggest that global temperature change will be greatest at high altitudes and high continental latitudes (Wang *et al.* 1991). High altitude temperatures are governed by evaporation of surface ocean water with transfer of 540 calories per gram of latent heat to higher latitudes. Albedo changes, as the result of changes in the ice and snow cover, produce high-latitude continental sensitivity to global climate forcing. Evidence for climatic events like the Medieval Warm Epoch and Little Ice Age come from North Atlantic border-lands and Europe above 45°N. It has been possible recently to demonstrate by study of montane glaciers that these climatic events are global rather than merely regional in scope (Rothlisberger 1986). Montane glaciers also provide perspective concerning the relative intensity of 20th Century global warming.

Hasenrath & Kruss (1992) have compared glacial maps for Mount Kenya (elevation 4.6 to 5 km). They estimate that the Mount Kenya Glacier retreated by 40% in area from 1963 to 1987. The retreat of the terminus of the Speke Glacier in the Ruwenzori Range of Uganda has accelerated from an average of $\approx$2 m/yr between 1958 and 1977 to more than 11 m/yr between 1977 and 1990 (Kaser & Noggler 1991). The areal extent of worldwide montane glaciation has been reduced by about 50% during the last century (Thompson 1991). Thompson *et al.* (1989) have studied the Dunde Ice Cap on the Qinghai-Tibetan Plateau at a summit elevation of 5235 m. According to these authors, in that region "the last 60 years were apparently one of the warmest periods in the entire record, equaling levels of the Holocene maximum between 6000 and 8000 years ago" (p. 474).

Evidence for high-latitude warming comes from dendroclimatologic studies and measurements of the temperature gradients in bore holes. Innes (1991) has documented high-altitude and high-latitude increases in 20th Century tree growth from about 30 studies of trees from Northern Europe and North America. In contrast to Thompson's conclusion, Innes states, "It appears that neither the northern extent nor the present growth rates of boreal forests have reached the extent of rates characteristic of the mid-Holocene, when the climate was demonstrably warmer." However, from their dendroclimatologic study of high-latitude boreal sites in Canada and Alaska, Jacoby & Arrigo (1989) conclude that "the global warming trend over the past century of increasing CO_2 has exceeded the recent (past three centuries) level of natural variability (pre-anthropogenic) of the climate system". In a study of old-growth Huon pines from a sub-alpine site in western Tasmania, Cook *et al.* (1992) concluded that, "Although the most recent 25-year period is warmer than any comparable period over the past 1090 years, it is not yet statistically unprecedented" (p. 215). Graybill & Shiyatov (1992) have reconstructed June-July temperatures for ad 961-1969 using tree-ring width variation of Siberian Larch from the Polar Urals (65-67°N). They concluded that "the relatively sustained rise in temperature from the 1880s to the 1960s has the greatest magnitude in the record for such an extended period (1.6°C difference between the 20-year intervals centered on 1870 and 1960)" (p. 411). Note that for this climate sensitive area, this recent regional warming event is threefold greater than 20th Century global warming of $\approx$0.5°C. Bore hole temperature gradients from Central and Eastern Canada (Beltrami *et al.* 1992; Shen & Beck 1992; Wang *et al.* 1992) as well as from the Alaskan Arctic (Lachenbruch & Vaughn 1986; Kakuta 1992) confirm the temperature rise from the late 19th Century through the late 20th Century. The temperature rise in Canada was typically 1–3°C and greater at Prudhoe Bay, Alaska where Kakuta (1992) estimates warming by 4°C from ad 1853 to ad 1973.

Summarizing, the modern instrumental record has made it possible to estimate 20th Century global warming of about 0.5°C (Folland *et al.* 1990). There is insufficient data to establish directly global climate change prior to the 19th Century. The existing data is biased toward relatively high-latitude, climatically-sensitive regions of the North Atlantic.

However, our comparison of the recent warming event with the paleoclimate record within climatically-sensitive regions suggests that this recent event exceeds in intensity natural global climate variability during the last millennium and is of comparable magnitude to the Holocene maximum between 6000 and 8000 years ago. Oxygen isotope stratigraphy also demonstrates the relative global stability of pre-anthropogenic climate. Solar forcing needs only account for global temperature variations of no more than ±0.5°C to account for all the natural climate variability during the last millennium. The Holocene maximum may exceed a global temperature increase of ±0.5°C. However, our previous estimate of ±0.6°C variability from the polar oxygen isotope record is probably a conservative upper limit for global temperature change during the Holocene.

2. Solar forcing of Holocene climate

2.1. *The Sun as a low-frequency harmonic oscillator*

The production of cosmogenic ^{14}C is modulated by solar activity via the magnetic field associated with the solar wind. The relationship is inverse and, so, the various historically observed solar minima such as the Maunder Minimum occur as ^{14}C maxima and the Medieval Solar Maximum from ad 1100-1250 occurs as a ^{14}C minimum in the spectrum of ^{14}C variations. The reader may refer to two papers for recent reviews of the subject (Damon & Sonett 1991; Stuiver *et al.* 1991). Damon & Sonett (1991) concluded that medium- and short-term variations with periods of ≈208 (Suess), 88 (Gleissberg), and 11 (Schwabe) years are in large part produced by solar wind modulation. A longer period of ≈2300 years (Hallstattzeit) seemed enigmatic but appears to 100% amplitude modulate the Suess cycle. The ≈2300-year period seems to be associated with recurrent Little Ice Ages (Schmidt & Gruhle 1988). Stuiver & Braziunas (1991), using the maximum entropy spectral method (MEM), chose a 420-year period as a fundamental with harmonics related to changes in the solar convective zone and noted that it splits into three periods at higher AR order. We note that the apparent splitting is an artifact of the MEM method. The peaks are always present but vary greatly with the autoregressive (AR) order. Too low an AR order results in oversmoothing. We deduced from both DFT and MEM power spectral analyses that the four periods (526, 424, 356 and 300 years) are the 4, 5, 6, 7 harmonics of the poorly resolved Hallstattzeit period that we estimated from its harmonics to be 2120 years (Damon & Jirikowic 1992a,b). Further analysis yielded 212 years for the Suess period with strong (≥95% confidence) harmonics at 106 years (2nd harmonic) and 53 years (4th harmonic).

The harmonics of the Suess period combined with the Gleissberg period appear to strongly modulate the 11.1-year (Schwabe) cycle. Furthermore, a reasonable fit to the annual Wolf Sunspot Indices can be obtained using a squared amplitude modulation model (after Sonett 1982) by allowing these periods (212, 106, 53, 88.3 years) to amplitude modulate the 11.1-year carrier (Damon & Jirikowic 1992a).

Subsequent to the work summarized above, one of the authors (Jirikowic) added Thomson's (1990) multi-taper power spectral analysis technique to the techniques made available to us by Prof. Charles Sonett here at the University of Arizona (periodogram, Bretthorst-Bayesian, Maximum Entropy [MEM], Discrete Fourier Transform [DFT]). The analysis of the latest version of the high-precision ^{14}C data set (Stuiver & Reimer 1993) is shown in Figure 2. The Hallstattzeit fundamental and first nine overtones (2–$10f_H$) clearly resolve at ≥95% confidence (harmonic F-test). The Suess period (f_S and $10f_H$) has unusually high power spectral density and appears to act as a fundamental as well as a harmonic of the Hallstattzeit. The accounts for the high spectral power

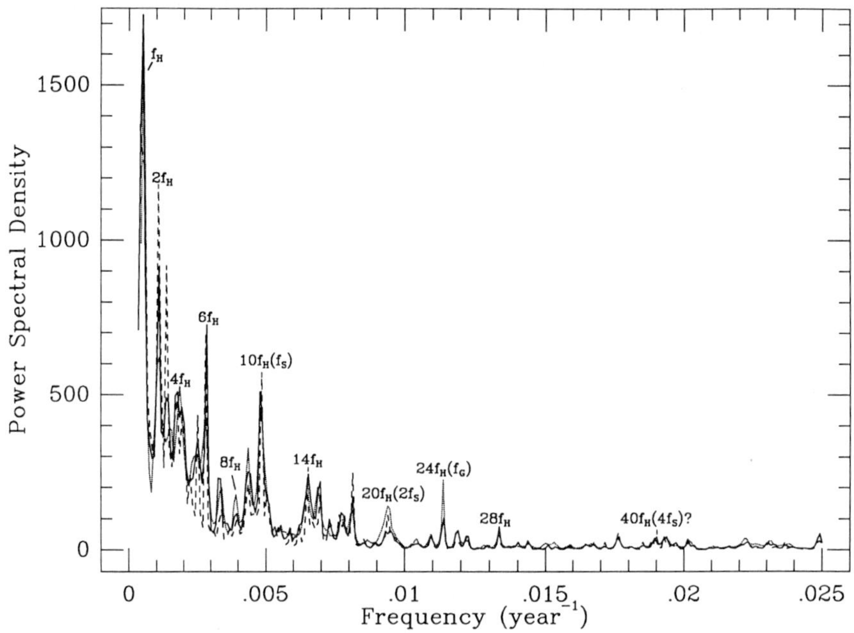

FIGURE 2. Power spectral density estimates using Thomson's Multi-taper Method (Thomson 1990) and three autoregressive modeling orders: 120 (dotted), 160 (solid) and 200 (dashed) from the 1993 Radiocarbon Calibration tree-ring δ^{14}C data (Stuiver & Reimer 1993). Annotated peaks show assignment to hypothetical harmonics to the fundamental Hallstattzeit (subscript H), Suess (subscript S) and Gleissberg (subscript G) variations. All these peaks have greater than 95% F-test confidence. Although the Hallstattzeit is poorly resolved, the annotated harmonics combine to suggest a fundamental period of 2108 $\pm$ 15 years. Earlier spectral analyses (Damon & Jirikowic 1992b) showed a strong fourth harmonic of the Suess that is not apparent here.

spectral density of its second harmonic ($2f_S$, $\geq$95% confidence). However, its fourth harmonic appears as a relatively weak doublet, unlike the MEM and DFT techniques ($\geq$95% confidence). The Gleissberg period is prominent at $\geq$95% confidence (88.1 years).

Neglecting the poorly resolved second Hallstattzeit harmonic but using harmonics 3-10 plus 14, 15, 17, 18 (all with $\geq$95% confidence) to calculate the Hallstattzeit fundamental, we obtain 2108$\pm$15 ($\bar{\sigma}$ years compared with 2115$\pm$18 ($\bar{\sigma}$) years previously obtained using the combined MEM and DFT methods (Damon & Jirikowic 1992b).

The enigmatic Hallstattzeit period may be divided into two episodes. For $\approx$1520 years, it is associated with high solar activity and weak periodicity followed by a $\approx$600–year episode in which solar activity is low on the average and the Suess period and its harmonics as well as the Gleissberg period are strong. These 600-year episodes are associated with Little Ice Ages. Thus, when the Medieval Solar Maximum ended at ad 1250, the Little Ice Age began accompanied by a series of profound solar minima and terminated after about 600 years toward the middle of the 19th Century when the Modern Solar Maximum began (Jirikowic & Damon 1994).

Instrumental analysis has confirmed a significant change in solar irradiance of 0.1% during the 21st solar cycle (Willson & Hudson 1988, 1991). Paleoclimate data suggests

Solar Condition	Change in Solar[1] Irradiance	Δ Forcing (tropopause level)	ΔT_{eq} (°C) Glacial[2] 2.4°C	ΔT_{eq} (°C) Present Glacial[2] 1.0°C	ΔT_{eq} (°C) Future[3] 0.6°C
Present Sun AD 1980–1986	0.0	0.0	0.0	0.0	0.0
No plage, spots or network	−1.5	−0.36	−0.9	−0.4	−0.2
No–cycling stars (average)	−2.4	−0.58	−1.4	−0.6	−0.3
Minimum possible Sun	−3.5	−0.84	−2.0	−0.8	−0.5

[1]From Lean *et al.* (1992).
2Calculated from Hansen *et al.*(1984).
[3]Calculated from Hansen *et al.*(1984) assuming variations are negligible on an ice–free Earth.

TABLE 2. Decreasing equilibrium temperature (T_{eq}) as a function of solar condition.

that longer periods such as the Hallstattzeit, Suess and Gleissberg periods may also be associated with changes in solar irradiance (e.g., see Eddy 1977). Further evidence comes from solar and astrophysics.

2.2. *The Sun as a variable star*

As discussed elsewhere in this colloquium's proceedings (see Lockwood 1994), 74 solar-type stars have been observed at the Mount Wilson Observatory. Of these 74, 13 have been studied monthly since 1966 and nightly since 1980 (Baliunas & Jastrow, 1990). Only sporadic measurements were made since 1978 for the other 61 stars. Nine of the 13 stars were observed to show a range of magnetic activity similar to the solar cycle whereas four that showed no variation of magnetic activity with time were assumed to be analogous to the Sun during the Maunder and other solar minimum. The nine magnetically active stars had regular activity cycles like the Sun. Their brightness varied by up to 0.4% compared to 0.1% for the Sun during cycle 21. Baliunas & Jastrow (1990) state, "Our results suggest that when a solar-type star enters or leaves a Maunder Minimum, its brightness may change by more that the 0.1% observed during the Sun's 11-year cycle" (p. 522). Lockwood *et al.* (1992) presented a compilation of eight years of observations of 33 stars at the Lowell Observatory. They report that the stars whose mean activities were closest to the Sun's varied in brightness by 2.5 to 10 times the variation in brightness of the Sun. They state, "This raises the question of whether the Sun's cycle irradiance variation is always so small or whether larger changes occur in other cycles or over longer time intervals" (p. 655).

2.3. *Solar variability and the Maunder Minimum*

Based on current knowledge of the Sun and Sun-like stars, Lean *et al.* (1992) estimated the change in solar irradiance and forcing at the top of the atmosphere for various conditions of the Sun (see Table 2). Assuming, as did Baliunas & Jastrow (1990), that the Maunder Minimum was equivalent in change of irradiance to the average solar-like non-cycling stars yields a 2.4 change in irradiance. Also, assuming that a global equilibrium temperature decrease of 1°C was required to explain Little Ice Age cooling, they conclude, "... that solar forcing may have contributed some, but not all, of the observed warming from the Little Ice Age to the present time" (p. 1594).

Keeping in mind that the ≈0.5°C rise in temperature in recent time began from a

Feedback	Gain	ΔT (°C)
Unity	0	-0.6
Water vapor–clouds	0.415	−2.2
Land ice	0.169	−0.9
Sea ice	0.132	−0.7
Vegetation	0.057	−0.3
CO_2	0.113	−0.6
	$\sum g_i = 0.886$	$\sum \Delta T_{\text{eq}} = -5.3$

TABLE 3. Gains (g_i) and temperature increments resulting from feedbacks at the Ice Age Maximum (18 ka).

relatively cool global temperature in the late 19th Century, and this rise approaches or exceeds previous such warming events during the Holocene (see 1.2 above), we suggest that the estimate of 1°C decrease in global cooling exceeds what is required to explain the Little Ice Age cooling during the Maunder Minimum. We concluded that less than ±0.5°C global cooling is required to explain climate variability during the last millennium. The global temperature rise from the Maunder Minimum to ad 1980-1986, based on the paleoclimatic record and instrumental measurements discussed, is greater than 0.5°C but less than 1°C.

In order to evaluate the temperature decrease resulting from a given change in solar irradiance (Table 2), various positive feedbacks must be considered. For our analyses, we refer to Hansen *et al.* (1984). Their paper included analysis of the various gains (g_i) and temperature increments (ΔT_i) including those for water vapor-clouds, land ice, sea ice, vegetation, and CO_2 (see Table 3).

From Table 3, the equilibrium temperature change, ΔT_{eq}=5.3°C and the temperature increment that would occur without feedbacks, ΔT_0=0.6°C. The total feedbacks, $\sum f_i$ is related to the total gain, $\sum g_i$, by:

$$\sum_i f_i = \frac{1}{1-\sum_i g_i} = \frac{1}{1-0.886} = 8.8 \tag{2.1}$$

and ΔT_0 is related to ΔT_{eq} by:

$$\Delta T_0 = \frac{\Delta T_{\text{eq}}}{\sum_i f_i} = \frac{5.3°\text{C}}{8.8} = 0.6°\text{C} \tag{2.2}$$

The relative increment in solar energy required to yield the increment ΔT_0 is:

$$\frac{\Delta S_0}{S_0} = 4\frac{\Delta T_0}{T_{\text{eq}}} = \frac{4 \times 0.6°\text{K}}{255°\text{K}} = 9.4 \times 10^{-3} \tag{2.3}$$

Note that the surface temperature of Earth is warmer by ≈33°K due primarily to the greenhouse gases. The change in solar energy required to obtain ΔT_0 is

$$9.4 \times 10^{-3} \times 1367.54\text{Wm}^{-2} = 12.8\text{Wm}^{-2}$$

that corresponds to a forcing at the tropopause level of 12.8 Wm^{-2}÷5.7 or 2.2 Wm^{-2}. This yields a climate sensitivity of 5.3/2.2 Wm^{-2} = 2.4°C/Wm^{-2}.

Likewise, Hansen *et al.* (1984) estimate $\sum f_i$ at present to be 3 to 4, we take f ≈3.6 that yields a climate sensitivity of 1°C/Wm^{-2}. If we assume in a world without ice, albedo effects are negligible, the temperature increase required to attain an ice-free state

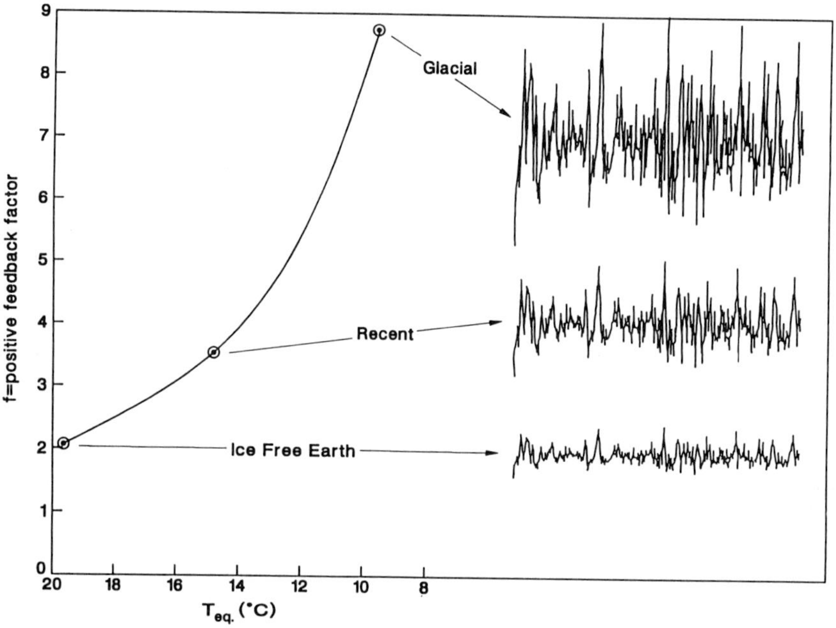

FIGURE 3. The relation between equilibrium surface temperature (T_{eq}), positive feedbacks ($\sum f_i$) and climate sensitivity. As a consequence of the increased positive feedbacks in colder climates, the same change in solar energy would have the greatest effect during glacial epochs and the least effect for an ice-free Earth.

would be a minimum of 4.8°C (Barron & Washington 1984), the sum of feedbacks would be 2.1 and the climate sensitivity would be $\approx$0.6°C/Wm^{-2}.

In order to go from the present climate to an ice-free (swamp-ocean) climate requires a forcing at the tropopause level of 8.6/Wm^{-2} or an increase of 3.5% in the solar constant whereas a cooling from present to the climax of the last ice age at 18 ka requires only a forcing decrease of 2.2 Wm^{-2} or a decrease in the solar constant of 9.4×10^{-3}‰. Neither of those changes are obtainable by changes in solar irradiance alone (Table 2). However, significant solar modulation of climate change within the two steady states, glacial-inter-glacial, can be obtained. Neglecting the increased positive feedback during a Little Ice Age, which we estimate to be only $\approx$10%, a peak-to-peak change of 0.6°C (Table 2) explains most, if not all, of the Holocene temperature variation. The same change in solar irradiance, 2.4‰, would produce a peak-to-peak variation of 1.4°C at 18 ka. Figure 3 illustrates the increased variability of climate during the ice ages and decreased variability at present and during an ice-free climate.

The evidence that the recent warming event approaches or perhaps exceeds past events sets limits on natural variability during the Holocene but we must keep in mind that warming since the Little Ice Age is the combined effects of increased solar activity and greenhouse forcing. According to Hansen & Lacis (1990), greenhouse forcing to 1989 was 2.04 Wm^{-2}, to which must be added our estimated solar forcing of 0.58 Wm^{-2} (Table 2) or about 2.6 Wm^{-2} total. This forcing at equilibrium, assuming the validity of the climate sensitivity of 1°C/Wm^{-2}, would yield an increase of 2.6°C since the Maunder Minimum.

However, heat is transported to the mixed layer of the ocean and from there to the deep sea resulting in a long lag time. From their three-dimensional (3D) model, Hansen *et al.* (1984) estimated "the time for the global area-weighted mixed layer temperature to reach 63 percent of its equilibrium response in 124 years" (p. 156). The 124-year e-folding time is applicable for the high climate sensitivity of $1°C/Wm^{-2}$. If we divide the 0.6°C (Table 2) equilibrium temperature increase equally between warming after the Maunder Minimum and additional warming during the Modern Solar Maximum, we calculate that 25% of the current warming of 0.5°C is from solar forcing and 75% is the result of greenhouse gas forcing. The temperature rise following the Maunder Minimum would attain its equilibrium value of 0.3°C to which 0.125°C must be added from the warming due to the current solar maximum. The resultant 0.425°C would then constitute a little more than half of the 0.8°C warming since the Maunder Minimum.

Our argument, thus far, has been based upon the high climate sensitivity of $1°C/Wm^{-2}$. As a result of uncertainties in the feedbacks, the range of possible climate sensitivities lies between $0.5°C/Wm^{-2}$ and $1.25°C/Wm^{-2}$ (Hansen & Lacis 1990). The e-folding time is also a function of the feedbacks ($\sum f_i$). For the lowest value of climate sensitivity, $0.5°C/Wm^{-2}$, the e-folding time is only about 40 years. This much shorter residence time slightly more than compensates for the lower greenhouse forcing climate sensitivity but the solar contribution falls to 13% of the current global warming. The total temperature rise since the Maunder Minimum becomes 0.65°C, of which only 33% is due to solar forcing.

It is of interest to apply this analysis to pre-anthropogenic climate change. For this, we take predicted temperature change between the Medieval Solar Maximum (ad 1100-1250) and the immediately following Wolf Minimum (ad 1250-1350). We assume, as above, that the equilibrium warming and subsequent cooling would each be 0.3°C. The predicted temperature fall must be calculated from the e-folding time of 124 years. It would be ≈0.4°C. This global change is not great but would be observable in climatically sensitive regions using appropriate paleoclimate indicators. It would probably be below detection levels in other regions. If the climate sensitivity is $0.5°C/Wm^{-2}$, the temperature decrease would only be ≈0.3°C but probably still observable. At high latitudes and at high elevations, the temperature might change by 1°C as compared to a 2°C warming during the current global warming event. There remains the possibility of greater cooling during prolonged periods of intense solar minima such as the Little Ice Ages (ad 1250-1850), for example. To completely account for Holocene temperature changes, however, it seems that all permissible solar and terrestrial factors must be maximized.

3. Conclusions and summary remarks

1. Pleistocene climate alternates between two steady states, glacial and interglacial, separated by transitional episodes. Climatic fluctuations around steady state are greater during glacial epochs than interglacial epochs by about a factor of three to four as observed in oxygen isotope stratigraphy.

2. Regional temperature changes are much more variable than and should not be confused with global temperature change. Since climatic events like the Little Ice Age, Medieval Warm Epoch and Altithermal were originally defined based upon paleoclimatic and historical events from Northern Europe and the North Atlantic regions, there has been a tendency to exaggerate global temperature changes during the Holocene.

3. Present global warming can serve as a frame of reference in evaluating Holocene global temperature changes from paleoclimate records within climatically sensitive regions such as high-latitude and high-altitude continental areas. Such comparison strongly

suggests that global temperature change did not exceed ±0.5°C during the last 10 ka or ±2.0°C during the last glacial period.

4. The Sun is a low frequency harmonic oscillator with solar activity fundamental frequencies of 1/2120 yr, 1/212 yr, 1/87 yr and 1/11 yr plus harmonics. The 1/2120 yr cycle includes two phases, episodes lasting 600 years during which the century-scale variation includes pronounced minima such as the Maunder Minimum separated by 1500-year episodes of high and more chaotic activity.

5. The Sun in one of a family of Sun-like stars most of which have cycles of brightness similar to the Sun's. These Sun-like stars vary in brightness up to four times the variation in brightness of the Sun. Some of the Sun-like stars are in quiescent states similar to the Sun during the Maunder Minimum (Baliunas & Jastrow 1990; Lockwood *et al.* 1992).

6. Lean *et al.* (1992) suggest that, during a solar minimum, the Sun decreases in brightness by 2.4‰ relative to the present Sun similar to the average variation of the non-cycling stars (Table 2).

7. Based on the 3D GCM analysis of Hansen *et al.* (1984), decreases in irradiance of 2.4‰ during solar minima relative to the present Sun would result in decrease of equilibrium global temperature by from 0.3°C to 0.6°C during the present interglacial and 0.9°C to 1.4°C during the last glacial maximum at 18 ka.

8. Applying paleoclimatic restraints on global temperature and using the 3D GCM analysis of Hansen *et al.* (1984) with a climate sensitivity of $1°C/Wm^{-2}$, we estimate that global temperature increased by ≈0.3°C following the Maunder Minimum and more than 0.1°C of the 0.5°C of the current global warming since the late 19th Century would be the result of increased solar irradiance during the Modern Solar Maximum. Thus, about 50% of warming since the Maunder Minimum would be the result of increase solar activity. If the lowest estimated climate sensitivity is assumed ($0.5°C/Wm^{-2}$), the solar contribution would be only about 33%.

9. A climate sensitivity of $1°C/Wm^{-2}$ requires that, if no more greenhouse gases were released to the atmosphere, global average temperature would continue to increase by 2.6°C due ≈20% to solar activity and ≈80% to the effect of greenhouse gases. Such an increase would be unprecedented during the last 2 Ma. With inevitable further release of greenhouse gases, global climate would tend towards an ice-free globe held in check only be the large thermal inertia of the global oceans and polar ice.

10. Assuming the validity of the astrophysics-solar physics analysis of Lean *et al.* (1992), changes in solar irradiance cannot result in a change from interglacial to glacial as suggested by Opik (1965). However, the century-scale and millennial changes in solar activity can account for much of the climate fluctuations around the glacial and interglacial steady states due to the existence of strong positive feedbacks.

11. In order to completely explain Holocene global temperature changes by solar forcing, all terrestrial and solar factors must be optimized. Since this is unlikely, we conclude that solar forcing of pre-anthropogenic global climate change is a significant and perhaps dominant factor, but other processes must also be significant.

12. Added note: The Vostok Antarctica core demonstrates relative climate stability during the present interglacial (Holocene) and also the penultimate (Eemian) interglacial (Dansgaard *et al.* 1993). The apparent instability during the Eemian suggested by the recent GRIP and GISP cores from Greenland (ibid.) appears to be caused by layer thinning and deformation rather than climate instability (e.g., Grootes *et al.* 1993).

Acknowledgments. We are grateful to Professor Charles Sonett for the use of his computer facilities and programs for power spectral analysis and for valuable advice. We also thank Dr. Thomas Crowley and James Hansen for their helpful comments.

This work was supported by NSF Grants ATM-8919535, EAR-9203383 and the State of Arizona.

REFERENCES

BALIUNAS, S. & JASTROW, R. 1990 Evidence for long-term brightness changes of solar-type stars. *Nature* **348**, 520–523.

BARRON, E. J. & WASHINGTON, W. M. 1984 The role of geographic variables in explaining paleoclimate results from Cretaceous climate model sensitivity studies. *J. Geophys. Res.* **89**, 1267–1279.

BELTRAMI, H., JESSUP, A. M. & MARESCHAL, J.-C. 1992 Ground temperature histories in eastern and central Canada from geothermal measurements: evidence for climate change. *Paleogeogr., Paleoclimatol., Paleoecol. (Global Planetary Section)* **98**, 167–184.

BODEN, T. A., SEPANSKI, R. J. & STOSS, F. W. (eds.) 1991 *Trends '91: A Compendium of Data on Global Change.* Pub. No. ORNL/CDIAC-46, ESD Pub. No. 3746, Carbon Dioxide Information Center, Oak Ridge, TN, pp. 665.

BRADLEY, R. S. 1985 *Quaternary Paleoclimatology: Methods of Paleoclimate Reconstruction.* Allen and Unwin, Boston, USA. pp. 472.

COOK, E., BIRD, T., BARBETTI, M., BUCKLEY, B., D'ARRIGO, R. & FRANCEY, R. 1992 Climatic change over the last millennium in Tasmania reconstructed from tree-rings. *The Holocene* **2**, 205–217.

CROWLEY, T. J. & NORTH, G. R. 1991 Paleoclimatology. Oxford Univ. Press, New York, pp. 339.

DAMON, P. E. & JIRIKOWIC, J. L. 1992a The Sun as a low-frequency harmonic oscillator. *Radiocarbon* **34**, 199–205.

Damon, P. E. & Jirikowic, J. L. 1992b Radiocarbon evidence for low frequency solar oscillations. In *Rare Nuclear Processes* (ed. P. Povinec). pp. 177–202. Proceedings of the 14th Europhysics Conference on Nuclear Physics, World Scientific Publishers, Singapore.

DAMON, P. E. & SONETT, C. P. 1991 Solar and terrestrial components of the atmospheric ^{14}C variation spectrum. In *The Sun in Time* (ed. C. P. Sonett, M. S. Giampapa & M. S. Matthews). pp. 361–388. Univ. of Arizona Press, Tucson, AZ, USA.

DANSGAARD, W. 1964 Stable isotopes in precipitation. *Tellus* **16**, 436–468.

DANSGAARD, W., CLAUSEN, H. B., GUNDESTRUP, N., HAMMER, C. U., JOHNSEN, S. J., KRISTENSDOTTIR, P. M. & REEH, H. 1982 A new Greenland ice core. *Science* **218**, 1273–1277.

EDDY, J. A. 1977 Climate and the changing Sun. *Climatic Change* **1**, 173–190.

EPSTEIN, S. 1959 The variations of the $^{18}O/^{16}O$ ratio in nature and some geologic implications. In *Researches in Geochemistry* (ed. P. H. Abelson). pp. 217– 240. John Wiley and Sons, New York, USA.

FAIRBANKS, R. G. 1989 A 17,000-year glacio-eustatic sea level record: influence of glacial melting rates on the Younger Dryas event and deep-ocean circulation. *Nature* **342**, 637–642.

FOLLAND, F. P., KARL, T. & BARNETT, T. P. 1990 Observed climate variations and change. In *Climate Change (The IPCC Scientific Assessment)* (ed. J. T. Houghton, G. J. Jenkins & J. S. Ephraums). pp. 195–238. Cambridge Univ. Press.

GRAYBILL, D. A. & SHIYATOV, S. G. 1992 Dendroclimatic evidence from the northern Soviet Union. In *Climate Change since a.d. 1500* (ed. R. S. Bradley & P. D. Jones). pp. 393–414. Routledge, London.

GROOTES, P. M., STUIVER, M. & WHITE, J. M. C. 1983 A long isotopic record at the Greenland Summit. *EOS* **74(43)**, 78 (abstract).

HANSEN, J. E. & LACIS, A. A. 1990 Sun and dust versus greenhouse gases: an assessment of the relative roles in global climate change. *Nature* **346**, 713–719.

HANSEN, J., LACIS, A., RIND, D., RUSSELL, G., STONE, P., FUNG, I., RUEDY, R. & LERNER, J. 1984 Climate sensitivity: analysis of feedback mechanisms. In *Climate Processes and*

Climate Sensitivity (ed. J. E. Hansen & T. Takahashi). pp. 130–163. Geophys. Monograph 29, Maurice Ewing, 5, Amer. Geophys. Union, Washington, D.C., USA.

HANSON, K., MAUL, G. A. & KARL, T. R. 1989 Are atmospheric "greenhouse" effects apparent in the climate record of the contiguous U.S. (1895-1987)? *Geophys. Res. Lett.* **16**, 49–52.

HASENRATH, S. & KRUSS, P. D. 1992 Greenhouse indicators in Kenya. *Nature* **355**, 503–504.

INNES, J. L. 1992 High-altitude and high-latitude tree growth in relation to past, present and future global climate change. *The Holocene* **1**, 168–173.

JACOBY, G. C. & D'ARRIGO, R. 1989 Reconstructed Northern Hemisphere annual temperature since 1671 based on high-latitude tree-rings from North America. *Climatic Change* **14**, 39–59.

JIRIKOWIC, J. L. & DAMON, P. E. 1994 The medieval solar activity maximum. *Climatic Change*, in press.

JOHNSEN, S. J., DANSGAARD, W., CLAUSEN, H. B. & LANGWAY, C. C. JR. 1972 Oxygen isotope profiles through the Antarctic and Greenland ice sheets. *Nature* **235**, 429–434.

KAKUTA, S. 1992 Surface-temperature history during the last 1000 years near Prudhoe Bay: applying control theory to the inversion of borehole temperature profiles. *Paleogeogr., Paleoclimatol., Paleoecol. (Global Planetary Section)* **98**, 225–244.

KASER, G. & NOGGLER, B. 1991 Observations on Speke Glacier, Ruwenzei Range, Uganda. *Jour. of Glaciology* **37**, 313–318.

LACHENBRUCH, A. & MARSHALL, B. V. 1956 Geothermal evidence from permafrost in the Alaska Arctic. *Science* **234**, 689–696.

LEAN, J., STUMANICH, A. & WHITE, O. 1992 Estimating the Sun's radiative output during the Maunder Minimum. *Geophys. Res. Lett.* **19**, 1591–1594.

LOCKWOOD, G.W. 1994 Irradiance variations of stars. In *The Sun as a Variable Star: Solar and Stellar Irradiance Variations* (ed. J.M. Pap, C. Frohlich, H.S. Hudson & S.K. Solanki). Cambridge Univ. Press, in press.

LOCKWOOD, G. W., SKIFF, B. A., BALIUNAS, S. L. & RADICK, R. R. 1992 Long-term solar brightness changes estimated from a survey of Sun-like stars. *Nature* **360**, 653–655.

MCINTYRE, A., MOORE, T. C. & 34 OTHER CLIMAP PROJECT MEMBERS 1976 The surface of the ice-age Earth. *Science* **191**, 1131–1144.

OPIK, E. J. 1965 Climatic change in cosmic perspective. *Icarus* **4**, 289–307.

ROTHLISBERGER, F. 1986 *10000 Jahre Gletschargeshicte der Erde.* Aarau, Verlag Sauerlander, 416 pp.

SCHMIDT, B. & GRUHLE, W. 1988 Klima, Radiokohlenstoffgehalt und Dendrochronologie. *Naturwissenshaftliche Rundschau* **5**, 177–182.

SHEN, P. Y. & BECK, A. E. 1992 Paleoclimate change and heat flow density inferred from temperature data in the Superior Province of the Canadian Shield. *Paleogeogr., Paleoclimatol., Paleoecol. (Global Planetary Section)* **98**, 143–165.

SONETT, C. P. 1982 Sunspot time series: spectrum from square law modulation of the Hale Cycle. *Geophys. Res. Lett.* **9**, 1313–1316.

STUIVER, M. & BRAZIUNAS, T. F. 1989 Atmospheric ^{14}C and century-scale solar oscillations. Nature **338**, 405–408.

STUIVER, M & REIMER, P. J. 1993 Extended ^{14}C data base and revised CALIB 3.0 ^{14}C age calibration program. *Radiocarbon* **35**, 215–230.

STUIVER, M., BRAZIUNAS, T. F., BECKER, B. & KROMER, B. 1991 Climatic, solar, oceanic, and geomagnetic influences on late-glacial and Holocene atmospheric ^{14}C/^{12}C change. *Quaternary Research* **35**, 1–24.

THOMSON, D. J. 1990 Quadratic-inverse spectrum estimates: applications to palaeoclimatology. *Trans. Roy. Soc. London* **A332**, 539–597.

THOMPSON, L. G. 1991 Private communication, Workshop on the Medieval Warm Epoch, Tucson, Arizona.

THOMPSON, L. G., MOSLEY-THOMPSON, E., DAVIS, M. E., BOLZAN, J. F., DAI, J., YAO,

T., Gunderstrup, N., Wu, X., Klein, L. & Xie, Z. 1989 Holocene-Late Pleistocene climatic ice core records from Qinghai-Tibetan Plateau. *Science* **246**, 474–477.

Wang, K., Lewis, T. J. & Jessup, A. M. 1992 Climatic changes in eastern and central Canada inferred from deep borehole temperature data. *Paleogeogr., Paleoclimatol., Paleoecol. (Global Planetary Section)* **98**, 129–141.

Wang, W. C., Dudek, M. P., Liang, X. Z. & Kiehl, J. T. 1991 Inadequacy of effective CO_2 as a proxy in simulating the greenhouse effect of other radiatively active gases. *Nature* **350**, 573–577.

Willson, R. C. & Hudson, H. S. 1988 Solar luminosity variations in solar cycle 21. *Nature* **332**, 810–812.

Willson, R. C. & Hudson, H. S. 1991 The Sun's luminosity over a complete solar cycle. *Nature* **351**, 42–44.

Response of the Middle Atmosphere to Solar Variability – Model Simulations

By THERESA Y. W. HUANG AND GUY P. BRASSEUR

National Center for Atmospheric Research, Boulder, CO 80307, USA

Solar flux variations could affect the middle atmosphere through modulating the photolysis of chemical series and solar heating rates. Indirect feedback effects from chemical, radiative, and dynamical interactions could provide additional sources for perturbations in the middle atmosphere. In this paper, recent developments in modeling the effect of solar variability on the middle atmosphere is described. For the 27-day solar rotational cycle, the temperature and ozone response in the stratosphere predicted by one- and two-dimensional models compares well with data analyses. For the 11-year solar cycle, model simulations suggest a non-negligible ozone/temperature response compared to changes produced by anthropogenic perturbations in the stratosphere. There is no sufficient long-term atmospheric dataset to establish a statistically significant correlation with the 11-year solar cycle. But in general, agreement between the observational analysis (for periods of one to two solar cycles) and model simulations of the long-term solar variability effect is unsatisfactory.

1. Introduction

Understanding how solar flux variations affect our atmosphere provides a way to learn about the interactive processes among the photochemistry, dynamics and radiation in the middle atmosphere. The long-term trends of temperature and ozone associated with anthropogenic perturbations can be more accurately estimated if the long-term changes caused by solar variability are well understood. Until recently, modeling the response of the middle atmosphere to solar variability had been impeded by the absence of reliable solar flux measurements, and by the lack of observational data required for meaningful model verification. With satellite measurements made during the past decade, more information is available regarding the cyclic variation of solar irradiance and relevant atmospheric data. This warrants a look into the effect of solar flux variation on the middle atmosphere from a model simulation perspective.

The electromagnetic flux emitted by the Sun varies with time predominantly in the 27-day and 11-year time period. Changes in solar irradiance on the 27-day solar rotational time scale arise from the uneven distribution of the active regions on the Sun. The long-term irradiance changes related to the 11-year solar cycle are caused by the changing emission of bright magnetic features, such as faculae and the magnetic network. Shown in Figure 1 is the time evolution of solar flux from observations by the Solar Mesosphere Explorer (SME), between the period 1982 – 1988 for the Lyman-α line (121.6 nm), O_2 Schumann-Runge bands (180–200 nm), and the O_2 Herzberg continuum (205 nm) (extracted from Brasseur 1993). Enveloped within the long-term variation of the solar flux are the short-term oscillations corresponding to the Sun's rotational period. The variability of the solar irradiance as a function of wavelength, as estimated by Rottman (1988), is shown in Figure 2a for the 11-year solar cycle and in Figure 2b for the 27-day solar variability. The largest change seen in the solar flux is in the Lyman-α line; the variability decreases substantially with increasing wavelength, with negligible variations beyond 300 nm. A large uncertainty in the solar flux variation shown in Figure 2a arises from the fact that SME data cover less than one solar cycle period. The long-term drift that may occur in solar detectors is also a cause of concern. In fact, for the 200-210 nm

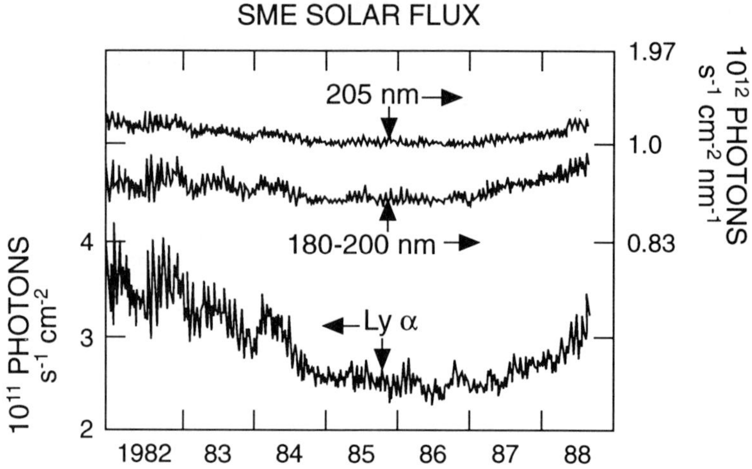

FIGURE 1. Solar actinic flux measure by the Solar Mesosphere Explorer at different wavelengths (121.6 nm, 180-200 nm, 205 nm) between 1982 and 1988, from Brasseur (1993).

region that is important to the photochemistry of the stratosphere, major uncertainties exist regarding the magnitude of the 11-year solar flux variability from a low 5% reported by Rottman (1988) to a high 10% described by Deland & Cebula (1993).

Observational studies that correlate the ozone/temperature temporal variability with variations in the solar irradiance suggest a connection between the solar cycle and the thermal and chemical structure of the middle atmosphere. This is especially true for the short-term variation, as revealed by analyses based on satellite observations in the 1980's (Gille *et al.* 1984; Hood 1986; Keating *et al.* 1987; Hood *et al.* 1991; and Hood & Jirikowic 1991). Convincing statistics are more difficult to obtain for the 11-year solar cycle response since reliable measurement records for the middle atmosphere span only slightly more than one solar cycle. Recent satellite data analyses of ozone, however, suggest a likely correlation with the long-term solar cycle (Chandra 1991; Stolarski *et al.* 1991; Hood & McCormack 1992; Hood *et al.* 1993), as do observations of stratospheric temperature measured by radiosondes (Labitzke *et al.* 1986; Angell 1988, 1991), rocketsondes (Angell 1991) and lidar (Hauchercorne & Chanin 1980; Chanin *et al.* 1987).

The effects of solar variability on the dynamics of the middle atmosphere still remain a matter of discussion. Labitzke & van Loon (1988) suggest an association between the 11-year solar cycle and the stratosphere, based on the analysis of more than three solar cycles of temperature and geopotential height data. Their correlation appears to be affected by the phase of the quasi-biennial oscillation (QBO). This study was followed by a series of statistical analyses on tropospheric temperature and pressure fields (van Loon & Labitzke 1988; Labitzke & van Loon 1989), in which relationships with solar activity were also found. Later studies such as those of Baldwin & Dunkerton (1989), and Salby & Shea (1991) argue against a physical explanation underlying these statistical correlations. Kodera (1993), however, calculated the correlation of the relationship between temperature and zonal wind and showed that a quasi-decadal variation can be seen without decreasing the sampling frequency. This implies that aliasing due to undersampling is not the primary cause of the correlation found by Labitzke & van Loon (1988). Another development that indicates the importance of dynamical feedback effect is the finding by Kodera & Yamazaki (1990) of large December zonal wind variations

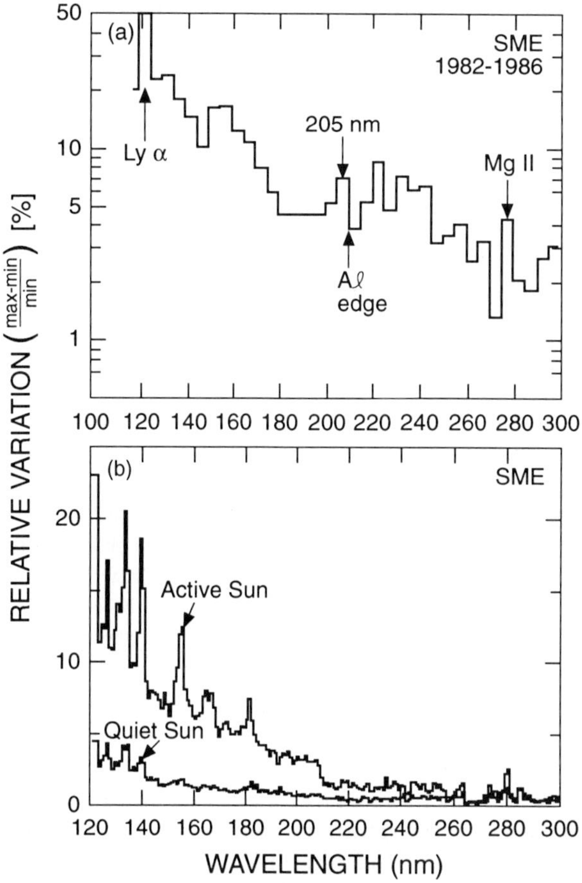

FIGURE 2. Variation (%) in solar ultraviolet flux as a function of wavelength derived from measurements by the Solar Mesosphere Explorer. (a) Variation between solar maximum and solar minimum conditions (11-year cycle); (b) Variation over a solar rotation period (27-day) for active and quite Sun, respectively, from Brasseur (1993).

tude) with solar activity in the upper stratosphere at 40°N. Hood *et al.* (1993) found a similar but somewhat smaller winter solstice zonal wind variation (14 m/s) for the Southern Hemisphere.

Theoretically, the modulation of the solar radiation can affect the structure of the atmosphere several different ways. The most direct is through changes in the solar heating rate and the photodissociation frequency of the photochemical species; these effects would be seen mainly in the higher altitude region (the mesosphere and thermosphere) since the larger changes in the solar flux occur at wavelengths shorter than 200 nm. Although these shorter wavelengths are unable to penetrate into the lower stratosphere, the intricate coupling of photochemistry, radiation, and dynamics in the middle atmosphere may make it possible for the solar variability to affect the lower altitudes of the middle atmosphere through some feedback mechanisms. For example, changes in solar irradiance in the Schumann-Runge bands and Herzberg continuum can affect the photodissociation of molecular oxygen in the mesosphere and stratosphere, which in turn influence ozone abundance. A change in the ozone concentration would affect both the long-wave and short-wave radiative heating which force changes in the temperature, and therefore could affect the meridional circulation and wave activity. The radiative/dynamical changes could then potentially feedback on the ozone balance to modulate the ozone response to the solar variability. Another factor that would allow a response to occur in a region of the atmosphere where the direct impact of solar variability is small is the transport of long lived trace gases from their source regions to other areas. One such example is provided by the odd nitrogen compounds, which are produced through ionic processes in the thermosphere (highly dependent on solar activity) and are transported down to the stratosphere in the polar night. This might directly contribute to the destruction of ozone in the middle to upper stratospheric region when sunlight is present. Another candidate for a feedback process is provided by the effect of wave-mean flow interactions. Zonal wind changes caused by solar variability may trigger changes in the pattern of wave propagation and the magnitude of wave forcing, which could affect the thermal structure of the atmosphere and the zonal wind.

With all the direct solar variability and potential feedback mechanisms operating in the atmosphere, a model simulation is valuable since it enables us to unambiguously test out the connections between solar variability and the atmospheric response within a chosen physical and chemical framework. Through the construction of a model with various degrees of complication, and with the increasing availability of data for comparison, we should be able to identify the major mechanisms that produce the atmospheric response to solar variability. Although it is too early to draw final conclusions about the solar-atmospheric relationship, considerable progress has been made towards our understanding of how solar variability may affect the middle atmosphere. A review will be presented here with the emphasis on model development.

2. Short-term solar variability

With the availability of satellite measurements in the last decade, a statistically significant relationship between 27-day variations in the solar flux and atmospheric quantities, such as ozone concentration and temperature, has been established. A simulation of the response of the atmosphere to the 27-day solar cycle is a good starting point for understanding the fundamental processes underlying the solar-atmospheric relationship.

Early modeling efforts have mainly been focused on using one-dimensional photochemical models to interpret the satellite data analyses (Eckman 1986; Brasseur *et al.* 1987; Summers *et al.* 1990). From a simplified, linearized continuity equation of ozone perturbation without considering temperature feedback, it can be shown that the ozone response is proportional to the O_2 photodissociation frequency (Hood 1986; Brasseur *et al.* 1987). The sensitivity of ozone and temperature relative to the solar flux change at 205 nm is therefore a good indicator of the stratospheric response to solar variability. The temperature and ozone response predicted by these one-dimensional models are in reasonable agreement with observational data for the stratosphere. For example, Figure 3 shows the temperature and ozone response and their phase lags relative to the 205 nm percentage change of solar irradiance calculated from the model of Brasseur *et al.* (1987) and compared with observations estimated by Hood (1986) and Keating *et al.* (1987). At 1 mb, the model predicts a 0.05% change of temperature relative to a 1% change in 205 nm flux, consistent with Hood's (1986) data analysis (0.05 to 0.07%). The calculated temperature phase lag (Figure 3b), however, is approximately a factor of 2 smaller than the values derived from observations. In an earlier work, Hood (1986), based on a simple analytic model, also showed a factor of 2 discrepancy between the predicted and observed phase lags. He concluded that the discrepancy is due to the neglect of dynamical coupling. The inability of the photochemical-radiative model to properly simulate the temperature phase lag explains why the ozone sensitivity and phase lag (shown in

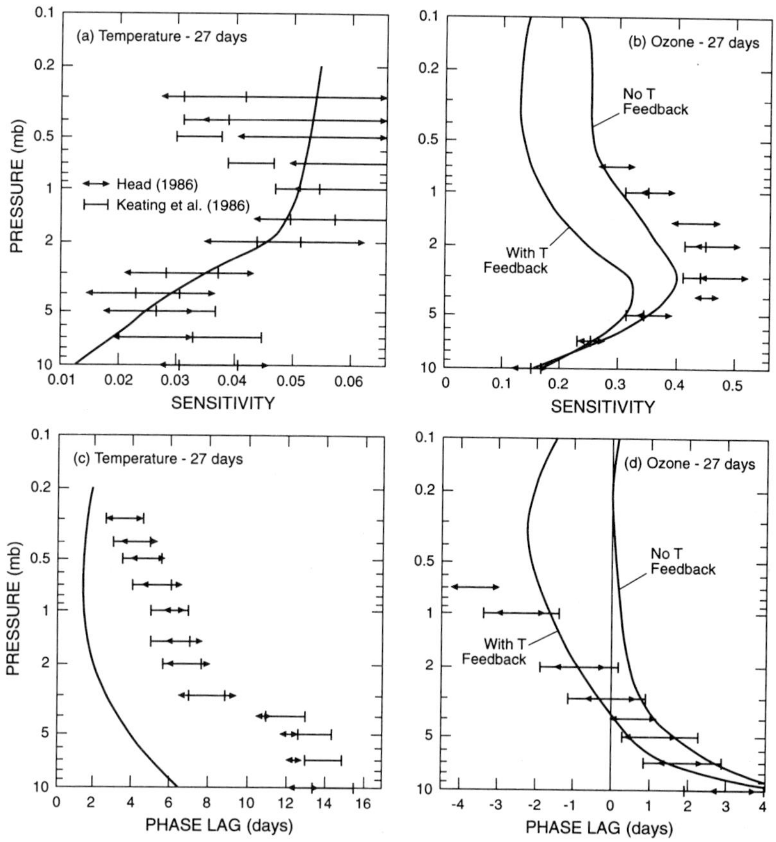

FIGURE 3. a. Amplitude of temperature variation (expressed in percent of the background temperature) calculated in response to a variation in the solar irradiance of 1% at 205 nm, with a period of 27-day. b. Same as a., except for ozone variation. c. Same as a., except for temperature phase lag (expressed in days) d. Same as a, except for ozone phase lag (expressed in days), from Brasseur *et al.* (1987).

Figure 3b and 3c) cannot be more accurately modeled. Nevertheless, the calculations illustrate the effect of temperature feedback on the ozone phase lag. Observations show a negative phase lag (e.g. ozone peak ahead of the solar flux peak) above the level of 3 mb (~40 km), and the model is able to simulate this behavior only when the temperature feedback effect is included. This is because the negative phase lag is mainly a result of the inverse temperature dependence of ozone due to HOx chemistry in the upper stratosphere.

In the mesosphere the ozone response to solar variability is primarily forced by the photolysis of water vapor at Lyman-α (121.6 nm), which produces hydroxyl radicals and hydrogen atoms that contribute to the ozone destruction. Summers *et al.* (1990) used a one-dimensional photochemical-radiative model to investigate the mesospheric response to solar flux variations. A comparison between the calculated ozone response (Figure 4a) with the results of Keating *et al.* (1987) reveals that the observed ozone response is well reproduced when the vertical eddy mixing coefficient used in the model is a factor of 5 smaller than the observationally inferred values. This suggests either a much smaller water vapor mixing ratio in the mesosphere or a less efficient HOx catalytic removal of

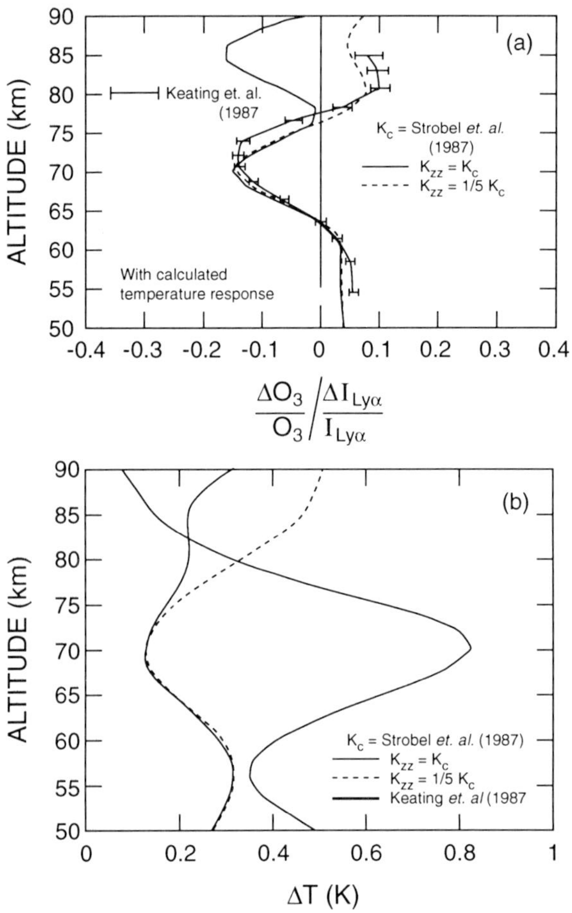

FIGURE 4. a. Ozone response (expressed in percent of background ozone) in response to a variation in the solar irradiance of 1% at 121.6 nm, for a 27-day sinusoidal variation of solar UV flux, using the baseline model Kzz profile, and Kzz decreased by a factor of 5. b. Temperature response (°K) for corresponding Kzz profiles, from Summers *et al.* (1990).

ozone than currently accepted. The most significant discrepancy between the models and observations in the mesosphere, however, is the inability of models to produce the positive correlation between temperature and solar activity in the vicinity of 70 km, opposite to that of ozone and solar activity, as seen in Figure 4b (Keating *et al.* 1987; Hood *et al.* 1991). Hood *et al.* (1991) suggested that HOx chemical heating (not included in the model of Summers *et al.* 1990) could explain this discrepancy, but Huang & Brasseur (1993) pointed out that the HOx chemical heating is weak near 70 km. Hood *et al.* (1991) also suggested that dynamical feedback, that is not adequately described in the models, may play a role.

The discrepancies between the one-dimensional model predictions and observations of the solar effects require that more elaborate models be used to establish the mechanisms responsible for the observed atmospheric response to solar variability. A two-dimensional model with a formulation of coupled chemical, radiative and dynamical processes could account for potential feedbacks not simulated by simple one-dimensional models. Brasseur (1993) used such a model to investigate the temperature and ozone

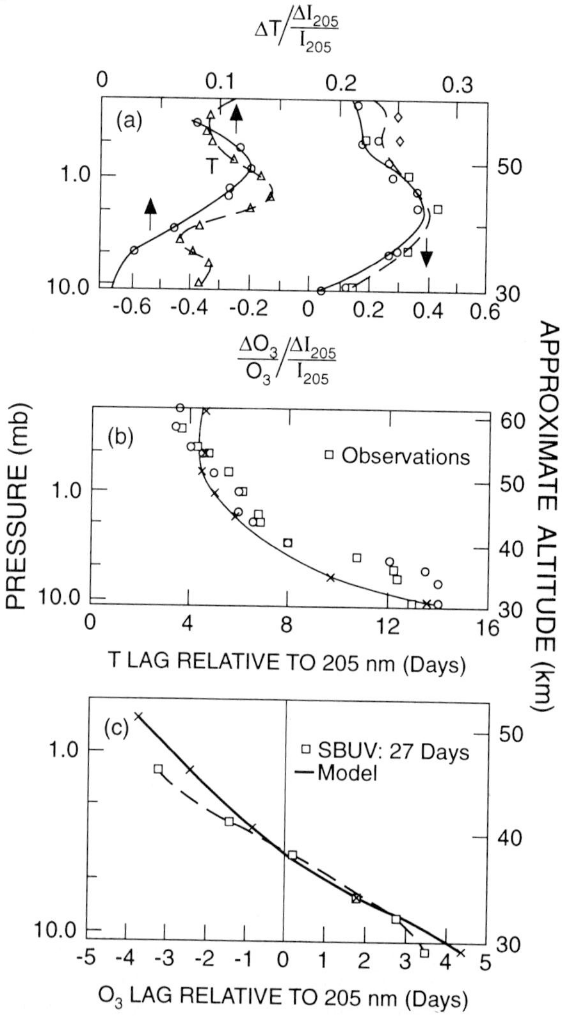

FIGURE 5. a. Ozone and temperature sensitivities relative to a 1% variation (27-day) in solar flux at 205 nm (expressed in units of percentage for ozone, and °K for temperature). Calculated values representative of equatorial regions are compared with data analyzed by Keating *et al.* (1987). b. Calculated time lag (days) of the temperature signal relative to the 27-day solar forcing, compared with observations data (Hood 1986; Keating *et al.* 1987). c. Calculated time lag (days) of ozone signal relative to the 27-day solar forcing, compared with observational data (SBUV) analyzed by Keating *et al.* (1987), from Brasseur (1993).

response associated with the solar rotational cycle. Shown in Figure 5a-c are the amplitudes and phases of the ozone and temperature responses in the stratosphere calculated at the equator by Brasseur (1993), compared to an analysis of data averaged over ±20° latitude (Hood 1986; Keating *et al.* 1987). The calculated ozone sensitivity is of the order of 0.06% relative to a 1% increase in solar irradiance at 205 nm at 30 km, 0.38% at 40 km, 0.25% at 50 km. The calculated temperature sensitivity is 0.01°K per 1% increase in the 205 nm solar flux at 30 km, 0.06°K at 40 km. The overall agreement with the data is good. The phase lags for the temperature and ozone (Figure 5b and 5c) calculated by the 2-D model are also in much better agreement with observations than those derived by 1-D models (Figure 3). The ozone response in the mesosphere (50 to 80

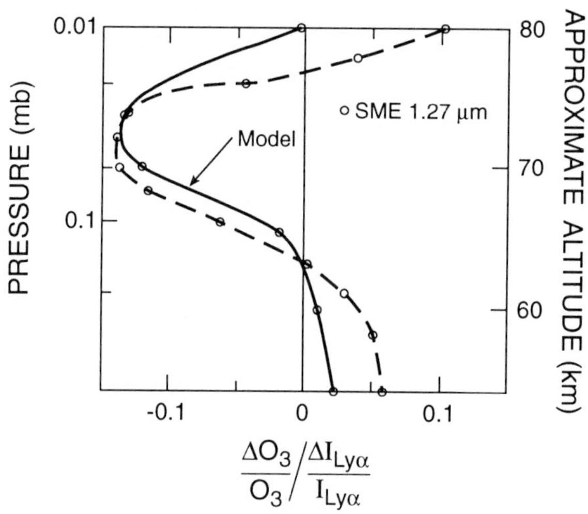

FIGURE 6. Ozone response in the mesosphere relative to a 1% variation (27-day) solar flux at 121.6 nm (expressed in percentage). Calculated values compared to the analysis by Keating *et al.* (1987), from Brasseur (1993).

km altitude) is shown in Figure 6. It is expressed relative to the solar flux at Lyman-α, since the influence of Lyman-α becomes dominant in this region of the atmosphere. The model and observation compares well for mesospheric ozone. Because the radiative and thermal processes important for the mesosphere are not included in the model used by Brasseur (1993), the response of the temperature in the mesosphere cannot be accurately represented by this model. However, Huang & Brasseur (1993) used an extended-altitude version of the model, with a more detailed description of the radiative processes of the mesosphere, and derived the steady-state ozone and temperature response to solar flux change. Even with this more detailed formulation, they could not reproduce the opposite response of temperature and ozone observed at 70 km. Although it was a steady-state simulation, it seems unlikely that using a sinusoidal variation of solar flux would alter the overall picture. Assuming that the observed responses in the mesosphere are indeed related to the solar cycle, it appears that either some important dynamical effect has been ignored, or the current understanding of the photochemical or radiative process is flawed and/or incomplete in the mesosphere. It is clear that much more remains to be learned.

3. Long-term solar variability

The hypothesis that the 11-year solar cycle modulates the ozone concentration and temperature structure has been discussed for decades, and many theoretical studies have been performed to estimate the effect of long-term solar variability on the middle atmospheric climatology (Callis & Nealy 1978; Penner & Chang 1978; Natarajan *et al.* 1981; Brasseur & Simon 1981; Garcia *et al.* 1984; Callis *et al.* 1985; Wuebbles *et al.* 1991; Brasseur 1993; Brasseur & Huang 1993, and references therein). Although the lack of data that covers a sufficiently long period of time precludes an evaluation of the accuracy of the numerical investigations, most models have consistently predicted a small,

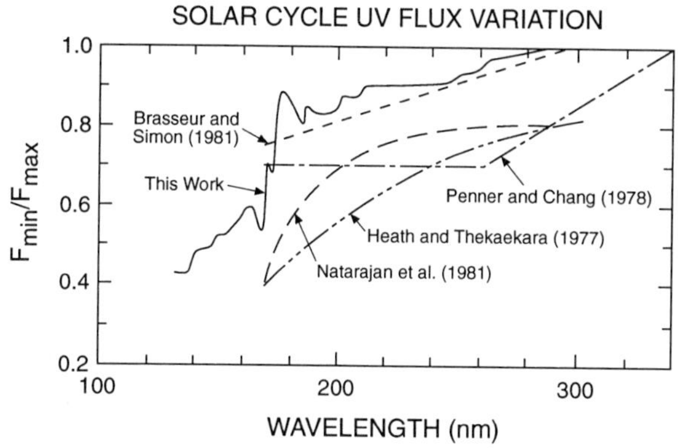

FIGURE 7. Solar flux variation ratio (flux solar minimum/flux solar maximum) adopted in Garcia *et al.* (1984) and several other studies, from Garcia *et al.* (1984).

but non-negligible response of the stratosphere to the 11-year solar cycle when currently accepted long-term solar flux variation values are used in the models.

A one-dimensional photochemical-radiative models have been used by Callis & Nealy (1978), Penner & Chang (1978), and Natarajan *et al.* (1981) to simulate the effect of the 11-year solar variability on the chemical and thermal structure of the stratosphere. In the study of Penner & Chang (1978), the ratio of solar flux between solar minimum and solar maximum at 210 nm ranges between 0.6 to 0.8; a similar value (around 0.7) is used in the work of Natarajan *et al.* (1981). Currently accepted values range from 0.9 to 0.95. As a consequence of these large changes that were assumed in their models, the change of ozone and temperature from solar minimum and maximum conditions can be quite large. For example, in Penner & Chang (1978), an ozone change of more than 10% is predicted at 30 km, and a temperature change of around 2 to 10°K is calculated in the stratosphere; a total ozone change of 6.5% is determined by Natarajan *et al.* (1981).

Using a photochemical-radiative two-dimensional model with specified dynamical conditions, Brasseur & Simon (1981) calculated the chemical and thermal response of the stratosphere to solar variation. An updated solar flux variation is used that is considerably larger than the solar minimum and maximum flux ratio assumed by Penner & Chang (1978) and Natarajan *et al.* (1981). Brasseur & Simon (1981) simulated an ozone response that is generally less than 10%, a total ozone variation on the order of 3%, and a temperature response of 2 – 4°K in the stratosphere. Garcia *et al.* (1984) used a two-dimensional model with coupled photochemistry, radiation, and circulation, and showed that the temperature and ozone response in the stratosphere to be even weaker, with a maximum temperature change of less then 2.5°K, and a maximum ozone change of less than 5% in the tropical stratosphere. This is partly a result of the even smaller solar UV flux variation used in their work, as evident from Figure 7 (extracted from Garcia *et al.* 1984). An exception to the weak stratospheric response is seen at the high latitude region in the southern hemisphere during the winter-spring transitional period (see Figure 8), where an ozone reduction of up to 40% is simulated. This is a direct consequence of the large variation of thermospheric NOx produced by auroral particle precipitation that is included in their model. Large amounts of NOx (during solar maximum condition), that are transported down from the thermosphere during winter, interact photochemically in the stratosphere to give a large ozone reduction after the return of solar insolation.

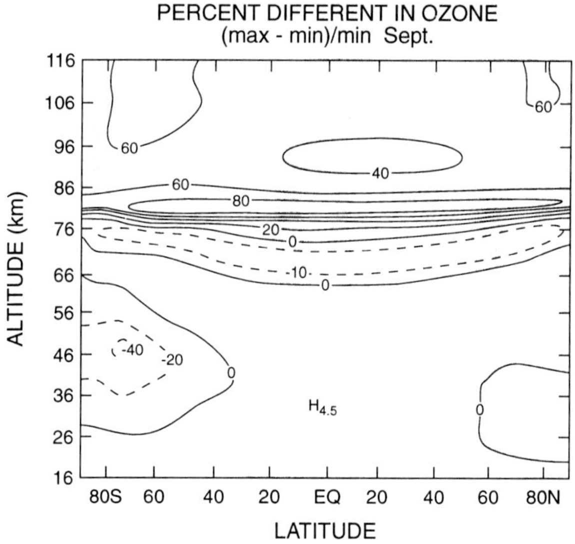

FIGURE 8. Ozone percentage change between solar minimum and maximum for September (end of Northern Hemisphere summer/Southern Hemisphere winter, from Garcia *et al.* (1984).

As mentioned in the introduction, a feedback mechanism that can be potentially important is the dynamical wave-mean flow interaction. This effect can most easily be examined by analyzing the zonal wind change. Zonal wind perturbation should change the wave propagation and dissipation structure, and consequently feedback into the thermal and chemical structure. Garcia *et al.* (1984) showed that the greatest changes in the zonal wind occur near the core of the polar night jet, with an increase of 3 m/s at solar maximum. This is approximately the same magnitude as the zonal wind change obtained by Callis *et al.* (1985), in which they calculated a zonal wind change of 4 to 6 m/s with a two-dimensional radiative equilibrium model, using specified profiles of ozone variation. The simulated zonal wind change varies between models, depending on the model itself and the specified variation in the solar flux. Huang & Brasseur (1993) obtained a zonal wind change of less than 0.5 m/s throughout the stratosphere. They attributed the weak response to the smaller solar flux variation used and the negative zonal wind feedback on gravity wave forcing that is included in their study. Regardless of the specified solar flux variation and the different models used, the less than 5 m/s zonal wind change simulated by most models is at odds with the larger than 20 m/s zonal wind change from solar minimum to solar maximum analyzed by Kodera & Yamazaki (1990), using combined rocket and National Meteorological Center derived winds spanning slightly more than one solar cycle. One solar cycle worth of wind data are hardly enough to conclusively establish a statistical correlation with solar variation, but if the zonal wind change observed is truly solar in origin, it would imply that feedbacks involving wave-mean flow interaction currently not described in models play a crucial role in this question.

Brasseur (1993) and Huang & Brasseur (1993) have used use a coupled radiative, chemistry, and dynamical two-dimensional model for the response simulation and the solar flux variability shown in Figure 2a. The total ozone column percentage change from solar minimum to maximum condition (Brasseur 1993) is shown in Figure 9. The change in ozone abundance is generally of the order of 1.1 – 1.3% at the equator, and increases with latitude. The increase with latitude is more pronounced during winter and early

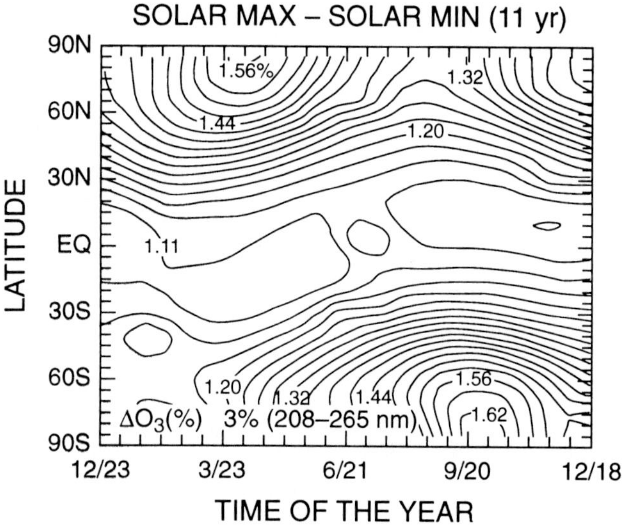

FIGURE 9. Ozone column response (expressed in percentage) as a function of latitude and time of year, from Brasseur (1993).

spring because strong planetary wave activity during this time of the year transports ozone-enhanced air masses from lower latitudes to higher latitudes. The latitude vs. height ozone concentration response is shown in Figure 10 (Huang & Brasseur 1993) for Northern Hemisphere fall equinox. A rather small increase (1–2%) in the stratospheric concentration of ozone is seen (except at higher latitudes) as a result of enhanced O_2 photolysis in the Herzberg continuum and Schumann-Runge bands. It is worth mentioning that a recent study by Deland & Cebula (1993) implies that the 6.6% change in the solar flux at 205 nm associated with the long-term solar cycle used in the model study of Huang and Brasseur may be on the low side. Thus the ozone change shown in Figure 10 probably represents the lower limit of the long-term solar cycle effect. The temperature difference between solar maximum and minimum conditions is shown in Figure 11 (Huang & Brasseur 1993). The temperature response is limited throughout the stratosphere, with a maximum temperature increase of around 1.4°K near the stratopause. This is caused by an increase in ozone and a larger solar flux available for absorption by ozone during solar maximum conditions.

Although caution has to be taken when comparing models with observations since the amount of data is insufficient, the agreement between the model results and available data is unsatisfactory. Using the result from Brasseur (1993) as an example, the observed ozone response near the stratopause in the tropics (Figure 12a) is a factor 2–3 higher than the calculated response, while the opposite is found near 30–35 km altitude (Keating *et al.* 1993; Hood *et al.* 1993). Similarly, the temperature response (Figure 12b) between the model and observation is very different. The calculated temperature response in the upper stratosphere (1–2°K) is much smaller than that derived from observations (5°K); in addition, the signs of temperature change below 35 km are different. Moreover, the variation with height of the model responses does not match the profiles retrieved from observations. It is not clear why these discrepancies exist; it could involve chemical, radiative, as well as dynamical processes which are not yet fully understood.

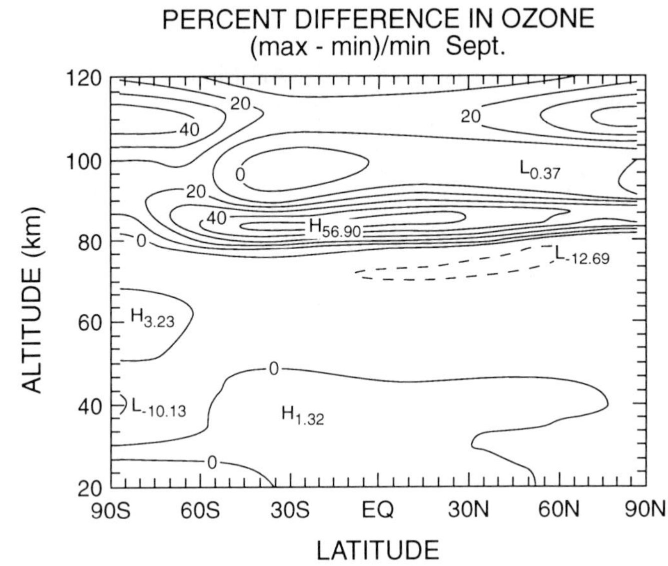

FIGURE 10. Ozone response (expressed in percentage) as a function of latitude and height for September, from Huang & Brasseur (1993).

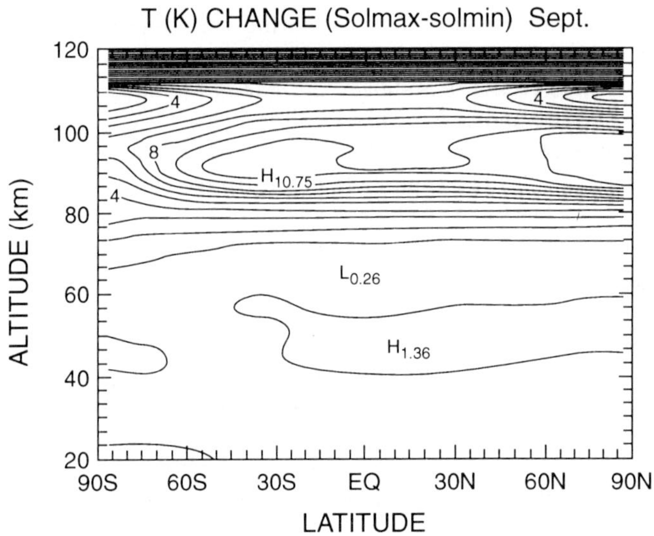

FIGURE 11. Temperature change (°K) between solar minimum and maximum conditions for September, from Huang & Brasseur (1993).

It is plausible that a three-dimensional model is needed to fully realize the true extent of the atmospheric response to solar flux change, especially in terms of the dynamical wave-mean flow interaction effect. A potentially important missing link could be the effect of tropical dynamics that causes the quasi-biennial and semi-annual oscillation phenomena, since several studies suggest that quasi-biennial oscillation could be involved in the

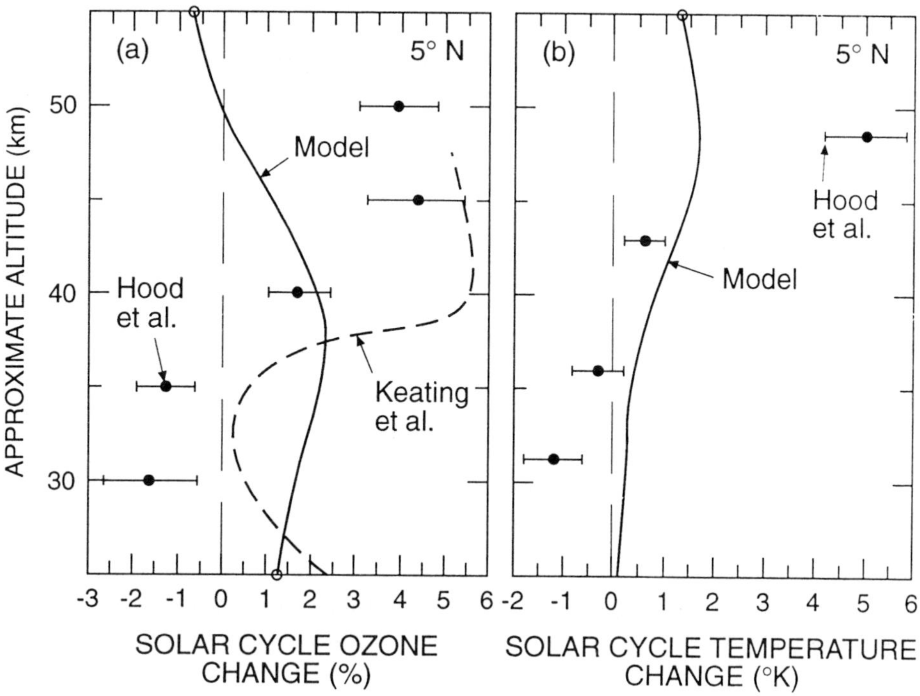

FIGURE 12. Comparison between model and observed variations in (a) ozone concentration and (b) temperature at 5°N between 25 and 55 km altitude, in response to the 11-year solar cycle. The ozone variation calculated by Keating *et al.* (1993) is based on SBUV and SAGE II data. Hood *et al.* (1993) based their analysis on SBUV observations for ozone and NMC analyses for temperature, from Brasseur (1993).

response of the stratosphere to the solar cycle solar cycle (Labitzke & van Loon 1988; Kodera 1991; Kodera *et al.* 1991).

REFERENCES

ANGELL, J. K. 1988 Variations and trends in tropospheric and stratospheric global temperatures, 1958–1987. *J. Climate* **1**, 1296–1313.

ANGELL, J. K. 1991 Stratospheric temperature change as a function of height and sunspot number during 1972–89 based on rocketsonde and radiosonde data. *J. Climate* **4**, 1170–1180.

BALDWIN, M. P. & DUNKERTON, T. J. 1989 Observations and statistical simulations of a proposed solar cycle/QBO/weather relationship. *Geophys. Res. Lett.* **16**, 863–867.

BRASSEUR, G. 1993 The response of the middle atmosphere to long-term and short-term solar variability: A two-dimensional model. *J. Geophys. Res.* **98**, 23,079–23,091.

BRASSEUR, G. & SIMON, P. C. 1981 Stratospheric chemical and thermal response to long-term variability in solar UV irradiance. *J. Geophys. Res.* **86** 7343–7362.

BRASSEUR, G., RUDDER, A.D., KEATING, G.M. & PITTS, M.C. 1987 Response of middle atmosphere to short-term solar ultraviolet variations: 2. Theory. *J. Geophys. Res.* **92**, 903–914.

CALLIS, L. B. & NEALY, J.E. 1978 Solar UV variability and its effect on stratospheric thermal structure and trace constituents. *Geophys. Res. Lett.* **5**, 249–252.

CALLIS, L. B., ALPERT, J.C. & GELLER, M.A. 1985 An assessment of thermal, wind, and planetary wave changes in the middle and lower atmosphere due to 11-year UV flux variations. *J. Geophys. Res.* **90**, 2273–2282.

CHANDRA, S. 1991 The solar UV related changes in total ozone from a solar rotation to a solar cycle. *Geophys. Res. Lett.* **18**, 837–84.

CHANIN, M.L., SIMRES, N. & HAUCHECORNE, A. 1987 Long-term variations of the temperature of the middle atmosphere at mid-latitude: dynamical and radiative causes. *J. Geophys. Res.* **92**, 10933–10941.

DELAND, M. T. & CEBULA, R.P. 1993 The composite Mg II solar activity index for solar cycles 21 and 22. *J. Geophys. Res.* **98**, 12809–12823.

ECKMAN, R. S. 1986 The response of ozone to short-term variations in the solar ultraviolet irradiance, 1., A theoretical model. *J. Geophys. Res.* **91**, 6695–6704.

GARCIA, R. R., SOLOMON, S., ROBLE, R.G/ & RUSCH, D.W. 1984 A numerical response of the middle atmosphere to the 11-year solar cycle. *Planet. Space. Sci.* **32**, 411–423.

GILLE, J. C., SMYTHE, C.M. & HEATH, D.F. 1984 Observed ozone response to variations in solar ultraviolet radiation. *Science* **225**, 315–317.

HAUCHERCORNE, A. & CHANIN, M.L. 1980 Density and temperature profiles obtained by lidar between 35 and 70 km. *Geophys. Res. Lett.* **7**, 565–568.

HOOD. L. L. 1986 Coupled stratospheric ozone and temperature responses to short-term changes in solar ultraviolet flux: An analysis of Nimbus 7 SBUV and SAMS data. *J. Geophys. Res.* **91**, 5264–5276.

HOOD, L. L., HUANG, Z. & BOUGHER, S.W. 1991 Mesospheric effects of solar ultraviolet variations: Further analysis of SME IR ozone and Nimbus 7 SAMS temperature data. *J. Geophys. Res.* **96**, 12989–13002.

HOOD, L. L. & JIRIKOWIC, J.L. 1991 Stratospheric dynamical effects of solar ultraviolet variations: Evidence from zonal mean ozone and temperature data. *J. Geophys. Res.* **96**, 7565–7577.

HOOD, L. L. & MCCORMACK, J.P. 1992 Components of interannual ozone change based on Nimbus 7 TOMS data. *Geophys. Res. Lett.* **19**, 2309–2312.

HOOD, L. L., JIRIKOWIC, J.L. & MCCORMACK, J.P. 1993 Quasi-decadal variability of the stratosphere: Influence of long-term solar ultraviolet variations. *J. Atmos. Sci.* **50**, 3941–3958.

HUANG, T.Y.W. & BRASSEUR, G.P. 1993 Effect of long-term solar variability in a two-dimensional interactive model of the middle atmosphere. *J. Geophys. Res.* **98**, 20413–20427.

KEATING, G. M., PITTS, M.C., BRASSEUR G. & RUDDER, A.D. 1987 Response of middle atmosphere to short-term ultraviolet variations: 1. Observations. *J. Geophys. Res.* **92**, 889–902.

KEATING, G. M., CHIOU, L.S. & HSU, N.C. 1993 Coupling between middle atmosphere trend estimates and solar effects in ozone vertical structure. *Adv. Space Res.*, in press.

KODERA, K. 1991 The solar and equatorial QBO influences on the stratospheric circulation during the early Northern-Hemisphere winter. *Geophys. Res. Lett.* **18**, 1023–1026.

KODERA, K. 1993 Quasi-decadal modulation of the influence of the equatorial quasi-biennial oscillation on the north polar stratospheric temperatures. *J. Geophys. Res.* **98**, 7245–7250.

KODERA, K. & YAMAZAKI, K. 1990 Long-term variation of upper stratospheric circulation in the Northern Hemisphere in December. *J. Met. Soc. Japan* **68**, 101– 105.

KODERA, K., CHIBA, M. & SHIBATA, K. 1991 A general circulation model study of the solar and QBO modulation of the stratospheric circulation during the Northern Hemisphere winter. *Geophys. Res. Lett.* **18**, 1209–1212.

LABITZKE, K., NAUJOKAT, B. & ANGELL, J.K. 1986 Long-term temperature trends in the middle stratosphere of the Northern Hemisphere. *Adv. Space. Res.* **6**, 7–16.

LABITZKE, K. & VAN LOON, H. 1988 Associations between the 11-year solar cycle, the QBO and the atmosphere. Part I: The troposphere and stratosphere in the Northern hemisphere in winter. *J. Atmos. Terr. Phys.* **50**, 197–206.

Labitzke, K. & van Loon, H. 1989 Association between the 11-year solar cycle, the QBO, and the atmosphere. Part III: Aspects of the association. *J. Climate* **2**, 554– 565.

Natarajan, M., Callis, L.B. & Nealy, J.E. 1981 Solar UV variability: Effects on stratospheric ozone, trace constituents and thermal structure. *Pure Appl. Geophys.* **119**, 750–779.

Penner, J. E. & Chang, J.S. 1978 Possible variations in atmospheric ozone related to the 11-year solar cycle. *Geophys. Res. Lett.* **5**, 817–820.

Rottman, G. J. 1988 Observations of solar UV and EUV variability. *Adv. Space. Res.* **8(7)**, 53–66.

Salby, M. L. & Shea, D.J. 1991 Correlations between solar activity and the atmosphere: An unphysical explanation. *J. Geophys. Res.* **96**, 22579–22595.

Stolarski, R. S., Bloomfield, P., McPeters, R.D. & Herman, J.R. 1991 Total ozone trends deduced from Nimbus 7 TOMS data. *Geophys. Res. Lett.* **18**, 1015–1018.

Summers, M. E., Strobel, D.F. Bevilacqua, R.M., Zhu, X., Deland, M.T., Allen, M. & Keating, G.M. 1990 A model study of the response of mesospheric ozone to short-term solar ultraviolet flux variations. *J. Geophys. Res.* **95**, 22523–22538.

van Loon, H. & Labitzke, K. 1988 Association between the 11-year cycle, the QBO, and the atmosphere. Part II: Surface and 700 mb in the Northern Hemisphere in winter. *J. Climate* **1**, 905–920.

Wuebbles, D. J., Kinnison, D.E., Grant, K.E. & Lean, J. 1991 The effect of solar flux variations and trace gas emissions on recent trends in stratospheric ozone and temperature. *J. Geomag. Geoelectr.* **43**, 709–718.

A Probable Connection between Solar and Atmospheric Decadal Variability

By KARIN LABITZKE[1] AND HARRY VAN LOON[2]

[1]Freie Universität, Berlin, Germany

[2]National Center for Atmospheric Research, Boulder, CO 80307, USA

The paper reviews briefly some recent investigations of possible effects of decadal solar variations on the stratosphere and troposphere, drawing attention to the stable, strong correlations between the height of constant pressure levels in the stratosphere and the sunspot cycle.

1. Introduction

The growing interest in the variability of climate has sharpened the realization that our knowledge of the natural variability and forcing of the atmosphere is imperfect. Partly as a result of the restricted insight into the causes of climate change, interest has revived in the possible effects of the Sun's variability on atmospheric variations. The popular belief of the past century was — to paraphrase Godson (1991) — that the variable solar activity had no influence on climate, or that if it did, the effects should always be the same with regard to sign and size. This scepticism is largely due to the fact that the statistical investigations, which dominate the research, although they often produce suggestive results, have been limited by their failure to explain how the Sun could produce the observed variations in the atmosphere. Some of the recently published papers reviewed below are no exception in this respect, but the statistics they present are of sufficient significance that they should be taken into account — even if their underlying cause should turn out not to be solar.

2. Effects of the 11-year sunspot cycle

2.1. *Summer*

If the 11-year solar cycle has a noticeable effect on the Earth's atmosphere, there is no self-evident reason why the outcome should be the same everywhere, or that it should appear as a purely radiative effect. The statistics presented below (based on the historical daily analyses at the Freie Universität, Berlin) suggest that dynamics play a major role in the spatial distribution of the atmosphere's response to the 11-year solar variability. This is evident in Figure 1, which shows the correlation between a 36-year series of 30-mb height in summer and the 10.7cm solar flux. This flux is a good index of the sunspot cycle. We have computed such correlations every year during the past five years, starting with a sample of 31, and they have remained stable with each additional year. The correlations over more than three quarters of the area are locally significant above the 1% level; and over one third of the area the local significance is above the 0.1% level. The map as a whole is field significant at better than the 1% level (see Livezey & Chen (1983), for an explanation of field significance). The shape of the correlation pattern, which is repeated in each month of the year, implies that the probable influence of the 11-year solar cycle is areally concentrated by atmospheric motions.

In order for the 30-mb geopotential height to vary with a period similar to that of the 11-year sunspot cycle, as suggested by Figure 1, the temperature of the air column below or the surface pressure, or both, must contain the same periodicity. As shown

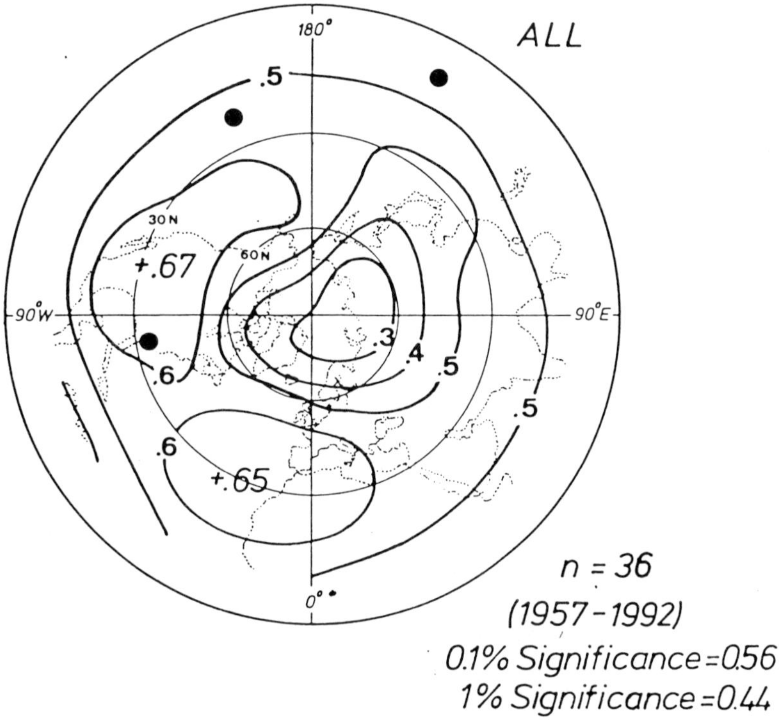

FIGURE 1. Correlations between 30-mb height in July-August and the 10.7cm solar flux. The 0.1% local significance level is 0.56. The dots mark the locations in Figure 2. Updated from van Loon & Labitzke (1990).

in the following, this periodicity does not exist in the surface pressure but is obvious in the temperature: Figures 2a,b show the difference in the temperature between solar maxima and minima from the ground to 30 mb (about 24 km) at two lower-latitude stations in areas of Figure 1 where the correlations are high. The two have in common that the troposphere is warmer in solar maximum, that the difference is negative in the levels about the tropopause (100 mb), and that on average the air in the stratosphere is warmer in the solar maxima. At the station nearest the axis of highest correlation, Charleston, the average temperature difference increases with height above 700 mb to 1.5°C just below the tropopause, at Truk the largest average difference is 0.8°C. The association between the 11-year solar cycle and the global temperature in the upper troposphere and lower stratosphere was studied by Sellers & Liu (1988). They found that the low frequency variation of the first EOF in the temperature is "primarily and strongly" correlated with the quasi-decadal solar cycle.

These correlations in the vertical do not prove that a solar influence exists which affects the stratosphere through the troposphere, either directly in the latitudes concerned or indirectly from low latitudes through the Hadley circulation. It is, for instance, also possible that the stratosphere warms at solar maximum through absorption of UV radiation by ozone (Hood *et al.* 1993), and that this changes the vertical distribution of temperature and thereby changes the height and temperature of the tropopause. The temperature in the upper troposphere would be affected in this way. Another mechanism

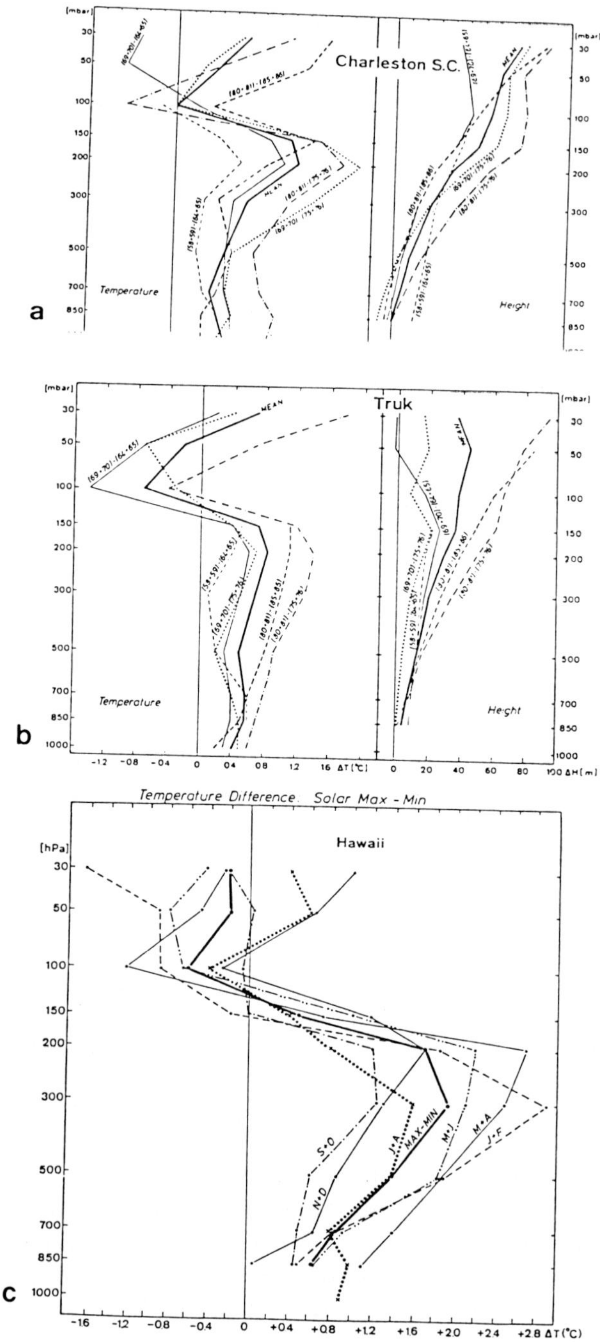

FIGURE 2. The change of temperature and geopotential height in July-August from solar minimum to maximum at Charleston (South Carolina) and Truk (7.5N, 152E). The heavy line is the average. From Labitzke & van Loon (1992). (c) The average temperature change from solar minimum to maximum at Lihue (22N, 159W), through the year and in the annual mean (heavy solid line). From Labitzke & van Loon (1993).

is that of forcing atmospheric changes due to variations in latent heat release from effects on cloud microphysics of solar wind induced changes in the global electric circuit (Tinsley & Deen 1991).

A more plausible idea is that the Sun at its highest 11-year activity heats the ocean surface at low latitudes more than at its lowest activity. This would intensify the Hadley circulation in solar maxima, causing warming in the subtropics by increased subsidence and thus affect the tropopause and the lower stratosphere. The objection to this idea is that the change of total solar irradiance from solar minimum to maximum, as measured by satellite during the recent solar cycle, amounts to only 0.1%. The warming of the tropical water in the solar maximum need not be large: the drastic change in the global atmospheric circulation in the years after 1977 (Labitzke & van Loon 1993) was accompanied by an average change in the tropical sea surface temperature of only 0.3°C (Bottomley *et al.* 1990), which is not far from observational accuracy.

2.2. *Midwinter*

The basic shape of the correlation pattern between 30-mb height and the 11-year solar cycle, a crescent of high correlations along 30°N on the water-dominated side of the hemisphere, is also present in January-February (Figure 3a) but the correlations are lower than in the 10 other months of the year (cf., Figure 1), reaching the 5% significance level only between 20°N and 45°N over the Pacific Ocean. It was found, however, that the stratosphere responds differently to solar changes in either extreme of the Quasi-Biennial Oscillation (QBO) in the 50–40 mb wind of the equatorial stratosphere (Labitzke 1987). In the east years (Figure 3b) the correlation pattern is the same as in the ten other months of the year, and over more than half of the map the correlations are locally significant above the 5% level. If one assumes that the solar effect expresses itself through changes in the atmosphere where the crescent of high correlations lies, then — owing to the marked teleconnections in winter (Shea *et al.* 1992) — the changes at high latitudes must be opposite to those in the tropics and subtropics and thus account for the large negative correlations in the Arctic (Figure 3b). In the east years of the QBO, the polar vortex tends, in other words, to be stronger at high and weaker at low solar activity, whereas theory (Holton & Tan 1980, 1982) and model experiment (e.g., (Dameris & Ebel 1990) indicate that the vortex should generally be weak in east years.

This relationship also emerges from the scatter diagram in Figure 4: the correlation between the zonally-averaged geostrophic wind at 60°N in the east years and the solar flux is 0.7, which is at the 1% significance level. The asterisks in the diagram mark the winters when major midwinter warmings (breakdowns of the polar vortex) took place in January or February; they are prevalently associated with lower values of the solar flux.

Conditions are different in the west years of the QBO (Figure 3c): in these years there are no statistically significant correlations over most of the map, the only noteworthy ones (positive) being over the Arctic, and there is no sign of the strong teleconnections between higher and lower latitudes which otherwise dominate the winter climate of the stratosphere (Shea *et al.* 1992). In the west years the Arctic stratosphere is positively correlated with the solar cycle (see the scatter diagram of north polar, 30-mb temperature versus the 10.7cm flux in Figure 5) because the major midwinter warmings in these years take place only at solar maximum.

Our correlations suggest that high solar activity brings on a disturbed state: at solar minimum the vortex is intense in the west and weak in the east years which, as mentioned above, is predicted by theory and confirmed by experiment; in contrast, at solar maximum the vortex weakens in the west and strengthens in the east years, contrary to theory.

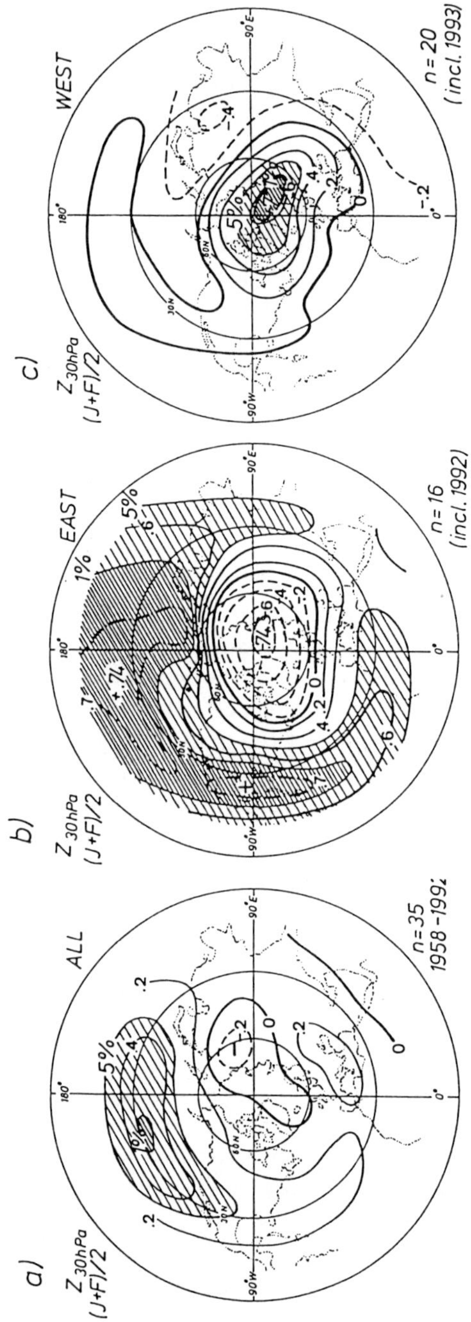

FIGURE 3. Correlations between the 10.7 cm solar flux and (a) the 30-mb height in January-February, $n = 35$ years; (b) in the east years of the QBO; (c) in the west years. The areas where the local statistical significance is above the 5% level are lightly hatched, those above the 1% level are densely hatched. Updated from Labitzke & van Loon (1988).

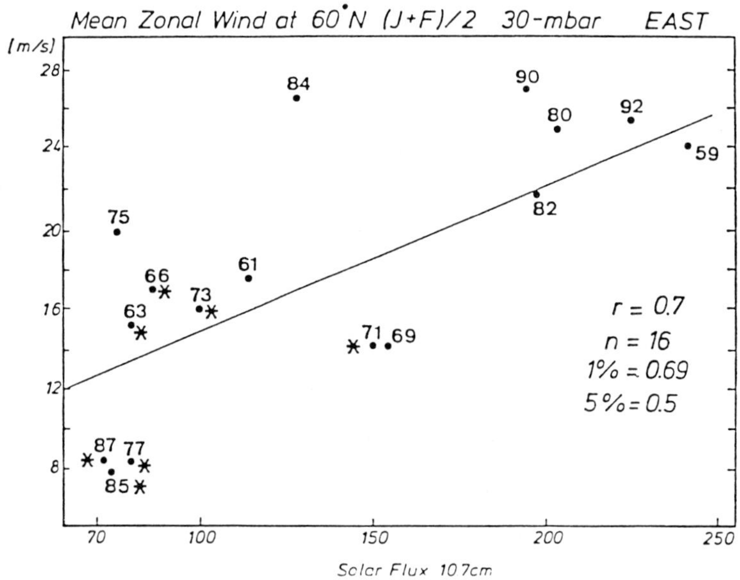

FIGURE 4. Zonally-averaged geostrophic wind at 60N in January-February at 30-mb in the east years plotted against the 10.7 cm solar flux. The asterisks denote major midwinter warmings.

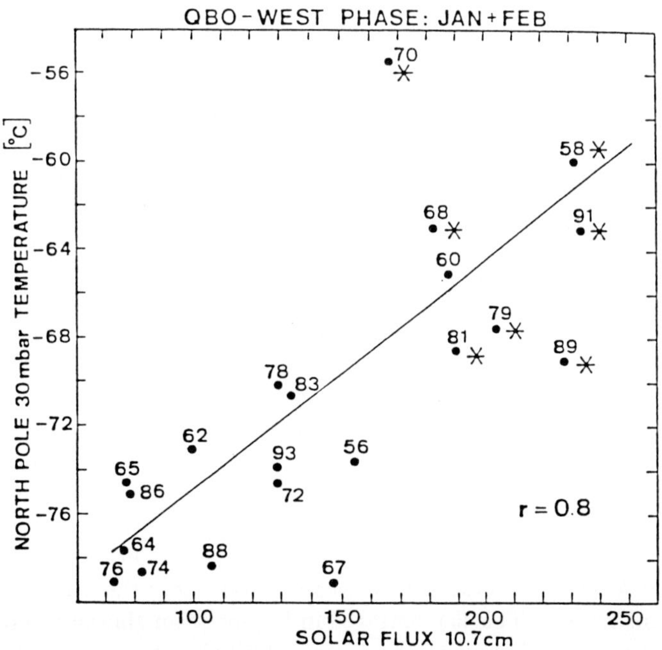

FIGURE 5. 30-mb temperature at the North Pole in January-February plotted against the 10.7cm solar flux in the west years of the QBO. Asterisks denote major midwinter warmings. Updated from van Loon & Labitzke (1990).

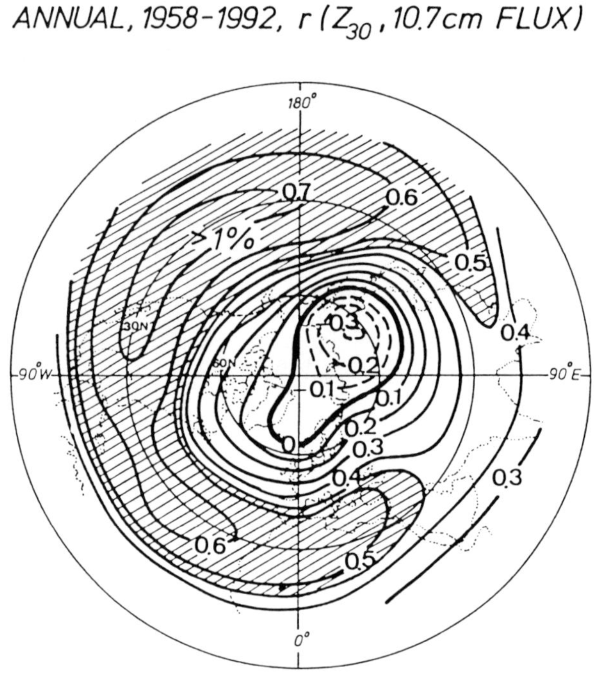

FIGURE 6. Correlations between the annual mean 30-mb height, 1958–1992, and the 10.7 cm solar flux. In the hatched areas the local statistical significance exceeds the 1% level. Updated from van Loon & Labitzke (1990).

2.3. *Year*

The map of correlation between the annual mean 30-mb height and the 10.7 cm solar flux (Figure 6) sums up the statistical analyses of this section. Not unexpectedly, the pattern is that of a crescent of strong, stable correlations. The latter exceed the 1% significance level over half the area of the map, and are above the 0.1% significance level from 150°E westward to 10°W, between 15°–20°N and 40°–45°N. Even though no physical mechanism has yet been proposed to explain the large and stable statistical association between the 11-year solar cycle and the changes in the lower stratosphere, it is reasonable to propose that correlations such as those in Figures 6 and 1 are not due to chance. It is also unlikely that the decadal variation which they reflect is due to an internal oscillation in the ocean-atmosphere system with a period close to that of the solar cycle, since it is unlikely that precisely in the time interval which was available to us such an internal oscillation should have been in phase with the solar cycle.

The crescent of high positive correlations in Figure 6 indicates that the atmosphere below the 30-mb level must in the annual mean be warmer at the peaks than in the valleys of the 11-year solar cycle. This is illustrated in Figure 2c for Hawaii where the correlation is between 0.6 and 0.7 in Figure 6: the heavy solid line is the mean annual temperature difference between maxima and minima in three solar cycles. The difference is positive in the troposphere, reaching almost two degrees at 300mb to 200mb, but slightly negative above the tropopause. *The large positive correlation coefficient between the 30-mb height and the solar flux is thus due entirely to the difference in the tropospheric temperature between the extremes of the 11-year sunspot cycle.*

The vertical structure of the temperature difference has the same shape throughout the year, but the values are smaller in summer and autumn than in winter and spring (Figure 2c). Evidently, the effect therefore is not *directly* linked with the local elevation of the Sun but must be associated with the interseasonal changes of the general circulation. The highest correlations in Figure 6 are in the belt of predominantly downward motion in the subtropical high; and the higher mean temperature in the solar maxima than minima is thus likely associated with differences in the strength of the subsidence. The subsidence is stronger in winter and spring than in summer and autumn because the convergence in the upper troposphere between the meridional flow in the Hadley and Ferrell circulations is stronger during the winter and spring when the temperature difference between equator and the North Pole is largest.

3. Conclusion

Satellites have in recent years provided a reliable picture of changes in the Sun's irradiance during a sunspot cycle; and papers have appeared with statistics which imply that the lower atmosphere responds to solar variability on several time scales. On the decadal time scale, the correlations of Labitzke & van Loon (1988), (1992) and van Loon & Labitzke (1990) have remained stable and increasingly statistically significant with the addition of 5–7 years of data. The basic pattern of our correlations with the sunspot cycle is shaped like a crescent with the largest correlations between 20°N and 45°N in the Pacific - Atlantic area, and it is present in every month. The correlations are statistically significant through most of the year, but the significance is especially high in summer and for the annual average. The statistically insignificant correlations in January-February improve when the data are grouped according to the wind near 50 mb above the equator (the QBO). The east-wind years then show a highly significant pattern, the same as that in other months, whereas the west-wind years have no significant correlations outside the Arctic; there they are positive in contrast to the east years. Kodera (1993) has demonstrated that changes in the relationship between equatorial and polar regions on the decadal scale are a fundamental aspect of the opposite signs in the arctic correlations with the sunspot cycle in January-February.

Acknowledgments. We thank the members of the Stratospheric Research Group, Freie Universität Berlin, for technical and professional assistance. Part of the research was funded by the Bundesministerium für Forschung und Technologie, Projekt 07KF2012.

REFERENCES

BOTTOMLEY, M., FOLLAND, C.K., HSIUNG, J., NEWELL, R.E. & PARKER, D.E. 1990 Global Ocean Temperature Atlas (GOSTA). *Joint Met. Office/MIT Project, Project supported by US Dept of Energy, US NSF and US Office of Nav. Res.* 20+iv pp and 131 Plates, HMSO, London.

DAMERIS, M. & EBEL, A. 1990 The quasi-biennnial oscillation and major atmospheric warmings: a three-dimensional model study. *Ann. Geophys.* **8**, 79–86.

GODSON, W.L. 1991 The recognition and assessment of decadal-scale variation of the effects of solar variability on climate. *J. Geomagn. Geoelectr.* **43**, 785–793.

HOLTON, J.R. & TAN, H.CH. 1980 The influence of the equatorial QBO in the global circulation at 50mb. *J. Atmos. Sci.* **37**, 2200–2208.

HOLTON, J.R. & TAN, H.CH. 1982 The quasi-biennial oscillation in the northern hemisphere lower stratosphere. *J. Met. Soc. Japan* **60**, 140–148.

HOOD, L.L., JIRIKOVIC, J.L. & MCCORMACK, J.P. 1993 Quasi-Decadal variability of the stratosphere: Influence of long-term solar ultraviolet variations. *J. Atmos. Sci.* **50**, in press.

KODERA, K. 1993 The quasi-decadal modulation of the influence of the equatorial QBO on the north-polar stratospheric temperatures. *J. Geophys. Res.* **98**, 7245–7250.

LABITZKE, K. 1987 Sunspots, the QBO, and the stratospheric temperature in the North Polar region. *Geophys. Res. Lett.* **14**, 535–537.

LABITZKE, K. & VAN LOON, H. 1988 Association between the 11-year solar cycle, the QBO and the atmosphere. Part I: The troposphere and stratosphere in the Northern Hemisphere in winter. *J. Atmos. Terr. Phys.* **50**, 197–206.

LABITZKE, K. & VAN LOON, H. 1992 Associations between the 11-year solar cycle and the atmosphere. Part V: Summer. *J. Clim.* **5**, 240–251.

LABITZKE, K. & VAN LOON, H. 1993 Some recent studies of probable connections between solar and atmospheric variability. *Ann. Geophysicae* **11**, in press.

LIVEZEY, R.E. & CHEN, W.Y. 1983 Statistical significance and its determination by Monte Carlo techniques, *Mon. Weath. Rev.* **111**, 46–58.

SELLERS, W.D. & LIU, W. 1988 Temperature patterns and trends in the upper troposphere and lower stratosphere. *J. Clim.* **1**, 573–581.

SHEA, D.J., VAN LOON, H. & LABITZKE, K. 1992 Point correlations of geopotential height and temperature at 30mb and between 500mb and 30mb. *NCAR Techn. Note*, NCAR/TN-368+STR.

TINSLEY, B.A. & DEEN, G.W. 1991 Apparent tropospheric response to MeV-GeV particle flux variations: A connection via electrofreezing of supercooled water in high-level clouds? *J. Geophys. Res.* **96**, 22283.

VAN LOON, H. & LABITZKE, K. 1990 Association between the 11-year solar cycle, the QBO and the atmosphere. Part IV: The Stratosphere, not grouped by the phase of the QBO. *J. Clim.* **3**, 827–837.

Solar Activity and Global Temperature

By EIGIL FRIIS-CHRISTENSEN
AND KNUD LASSEN

Solar-Terrestrial Physics Division, Danish Meteorological Institute, Lyngbyvej 100, DK-2100, Copenhagen, Denmark

A major problem in the determination of the magnitude of a possible solar effect on climate is that no physical parameter of solar energy output exists that has been observed long enough to be used for long-term analyses. Therefore, a number of indirect parameters have been proposed, with the sunspot number as the most commonly used parameter. Recently it has been suggested that climatic effects may be more directly associated with the length of the solar cycle. Whereas the magnitude of the sunspot number is only believed to be reliable back to 1750, determination of solar activity minima may be based on other types of data. A recent reconstructed series of solar cycle lengths back to 1500 gives new information about solar activity in particular before and during the Maunder Minimum. A comparison with reconstructed temperature records has revealed that the good agreement between the solar cycle length and the global temperature found for the modern instrumental temperature record is also characteristic for the total series of reconstructed temperature data. A further result is that the response of the temperature during the pre-instrumental era is the same as for the modern temperature record. This finding confirms the close association beween terrestrial temperature and solar activity measured in terms of the solar cycle length.

1. Introduction

An observed rise in the global average temperature of about 0.5° C since 1890 has caused concern in the public because it could indicate the climatic response to a changing composition of the atmosphere due to an increase of greenhouse gases caused by anthropogenic activity. Therefore an Intergovernmental Panel on Climate Change (IPCC) was established in 1988 by the World Meteorological Organization (WMO) and the United Nations Environmental Programme (UNEP). In the report published by this panel it is concluded that it is presently not possible to ascribe unequivocally the found changes in climate to the enhanced greenhouse effect since the size of the global warming is of the same magnitude as natural climate variability.

One of the reasons for the uncertainty regarding the cause of the observed increase in temperature is that the temperature has not followed the monotonic rise that observations of greenhouse gases and theoretical calculations predict. On the contrary, there have been intervals, in particular from around 1940 to around 1970, where the global temperature was decreasing during a period of rapid build-up of CO_2 in the atmosphere. These apparently contradictory observations are usually ascribed to a high degree of internal variability in the climate system, which may be reproduced in the results of climate models. A different explanation of the long-term variations of the temperature could, however, be a possible connection between observed climate changes and changes in solar activity. Thus, in a detailed study of long-term solar variations Eddy (1976) drew attention to the coincidence of the so-called Maunder Mimimum in solar activity with the coldest excursion of the "Little Ice Age", already noted by many who had looked at the possible relationships between the Sun and terrestrial climate.

Reid (1987, 1991) noticed a striking similarity between the globally averaged sea surface temperature (SST) and the long-term variation of solar activity represented by the 11-

year running mean Zürich sunspot number. In accordance with Eddy (1976) he suggested that the solar irradiance may have varied by approximately 0.6% from 1910 to 1960 in phase with the 80-90 year cycle (the Gleissberg period) of solar activity represented by the envelope of the 11-year solar activity cycle. Friis-Christensen & Lassen (1991) pointed out a major difficulty with this interpretation. They examined the Northern Hemisphere land air temperature and noted that this record was leading both the SST record and the sunspot record, in fact by as much as 20 years. From this discrepancy they concluded that it is unlikely that long-term variations of solar activity can be sufficiently well represented by some average value of the sunspot number itself. They further pointed out that the long-term variation of the variable length of the '11-year' sunspot cycle was well correlated with the Northern Hemisphere land surface temperature during the entire interval of systematic temperature measurements and suggested that the solar cycle length could provide an indicator of long-term variations in solar luminosity.

Thus, there appears to be an increasing approval to the view that the long-term variation of the solar activity contributes appreciably to the long-term variations of the mean global temperature. An elucidation of the magnitude of a possible solar contribution to the temperature increase of the present century crucially influences the estimate of the magnitude of the anthropogenic greenhouse forcing and consequently the predictions of the climate models.

2. Solar activity

It is well-known that solar activity varies with an approximately 11-year cycle. This cycle is a well documented feature first found by analyzing the observational record of the number of sunspots on the solar disk. Later the existence of the cycle was noticed in a number of different observations of the Sun as well as of the upper atmosphere, which is influenced by the varying flux of solar energetic particles and ultraviolet radiation. The amplitude of the 11-year cycle appears to be modulated by a 70-90 years undulation, the so-called Gleissberg period.

This is illustrated in Figure 1 which shows the monthly mean sunspot number from 1750 to 1991. It is evident that the monthly mean sunspot number is affected by noise that is superposed on an assumed quasi-regular 11-year period. Apparently the sunspot number returns to a rather low minimum value during each cycle but this does not necessarily mean that this activity level is close to a possible lower threshold of solar activity as such. The solar activity at different sunspot minima may be rather different, even if it does not manifest itself in the observed sunspot number directly. This fact may be clearly demonstrated by looking at a quite different parameter of solar activity. During more than 120 years systematic observations of the geomagnetic field on the surface of the Earth have been performed. The geomagnetic activity shows a dominating approximately 11-year cycle, although not as evident as the annual sunspot number. Contrary to the sunspot variation, however, the geomagnetic cyclic variations are superimposed on a long-term variation of similar amplitude including a nearly monotonic increase from 1900 to 1960. From the statistical relation between geomagnetic activity and satellite measurements of the solar wind, Feynman & Crooker (1978) conclude that solar activity as represented by the solar wind velocity or that the southward component of the interplanetary magnetic field has been increasing since the beginning of the century. The long-term variation in solar output through the solar wind measured by means of the geomagnetic aa-index suggests that similar longterm changes in other manifestations of solar energy output may also have occurred.

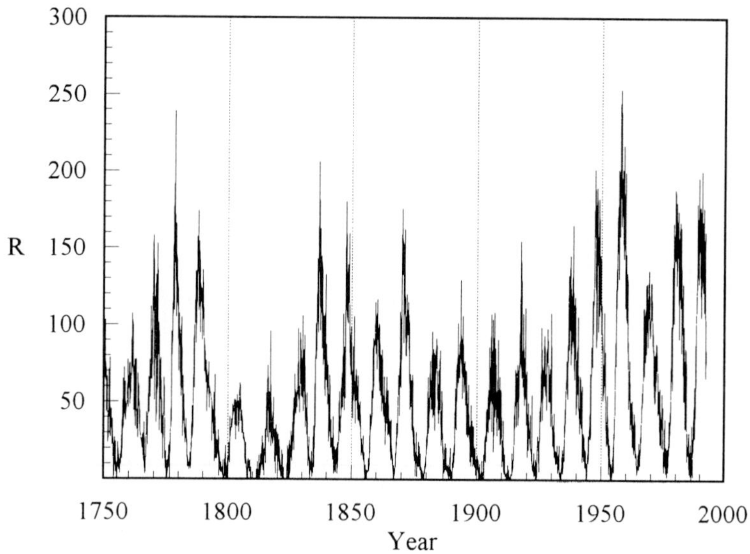

FIGURE 1. The monthly mean of the sunspot number 1750 to 1991.

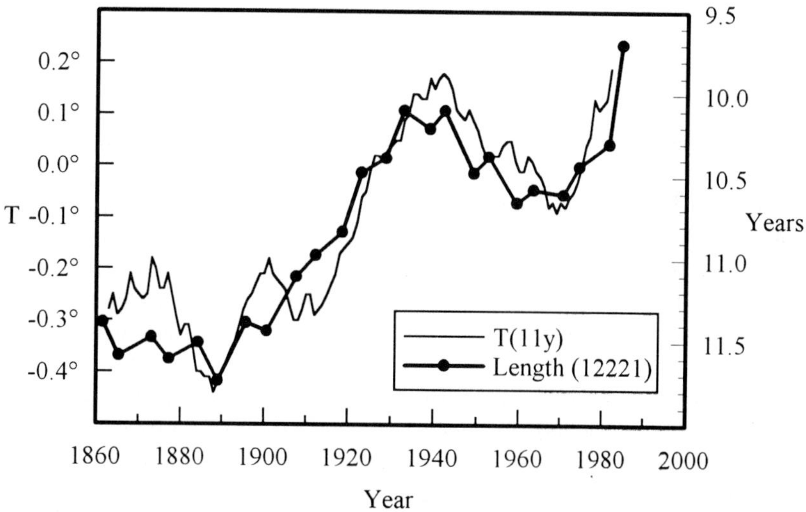

FIGURE 2. 11-year running mean of the annual average Northern Hemisphere land air temperature by Jones *et al.* (1986) and Jones (1988) relative to the average temperature 1951-1980, and the filtered length of the sunspot cycle.

3. Solar cycle length versus temperature

The most acknowledged virtue of the sunspot number is that this parameter has been observed over a very long time period. However, as discussed by Friis-Christensen & Lassen (1991) the long-term behaviour of solar activity may be better represented by the varying length of the solar cycle. Support for this suggestion was found in a very high correlation between this parameter and the long term variation of the Northern Hemisphere land air temperature as given by Jones *et al.* (1986) and Jones (1988).

Figure 2 shows the variation of the sunspot cycle length together with the 11-year

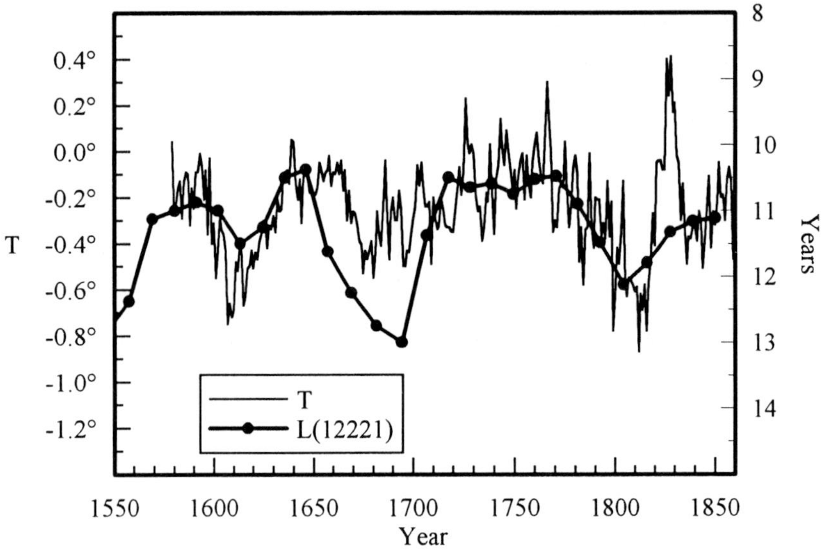

FIGURE 3. Annual mean values of the Northern Hemisphere land air temperature 1579-1860 relative to the average temperature 1881-1975, reconstructed by Groveman & Landsberg (1979). Also plotted is the filtered value (1,2,2,2,1 filter) of the sunspot cycle length.

running mean of the Northern Hemisphere land air temperature from 1860. The 11-year running mean filter was chosen in order to suppress the temperature variations within a solar cycle which obviously cannot be ascribed to the possible long-term external forcing.

The limited extent of instrumental temperature records and the possible complex behaviour of a Sun-climate relationship that makes a filtering of the time series necessary, also means that the statistical significance of the correlation is not sufficiently large to prove that the correlation is not just a coincidence. Since the shorter term variations in solar activity and temperature may have quite different relationships the only possible way to obtain an increased significance is to try to use a longer time series of temperature data.

Although reconstructed temperature series are available back to the second half of the 16th century, solar activity has not been regularly monitored prior to 1750. However, several efforts have been made to reconstruct the sunspot variation during the preceding centuries by means of sporadic observations in combination with proxy data. Lassen & Friis-Christensen (1993) have recently made a critical assessment of such published solar data covering the period 1500-1750. Based on this they presented a revised series of epochs of minimum solar activity and compared this with reconstructed temperature time series.

A comprehensive reconstruction of the Northern Hemisphere temperature covering a 400 year period was performed by Groveman & Landsberg (1979). They used several local temperature measurements together with proxy data from many places in the Northern Hemisphere and performed a multiregressional analysis of the data that resulted in a set of empirical formulas relating each proxy data series to the measured Northern Hemisphere temperature. Using this set of empirical relations they then calculated the temperature for the Northern Hemisphere for the last 400 years.

In Figure 3 is plotted annual mean values of their reconstructed temperatures from 1579 to 1860 on the same graph as the variation of the sunspot cycle length. The reconstructed temperature values are given as departures from the average Northern Hemisphere temperature 1881-1975.

Naturally, the reconstruction of the Northern Hemisphere temperature must be less confident than the modern record. Apart from this there exist, of course, year-to-year variations due to internal oscillations in the climate, El Niño effects, volcanic eruptions etc. Taking these variations into consideration, the comparison between the temperature record and the solar activity during the pre-instrumental period reveals that there is a good association between the trends in the temperature and in the solar cycle length record.

An obvious disagreement between the two time series in Figure 3 is noticed in the second half of the seventeenth century. Whereas the solar activity here reaches its lowest level of the five centuries the reconstructed temperature remains at a level remarkably higher than during the two surrounding minima near 1610 and 1805. This is the more surprising because this period of low solar activity - the so-called Maunder Minimum - is often regarded as being associated with the culmination of the Little Ice Age (Eddy 1976). However, it is possible that the restricted number of localities used in the reconstruction of the temperature may have resulted in an average temperature variation that is biased by a local effect at a single station.

In addition to the Groveman-Landsberg series there exist a few independent series of proxy temperature data representing more restricted regions of the Northern Hemisphere.

Thus, it has been demonstrated earlier (Friis-Christensen & Lassen 1991) that an index of the North-Atlantic sea ice in the waters around Iceland oscillated in concert with the longterm variation of the solar activity during the years 1750-1970. Although the individually measured extensions of sea ice suffer from a number of different influences and probably cannot be used directly as a temperature index, it seems reasonable that the absence of considerable amounts of ice would be associated with relatively high hemispheric average temperatures. Lassen & Friis-Christensen (1993) show a comparison between the smoothed sunspot cycle lengths and a 22-year running mean of the extent of sea ice around Iceland from 1600 to 1780. The comparison supports and improves their earlier result that maxima in the long-term solar activity have been accompanied by minima in the 22-year running mean of the extent of sea-ice around Iceland, i.e. by increased average temperatures at high northern latitudes.

A similar conclusion was reached by Hameed & Gong (1993). These authors combined data of blossoming of plants noted in personal diaries and other documents originating from the area of the Middle and Lower Yangtse River Valley with records of the last day of snow event in the spring season of each year between 1720-1800 kept in the Palace Museum in the Forbidden City. The combined data sets made it possible to estimate the long-term variations of spring temperature in the region from 1580 to 1920. Their results show a strong co-variation between the spring temperature in Central China and the length of the solar cycle since 1750. In Figure 4 we have redrawn their smoothed temperature curve and show it together with the revised (extended) time series of the solar cycle length. Note that the proxy temperature values are given in the unit "days before or after an average date of the event". It is clearly demonstrated that the climate in this region of Asia has oscillated in concert with solar activity at least during the past 300 years. Elsewhere it has been argued that regional Chinese time series of climate data represent large-scale temperature changes rather well (Bradley & Jones 1993).

In common the above-mentioned three examples reveal that the good agreement between the solar cycle length and the global temperature found for the modern instrumental temperature record is also characteristic for the total series of reconstructed temperature data. The only essential exception is the disagreement between the two

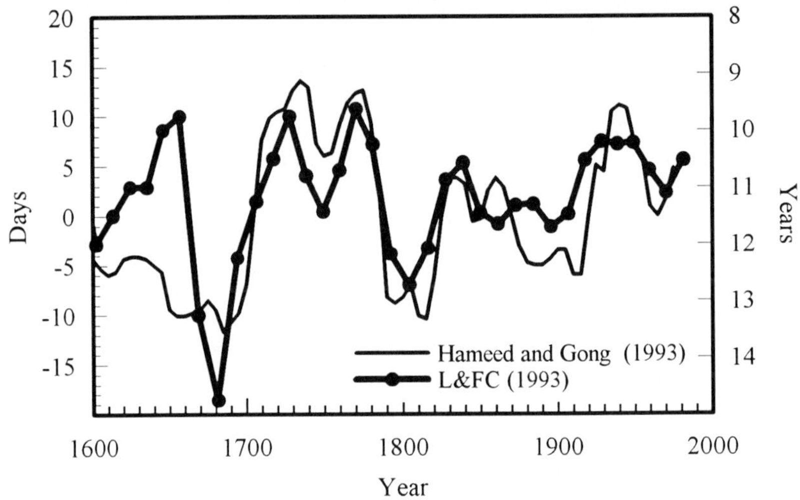

FIGURE 4. Variation of spring temperature in the Middle and Lower Yangtse River Valley deduced from phenological data. The curve is redrawn after Hameed & Gong (1993). The ordinate indicates "days before or after an average date". Also plotted is the filtered value (1,2,2,2,1 filter) of the solar cycle length (Lassen & Friis-Christensen 1993).

time series in the second part of the seventeenth century (Figure 3). However, no counterpart to this is apparent in the series from Iceland and China. This seems to confirm our assumption that the reconstructed temperatures of Groveman and Landsberg are unreliable (too high) during the interval 1650-1700. The remaining part of the figure shows a strong association between the trend of the reconstructed temperature curve and the long-term variation of the length of the solar cycle. There appears to be a linear relation between the two quantities. This is illustrated in Figure 5 which also shows the regression line with a regression coefficient ($\pm\sigma$) of $(0.23 \pm 0.04)^{\circ}$C/y.

Essentially, the reconstructed temperature series 1579-1880 covers a period that is situated prior to the rapid increase of industrial production and resulting air pollution. It is therefore of interest to compare the magnitude of the regression coefficient for this period with a similar coefficient for the latest 130 years as shown in Figure 6. Since the solar cycle length in Figure 2 is essentially decreasing with time it might be expected that the effect of an increasing greenhouse warming through the 130 years would show up as a curvature of the function combining temperature and solar cycle length, or in an increase in the slope of the regression line. Neither of these possibilities is apparent in the figure. Actually, the regression coefficient has been found to be $(0.30 \pm 0.03)^{\circ}$C/y. The difference between the regression coefficients is barely significant. Thus we must conclude that the data used in our study do not indicate any change in the relationship between Northern Hemisphere temperature and long-term solar activity through the last 400 years. In consequence, influence of anthropologic activities cannot be traced with certainty in our data up to the 1980's.

4. Discussion

Research in possible Sun-climate relationships has been dominated by the unsuccessful attempts to identify a physical mechanism that could account for the hypothesized effects on climate based on a large number of correlations between various solar and climate

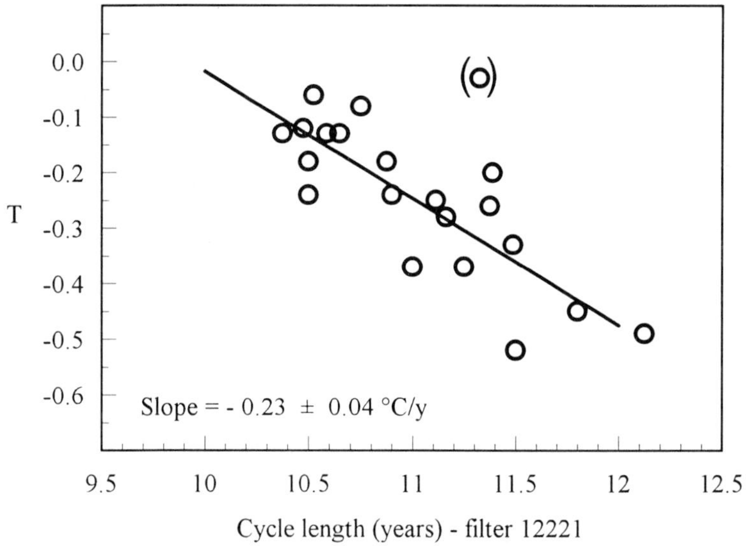

FIGURE 5. Annual mean values of the Northern Hemisphere land air temperature from Figure 3 (1656-1696 not included, see text) versus filtered value (1,2,2,2,1 filter) of the solar cycle length. The value labelled () is not included in the calculation of the regression coefficient.

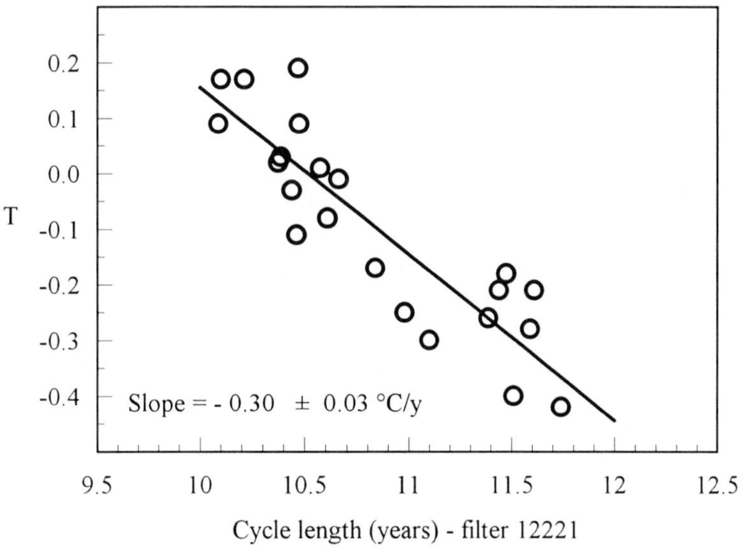

FIGURE 6. Annual mean values of the Northern Hemisphere land air temperature from Figure 2 versus filtered value (1,2,2,2,1 filter) of the solar cycle length.

parameters. The simplest conceivable physical mechanism would be a long-term variation of total solar irradiation that is sufficiently large to have an observed temperature effect on climate. Satellite measurements over approximately one solar cycle have shown that the irradiance is not constant, but model calculations show that it varies too little (less than 0.1%) during a solar cycle to be of major importance for climate (Wigley & Raper 1990). However, no measurements yet exist that do exclude the possibility of larger variations in total irradiance over a longer period of time. On the contrary, Lean *et al.*

(1992) from correlations between total solar irradiance measurements and Ca II solar emissions estimate that in the absence of surface solar magnetism (which may have been the case during the Maunder sunspot minimum period around 1700) the total solar irradiance may have been reduced by about a quarter per cent.

The hypothesis of a solar forcing through total irradiance variations may be tested using simple energy-balance climate models. Schlesinger & Ramankutty (1992) find strong circumstantial evidence that there have been intercycle variations in solar irradiance which have contributed to the observed temperature changes since 1856, but they also, in agreement with Kelly & Wigley (1992), find that greenhouse gases have been since the nineteenth century the dominant contributor to the observed temperature changes. Kelly & Wigley (1992) conclude that the combination of greenhouse forcing and solar cycle length forcing can explain many features of the observed temperature record. They do call for caution, however, regarding the interpretation of this result because the cycle-length history is not uniquely defined and because they regard the implied irradiance changes prior to 1900 as unrealistically large. Schlesinger & Ramankutty (1992) simulate the temperature anomalies from 1638 to 1985 for different values of the various forcing parameters. Their simulated temperatures may be compared with the temperature record shown in Figure 2 and Figure 3. Such a comparison shows the best agreement when greenhouse gases as well as solar forcing and anthropogenic sulphate aerosol radiative forcing are taken into account. In all cases, however, the inclusion of solar influence in the simple climate model does imply a large reduction, at least 48%, of the estimated sensitivity of the climate to an increased CO_2 content of the atmosphere.

The energy-balance models are simple and inexpensive in terms of computer resources. The problem with these models is that it is assumed that the global climate can be calculated solely by means of global averages of radiative forcings. The different radiative forcings that exist, including solar irradiance, greenhouse gases, and anthropogenic sulphate aerosols, all have different spatial distributions and, in case of the latter, a particularly inhomogeneous distribution that makes it rather dubious if a one-dimensional climate model is at all appropriate in a quantitative distinction of the effects of these different forcings. Whether the cause of a temperature trend is *solar*, through changes in the Sun's luminosity or *human*, through an increase in greenhouse gases, the effect is unlikely to be through direct radiation, but more likely through changes in the general circulation. It is conceivable that a comparatively small rise in the Sun's irradiation can effect a steeper tropical to polar temperature gradient, because of the different solar zenith angle at low and high latitudes (Labitzke & van Loon 1993). If so, an increased Hadley circulation may create regional temperature trends. In particular the fact that most of the rise in global temperature during the present century had its origin in a very fast increase in the temperature in the Northern polar regions during 20 years from 1920 to 1940, simultaneous with a similarly fast rise in the solar activity expressed in the variation of the length of the sunspot cycle, suggests a more direct solar effect than the climate models indicate.

REFERENCES

EDDY, J. A. 1976 The Maunder Minimum. *Science* **192**, 1189–1202.

BRADLEY, R. S. & P. D. JONES 1993 'Little Ice Age' summer temperature variations: their nature and relevance to recent global warming trends. *The Holocene* **3**, 387–396.

FEYNMAN, J. & N. U. CROOKER 1978 The solar wind at the turn of the century. *Nature* **275**, 626-627.

FRIIS-CHRISTENSEN, E. & K. LASSEN 1991 Length of the solar cycle: an indicator of solar activity closely associated with climate. *Science* **254**, 698–700.

GROVEMAN, B. S. & H. E. LANDSBERG 1979 Simulated Northern Hemisphere temperature departures 1579-1880. *Geophys. Res. Lett.* **6**, 767–769.

HAMEED, S. & G. GONG 1993 Long term variations of phenological dates in the Middle and Lower Yangtse River Valley. Preprint.

JONES, P. D., S. C. B. RAPER, R. S. BRADLEY, H. F. DIAZ, P. M. KELLY, & T. M. L. WIGLEY 1986 Northern Hemisphere surface air temperature variations: 1851–1984. *J. Clim. Appl. Met.* **25**, 161–179.

JONES, P. D. 1988 Hemispheric surface air temperature variations: Recent trends and an update to 1987. *J. Climate* **1**, 654–660.

KELLY, P. M. & T. M. L. WIGLEY 1992 Solar cycle length, greenhouse forcing and global climate. *Nature* **360**, 328–330.

LABITZKE, K. & H. VAN LOON 1993 Some recent studies of probable connections between solar and atmospheric variability. *Ann. Geophysicae*, in press.

LASSEN, K. & E. FRIIS-CHRISTENSEN 1993 Solar variability during the past five centuries and the apparent association with terrestrial climate. Submitted to *J. Atm. Terr. Phys.*

LEAN, J., A. SKUMANICH, & O. WHITE 1992 Estimating the Sun's radiative output during the Maunder Minimum. *Geophys. Res. Lett.* **19**, 1591–1594.

REID, G. C. 1987 Influence of solar variability on global sea surface temperatures. *Nature* **329**, 142–143.

REID, G. C. 1991 Solar total irradiance variations and the global sea surface temperature record. *J. Geophys. Res.* **96** (D2), 2835–2844.

SCHLESINGER, M. E. & N. RAMANKUTTY 1992 Implications for global warming of intercycle solar irradiance variations. *Nature* **360**, 330–333.

WIGLEY, T. M. L., & S. C. B. RAPER 1990 Climate change due to solar irradiance changes. *Geophys. Res. Lett.* **17**, 2169–2172.

LIST OF CONTRIBUTED (POSTER) PAPERS

Observational Programs for Solar and Stellar Irradiance Variability

Cryogenic Absolute Radiometer – CSAR
J. E. Martin and N. P. Fox

A Review of the Nimbus-7 ERB Solar Dataset
H. L. Kyle, D. V. Hoyt, and J. R. Hickey

Variability of the Solar Constant Measured at Earth's Surface and by Artificial Satellite Nimbus-7
M. A. M. Shaltout

Solar Total Irradiance Variations Observed during Minimum through Maximum of Solar Activity Cycle
A. Mecherikunnel

Long-Term Variations in Total Solar Irradiance
J. M. Pap, R. C. Willson, C. Fröhlich, R. F. Donnelly, and L. Puga

Observations of the Solar Constant with SOLCON (ATLAS 1 & 2) and SOVA 1 (EURECA)
D. Crommelynck, A. Fichot, B. Penelle, J. Romero, C. H. Wehrli, C. Fröhlich, and V. Domingo

Solar Total Irradiance Variability from SOVA 2 on Board EURECA
J. Romero, C. H. Wehrli, and C. Fröhlich

Fraunhofer Line Variability in the Solar Irradiance Spectrum
W. C. Livingston

Long-Term Changes of some Fraunhofer Lines
I. Vince, J. Skuljan, L. Popovia, and M. Karabin

Imaging Solar Spectral Irradiance Variations
J. Harvey, D. Koolbeck, T. Duvall Jr., and S. Jefferies

Active Region Effects on Solar Irradiance at NaI D Lines
C. Marmolino, V. Andretta, M. T. Gomez, M. Oliviero, and G. Severino

Photometer for the Study of Solar Brightness Variations
E. A. Gurtovenko, I. G. Kesel'man, R. I. Kostyk, S. N. Osipov, Yu. D. Zhugzhda, and N. I. Lebedev

Absolute Measurements of the Solar Spectrum in the Wavelength Range 310 – 680 nm
K. A. Burlov-Vasiljev, E. A. Gurtovenko, and Yu. B. Matvejev

The Brightness on the Solar Disk in the Continuum for the Spectral Range 1.0 – 2.4 μm
E. A. Makarova, E. M. Roshchina, and A. P. Sarychev

Solar Spectral Irradiance at 335, 500 & 862 nm from SOVA 2 on Board EURECA
C. H. Wehrli, J. Romero ,and C. Fröhlich

Comparisons of the Mg II Index Products from the NOAA-9 and NOAA-11 SBUV/2 Instruments
R. Cebula and M. T. Deland

The NOAA/SEL Modified Mg II Center-to-Wing Ratio, 1978 – 1992
L. C. Puga and S. D. Bouwer

The Solar Ca II K Index and the Mg II Core-to-Wing ratio
R. F. Donnelly, O. R. White, and W.C. Livingston

The Solar Irradiance between 110 nm and 410 nm Measured by the SUSIM Space Shuttle Instrument
M. E. VanHoosier, G. E. Brueckner, D. K. Prinz, and J. Lean

An Ultraviolet Spectral Atlas of the Sun between 1190 – 1730 Å
P. Brekke

Solar Spectral Irradiance Measurements from the HIRES Sounding Rocket Payload
J. G. Timothy, P. Bergamini, T. E. Berger, and D. C. Slater

Using the Voyager Spacecraft for Solar EUV Spectral Radiometry
P. L. Smith, W.H. Parkinson, B.R. Sandel, and J.B. Holberg

Analysis of Voyager EUV (500 – 1700 Å) Solar Spectra during the period 1981 – 1992
G. de Toma, R. Vervack, B.R. Sandel, and R. Stalio

Solar cycle Variations in the Extreme Ultraviolet
W.D . Pesnell and W. R. Hoegy

Solar Cycle 21 and 22 Observed by Pioneer Venus Orbiter
W. R. Hoegy and W. D. Pesnell

The Observations of Solar EUV Variations near Activity Maxima
T. V. Kazachvskaya, A. I. Lomovsky, and A. Nusinov

Observed Variability of the Solar EUV Irradiance
T. N. Woods, J. Worden, G. J. Rottman, S. C. Solomon, and G. Schmidtke

Solar EUV Variability Measurements from Lunar Observations with EUVE
P. Judge

X-ray Irradiance Variability Observed by the GOES Satellite
M. J. Aschwanden

Variations in Photometric and Spectrophotometric Parameters of Solar Type Stars: Comparison with the Sun
I. N. Glushneva and A. V. Kharitonov

The List of Stars Recommended as Spectrophotometric Standards
E. I. Terez

Variable Stars with Hipparcos
L. Eyer, M. Grenon, J-L Falin, M. Froeschle, and F. Mignard

Variations during Stellar Cycles
D. V. Gray

Spectroscopic and Photometric Variability in T Tauri Stars: Space and Ground-Based Observations
M. T. V. T. Lago, C. Castro, and J. F. Gameiro

Search for Rapid, Small Scale Variability in Optical Emission of Late-Type Flare Stars
R. Ventura, I. Pagano, G. Peres, and M. Rodono

A Stellar Photometer for Doppler Shift Measurements on Oscillating Stars
P. Costa, S. Catalano, M. Rodono, R. Ventura, and A. Cacciani

Variation of the Solar Diameter from Solar Eclipse Observations, 1715 – 1991
A. D. Fiala, D. W. Dunhan, and S. Sofia

Latitudinal Variation of the Solar Limb-Darkening Function
R. J. Kroll

Variability of Solar and Stellar Irradiance Related to the Network, Active Regions (Sunspots and Plages), and Large-Scale Magnetic Structures

Improvement of the Photometric Sunspot Index and Changes of the Sunspot Contrast within the Solar Cycle
C. Fröhlich, J. M. Pap, and H. S. Hudson

Modelling Solar Irradiance Variations with Area Dependent Photometric Sunspot Index
P. N. Brandt, M. Stix, and H. Weinhardt

Facular Excess Radiation and the Energy Balance of Solar Active Regions
M. Steinegger, P. N. Brandt, H. Haupt, and W. Schmidt

On the Cause of Total Irradiance Variations Observed by the CCD Solar Surface Photometer
J. Nishikawa

Contrast Functions of Active Region versus Quiet-Sun Faculae
J. K. Lawrence and K. Topka

Cycle and Latitude Variation of the Number of Network Bright Points
R. Muller and Th. Roudier

Measures of the Solar Magnetic Field
A. Ruzmaikin, J. Lawrence, and C. Cadavid

Ca II K Line Variations in the Sun as a Star during Solar Cycles 21 and 22
K. R. Sivaraman, S. S. Gupta, R. Kariyappa, P. S. M. Aleem, and K. Sundararaman

Variability of the Solar Chromospheric Network over a Solar Cycle
R. Kariyappa and K.R. Sivaraman

Ca II K Line Profile of the Truly Quiet-Sun
K. R. Sivaraman, S. S. Gupta, and R. Kariyappa

Long-Term Magnetic Activity in the Sun and Solar-Type Stars
W. H. Soon, S. L. Baliunas, and Q. Zhang

Stellar Irradiance Variations due to Surface Temperature Inhomogeneities
K. G. Strassmeier

A Diagnostics of Solar Atmospheric Models: Search for "Black Flares"
L. van Driel-Gesztelyi, H. S. Hudson, B. Anwar, E. Hiei, and S. Tsuneta

Cyclic Variations of Solar Prominences Magnetic Fields
I. S. Kim

Oscillating Loops in Dwarf *M* Stars: A Pointer to a Coronal Heating Mechanism?
J. G. Doyle and A. D. Andrews

On Correlation between the Total Solar Irradiance and CI
P. Gašpar, V. Rušin, and M. Rybanský

Rotational Characteristics of the Solar Corona: Emission Line Fe XIV 5303 Å Observations 1964 – 1991
J. Rybák and V. Rušin

Solar Cycle Variation of the Microwave Spectrum and Total Irradiance
E. J. Schmahl and M. R. Kundu

Total Flux at 115 cm and 10.7 cm during Solar Cycle 22
A. Tlamicha

Solar Brightness Distribution and its Variability at 3 Millimeter Wavelength
V. G. Nagnibeda and V. V. Piotrovitch

Latitudional Variability of Large-Scale Coronal Temperature and Density Distributions, 1984 – 1992
M. Guhathakurta and R. R. Fischer

Variability of Solar Irradiance Related to the Large-Scale Convective Patterns
P. Ambrož

New Observations of Solar Cycle Variability
P. S. McIntosh

Are Sunspot Magnetic Field Strengths Increasing?
J. A. McKinnon, J. H. Allen, and C. C. Abston

Empirical Models of Solar Total and Spectral Irradiance Variability

Long-Term Irradiance Variations of the Sun: What do Solar-Type Stars Tell Us?
Q. Zhang, W. H. Soon, S. L. Baliunas, G. W. Lockwood, B. A. Skiff, and R. R. Radick

Modeling the Solar Atmosphere and Irradiance Spectrum
E. H. Avrett

Active Region Evolution and Solar Flux Variations
T. P. Hartsell

Chromospheric UV Radiation, the Modeling of Observed Spectral Variability
J. M. Fontenla

On Variability of some Characteristics of Solar Flux
E. A. Makarova, T. V. Kazachevskaya, and A. V. Kharitonov

Lyman-alpha Line Intensity as a Solar Activity Index in the Far Ultraviolet Range
A. A. Nusinov and V. V. Katyushina

SERF3 Model of EUV and UV Temporal Variations
R. F. Donnelly, D. F. Heath, and B. M. Schlesinger

Indices for Estimating Long-Term Coronal Flux Variations
R. F. Donnelly

A Method to Improve Estimates of Solar EUV Flux
S. D. Bouwer and W. K. Tobiska

Interplanetary Hydrogen and the Variable Sun
E. Kyrölä, T. Summanen, and P. R. Raback

Cosmic Rays – an Indicator of Solar Activity
V. Rušin and M. Rybanský

Meridional Motions of Sunspot Groups Depending on Different Activity Cycles Determined from Two Independent Sets of Data
G. Lustig and H. Wöhl

A New Method of Analysis of the Photometric and Spectral Data of Stars with Surface Inhomogeneities
D. P. Kjurkchieva

Spot Parameters of CC ERI during 1956/57
D. P. Kjurkchieva and V. G. Shodrov

Active Region Area Coverages on Solar-Type Stars
A. Giampapa and V. Andretta

The Morphology and Evolution of Stellar Dynamo Variations
S. Saar, A. Brandenburg, S. Baliunas, and R. Donahue

Eclipse Imaging and Spot Models of YY Geminorum
C. J. Butler, E. Budding, J. G. Doyle, B. J. Kellett, G. E. Bromage, T. Simon, and B. Foing

Periodicities in Solar Luminosity Data
T. Bai

152-Day Periodicity during Solar Cycle 22
A.Özgüz and Atac

The Periodicity of Solar Activity Cycle
Y. V. S. Romanon and N. S. Zgonyaiko

Solar and Stellar Oscillations, Irradiance Variations and their Interpretation

Sunspot Umbral Oscillations
J. Staude

Radiation-Hydrodynamic Waves and Global Solar Oscillations
N. S. Dzhalilov, Y. D. Zhugzhda, and J. Staude

Excitation of Solar Gravity Waves
B. N. Andersen

Solar Noise Simulations in Radiance and Irradiance
B. N. Andersen

Secular Variations in the Spectrum of Solar P-Modes
A. Jiménez, P. L. Pall, C. Régulo, T. Roca Cortés, and L. S. Duarte

Phase Differences between Irradiance and Velocity Low Degree Solar Acoustic Modes Revisited
A. Jiménez

P-Mode Frequency Variation in Relation to Global Solar Activity
K. T. Bachmann and T.M. Brown

Comparison of Helioseismic Parameters of High Degree Solar P-Modes for Different Epochs during Solar Cycle 22
E. J. Rhodes, Jr., S. G. Korzennik, and A. Cacciani

Solar Cycle Variations in P-Modes and Chromospheric Magnetism
R. Jain and B. Roberts

Wavelet versus Fourier Analysis of Solar Radius Variability
A. Vigouroux and P. Delache

On P-Mode Oscillations in Stars from 1 to 2 *M*
N. Audard and J. Provost

Asteroseismology of Rapidly Oscillating Peculiar *A* Stars
D. W. Kurtz

"Star as the Sun" Observations in Seismology of Distant Stars
D. E. Mkrtichian

Time Correlation between Solar Structural Parameters: P-Mode Frequencies, Radius and Neutrino Flux
Ph. Delache, V. Gavryusev, E. Gavryuseva, F. Laclare, C. Régulo, and T. Roca Cortés

Time Variation of the Solar Neutrino Flux
V. Gavryusev and E. Gavryuseva

Solar Structure as Constrained by Observed Solar Oscillation Frequencies
A. N. Cox and J. Guzik

Using Solar Oscillations to Study the Implications of Element Diffusion for Solar Interior Structure
J. A. Guzik and A. N. Cox

The Solar Differential Rotation, Meridional Circulation, and Irradiance Variations
K. L. Chan and H. G. Mayr

Differentional Rotation of Stellar Convective Zones and Magnetic Dynamo
G. S. Bisnovaty-Kogan

Luminosity Variations in a Mean-Field Dynamo
A. Brandenburg, D. Moss, and I. Tuominen

On the Magnetic Fluctuations Produced by a Large-Scale Magnetic Field
A. Brandenburg, F. Krause, A. Nordlund, A. Ruzmaikin, R. F. Stein, and I. Tuominen

On the Relation of X-ray and H, K Variability to Stellar Dynamo
N. Kleeorin, I. Rogachevskii, and A. Ruzmaikin

A Modification of the Torsional Oscillations by MHD Turbulence of the Solar Convective Zone
N. Kleeorin, I. Rogachevskii, and A. Ruzmaikin

A New Type of Magnetic Buoyancy as a Trigger of the Sunspot Formation and Solar Oscillations
N. Kleeorin, M. Mond, and I. Rogachevskii

The Response of the Earth's Atmosphere to Solar Irradiance Variations and Sun-Climate Connections

Spectral Extrapolation of Solar Irradiance Variability from Nimbus Irradiance Data
W. Mende

Variations of the Magnetic Fields of the Sun and the Earth within 7 – 50 Days Period
V. P. Bobova and N. N. Stepanian

The Investigation on Factors Effecting Solar Irradiance and Climate Variability
Chen Xie-zhen

Ionospheric Ground Patrol of the Solar EUV Irradiance Variability
L. A. Antonova, G. S. Ivanov-Kholodny, and V. E. Chertoprud

Thermospheric Airglow Response to the Solar Ultraviolet Irradiance
S. C. Solomon, S. M. Bailey, and T. N. Woods

On the Dependence of Total Ozone on the Long-Term Period of Solar Activity
G. A. Terez and E.I. Terez

Parallel Solar and Terrestrial Variations: Coincidence or a New Paradigm?
D. V. Hoyt

Distinction between the Climatic Effects of the Solar Corpuscular and Electromagnetic Variations
T. Baranyi and A. Ludmány

EUDOSSO: A Space Project for Solar Oscillations and Long-Term Variability Relevant to Climatic Changes
L. Paterno and S. Sofia

Long-Term Persistence of the Solar Activity
A. Ruzmaikin, J. Feynman and P. Robinson

Solar Magnetic and Bolometric Cycles Recorded in Sea Sediments
G. C. Castagnoli, G. Bonino, A. Provenzale, and C. Taricco

The Solar Irradiance Variations and Excited Emission of the Organic Molecules in the Flowstones' Microzonal Structure
V. N. Dermendijev, Y. Y. Shopov, and M.-E. S. Mihailov

Maximum and Minimum Temperatures at Armagh
C. J. Butler